Soils

Khan Towhid Osman

Soils

Principles, Properties and Management

Springer

Khan Towhid Osman
Department of Soil Science
University of Chittagong
Chittagong, Bangladesh

ISBN 978-94-007-5662-5 ISBN 978-94-007-5663-2 (eBook)
DOI 10.1007/978-94-007-5663-2
Springer Dordrecht Heidelberg New York London

Library of Congress Control Number: 2012951125

Printed on acid-free paper

Springer is part of Springer Science+Business Media (www.springer.com)

For my teacher
Professor Aminul Islam

Preface

I obtained my M.S. degree in Soil Science in 1976, and I had been a teacher in the Department of Botany, University of Chittagong, Bangladesh, for 25 years since 1977. I taught undergraduate and graduate students of Botany the origin and development of soils, properties of soils, growth and distribution of plants in relation to properties of soils, soil fertility and productivity management, and soil conservation. I also gave lectures on different aspects of ecology, particularly edaphic factor of vegetation development and distribution, and agronomy, especially soil and crop management for sustainable yield. For sometime in the 1990s, I worked off and on as a guest faculty in the Institute of Forestry and Environmental Sciences, where I lectured on forest soils and forest soil management to the undergraduate students of forestry. Meanwhile, I obtained my Ph.D. in 1996 on the growth of teak (*Tectona grandis*) in relation to soil properties in the southeastern hilly areas of Bangladesh. I joined the Department of Soil Science in 2000 as the founding chair and have been working there as a professor since. Versatile as my academic experiences have been, I could see the connections of soil science with other relevant branches of knowledge and felt the necessity of integrating them in one volume.

During my studentship and teaching life, I had the opportunity to study some good books on soil science. I enjoyed much the works of H. O. Buckman and N. C. Brady (*The Nature and Properties of Soils*, 10th edn.); N. C. Brady and R.R. Weil (*The Nature and Properties of Soils*, 14th edn.); L. M. Turk and H. D. Foth (*Fundamentals of Soil Science*); M. J. Singer and D. N. Munns (*Soils: An Introduction*); M. E. Sumner (*Handbook of Soil Science*); R.L. Donahue, R. W. Miller, and J. C. Shikluna (*Soils: An Introduction to Soils and Plant Growth*); E. A. Fitzpatrick (*An Introduction to Soil Science*); C. A. Black (*Soil Plant Relationships*); E. J. Russel (*Soil Conditions and Plant Growth*); J. S. Joffe (*Pedology*); H. Jenny (*Factors of Soil Formation*); R. J. Schaetzl and S. Anderson (*Soils Genesis and Geomorphology*); R.E. Grim (*Clay Mineralogy*); F. E. Bear (*Chemistry of the Soil*); H.L.Bohn, B. L. McNeal, and G. A. O'Connor (*Soil Chemistry*); G. Sposito (*The Chemistry of Soils*); D. L. Sparks (*Environmental Soil Chemistry*); K. H. Tan (*Principles of Soil Chemistry*); USDA (*Soil Survey Manual, Soil Taxonomy*); USDA Salinity Laboratory Staff (*Diagnosis and Improvement of Saline and Alkali Soils*); L. D. Baver (*Soil Physics*); D. Hillel (*Introduction to Soil Physics*); M.B. Kirkham (*Principles of Soil and Plant Water Relations*); P. J. Kramer (*Plant and Soil Water Relationships: A Modern Synthesis*); H. Marschner (*Mineral Nutrition of Higher Plants*); H. D. Chapman (*Diagnostic Criteria for Plants and Soils*); M. M. Kononova (*Soil Organic Matter*); H. H. Benett (*Soil Conservation*); S. L. Tisdale, W. L. Nelson, and J. D. Beaton (*Soil Fertility and Fertilizers*); K. Kilham (*Soil Ecology*); R. P. C. Morgan (*Soil Erosion and Conservation*); N. van Breemen and P. Buurman (*Soil Formation*); A. Martin (*Introduction to Soil Microbiology*); F. J. Stevenson (*Cycles of Soil: Carbon, Nitrogen, Phosphorus, Sulfur, Micronutrients*); N. Juma (*The Pedosphere and Its Dynamics*); R. F. Fisher and D. Binkley (*Ecology and Management of Forest Soils*); K. A. Armson (*Forest Soils: Properties and Processes*), J. B. Jones, Jr. (*Agronomic Handbook*), and so on. These authors have stimulated my interest in learning soil science and delivering my acquired knowledge to my students in a systematic way. I have keenly noticed the responses of my students, whom I have seen to have hard times with information extraction and interpretation of texts, which led me to conclude that despite the availability of plenty of good texts, there is still scope of new books with novel styles of presentation, updated information, and new interpretations. The present work is an attempt toward this.

Chapters 1, 2, 3, and 4 of this book emphasize soil as a natural dynamic body, its origin and development, and its systematic study for a better understanding of its properties and management. Chapter 1 deals with the concepts of soil—soil as it occurs in nature, its makeup, and ecosystem functions. Chapter 2 gives an account of the elemental, mineralogical, and petrological composition of the lithosphere and weathering of rocks and minerals including biogeochemical weathering and its products. Chapter 3 is a brief account of soil-forming factors and processes. Chapter 4 contains modern soil classification systems. Chapters 5, 6, 7, 8, and 9 describe physical, chemical, and biological properties of soils in relation to plant growth. Chapter 6 deals with soil water—water as a component of soil, its hydrological properties, moisture constants and potentials, water movement through soil and plant, water stress, and waterlogging. Irrigation and drainage methods have also been treated in considerable detail. Chapter 9 deals with biological properties of soils with a good account on soil fauna, which is not generally stressed in basic soil science books despite its significant role in determining soil characteristics. Chapter 10 addresses plant nutrients and soil fertility management. Physiological functions of nutrients in plants, behavior, and availability of nutrients in soil, plant nutrient requirements, nutrient interactions in plants and soils, soil fertility evaluation, organic and inorganic fertilizers, and methods of fertilizer application have been discussed with sufficient details. Problem soils and their management have been treated in Chap. 11. Chapter 12 reviews soil resources and soil degradation. Recent data and literature have been consulted to incorporate most recent developments in the field. This book has accommodated one chapter for wetland soils (Chap. 13) and another for forest soils (Chap. 14) unlike most fundamental soil science books. In spite of the fact that forests occupy almost one-third land area of the world and forest soils have tremendous ecological roles, they are not generally included in discussions of common basic soil science texts, which, in my view, is a major exclusion. Therefore, I have attempted to give a comprehensive yet concise account of forest soils. Chapter 15 emphasizes soil study in a changing climate, an issue that has attracted much attention in recent decades.

This book is intended for undergraduate and graduate students of soil science, and agricultural, biological, and environmental sciences, who study soil as a natural resource. Professionals, including agronomists, horticulturists, gardeners, geologists, geographers, ecologists, biologists, microbiologists, and silviculturists, may find something of their interest as well.

In this text, soil processes and properties have been explained with adequate examples, tables, and figures. In order to make matters comprehensive, necessary generalization and simplification were done, though bearing the danger of oversimplification of soil as a complex entity in mind. Students, I have noticed often, get overwhelmed by throngs of information without being able to have a complete understanding of the central concept. This is why I inserted meaningful messages in section headings so that even before going over a particular section, one might get the gist right away.

Chittagong Khan Towhid Osman

Acknowledgements

I sincerely appreciate the inspiration of my students during the preparatory phase of this book. Their queries and responses made me eager to learn more, systematize topic organization, and present materials in an interesting way.

My colleague Mr. Mohammad Enayet Hossain went painstakingly and meticulously through the manuscript and suggested corrections on many occasions. I cannot be more grateful to him. My colleagues Dr. Abul Kashem, Mr. Enamul Haque, and Mr. Zafar Afsar extended their cooperation during the finalization of the draft. I would also like to thank Professor Brian Alloway, University of Reading, UK, for his general advice on how to make the text better.

My special thanks go to Dr. Kamrul Huda, Dr. Animesh Biswas, Mr. Md. Mueed Ul Zahan, Mr. Imam Hossain, Ms. Priyanka Chak, Mr. Khan Md. Rafee, and Mr. M. A. Hannan, who lent themselves freely in image processing. Many organizations also were kind enough to permit me to use their resources. I am indebted to them all.

I am grateful to my sons, Shovon, Shajib, and Shourabh, for their encouragement. My wife, Mrs. Taslima Begum, managed the household and allowed me the time necessary for writing.

Khan Towhid Osman

Contents

List of Abbreviations

AEC	Anion exchange capacity
ADP	Adenosine diphosphate
ATSDR	Agency for Toxic Substances and Disease Registry
ATP	Adenosine triphosphate
AWC	Available water capacity
BSP	Base saturation percentage
BTEX	Benzene, toluene, ethylbenzene, and xylenes
CCE	Calcium carbonate equivalence
CEC	Cation exchange capacity
CEICC	Council-certified Environmental Infection Control Consultant
CNR	Critical nutrient range
CNS	Central nervous system
CRF	Controlled release fertilizer
CTF	Controlled traffic farming
CWR	Crop water requirement
CyDTA	Cyclohexane diamine tetraacetic acid
DAP	Di-ammonium phosphate
DCE	Dichloroethylene
DDT	Dichlorodiphenyltrichloroethane
DNA	Deoxyribonucleic acid
DTPA	Diethylene triamine penta-acetic acid
DRIS	Diagnosis and Recommendation Integrated System
dS	Decisiemens
EDTA	Ethylene diamine tetraacetic acid
EDDA	Ethylene diamine diacetic acid
ENV	Effective neutralizing value
ESP	Exchangeable sodium percentage
ET	Evapotranspiration
FAO	Food and Agricultural Organization
FC	Field capacity
GCM	Global Circulation Model
GLASOD	Global assessment of soil degradation
ICIMOD	International Center for Integrated Mountain Development
IPCC	Intergovernmental Panel on Climate Change
IRRI	International Rice Research Institute
IWR	Irrigation water requirement
kPa	Kilopascal
MAP	Mono-ammonium phosphate
MAST	Mean annual soil temperature
MP	Muriate of potash
MWHC	Maximum water-holding capacity

NPEs	Nonylphenol-ethoxylates
NRCS	Natural Resources Conservation Service
OSP	Ordinary super phosphate
PAEs	Phthalate Esters
PAHs	Polyaromatic hydrocarbons
PCBs	Polychlorobiphenols
PCE	Polychloroethylene
POPs	Persistent organic pollutants
PWP	Permanent wilting point
RNA	Ribonucleic acid
RUSLE	Revised Universal Soil Loss Equation
SALT	Sloping Agricultural Land Technology
SAR	Sodium adsorption ratio
SOM	Soil organic matter
SPAC	Soil-plant-atmosphere continuum
SRF	Slow-release fertilizer
TCE	Trichloroethylene
TDS	Total dissolved salts
TEB	Total exchangeable bases
TSP	Triple superphosphate
UNCCD	United Nations Convention to Combat Desertification
UNESCO	United Nations Educational, Scientific and Cultural Organization
USDA	United States Department of Agriculture
USEPA	United States Environmental Protection Agency
USLE	Universal Soil Loss Equation
VC	Vinyl chloride
WRB	World Reference Base
WRI	World Resources Institute
WUE	Crop water use efficiency

Concepts of Soil

1

Soils are natural unconsolidated materials on the surface of the earth and are composed of solid, liquid, and gas. They have organic as well as inorganic matter, which are intimately mixed together by natural processes. By this mixing and transforming, they are aggregated into a porous body. The pores accommodate air and water. Thus, there are four major components of soil—mineral matter, organic matter, water, and air. The soil has evolved through pedogenic processes as a dynamic and a three-dimensional body. Soils have attained the capacity of supporting various ecosystems on the earth. They provide plants with anchorage, nutrients, water, air, and warmth and protect them from toxins. Soils have crucial ecological functions: It is a transformer of energy, it is a recycler of materials, it is a purifier of water, and above all, it functions as an ecosystem component.

1.1 Different People Have Different Concepts of Soil

Everybody is familiar with soils, and everybody has his own ideas about soil. These ideas differ from people to people because they think of soils in different ways, or they use soils for different purposes. To a farmer, soil and land have the same meanings. He understands soil where crops can be grown. To him, only the root zone of the crop plants is important. A potter uses mud for pottery; to him, there is no difference between mud or clay and soil. Professionals also differ in their concepts of soil. For example, a geologist views soil as the biologically modified upper part of the regolith, a term to mean unconsolidated materials on the earth's surface over bedrock and formed by weathering of rocks and minerals. To a geographer, soil is the discontinuous thin envelope of loose material on the surface of the earth. An engineer considers soil as any kind of loose, unconsolidated earth material on which foundations of roads, buildings, and other structures are constructed. He does not distinguish between soil and other loose materials on land such as sediment.

All these views partly reflect the nature of soils, but none gives a complete idea about soil as it occurs and functions in nature. To a soil scientist, soil is a three-dimensional body which has characteristics that distinguish it from other natural materials. The soil is an unconsolidated material on the earth's surface that has evolved through complex pedogenic processes by the natural amalgamation of mineral and organic matter; that has achieved distinct morphological, physical, chemical, and biological characteristics; and that has attained the capacity of supporting vegetation and other life forms. It is, in fact, an ecosystem itself and, at the same time, a part of the greater terrestrial ecosystems.

1.2 There Are Many Different Definitions of Soil

Different scientists have defined soil in different ways. Some of these definitions are given below to show the evolution of the modern concepts of soil.

> Soils are applied solely to those superficial or nearly superficial horizons of rocks, that have been more or less modified naturally by the interaction of water, air and various kinds of organisms, either living or dead; this being reflected in a certain manner in the composition, structure and color of such formations. Where these conditions are absent, there are no natural soils, but artificial mixtures of rock.
>
> Dokuchaev (1879)

> Soil is the uppermost weathered layer of the earth's crust; it consists of rocks that have been reduced to small fragments and have been more or less changed chemically together with the remains of plants and animals that live on it and in it.
>
> Ramann (1905)

> Soil is, more or less, loose, friable material in which, by means of their roots, plants may or do find foothold, nourishment as well as all other conditions of growth.
>
> Hilgard (1914)

> The soil is a natural body of mineral and organic constituents, differentiated into horizons, of variable depth, which differs from the material below in morphology, physical makeup, chemical properties and composition, and biological characteristics.
>
> Joffe (1949)

> Soil is a collection of natural bodies occupying a portion of the earth surface that supports plant growth and that has properties

K.T. Osman, *Soils: Principles, Properties and Management*,
DOI 10.1007/978-94-007-5663-2_1, © Springer Science+Business Media Dordrecht 2013

> due to the integrated effect of climate and vegetation acting upon parent material, as conditioned by relief, over a period of time.
>
> Kellogg (1960)

> Soil is a natural body comprised of solids (minerals and organic matter), liquid, and gases that occurs on the land surface, occupies space, and is characterized by one or both of the following: horizons, or layers, that are distinguishable from the initial material as a result of additions, losses, transfers, and transformations of energy and matter or the ability to support rooted plants in a natural environment.
>
> USDA, NRCS (2003)

1.3 All Loose Materials on the Surface of the Earth Are Not Soils

There are many loose materials resembling soils on the surface of the earth; they are not soils, although they may have limited capacity of supporting plant growth. For example, desert sands are not soils; normally, they lack organic matter, available water, and capacity to support plants. Sands are discrete particles; they are not aggregated like soil. Sandy beaches and dunes are not soils; they may have some organic matter, but they do not show profile development. Similar is the case with sediments on riverbeds. They are actually parent materials. Under favorable conditions and over time, these parent materials will develop into soils. Thus, parent materials below the solum (the depth to which pedogenesis has occurred; Sect. 3.1) and above the solid bedrock are not soils. Unconsolidated mineral or organic material thinner than 10 cm overlying bedrock, unconsolidated material covered by more than 60 cm of water throughout the year, and organic material thinner than 40 cm overlying water are not soils. Some soils may have water covering its surface to a depth of 60 cm or less either at low tide in coastal areas or during the driest part of the year in areas inland. Subaqueous soils (soil materials, mud/sediments, found underwater; Sect. 13.7.4) occur immediately below a water depth of <2.5 m (USDA, NRCS 2003).

1.4 Soil Is Not Land Itself; It Is a Part of Land

Often land is used synonymously with soil, but the two are not the same. Land is the non-water part of the earth's surface, while soil occupies only a thin upper part of some land. There is no soil on some other lands. Some land has rocky surfaces; some are covered by desert sands and some by glaciers and ice over the solid rocks. On some land, there are deposits of fresh sediments that have not been altered enough to be called soil. Soil has a lower limit (usually above the parent material—the material from which soil has developed), while land has none. Whatever deep we dig the land, it remains the same land. At a depth of the land, there is no soil. Land does not bear significant vegetation unless it has soil on it. The part of the land that can be seen at a glance is called landscape. The part of the landscape that has similar slope, vegetation, or cropping potential is known as a land. The same land may contain different soils at different positions. Both land and soil are natural resources, but soil is not normally owned as a property. A land can be owned as a property. Land may have a variety of uses depending on soil and other characteristics.

1.5 Soil Is a Natural Body

The soil is a natural entity. It has evolved over a long time in a place through natural processes under natural conditions from natural materials. The materials from which the soils are formed are called parent materials (Sect. 3.6). Parent materials may be organic and inorganic, although most soils (more than 99% of world soils) develop from inorganic or mineral parent materials. They are disintegrated and decomposition products of rocks and minerals (Chap. 2). Some soils develop from organic parent materials which are residues of past vegetation, usually accumulated under wet conditions. The natural processes involved in disintegration and decomposition of rocks and minerals are collectively known as weathering (Sect. 2.14). At a point of disintegration and decomposition during evolution of soil on a bare surface, organisms including plants, animals, and microbes take hold and add organic matter to the parent materials. Climate and organisms act upon it at a topography and transform it into soil after a long period of time (Jenny 1941, 1980). The processes involved in this transformation are collectively called pedogenesis (factors of soil formation and pedogenic processes are discussed in detail in Chap. 3).

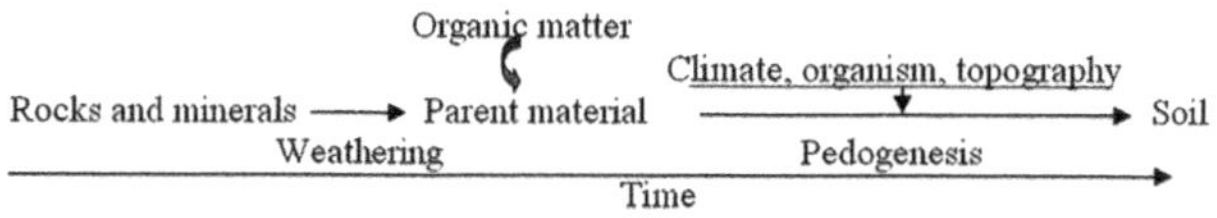

So, the soil is a natural body. Brady and Weil (2002) says, "Soil is a natural body in the same sense as that a mountain, a lake or a valley is."

1.6 Soil Is a Three-Dimensional Body

"A soil" is a natural unit that has a definite range of physical, chemical, and biological properties. But there are variations in soil properties at all directions of a landscape. After some distance in the lateral direction, we find a different soil having different properties. Soil has a lower limit too. The upper part of the regolith (unconsolidated and loose materials above bedrock; Sect. 3.1) is soil. So, a soil has three dimensions,

Fig. 1.1 A pedon; exposed sides of the pedon represent soil profile

two laterals and one vertical, that is, length, width, and depth. A soil has a volume too. The smallest volume of a soil having similar characteristics throughout the lateral dimensions is called a pedon (Soil Survey Staff 1975, 1999). Simonson (1978) defined pedon as the smallest effective unit of soil. A pedon is actually an arbitrary unit chosen for examination of soil in field and sampling for study. A pedon may have an area of 1–10 m^2, and it is usually hexagonal in shape (Fig. 1.1). It extends in depth into the parent material from which the soil has developed. A group of similar pedons that occur together in a landscape is called a polypedon. A polypedon represents a soil individual. Different soils in a landscape have polypedons of different pedon characteristics.

The pedon characteristics are exhibited by the exposed sides of the pedon. An exposed side of the pedon actually represents the soil profile (Sect. 3.1) which is defined as the vertical section of soil from surface down to the parent material. The horizontal layers in a soil profile roughly parallel to one another and formed by pedogenic processes are known as horizons. Horizons are, however, different from layers. A soil layer is any stratum which is not formed by pedogenic processes. A layer is generally formed by geological processes or by human activity. There are five master horizons denoted by O, A, E, B, and C. Below the C horizon, there is usually the bedrock in soils formed from residual parent materials and unaltered sediments in transported parent materials.

1.7 Soil Is a Dynamic Body

Soils continually change at different rates and along different pathways. They continually evolve and are never static for more than short periods of time (Schaetzl and Anderson 2005). Daniels and Hammer (1992) stated that soils were four-dimensional systems. Time is the fourth dimension because soils change with time. These changes are physical, chemical, and biological; most changes are biogeophysical and biogeochemical in nature. There is always exchange of gases (oxygen, carbon dioxide, water vapor) between the soil and the atmosphere, infiltration and percolation of water, leaching of materials in suspension and solution, and aggregation of particles. Soil minerals are continuously weathered releasing soluble substances into the soil solution. Plants absorb soluble cations and anions and alter the composition of soil solution. Soil organisms act relentlessly in changing the soil environment. There are always decomposition of organic matter and release of carbon dioxide, mineralization of nitrogen, phosphorus, sulfur, and bases, formation of humus and organo-metal complexes, fixation of nitrogen, etc. Despite these changes, the state of the soil always tends to remain at equilibrium with the environment it is exposed to. Thus, there are always complex interactions between the soil and the environment.

1.8 Soil Is a Transformer of Energy

Soil acts as a receptor and transmitter of solar radiant energy (Fig. 1.2). This energy is utilized in photosynthesis, evaporation, transpiration, heating of soil particles, water and gases, chemical reactions, physiological functions of soil organisms, etc. Energy is emitted from the soil by exothermic chemical reactions, organic matter decomposition and soil respiration, etc. Soil transmits energy through particle contacts, movement of water and gases, etc. Energy is absorbed and released in soil by wetting and drying, heating and cooling, and weathering.

Thus, energy is transformed in soil through physical, chemical, and biological processes. Energy is always at a state of flux within the soil.

1.9 Soil Is a Recycler of Materials

Almost all materials in soil, inherent to it or added to it, undergo diverse transformations. These transformations may be physical, chemical, biochemical, biological, or all. The most remarkable transformation that occurs in soil is the decomposition of organic matter. All organic substances added naturally or as manures and wastes are altered and decomposed by soil flora and fauna. The bulk is lost as CO_2 to the atmosphere; a part is retained as humus mixed with mineral soil particles, and the bases are released to the available soil nutrient pool. They are absorbed by growing plants.

We dispose a large quantity of domestic wastes daily to our home yard or gardens. If that had not decomposed within a short time, how noxious our life would have been. Thus, soil is a great recycler of materials. Soil has also a huge storage of

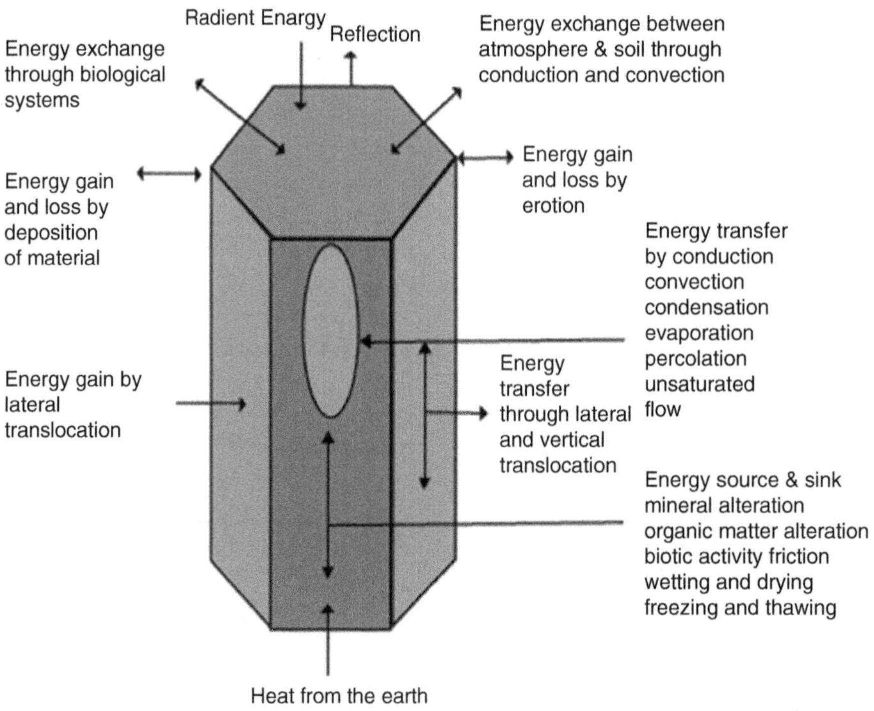

Fig. 1.2 Transformation of energy in soil (After Buol et al. 1997)

organic matter. Organic matter is continually being added and decomposed keeping the storage in a balance. Soil may be a significant C-sink if managed properly. Therefore, soil acts as an environmental buffer. Inorganic substances are also transformed by chemical or biological processes and are rendered soluble/insoluble, mobile/immobile, and active/inactive.

1.10 Soil Is a Purifier of Water

When water, rain or wastewater, falls on soil, some is evaporated and transpired, some runs off, and the most part infiltrates. Soil stores some water, but a considerable amount passes through the soil profile into the groundwater. During temporary storage and transmittance of this water in soil, suspended and dissolved organic matter is decomposed, the sediments are deposited in soil matrix, the contaminants are removed by reaction with soil constituents or exchanged in the colloidal phase, and the purified water enters into the groundwater or flows as stream water. In a healthy soil ecosystem, soil organisms reduce the impacts of pollution by buffering, detoxifying, and decomposing potential pollutants. Most surface water travels a long way over or through the soil before it reaches lakes and rivers. During the travel, sediments are deposited, and contaminants are removed by soil. Quality of water in the watershed largely depends on the quality of soil around. So, soil is a natural filter.

1.11 Soil Is an Ecosystem

An ecosystem is a unit of organisms interacting among themselves at a space and with their abiotic environmental components in a way that flow of materials and energy within the system tends to remain in a dynamic equilibrium. Ecosystems may be broadly classified as terrestrial ecosystems and aquatic ecosystems. Further divisions as forest ecosystem, grassland ecosystem, marine ecosystem, lake ecosystem, pond ecosystem, etc., may also be done. The soil is a component of all terrestrial ecosystems. However, the soil is itself an ecosystem in that it harbors a large number of organisms which interact among them and with the physical and chemical soil environment (Fig. 1.3).

Among the biotic components of the soil are the producers such as green plants, algae, and autotrophic bacteria; the consumers including soil animals, for example, detritus feeders and predators; the decomposers such as fungi and heterotrophic bacteria; and the transformers such as nitrifiers, denitrifiers, and sulfur bacteria. Their populations and functions depend on physical and chemical environment prevailing in the soil. A particular soil acts as a large number of habitats for a diverse group of organisms which have their own niches that overlap with others. The organisms have both competitive and cooperative relationships among themselves for functional and structural habits.

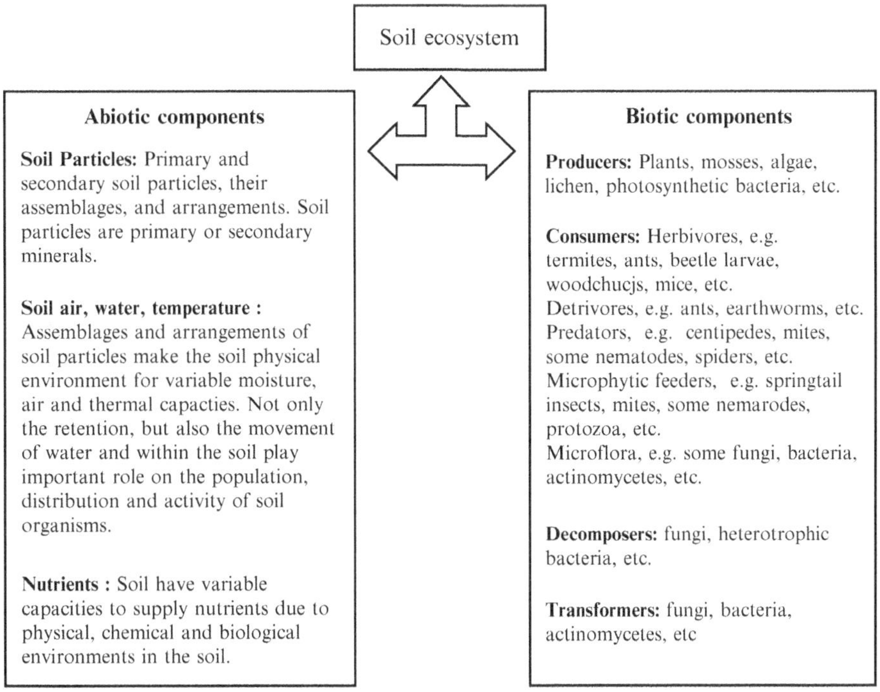

Fig. 1.3 Components of soil ecosystem

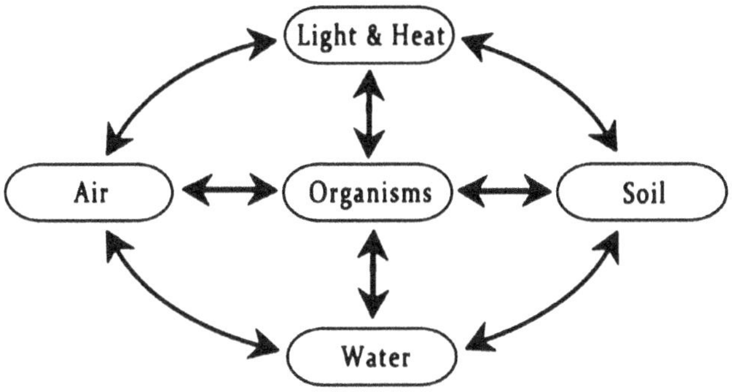

Fig. 1.4 Environmental components that surround organisms

1.12 Soil Is a Component of the Environment

Lodha (1996) defines environment as materials and conditions "that surround an individual or a community at any point in its life cycle." These include physical and cultural surroundings. Actually, environment consists of the complex of physical, chemical, and biotic factors that act upon an organism or an ecological community and ultimately determine its form and survival. An organism or a biotic community interacts with its environment. These interactions between individuals, between populations, and between organisms and their environment form ecological systems, or ecosystems.

Environment includes climate (light, temperature, air, humidity, precipitation, and wind), lithosphere (rocks and soils), hydrosphere (lakes, streams, groundwater, and ocean), and biosphere (organisms—flora and fauna) (Fig. 1.4). These environmental components interact to reach equilibrium and form the ecosystems. A permanent change in any of these conditions, such as a change in temperature or intensity of light, is a change in the environment. Change in the environment may be natural, as experienced through the evolution of atmosphere and life, or may be caused by development activities of human. Human-induced environmental changes are abrupt and tax on the health and survival of human itself.

Exchange of materials and energy takes place between the soil and its environment. For example, climatic elements, such as precipitation and temperature, have the most pronounced effect on the formation and properties of soils. For example, Oxisols (highly leached soils with accumulation of quartz, Fe and Al oxides) and Ultisols (strongly leached soils with accumulation of clay in subsoil) are associated with humid tropical climate and Spodosols (soils with accumulation of humus and oxides of Al and Fe in subsoil) with humid temperate climate (Chap. 3 for formation and Chap. 4 for characteristics of these soils). Climate affects weathering and determines to a large extent the contents of clay, weatherable minerals, soluble salts, organic matter, and nutrients in soil.

On the other hand, soil acts both as a great source and sink of CO_2. Further, wetland soils cause emissions of CH_4, H_2S, and NH_3 gases to the atmosphere. Soil properties affect the distribution of natural plants and growth and yield of crop plants.

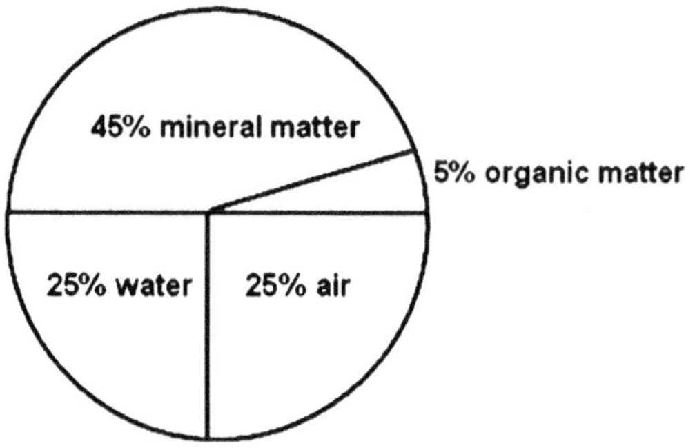

Fig. 1.5 Average volume composition of a loam-textured surface soil

Soil affects qualities of surface and groundwaters. Better managed watershed areas yield good quality water. Soils may become contaminated with heavy metals. These contaminants may pollute water bodies and affect human health.

1.13 Major Components of Soils Vary in Volume Proportions

On an average, an ideal loam-textured surface mineral soil contains 45% mineral matter, 5% organic matter, 25% water, and 25% air by volume (Fig. 1.5). However, the volume composition highly varies with soil types. For example, organic matter in soils of warm tropical areas, Ultisols, Oxisols, and other soils, is very low (<2%) due to rapid decomposition rates. Histosols (organic soils) and holistic horizons in other soils may contain very high (>80%) organic matter. Mollisols (grassland soils) have intermediate organic matter content (Chap. 4). Air and water contents in soils are more variable. Some soils are wet at all times such as hydric soils (soils saturated with water and conducive to the development of hydrophytic vegetation; Chap. 13), requiring artificial drainage if they are to be used for agriculture except wetland rice. Some soils such as Aridisols (soils of the dry regions) are continuously dry, because of inadequate rainfall or excessive drainage, and will grow few plants without irrigation. Most agricultural soils have adequate water to meet vegetation requirement during a considerable part of the year, although plants often suffer during periods of drought.

1.14 Soil Is a Medium of Plant Growth

Soil is one of the factors of plant growth. Terrestrial plants have the following demands on soil: anchorage, water, air, nutrients, and protection from toxins. Plants need adequate air, water, and nutrients in their root zone for optimum growth and yield. For example, plants require some hundred to some thousand grams of water to produce 1 g dry matter (Foth 1990). The capacity of soils to provide air, water, and nutrients depends on their physical, chemical, and biological properties. We find that plants suffer from poor air, water, and nutrient supply in some soils. There are soil management practices to overcome such soil problems. Soil air, water, and nutrients and their management are discussed in detail in Chaps. 5, 6, and 10, respectively.

Plant species differ in their soil requirements because of their evolution in different environments and due to differences in their genetic makeup. The demands of a particular plant even differ at different growth stages. We find *Lantana camara* in acid soils and *Casuarina equisetifolia* on saline soils. Soils suitable for rice may not be so for potato. In other words, all soils are not suitable for all plants. So, selection of crop plants according to the characteristics of the soil along with necessary amendments lies behind the success of crop production.

The capacity of soils to supply nutrients in available forms in adequate amounts and proper balance, and absence of any toxicity is known as soil fertility. On the other hand, the capacity of soil to produce optimum yield under optimum management is known as soil productivity.

Study Questions

1. Define soil. Distinguish between soils and non-soils.
2. Discuss soil as a natural body that undergoes continual changes.
3. What is an ecosystem? Explain that soil is itself an ecosystem besides being part of all terrestrial ecosystems.
4. Why does volume composition of soil differ widely? When does a soil contain much air or much water?
5. How does soil recycle organic matter and purify water?

References

Brady NC, Weil RR (2002) The nature and properties of soils, 13th edn. Pearson Education Inc., New Delhi
Buol SW, Hole FD, McCracken RJ, Southard RJ (1997) Soil genesis and classification, 4th edn. Iowa State University Press, Ames
Daniels RB, Hammer RD (1992) Soil geomorphology. Wiley, New York
Dokuchaev VV (1879) Short historical description and critical analysis of the more important soil classifications. Trav Soc Nat St Petersb 10:64–67 (in Russian)
Foth HD (1990) Fundamentals of soil science, 8th edn. Wiley, New York
Hilgard EW (1914) Soils. The Macmillan Company, New York
Jenny H (1941) Factors of soil formation a system of quantitative pedology. McGraw-Hill Book Company Inc., New York
Jenny H (1980) The soil resource: origin and behavior. Springer, New York
Joffe JS (1949) Pedology, 2nd edn. Somerset Press, Somerville
Kellogg CE (1960) Foreword. Soil Classification, a comprehensive System 7th Approximation. J Soil Sci 11:172–185
Lodha RM (1996) Environment and industry: an alarm. Shiva Publishers Distributors, Udaipur
Ramann E (1905) Bodenkunde, 2nd edn. Julius Springer, Berlin

Schaetzl RJ, Anderson S (2005) Soils: genesis and geomorphology. Cambridge University Press, Cambridge

Simonson RW (1978) A multiple-process model of soil genesis. In: Mahaney WC (ed) Quaternary soils. Geo Books, Norwich

Soil Survey Staff (1975) Soil taxonomy: a basic system of soil classification for making and interpreting soil surveys. U.S. Department of Agriculture handbook 436. USDA, Government Printing office, Washington, DC

Soil Survey Staff (1999) Soil taxonomy: a basic system of soil classification for making and interpreting soil surveys, 2nd edn, Agriculture handbook number 436. United States Department of Agriculture Natural Resources Conservation Service, Washington, DC

USDA, NRCS (2003) Keys to soil taxonomy, 9th edn. United States Department of Agriculture/Government Printing Office, Washington, DC

Soil as a Part of the Lithosphere

2

The lithosphere is the upper part of the earth. It includes the crust and the solid portion of the mantle. Lithosphere interacts with atmosphere, hydrosphere, and biosphere and produces the pedosphere (the soil with its biotic and abiotic components). The lithosphere contains rocks, minerals, and soils. It is made up with more than 100 chemical elements, but most of them are rare. Only eight elements—oxygen (O), silicon (Si), aluminum (Al), iron (Fe), calcium (Ca), sodium (Na), potassium (K), and magnesium (Mg)—constitute more than 99% of its volume. In the earth's crust, these elements generally form crystalline solid compounds of definite chemical composition which are known as minerals. Chemically, minerals may be sulfides, sulfosalts, oxides and hydroxides, halides, carbonates, nitrates, borates, sulfates, phosphates, and silicates. Most rock-forming minerals are, however, aluminosilicates of Ca, Mg, Na, and K because these elements are most abundant. Minerals are aggregated into rocks. Rocks may be igneous, sedimentary, and metamorphic. Igneous rocks are formed by solidification of magma or lava, sedimentary rocks are formed by lithification of sediments or by precipitation from solution and consolidation of remnants of plants and animals, and metamorphic rocks are formed from preexisting rocks by the change temperature and pressure in the solid state. By the action of natural forces over geological time, rocks and minerals are disintegrated and decomposed into new minerals and new compounds such as salts, acids, bases, and soluble substances. The processes are collectively known as weathering. However, the effects of rocks and minerals on mature soils are usually temporary. Their effects are profound in young and immature soils. Eventually, similar soils may develop from dissimilar rocks depending on other soil-forming factors.

2.1 Lithosphere Is the Outermost Part of the Earth

There are four concentric zones in the earth's interior (Fig. 2.1). The innermost zone is called the inner core which is a solid, spherical mass of iron, having a radius of 1,216 km. Above it is the outer core which is a layer of molten liquid rich in nickel and iron. It is about 2,270 km thick. The outer core is overlain by the mantle, which is solid but soft. The mantle is about 2,900 km thick. The crust is the outermost zone and is the hardened exterior of the earth. It varies in thickness from about 5–50 km. The uppermost part of the earth, the crust, and a part of the mantle which is hard, consolidated, and brittle together constitute the lithosphere (from Gk "*lith*" meaning rock). It extends about 100 km down the continental surface (Holmes 1979).

2.2 Lithosphere Interacts with Atmosphere, Hydrosphere, and Biosphere to Form the Pedosphere

Due to its unique position in space, the lithosphere continually interacts with atmosphere, hydrosphere, and biosphere and produces the pedosphere (Juma 2001). Pedosphere includes soils and their living and nonliving components. Figure 2.2 represents the interactions among these spheres.

Atmosphere is the layer of gases and vapor that surrounds the earth. It consists of a mixture of the gases, nitrogen (78.084% by volume), oxygen (20.9476%), water vapor (variable), carbon dioxide (0.0314%), and inert gases such as argon (0.934%) and some other rare gases (Lide 1996). Carbon dioxide critically controls the temperature and the climate on the earth. Biosphere is part of the earth's space in which organisms live, interact, and form ecosystems. It includes the primary producers, consumers, decomposers, and transformers. Hydrosphere is the total body of water that exists on or close to the earth's surface. Hydrosphere includes surface water and groundwater. The lithosphere and biosphere, respectively, impart mineral and organic inputs to soil. Lithosphere, biosphere, hydrosphere, and atmosphere together also provide other material and energy inputs, and set conditions for soil development.

K.T. Osman, *Soils: Principles, Properties and Management*,
DOI 10.1007/978-94-007-5663-2_2, © Springer Science+Business Media Dordrecht 2013

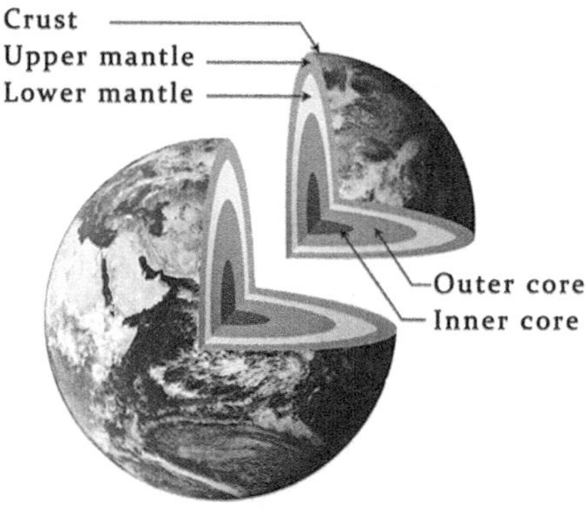

Fig. 2.1 Zones in the earth's interior

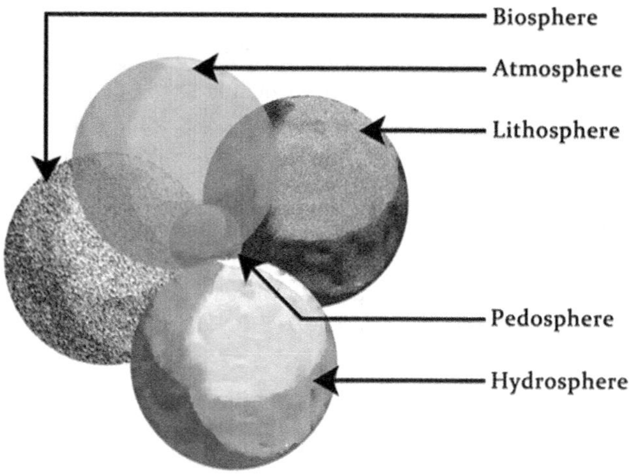

Fig. 2.2 Interactions of lithosphere, biosphere, atmosphere, and hydrosphere to produce the pedosphere

2.3 Eight Chemical Elements Constitute the Bulk of the Earth's Crust

Almost a 100 chemical elements are found in the earth's crust, but only a few makes up its bulk. Table 2.1 shows the abundance of some common elements of the earth's crust.

2.4 Chemical Elements in the Earth's Crust Form Minerals Under Natural Conditions

Only a few elements may be present in pure elemental forms in the earth's crust. They are called native elements. They include copper, gold, lead, mercury, nickel, platinum, and silver. They are found in ores. Most of the elements remain in different combinations as minerals. A mineral is defined as a naturally occurring, inorganic, crystalline solid that has a definite chemical composition (or a definite compositional range). Many minerals are composed of only one element; for example, diamond and graphite are composed of carbon alone. Most

Table 2.1 Elemental composition of the earth's crust

Element	Percent by weight[a]
Oxygen	46.6
Silicon	27.7
Aluminum	8.1
Iron	5.0
Calcium	3.6
Sodium	2.8
Potassium	2.6
Magnesium	2.1

[a]Source: http://hyperphysics.phy-astr.gsu.edu/hbase/tables/elabund.html

minerals are formed by two or more elements. On the basis of their origin, minerals may be divided into primary and secondary minerals. Minerals formed during solidification of magma or lava are called primary minerals, and minerals formed by alteration of the primary minerals and by resynthesis and recrystallization during weathering are secondary minerals. Chemically, the broadest groups of minerals are sulfides, sulfosalts, oxides and hydroxides, halides, carbonates, nitrates, borates, sulfates, phosphates, and silicates (Bishop et al. 2001). For their relative abundance in the earth's crust, silicon and oxygen are found as the major constituents of most minerals. The simplest form is the oxide of silicon, SiO_2. The most common mineral with this formula is quartz. However, the most abundant minerals in rocks of the earth's crust are the silicates. Since the third abundant element in the earth's crust is aluminum, most minerals are aluminosilicates of one or more of the remaining elements.

2.5 Silicate Minerals Are Important Rock and Soil Constituents

Important silicate minerals of rocks and soils include quartz, feldspars, pyroxenes, amphiboles, mica, and olivines. Quartz (SiO_2) is dominant in all types of rocks and coarse fraction of soils; feldspars include framework silicates, which, in addition to silicon, aluminum, and oxygen, contains potassium, sodium, or calcium. Those that contain potassium are called potassium feldspars such as orthoclase ($KAlSi_3O_8$) and microcline. Sodium and calcium containing feldspars are plagioclase feldspars, such as albite ($NaAlSi_3O_8$) and anorthite ($Ca_2Al_2Si_2O_8$) and their intergrades. Pyroxenes have the general formula $(Na,Ca,Mg,Fe,Al)_2Si_2O_6$. Augite, enstatite, hypersthene, diopside, rhodonite, etc., are important members of the pyroxene family. Amphibole is the family name of another very large and complex group of minerals. Hornblende $[Ca_2,Na(Mg,Fe^{2+})_4(Al,Fe,Ti)\ (Al,Si)_8O_{22}(O,-OH)_2]$ is the main rock-forming mineral of the family. Mica includes minerals with a sheet silicate structure. The two most common micas are biotite or black mica $[K(Mg,Fe)_3(AlSi_3O_{10})$

$(OH)_2$] and muscovite or white mica [$KAl_2(AlSi_3O_{10})(OH)_2$]. Olivines are magnesium and magnesium–ferrous-containing orthosilicates. They have the general formula $(Mg,Fe)SiO_4$. Silicate minerals containing Mg and Fe such as biotite, pyroxene, amphibole, and olivine are called ferromagnesian or mafic minerals. They are usually dark colored (dark greenish to black). Others, such as quartz, feldspars, and muscovite, are nonferromagnesian minerals. Silicate minerals are important sources of plant nutrients; weathering of these minerals releases soluble-base cations such as K^+, Ca^{2+}, and Mg^{2+}.

2.6 Some Other Minerals Are Also Abundant in Soil

Other kinds of minerals include carbonate minerals such as calcite ($CaCO_3$), magnesite ($MgCO_3$), and dolomite ($CaCO_3{\cdot}MgCO_3$); hydroxide minerals such as gibbsite [$Al(OH)_3$], boehmite [$AlO(OH)$], and goethite [$FeO(OH)$]; oxide minerals such as hematite (Fe_2O_3), magnetite (Fe_3O_4), ilmenite ($FeTiO_3$), pyrolusite (MnO_2), and zircon (ZnO_2SiO_2); phosphate minerals such as apatite [$Ca_5(PO_4)_3{\cdot}(F,Cl,OH)$]; sulfate minerals such as gypsum ($CaSO_4{\cdot}2H_2O$); and sulfide minerals such as chalcopyrite ($CuFeS_2$), galena (PbS), and pyrite (FeS_2). They are found in soils and in abundance in some rocks. For example, limestone is dominated by calcite, and the rock phosphate is mainly composed of apatites. Limestone parent materials produce calcareous soils, and apatite is the most important natural source of phosphorus. Most commercial phosphate fertilizers are manufactured from rock phosphate-containing apatite. The most important secondary minerals are the clay minerals such as kaolinite, smectite, vermiculite, illite, and chlorite (Sect. 8.3.1). Clay minerals are abundant in the finer fraction of soil and influence almost all its physical and chemical properties. Soils rich in clay minerals, especially smectites and vermiculites, are fertile and may be productive if properly drained.

2.7 A Few Minerals Constitute the Bulk of the Earth's Crust

Most of the several thousands of minerals that have so far been identified are rare. Only about 100 minerals constitute the major mineral components of rocks. Most of these minerals are also relatively uncommon. The bulk of the earth's crust is made up with a few silicate minerals (Fig. 2.3).

Feldspars constitute more than 50%, including 39% as plagioclase and 12% as orthoclase (Plummer et al. 2001). Quartz and pyroxene also contribute similar proportions to orthoclase. Olivine constitutes only 3%. For this, high proportions of quartz and feldspars are found in soils.

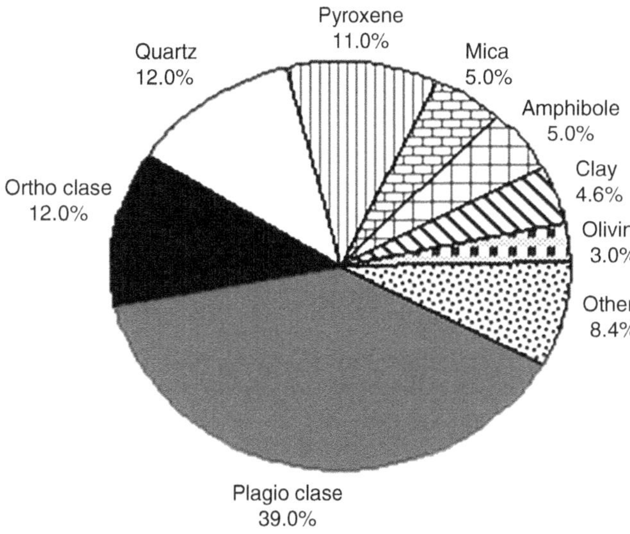

Fig. 2.3 Relative abundance of minerals in the earth's crust

2.8 Rocks Are Aggregates of Minerals

Rocks are the naturally occurring hard and consolidated inorganic materials of the earth. Coal and some limestone have, however, developed from plant and animal remains. Igneous rocks originate through the solidification of lava on the surface or magma in the interior of the earth. Flow of magma through preexisting igneous or sedimentary rocks gives rise to metamorphic rocks. Igneous and metamorphic rocks are weathered and form sediments which are deposited and lithified later into sedimentary rocks usually at a place different from its origin.

A rock can be composed of one mineral or of a large number of different minerals. Limestone and marble are composed mainly of calcite ($CaCO_3$) and rock phosphate of apatite [$Ca_5(PO_4)_3{\cdot}(F,Cl,OH)$] with some other impurities. Most rocks are, however, assemblages of several different minerals. Igneous rocks are composed of primary minerals, predominantly silicate minerals. Sedimentary rocks contain some original primary minerals and at the same time possess some altered and newly synthesized secondary minerals.

2.9 Igneous Rocks Are Formed by Solidification of Magma and Lava

Hot, molten, igneous material within the earth's interior is called magma. When magma erupts on the earth's surface, it is called lava. Magma originates generally in the asthenosphere and rises toward the surface of the earth through lines of weakness, fractures, fissures, joints, etc. On its way to the surface through the crust, magma exerts tremendous pressure and heat, and modifies (or metamorphoses) the adjacent rocks. However, magma may solidify at very great depths or near the surface. When solidification occurs near the surface, the rock is called hypabyssal rock. When solidification occurs deep within the earth, the rock is called plutonic or intrusive rock. Granite, gabbro, and diorite are examples of intrusive rocks. Igneous rocks are sometimes found to overlap sedimentary and metamorphic rocks. Many intrusive rocks, such as granite, have been exposed at the surface of the earth by uplift, erosion, and weathering. At present, there are many fertile soils that have developed from weathering products of granite.

When lava erupts on the earth's surface, volatile materials present in it generally escape, and the molten material forms the type of igneous rocks known as volcanic or extrusive rocks. Examples of extrusive rocks are basalt, rhyolite, and andesite; they are all fine-grained rocks. Lava cools relatively rapidly on the surface and very slowly beneath. The rate of cooling has a profound influence on the grain size of the minerals in rocks. Larger minerals form due to slow cooling, and smaller minerals develop by rapid cooling. Therefore, extrusive rocks are generally fine textured, and intrusive rocks are coarse textured. Some rocks may be of mixed textures. Sometimes lava may cool so rapidly that crystallization process does not take place. The resulting rocks are glassy or obsidian. Igneous rocks having silica content <45% are called ultramafic (serpentine, periotite), between 45 and 55% mafic (basalt, gabbro), between 55 and 65% intermediate (amphibolite, andesite), and >65% silicic or felsic rocks (granite, rhyolite); soils have developed from all these igneous rocks and volcanic ashes.

2.10 Sedimentary Rocks Are Mainly Formed by Lithification of Sediments

There are three modes of sedimentary rock formation: (1) lithification of sediments containing fragments of preexisting rocks, (2) precipitation from solution (i.e., chemically formed), or (3) consolidation of remnants of plants and animals (i.e., organically formed). Sedimentary rocks containing fragments of earlier rocks are known as clastic sedimentary rocks. The fragments may be rock pieces, such as pebbles of granite or shale, and individual mineral grains, such as sand-sized quartz and feldspar crystals. Clastic sedimentary rocks may be classified as coarse-grained, medium-grained, and fine-grained sedimentary rocks. Coarse-grained sedimentary rocks include conglomerate formed by cementation of rounded gravels and pebbles, and breccia formed by cementation of angular fragments. Medium-grained sedimentary rocks include sandstone formed by cementation of sand grains and siltstone formed by cementation of silt-sized particles. Examples of fine-grained sedimentary rocks are shales which are formed by cementation of clay-sized particles. Mudstone is formed by cementation of silt- and clay-sized particles together. Chemically formed sedimentary rocks include carbonate-containing rocks, such as limestone, which is composed of the mineral calcite ($CaCO_3$), dolostone composed of the mineral dolomite ($CaCO_3{\cdot}MgCO_3$), chert composed of hard reprecipitated silica, and rock salt composed of halite (NaCl). Many limestones are formed from consolidation of $CaCO_3$-rich shells of marine organisms. Other organically formed sedimentary rocks include coal and peat. Plants, animals, and their remnants are sometimes preserved and silicified beneath earth's surface. These are called fossils.

2.11 Metamorphic Rocks Are Formed from Preexisting Rocks by Change in Solid State

Metamorphic rocks are formed from preexisting rocks (igneous, sedimentary, or other metamorphic rocks) by the change in solid state either in texture or mineralogical composition through the action of heat, pressure, chemically active solution, etc. Thermal or contact metamorphic rocks are formed during intrusion of magma and found along margins of intrusive bodies, while regional metamorphic rocks occupy large areas and reveal much uplift and erosion after their formation. All types of preexisting rocks may undergo metamorphism.

2.12 Sedimentary Rocks Predominate in the Earth's Surface While Igneous Rocks in the Crust

Sedimentary rocks predominate in the petrological composition of the surface of the earth, but igneous rocks are more abundant when the entire crust is considered. Figure 2.4a shows relative abundance of different rocks within 5 km of the earth's surface, and Fig. 2.4b shows rock composition of the earth's crust.

2.13 Soil Characteristics Differ on Rock and Mineral Sources

Rocks and minerals produce parent materials, and they may have influences in determining characteristics of the resulting soils. This has been discussed well in Sect. 3.5 in relation to factors of soil formation. Now, some characteristics of soils likely to develop from different rocks are mentioned:

- *Volcanic Ash*: Volcanic ash deposits are fine grained and weather relatively easily. Plants may invade a new deposit of volcanic ash within only a few years. The resulting soils are fine textured and usually rich in organic matter and plant nutrients. They are fertile soils. Volcanic ash

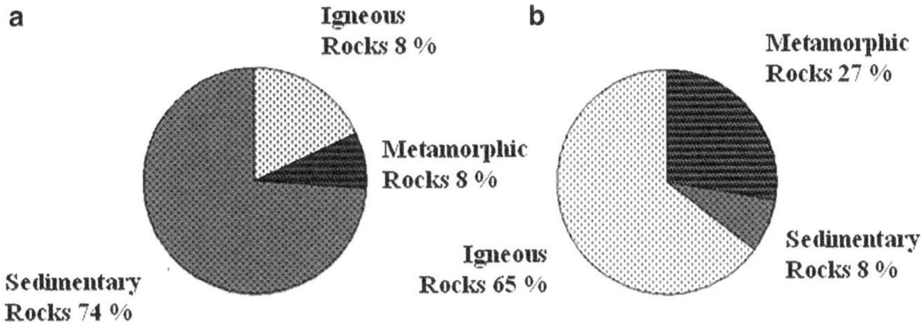

Fig. 2.4 Abundance of rocks in the earth's surface (**a**) and earth's crust (**b**)

soils are named Andisols in Soil Taxonomy of the USDA (Chap. 4). These soils are found exclusively in regions where active and recently extinct volcanoes are located. The soils cover about 124 million ha or 0.84% of the world's land surface (Leamy 1984).

- *Granite*: Granite is a coarse-grained rock and contains on an average about 25% quartz and 65% orthoclase with some small amounts of mica and hornblende. Soils developing from granite are usually sandy in the initial stage. After further soil development, feldspar grains weather into fine-grained clay minerals. Most soils from granite are friable, permeable, acidic, and low in base status with low nutrient contents. However, Jusop et al. (1995) observed that soils from granite in wet zone are low in pH and exchangeable cations, while soils in the dry zone are high in pH and exchangeable bases. Nielsen (2004) observed that most soils developed from granite in Australia are loose and deep, but with little cohesion or consolidation and susceptible to erosion.
- *Basalt*: Basalt is a fine-textured rock. Basalt is rich in ferromagnesian and calcic plagioclase minerals. These minerals weather relatively easily to fine-grained clay minerals. The resulting soils are usually fine textured. Under moderate intensity of weathering, for example, in temperate regions, a good amount of the minerals persist and maintain a high base status in soil. In well-drained and humid conditions, 1:1-type clay minerals (e.g., kaolinite, halloysite) and sesquioxides predominate. Soils developed from tertiary basalts in New South Wales were found to be deep, well-drained, red and with variable amounts of hematite, goethite, and gibbsite (Norrish and Pickering 1983). Soils have low CEC and high phosphate sorption (Jenkins and Morand 2002). Most of them are Ultisols; some may be Oxisols (Chap. 4).
- *Sandstone*: In sandstone, more than 50% materials are sand sized, mainly quartz, with impurities including feldspar and mica, and such cementing agents as silica, iron, and lime. Soils formed from sandstone are coarse textured, acidic, and poorly fertile. However, the characteristics of soils depend on the types of sandstones (grain size, mineralogical composition). Fine-grained sandstones are difficult to weathering, whereas coarse-grained sandstones are less resistant to weathering. So, soils developed from fine sandstones are shallower than soils developed from coarse-grained sandstones (Szafranek and Sklodowski 2006).
- *Limestone*: Limestone rocks contain mainly calcite ($CaCO_3$) with considerable impurities of other carbonates, silt, clay, quartz, and iron. Resulting soils are generally clayey. In warm humid climates, the Ca and Mg and bicarbonate ions are almost completely removed. USDA (1997) observed that most soils developed from limestone in the Limestone County of Texas were clay and clay loam with some sandy loams.

2.14 Weathering Is the Disintegration and Decomposition of Rocks and Minerals

The processes involved in the disintegration (fragmentation, splitting, detachment, etc.) and decomposition (a chemical change leading to the formation of simpler products) of rocks and minerals are collectively called weathering. Rocks are weathered by the action of such natural forces as heat, water, wind, glacier, organisms, chemical reactants, etc. The changes are either physical or chemical, or both. When rocks are only changed physically without any alteration of the chemical composition, the process is known as physical weathering. When the chemical and mineralogical composition of rocks are altered, or when a mineral is decomposed into soluble and insoluble products or gives rise to a different mineral, the process is called chemical weathering.

The principal agents of physical weathering are heat, water, wind, glaciers, and organisms. The most remarkable physical weathering is caused by temperature. It may be regarded as thermal weathering. Other physical weathering processes may be included in mechanical weathering. Chemical weathering occurs by reactions of minerals with acids, bases, salts, ions, etc., or even with water. Organisms produce carbonic acid and other

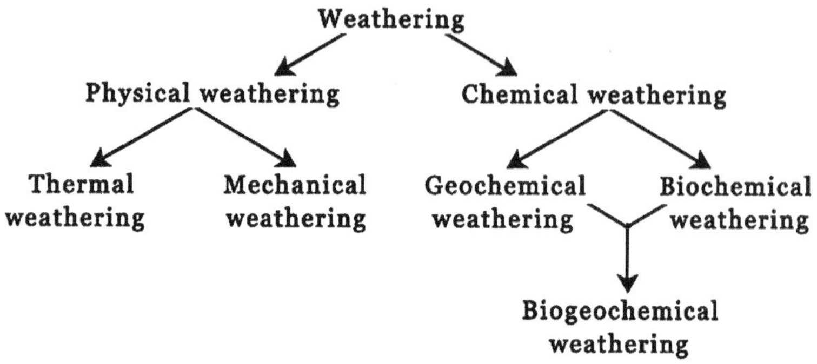

Fig. 2.5 Types of weathering

organic acids which actively participate in chemical alteration of rocks. Chemical weathering is, therefore, sometimes divided into geochemical weathering and biochemical weathering. Together they bring biogeochemical weathering (Fig. 2.5). Weathering of rocks and minerals produces parent materials of mineral soils.

2.14.1 Thermal Weathering Is Caused by Variation in Temperature

Expansion and contraction of rocks due to fluctuation in temperature cause thermal weathering. Rocks and minerals expand when heated by solar radiation and contract when cooled. Rocks expand during the day and contract at night. More expansion occurs during summer than in winter. But not all parts of the rock body are equally exposed to heat. So, different parts of rock expand and contract at different rates. Different minerals expand and contract at different rates as well. Since a rock is usually composed of a variety of minerals, and they expand and contract differently, internal pressures are created when rocks expand or contract. Due to this internal pressure, a rock on the earth's surface is ultimately fragmented after prolonged exposure to fluctuations of temperature.

Exfoliation is a process of thermal weathering which occurs in layered rocks, rocks with approximately homogenous composition at different zones, or rocks expanding almost at similar rates up to a depth. The outer part of a boulder is expanded and contracted usually at higher rates than the inner parts and may be detached in thin layers. In this process, a thin layer of rock is detached from the rock body, resembling the peeling up of onions.

2.14.2 Mechanical Weathering Is Caused by Water, Glacier, Wind, and Organisms

Alternate wetting and drying is an important process of mechanical weathering. Many rocks absorb water and swell when wetted and shrink when dried. Expansion and contraction in volume due to alternate wetting and drying create stresses that break up rocks into pieces. Water may accumulate in fractures and joints of rocks. When the temperature is lowered below the freezing point, ice is formed. Volume of ice increases by 9% and exerts tremendous lateral pressure on surrounding rocks. For this pressure, already fractured rocks are divided into several fragments. It is sometimes called frost wedging. Another type of mechanical weathering by freezing water is known as frost heaving, by means of which rock fragments are lifted upward. Flowing water has a considerable cutting action. Presence of suspended materials and rock fragments in water highly increases the capacity of this cutting action. By the water flow, the suspended materials roll and collide and disintegrate themselves as well as the walls and beds of rocks through which they travel. Large bodies of ice slowly moving along slopes are called glaciers. Glaciers forming on already weathered surfaces contain variously sized rock fragments including boulders. When glaciers flow down slope, ice begins to melt, and the rock fragments shatter one another. The flowing glaciers tend to or do smash whatever rock they find obstructing their way.

When an overlying rock is removed due to weathering and erosion, the underlying rock expands due to release of pressure. This might cause the underlying rock split. Another weak weathering agent is the wind. However, the presence of suspended particles increases the abrasive power of wind. Abrasion by wind over a long time may considerably weather rock surfaces. Large rock masses at higher elevation are pulled down by the gravity, and they may be detached. Deposition of such materials on piedmont and foothill are called colluviums. Plants may extend their roots in rock fractures and crevices. When they grow, the roots exert lateral pressure. Many burrowing animals ingest sand and gravel, and as this material passes through their gut, it gets physically comminuted (Suzuki et al. 2003).

2.14.3 Chemical Weathering Brings Chemical Changes in Rocks and Minerals

The main processes of chemical weathering are solution, hydration, hydrolysis, oxidation, reduction, carbonation, and other acid reactions. Minerals are not ordinarily soluble in water. For hundreds of years of wetting and for the presence of natural acids in water, which accelerates dissolution, minerals are slowly dissolved in natural water. However, many minerals particularly iron and aluminum oxides are hydrated by the binding of water to their molecules. Hydrolysis is probably the most effective chemical weathering process by which minerals are decomposed. During hydrolysis, water molecule splits into hydrogen and hydroxyl ions, and the hydrogen replaces a cation from the mineral structure. Due to this replacement, the mineral becomes susceptible to decomposition by further hydrolysis or other chemical reactions. Some minerals are oxidized; some others are reduced depending on the conditions. Carbonation is a very active and common chemical weathering process. Most carbonation processes are biochemical in nature. Carbonic acid is naturally produced by dissolution of CO_2 in water. Carbon dioxide is available from the atmosphere and as products of microbial and root respiration. Carbonic acid dissolves many minerals. Many other inorganic and organic acids are available in nature. Nitric acid may be present in rainwater. Sulfuric acid is produced by oxidation and hydrolysis of pyrite.

Plants and their associated microbiota directly impact chemical weathering by generating chelating compounds, by modifying pH through production of CO_2 and organic acids, by altering the exposed surface areas of minerals via nitrification, and by affecting the residence time of water (Drever 1994). Of these, the production of organic acids may be the most important contribution to biochemical weathering in soils. These acids can either be released directly from the organism or be derived as by-products of organic decomposition.

2.14.3.1 Examples of Geochemical/Biochemical/Biogeochemical Weathering

Solution

$$\underset{\text{(Gypsum – insoluble)}}{CaSO_4.2H_2O} \xrightarrow[\text{Water}]{} \underset{\text{(Soluble calcium and sulfate ions)}}{Ca^{2+} + SO_4^{2-} + 4H_2O.}$$

Hydration

$$\underset{\text{(Hematite – red)}}{Fe_2O_3 + 3H_2O} \xrightarrow[\text{Hydration}]{} \underset{\text{(Limonite – yellow)}}{Fe_2O_3.3H_2O}$$

Hydrolysis

$$\underset{\text{(Orthoclase – insoluble)}}{KAlSi_3O_8 + H_2O} \xrightarrow[\text{Hydrolysis}]{} \underset{\text{(Hydrogen aluminum silicate – insoluble)}}{HAlSi_3O_8 + KOH}$$

On further hydrolysis,

$$\underset{\text{(Hydrated Al oxide – insoluble)}}{HAlsi_3O_8 + H_2O} \xrightarrow[\text{Hydrolysis}]{} \underset{\text{(Silicic acid – soluble)}}{Al_2O_3.3H_2O + H_2SiO}$$

Oxidation

$$\underset{\text{(Pyrite – insoluble)}}{FeS_2 + O_2} \xrightarrow[\text{Oxidation}]{} \underset{\text{(Ferrous sulfate – soluble)}}{FeSO_4}$$

Reduction

$$\underset{\text{(Goethite – insoluble)}}{FeOOH} \xrightarrow[\text{Reduction}]{} \underset{\text{(Ferrous oxide)}}{FeO + O + H_2O}$$

Carbonation

$$\underset{\text{(Calcite – insoluble)}}{CaCO_3 + H_2CO_3} \xrightarrow[\text{Carbonation}]{} \underset{\text{(Soluble)}}{Ca^{2+} + HCO_3^-}$$

The processes of chemical weathering may operate simultaneously, or one process may follow another.

For example, hydration and carbonation may occur at the same time, or oxidation may follow hydrolysis as

$$\underset{\text{(Olivine)}}{MgFeSiO_4 + H_2O} \xrightarrow[\text{Hydrolysis}]{} \underset{\text{(Serpentine)}}{H_4Mg_3Si_2O_9 + SiO_2 + FeO} \xrightarrow[\text{Oxidation}]{} \underset{\text{(Hematite)}}{Fe_2O_3}$$

2.14.4 Physical and Chemical Weathering May Enhance Each Other

In nature, physical and chemical weathering operates individually or concurrently. In most cases, one accelerates the other. For example, when a large rock body is fragmented, it is more easily acted upon by chemical reactants. When some minerals in a rock are dissolved, it is more easily fragmented (Fig. 2.6).

However, the kind of weathering process that would dominate depends on the rocks and minerals themselves and on the environmental conditions at which weathering is taking place. Physical weathering dominates in extremely dry and cold conditions, and chemical weathering predominates in warm humid conditions.

2.14.5 The Rate of Weathering Varies with Minerals and Climate

Weathering of rocks generally begins slowly and accelerates as the rock breaks up and more surface area is exposed to attack. As weatherable minerals disappear by decomposition, weathering becomes slower again. It is accelerated by biological activity and due to removal of weathering products through leaching. Under similar conditions, different minerals weather at different rates. Generally, the more complex the mineralogical composition is, the more easily the mineral weathers. Plagioclase weathers more rapidly than orthoclase. Hornblende weathers fairly rapidly. Quartz is very resistant to weathering. Jackson et al. (1948) suggested the sequence of susceptibility of common minerals to

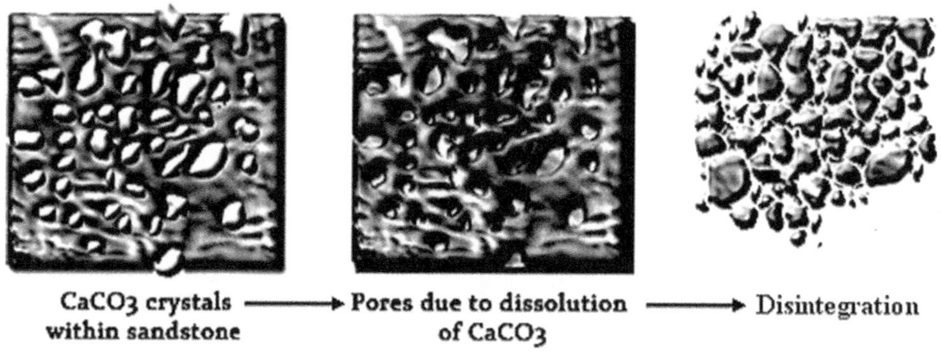

Fig. 2.6 Chemical weathering favors physical weathering

weathering as gypsum, halite > calcite, dolomite, aragonite > olivine–hornblende, diopside > biotite, glauconite, nontronite > albite, anorthite, microcline > quartz > illite, muscovite > hydrous mica > montmorillonite (smectite) > kaolinite, halloysite > gibbsite, boehmite > hematite, goethite, and limonite > anatase. A particular rock or mineral, however, weathers at different rates in different climatic conditions. Chemical weathering is low in areas where water and temperature are low. The rate of weathering is higher in warmer and moister areas. In the humid tropics, not only the rate of weathering is rapid but also the weathering products are lost quickly. Alfisols, Oxisols, and Ultisols (Chap. 4) are highly weathered and leached soils of the humid tropics. Also the products of weathering to be found in a soil depend on minerals themselves and the conditions of weathering (Ollier 1975). However, weathering is a continuous process; it proceeds even after the formation of the parent material and during pedogenesis.

Study Questions

1. Distinguish between primary and secondary minerals with examples. Mention the most abundant mineral groups of the earth's crust
2. Which minerals are found abundantly in the coarser and finer fractions of the soils?
3. Give examples of two mineral sources for each of the plant nutrients K, Ca, Mg, Fe, P, and S
4. Distinguish between plutonic and volcanic rocks. Why do the minerals of these two types differ in their grain size?
5. Explain biogeochemical weathering. With examples, distinguish between hydration and hydrolysis

References

Bishop A, Woolley A, Hamilton W (2001) Cambridge guide to minerals, rocks, and fossils, 2nd edn. University of Cambridge, Cambridge

Drever JI (1994) The effect of land plants on weathering rates of silicate minerals. Geochem Cosmochem Acta 58:2325–2332

Holmes A (1979) Principles of physical geology. Nelson, London

http://hyperphysics.phy-astr.gsu.edu/hbase/tables/elabund.html. Accessed 5 June 2011

Jackson ML, Tyler SA, Willis AL, Bourbeau GA, Pennington RP (1948) Weathering sequence of clay-size minerals in soils and sediments I. Fundamental generalizations. J Phys Chem 52(7):1237–1260

Jenkins B, Morand D (2002) A comparison of basaltic soils and associated vegetation patterns in contrasting climatic environments. In: Roach IC (ed) Regolith and landscapes in eastern Australia. CRC-LEME (Cooperative Research Center for Landscape Environments and Mineral Exploration), Australia

Juma NG (2001) The pedosphere and its dynamics: a systems approach in soil science. Salman Productions, Edmonton

Jusop S, Salleh R, Hanif M, Husni A, Awang K (1995) The mineralogy and chemical properties of soils on granite gneiss in three climatic zones in Sri Lanka. Pertanika J Trop Agric Sci 18(1):45–56

Leamy ML (1984) International Committee on Classification of Andisols (ICOMAND) Circular letter no. 6. New Zealand Soil Bureau. DSIR Lower Hutt

Lide DR (1996) Composition of the earths atmosphere, 77th edn, Handbook of chemistry and physics. CRC Press, London

Nielsen WA (2004) Soil survey of the granite soils of North East Tasmania. Division of Forest Research and Development, Technical report 06/2004, Forestry Tasmania, Hobart

Norrish K, Pickering (1983) Clay minerals. In Soils: An Australian viewpoint. Division of Soils, CSIRO, Melbourne, Academic Press, London

Ollier CD (1975) Weathering. Longman, London

Plummer CC, McGery D, Carlson DH (2001) Physical geology. McGraw Hill, Boston

Szafranek A, Sklodowski P (2006) Properties of acid brown soils developed from sandstones of the Oewiêtokrzyski region. Pol J Soil Sci 39:65–71

Suzuki Y, Matsubara T, Hoshino M (2003) Breakdown of mineral grains by earthworms and beetle larvae. Geoderma 112:131–142

USDA (1997) Soil survey of Limestone County, Texas. United States Department of Agriculture, Washington, DC

3 Factors and Processes of Soil Formation

Comprehensive field studies were carried out in Russia in the latter part of the nineteenth century by Dokuchaev (1879, 1883) and his followers who described the occurrence of different kinds of soils thoroughly using soil morphological and environmental features. They noticed that different environmental conditions were responsible for the development of different kinds of soils. Dokuchaev pointed out close relationship of soil properties with climate and vegetation of Russia. In the United States, Hilgard (1892) also emphasized the relationship between soil properties and climate. Latter, Jenny (1941) explained that any soil property is the function of five soil-forming factors such as climate, organism, relief, parent material, and time. Meanwhile, it was also established that the soil is a three-dimensional body. The vertical dimension which represents the soil profile is differentiated into several horizons by pedogenic processes. There are some basic and some specific soil-forming processes. Simonson (1959) stressed that many genetic processes are simultaneously and/or sequentially active in the genesis of a soil.

3.1 A Soil Profile May Be Differentiated into Several Horizons

A soil is said to have formed when it attains, through natural processes, an appearance (morphology) that is significantly different from the underlying parent material. The parent material is also unconsolidated but is not aggregated and differentiated into horizons. Generally, the soil has a profile which can be defined as the vertical section from surface down to the parent material. However, the identification of the parent material below is not always easy; rather, it becomes sometimes difficult, such as in deep alluviums. Soil Survey Staff (1993) have defined soil profile as "a vertical cut in the soil that exposes the genetic layers or horizons for examination, description, and recording of key features observed as well as for sampling for laboratory analysis if desired." Thus, most soil profiles are made up of a set of distinct horizons.

A horizon is a layer of soil approximately parallel to the land surface which can be differentiated from adjacent layers, or horizons, by identifiable physical, chemical, and biological characteristics. A soil horizon develops over a long time in a place through complex interacting pedogenic processes under a set of environmental conditions. There are five master horizons denoted by the capital letters O, A, E, B, and C (Fig. 3.1). Characteristics of these horizons are given below.

The O horizon is an organic horizon developed on either mineral soil surface or over an organic soil profile. This horizon develops from dead plant and animal residues, particularly under forest vegetation in humid temperate regions. It is seldom found in grassland vegetation. The entire soil profile in an organic soil such as peat soil may constitute the O horizon. Organic matter in some profiles may translocate to a mineral soil layer deep in the profile. This is not designated as O horizon. O horizons may have three subordinate horizons, Oi, Oe, and Oa horizons. Oi is characterized by fresh and largely undecomposed plant and animal residues. This layer is composed of leaves, twigs, and other dead parts of plants. Oe represents a layer of partially decomposed plant and animal material. Oa horizon contains finely divided amorphous material known as humus. These layers are correspondingly designated as L (litter), F (fragmented), and H (humus) layers in Canadian soil classification and in forest soil nomenclature (Chap. 14).

The A horizon is a mineral horizon formed on the surface of mineral soil or below the O horizon where it is present. It contains finely divided humus materials intimately mixed with the mineral matter giving it usually a darker color than the underlying horizons. Finer particles are often translocated from this horizon downward in soils developed from residual parent materials in humid climate. In such soils, the texture of the A horizon is somewhat coarser than the lower horizons.

The E horizon is a highly leached horizon formed below the A horizon in humid climates. Clay, iron and aluminum oxides, and soluble salts, and sometimes humus,

K.T. Osman, *Soils: Principles, Properties and Management*,
DOI 10.1007/978-94-007-5663-2_3, © Springer Science+Business Media Dordrecht 2013

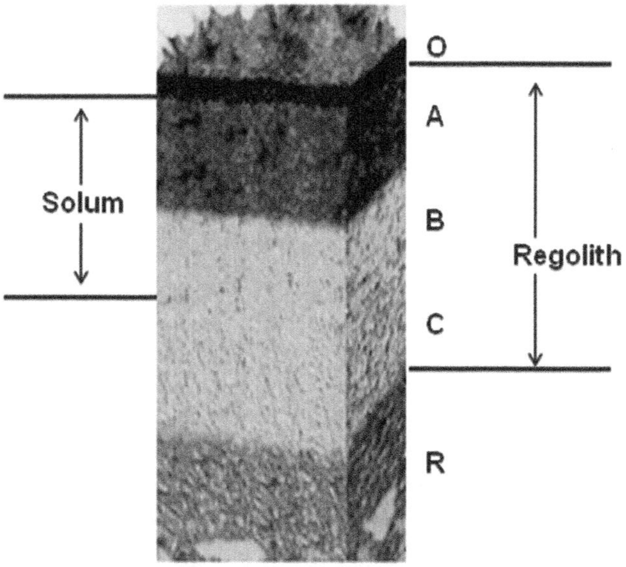

Fig. 3.1 A hypothetical soil profile

are translocated downward giving it a bleached appearance. This process of translocation is known as eluviation. Some eluviation may also take place from the A horizon, but the E horizon is the zone of maximum eluviation. E horizon usually retains resistant minerals in the sand and silt fractions. The color of the E horizon is, therefore, usually grayish to whitish. Prominent E horizons develop in soils under forest vegetation in humid temperate regions.

The B horizon is the zone of maximum illuviation, a process of accumulation of materials eluviated from upper horizons. B horizon usually develops below the A or E horizon, but sometimes E may be absent. In highly eroded soils, the B horizon may be exposed to the surface. Commonly B horizon exists in the middle part of the soil profile. Accumulation of silicate clays, iron and aluminum oxides, and sometimes humus takes place in humid climate. In arid regions, calcium carbonate, calcium sulfate, and other salts may accumulate in the B horizon.

The C horizon is the unconsolidated material, usually below the B horizon. It is the parent material that has not yet been altered by pedogenic processes. The upper horizons A, E, and B of mineral soils develop usually from the parent material by soil-forming processes. However, C sometimes represents a different material from the actual parent material. In sketches of soil profiles, sometime an R horizon is shown. This is confusing because R represents the solid rock body which is not a part of the soil and has not even been altered by the weathering processes. However, it may exhibit the kind of rock from which the residual parent material has originated.

In some soil profiles, a horizon may occur in between two master horizons and having characteristics of both the overlying and underlying horizons. Such horizons are transitional horizons and are designated by capital letters of the two master horizons in between those it occurs. Usually, characteristics of one horizon dominate over the other. The first letter indicates the characteristics of the dominating horizon and the second by the subordinate horizon characteristics. Common examples are AB, AE, EB, BE, and BC horizons. The AB horizon, for instance, contains characteristics of both A and B horizons, and it is more like A than B.

3.1.1 There May Be Subordinate Distinctions of Master Horizons

During and after examination of the soil profile in the field and analysis of samples in the laboratory, it becomes necessary to indicate the physical and chemical features of each horizon. For example, two soil profiles may have the same sequence of the master horizons, but differing in the accumulation of the kinds of materials in the B horizon. This subordinate distinctions are made by specific (as described by Soil Survey Staff 1993, 1994) small letters, followed by the master horizon capital letters. Thus, Ap is used to denote A horizon modified by plowing, Bk is a B horizon having accumulation of alkaline earth carbonates, and so on. The letters used for subordinate distinction of master horizons are listed in Table 3.1.

Under special circumstances, such as variation in color, texture, and structure, and for close examination of the soil profile, the B and C horizons often need to be subdivided even after using the subordinate distinctions. These subdivisions are denoted by Arabic numerals such as B2t. However, a soil profile may contain all the master horizons or may lack one or more of them. There are ample examples of A–C, A–B–C, A–E–B–C, and B–C profiles. In some very steep slopes, all the genetic horizons may be removed by erosion exposing the C or R to the surface.

3.2 Soil-Forming Factors Are Framed in the Fundamental Soil-Forming Equation

Jenny (1941) proposed a soil-forming equation to show relationships of a soil property with the soil-forming factors. In this equation, a soil property is regarded as a dependent variable, which depends on the soil-forming factors—the independent variables. The equation is known as fundamental equation of soil formation which may be written as

$$S = f(Cl, o, r, p, t)$$

where S = any soil property, Cl = climate, o = organism, r = relief or topography, p = parent material, t = time, and f = function.

Table 3.1 Letters to designate subordinate horizons

Letter	Features of distinction
a	Highly decomposed organic matter
b	Buried genetic horizon
c	Presence of concretions or nodules
d	Physical root restriction
e	Intermediately decomposed organic matter
f	Frozen soil or water (permafrost)
ff	Dry permafrost
g	Strong gleying
h	Accumulation of illuvial organic matter
i	Slightly decomposed organic matter
j	Accumulation of jarosite
jj	Evidence of cryoturbation
k	Accumulation of carbonates
m	Cementation or induration
n	Accumulation of sodium
o	Residual accumulation of sesquioxides
p	Disturbance due to plowing
q	Accumulation of silica
r	Soft or weathered bedrock
s	Accumulation of illuvial sesquioxides and organic matter
ss	Presence of slickensides
t	Accumulation of silicate clay
v	Plinthite
w	Development of structure or color
x	Presence of fragipan
y	Accumulation of gypsum
z	Accumulation of salts more soluble than gypsum

Climate and organisms are said to be active factors since their effects are direct and more remarkable. Relief, parent material, and time are called passive factors because their effects are less apparent. The fundamental equation as shown above appears to be a quantitative expression. It gives the impression that any soil property can be estimated from the variables—climate, organism, parent material, relief, and time. But some of these factors cannot be quantified. So, the above equation remains to be indicative rather than determinative.

3.3 Soil Formation Depends on the Interaction of Soil-Forming Factors

Soil-forming factors exert their effects individually and combinedly at the same time. Any soil property should then depend on the resultant of the interactions of these factors. For example, similar soils will develop under similar set of the factors. On the other hand, under similar sets of organism, parent material, relief, and time, differences in properties between two soils would be brought about by climate alone. It also indicates that similar soils would develop under dissimilar sets of conditions provided that the resultant effects of the factors are the same. For explanation, let us consider three different settings of the soil-forming factors below. In the first and third settings, the resultant effect is R_1 despite differences in organism, parent material, and time between two spaces. Theoretically and according to the meaning of the soil-forming equation, similar soil (S_1) will develop in these two different spaces. A different soil (S_2) will be formed in space 3 because the resultant effect is different.

Space	Factor interactions	Resultant effect	Resultant soil
1	$Cl_1 \times o_1 \times p_2 \times r_1 \times t_1$	R_1	S_1
2	$Cl_2 \times o_2 \times p_1 \times r_1 \times t_2$	R_2	S_2
3	$Cl_1 \times o_2 \times p_3 \times r_1 \times t_2$	R_1	S_1

Jenny (1980) states that for a given combination of Cl, o, r, p, and t, the state of the soil system is fixed; only one type of soil exists under these conditions. Any particular combination of these factors will give rise to a certain soil-forming process, a set of physical, chemical, and biological processes that create a particular soil (van Breemen and Buurman 2002). If resultant effects of the combination of factors are regarded as causal, distinguishing soil-forming factors into active and passive appears to be impractical. However, the above examples showing resultant effects in different spaces are arbitrary and may not be obtained under natural conditions. Usually, organism as a soil-forming factor (vegetation in particular) tends to be associated with climate. Since particular soil and vegetation zones are commonly associated with particular climatic zones, they may be thought as interdependent rather than being independent variables. Jenny (1980) regarded that soil and organisms at a space interact and are components of the same ecosystem. Components of such coupled systems evolve together. Buol et al. (1997) regarded climate, parent material, topography, and land use as the driving variables in soil formation. On the other hand, vegetation, animals, microbes, and physical and chemical properties of soil constitute the state variable. Driving variables interact with the state variables over time so that soil and vegetation development proceed together.

3.4 There Are Diverse Effects of Climate on Soil Formation

Conditions of rainfall, temperature, humidity, day length, etc., at a place over a long period are called climate. Climatic conditions at a place depend on geographic location; elevation; velocity and direction of prevalent wind; position of land in relation to sea, ocean, and mountains; characteristics of oceanic current; etc. Thus, climate is a complex phenomenon and exhibit considerable variation within a particular geographic region. For example, there are wet, dry, and desert climates within the tropics.

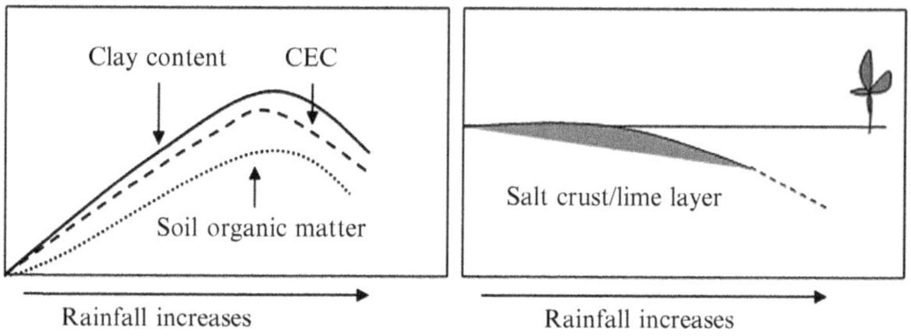

Fig. 3.2 Effect of rainfall on soil properties

The most important components of climate affecting soil formation are rainfall and temperature. Rainfall and temperature affect the kind, degree, and depth of weathering; nature of weathering products that accumulate or are lost; production and decomposition of organic matter; biogeochemical processes operating in the parent materials; etc. The effects of climate on soil formation are diverse and complex, partly due to its interaction with other factors, but some generalization can be made. The most important role rainfall plays in soil development is through the addition and loss of water to and from the parent material. Water acts as a reactant and a medium of translocation. As rainfall increases, rate of weathering, depth of weathering, diversity of weathering products, formation of clay, release of soluble substances, and organic matter production increase. However, leaching of materials in solution and suspension also increases.

In very dry conditions, although physical weathering can proceed well, chemical weathering is negligible. The resulting parent material is sandy with little signs of soil development. As one moves toward regions where there is some rainfall but the potential evaporation is higher than precipitation, one should experience that there is some chemical weathering and formation of soluble salts. Due to insufficient water to leach them downward and since some salts reach the surface with capillary rise of water, a salt crust may develop on thesurface soil. Sometimes, $CaCO_3$ may also accumulate on the surface (lime layer). Proceeding to further moister areas would reveal that the lime and salt layers are present at higher depth below the surface. These soils are classified as Aridisols (Chap. 4). On further increase in rainfall, the salt crust or the lime layer may be lacking. In very moist conditions, both formation and leaching of clay, soluble salts, humus, etc., are high. Contents of clay, organic matter, CEC, and nutrients increase as rainfall increases. Under very humid condition, contents of clay, humus, and available nutrients may decrease at least in the upper portion of the soil. Due to leaching of bases, soils become more acidic. Figure 3.2 shows the effect of rainfall on some soil properties. Under intense weathering in humid conditions, quartz tends to increase due to decomposition and removal of feldspars and other easily weatherable minerals. Some soluble Fe/Al may reprecipitate as their hydrous oxides.

3.4.1 Pedoclimate May Be More Important than Atmospheric Climate

Texture of parent material and soil greatly influences the proportion of rainfall that would be effective in soil formation. A coarse-textured parent material absorbs water rapidly, but it retains less water and allows more water to percolate than fine-textured parent material. Percolating water helps in leaching of colloidal and soluble materials downward. On the other hand, retention of higher amount of water should lead to higher chemical weathering. On a sloping land, a part of the rainfall moves down the slope and reaches nearby plains. So, soil on a sloping land has less effective rainfall for soil development. Vegetation creates microclimate through interception, stemflow, and transpiration. Soil temperature is greatly influenced by vegetation cover, humus, and moisture contents. Therefore, pedoclimate is another component of the climatic factor. Figure 3.3 gives the examples of different pedoclimate created by topography and soil texture.

Soil development requires geological time, and climate of an area may have changed several times during the past. It often becomes difficult to correlate the present soil properties with the present-day climate. It may also be difficult to understand the resultant effect of the past climatic changes from present-day soils. Many previous soils have been buried under sediments from which new soils have developed.

Soils in Different Climates

High temperature and high rainfall: In humid tropics, average mean temperature lies around 20 °C, and precipitation, mainly rainfall, varies from 1,000 to more than 3,000 mm. There may be continuous precipitous or seasonal rainfall areas. Evapotranspiration is less than precipitation so that there is plenty of available soil moisture for weathering, leaching, and plant growth. Biomass production is high, but decomposition is also high. Parent material is deeply weathered, and most minerals except quartz and some feldspars are removed. Bases are leached out, and the surfacesoil is usually moderately acid. Hydrous oxides of iron

(continued)

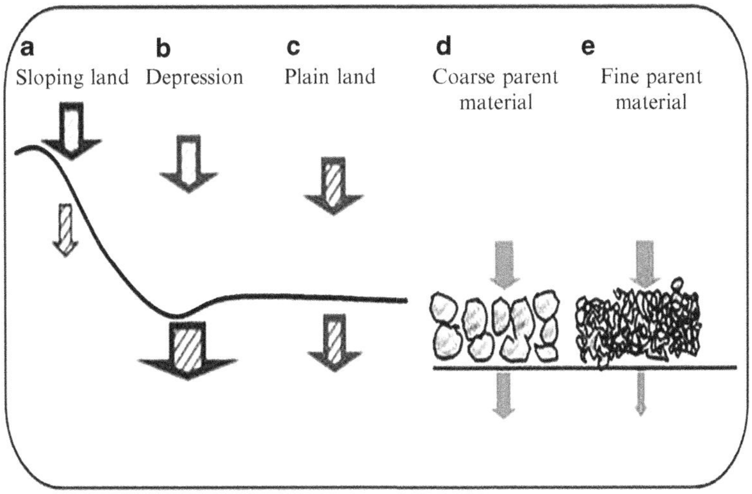

Fig. 3.3 Rainfall characteristics are similar, but effective moisture for soil development is different; (*a*) in sloping land, less water enters the soil; (*b*) in depression, amount of infiltration exceeds amount of rainfall. More water percolates in coarse-textured (*d*) than in a fine-textured (*e*) soil. In the above figure, (*a*), (*b*), and (*c*) show how topography modifies the effects of rainfall and determines the effective moisture for soil development; (*d*) and (*e*) show how parent material does the same

(continued)

and aluminum accumulate. Mineralogically thereare four soil types: Kaolinitic, Oxidic, Allophanic, and Smectitic soils (Juo and Franzluebbers 2003). Taxonomically the most residual soils are Alfisols, Oxisols, and Ultisols with some other soil orders.

Low temperature and high rainfall: In temperate regions, the average temperature of the coldest months is between 18 and −3 °C, and the average temperature of the warmest months is >10 °C. Evapotranspiration is less than precipitation so that there is enough available moisture for intense physical and chemical weathering and leaching. Decomposition of organic matter is less than organic matter production, and as a consequence, an O horizon commonly develops in undisturbed forest soils. Decomposition of organic matter produces organic acids which favor formation of extreme surface soil acidity. Prolonged eluviation produces a prominent E horizon below the A horizon. Illuviation of clay, humus, Fe/Al, etc., may take place in the B horizon (spodic horizon). The resulting soils are Spodosols. There are many Histosols and Alfisols as well.

Short cold summers and long extremely cold winters: Soil develops in regions with mean temperature ranges from 3 to 10 °C in the warmest and below −3 °C in the coldest months. Precipitation is low (60–160 mm) and occurs mostly as snow. Vegetation is shrub tundra in the south, and dwarf shrubs, herbs, mosses and lichens in the north (Margesin 2009). Extensive areas in northern Canada are covered with boreal forests. The soils are frequently covered with snow and remain frozen in the winter months. In vast areas, the soils below a depth remain permanently frozen (permafrost). The upper part of soil freezes in the colder months and thaws in warmer months. This part of soil is known as the active layer. The dominant soil-forming processes are strongly influenced by the occurrence of permafrost affected by cryoturbation and in some areas gleying in the unfrozen zone (Nortcliff 2009). Most soils are Gelisols intermittently, there are Histosols and Spodosols.

3.4.2 Climate Changes with Time

Soils of the Past Climate

Ancient soils or Paleosols developed under a past and different climate. They have no relevance in their physical and chemical characteristics to the present-day climate or vegetation. Such soils have formed on extremely old continental landforms in outliers of ancient rocks. Many Alfisols, Oxisols, and Ultisols developed under tropical rainforests and even Spodosols developed under temperate broadleaf forests have been subjected to the impacts of a different climate, for example, arid, for a successive long period. Paleosols are found in many parts of the world including Australia, Africa, Asia,

(continued)

(continued)

and America. Paleosols are sometimes found embedded in sedimentary rocks (Retallack 2001). Many pre-existed soils have been converted into soil fossils. In cases where volcanoes have been active, some soil fossils occur under the volcanic ash. Paleosols are of considerable archeological importance. Properties of such soils may indicate the agrological and vegetation characteristics of those soils and the nature of the climatic conditions under which they developed.

3.5 Organisms Provide Organic Inputs and Biochemical Transformations

Organisms include vegetation, animals, and microorganisms living in and on the soil or exploit it. Organisms add organic matter; transform organic and mineral matter; translocate ions, compounds, and particles; enhance soil erosion or protect soil; and remove some substances during soil formation. Vegetation penetrates roots and creates channels that affect movement of water within the soil. Plants shed leaves and other litter, which on decomposition form humus. Humus becomes admixed with mineral matter in soil. Plants absorb water and nutrients and secrete metabolites through their roots. Plants circulate nutrients in soil by absorbing them from depth and returning them on surface. The amount and proportion of nutrients absorbed from the soil differs with plant species. Some plants absorb and return more bases than others. Litter from such vegetation tends to prevent soils against becoming more acidic. Some plants may create soil acidity, too.

Soil microorganisms are important transformers. They mineralize organic matter; transform nitrogen, phosphorus, sulfur, iron, etc.; and produce organic substances, which form complexes and chelates with metals. Soil insects and animals add organic matter and mix soil materials laterally and vertically. Some insects consume soil particles and transform them biochemically. Worm and beetle casts are remarkable examples of transporting and transforming action of insects. Burrowing animals have tremendous capacity of mixing soil materials. A set of soils developed under similar climate, parent material, relief; and time along a vegetation gradient (such as grassland—grassland–forest transition—forest) is known as biosequence.

Soils Under Different Natural Vegetations

Grasslands: High organic matter content to a considerable depth below the surface, continuous root production, and high interception of rain. Typical soil order—Mollisols

Temperate coniferous forests: Low retention of bases (Ca, Mg, K) in vegetation, low recycling of nutrients, highly leached, and acidic soils. Typical soil order—Spodosols

Temperate deciduous forests: High retention of bases in biomass, high recycling of nutrients, and slightly to moderately acidic soils. Typical soil order—Alfisols

Tropical broadleaf forests: Intensive weathering and leaching, enrichment of oxides and hydroxides of Fe and Al, low organic matter accumulation, and moderately acidic soils. Typical soil order—Oxisols

3.5.1 Human Affects Soil Formation

In managed ecosystems, human modifies the soil-forming factors, alters the natural course of soil development, and directs it to a new course. Human interventions are either direct—through plowing, liming, manuring, and fertilizing—or indirect through changing the natural soil-forming factors, changing the vegetation by deforestation, changing the relief by leveling and terracing, changing the moisture regime through irrigation and drainage, and changing the parent material through transport, dumping, or the exploitation of peat and rocks, through erosion and sedimentation (Dudal 2004).

According to Amundson and Jenny (1991), it was not appropriate that human impact be integrated in the soil-forming factor "organisms." Humans differ from other species in the animal kingdom by their capacity for abstract reasoning and the development of culture. Human impact is goal directed, and differences between societies produce different effects on the environment. Dudal (2004) suggested that human be considered as the sixth factor of soil formation. Human contributes to global soil change (Arnold et al. 1990).

For managed soils, Rowell (1994) proposed the following soil-forming equation:

$$S = f(Cl, o, r, p, t_1) + f(m, t_2)$$

where Cl, o, r, and p have the usual meanings; t_1 stands for time for which the soil developed under natural conditions; m stands for management during and after bringing soil into cropping or other use; and t_2 is the time for which the soil remains under such management. It is, however, difficult to estimate the effects of the management practices for long periods of time. Cropping patterns and cultural practices certainly changed several times during the past. However, in many soils, the results of management, ill or well, are spectacular. Many soils have been degraded and left barren afterward.

Anthropogenic Soils

Anthropogenic soils are soils that bear the signs of long-term human impacts in their characteristics. They are human-altered and human-transported soils (USDA-NRCS 2007). These soils are classified as Anthrepts (Inceptisols having an anthropic or a plaggen epipedon) and in other categories having the formative elements Anthraltic, Anthraquic, Anthropic Humic, Anthropic, Anthroportic, Plaggic, and Plagghaplic in USDA Soil Taxonomy (Chap. 4). In WRB 2006 (FAO 2006), two reference soil groups of anthropogenic soils were recognized—Anthrosols (soils that have been formed or profoundly modified through long-term human activities, such as addition of organic materials or household wastes, irrigation, or cultivation) and Technosols (soils dominated by artifacts and having properties substantially different from those of natural soil materials and recognizable as such in the soil).

3.6 Parent Material Provides Raw Materials for Soil Development

Parent material is the initial state of the soil system. Parent materials provide the main raw materials for soil formation. Properties of parent materials, such as texture, acidity, lime and salt content, and mineralogical composition, are inherited to soils under certain conditions of soil genesis. Coarse-textured parent materials usually produce coarse-textured soils in the initial stage. Acid parent materials tend to produce acid soils. Lime-rich parent materials produce calcareous soils. Mineralogical composition of parent materials greatly influences the mineralogical composition of soil, the type of soluble products, and the kind and amount of clay minerals. However, the influence of parent materials on soils may not be long lasting. Under warm and humid climate, the effects of parent materials largely disappear as the soils grow older. Actually, many different parent materials may produce similar soils depending on other soil-forming factors, particularly climate. The effects of parent materials are very dominant on young and immature soils.

3.6.1 Parent Materials Are Diverse in Origin and Characteristics

Organic parent materials originate from deposition of residues of many generations of vegetation usually under wetland conditions, such as swamps and marshes. The organic deposition is called peat if the residues are sufficiently intact to permit the fibers to be identified. If most of the material has Continuedrial is called muck. Examples of organic parent materials include (1) moss peat, the remains of mosses such as sphagnum; (2) herbaceous peat, residues of herbs such as sedges and reeds; and (3) woody peat, the remains of woody plants. Most organic parent materials are mixtures of different types of peat arranged in successive or alternate layers. Organic parent materials may also occur in non-wetland positions in upland sites where leaf litter accumulates, in coastal rainforests of Canada and elsewhere where precipitation levels can reach 3,000 mm per year where great thicknesses of leaf litter and woody forest material accumulate, and in other forested regions where a layer of leaf litter directly overlies the bedrock surface. Soils developed from organic parent materials are classified as Histosols in Soil Taxonomy (Chap. 4). Histosols are further divided in to four suborders: folists, which are saturated with water for less than 30 cumulative days during normal years and not drained artificially; fibrists, which consist of slightly decomposed organic materials and are artificially drained at times; saprists, which are wet with highly decomposed organic materials; and hemists, which are moderately decomposed organic materials.

3.6.2 Mineral Parent Materials May Be Residual or Transported

When the regolith rests on the underlying rock from which it was produced and soils develop there, the parent material is called residual parent material or sedentary parent material. In many instances, weathered materials are carried away from its place of origin and are deposited in many different other places. If soils develop on these transported materials, the parent material is called transported parent material. Most present-day soils have developed from trans ported parent materials. Thus, mineral parent material is of two

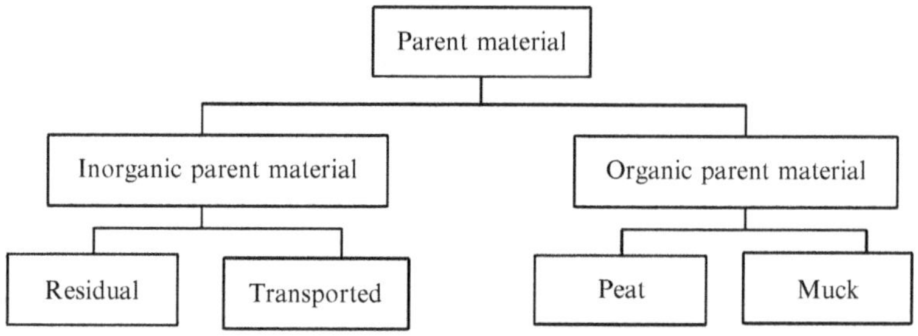

Fig. 3.4 Classification of parent materials

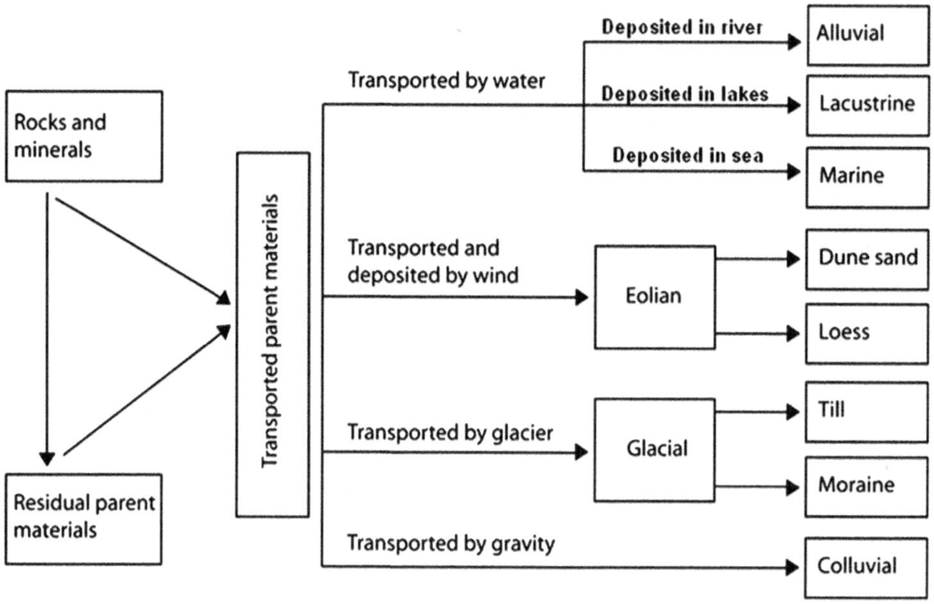

Fig. 3.5 Different types of transported parent materials

types: residual and transported, and organic parent materials are of two types: peat and muck (Fig. 3.4).

There are many kinds of transported parent materials depending on the agents of transportation and places of deposition. The agents of transportation may be water, wind, glacier, or gravity. Water transported materials deposited in riverbeds, lakes, and oceans are, respectively, called alluvial, lacustrine, and marine parent materials. Windblown transported parent materials are eolian parent materials; parent materials carried by glaciers are glacial parent materials, and those transported by gravity are colluvial parent materials (Fig. 3.5).

In many geographic regions as in Bangladesh in Southeast Asia, most soils (>80%) have developed from alluvial parent materials composed of Himalayan sediments. Gangetic alluvium, Brahmaputra alluvium, Meghna alluvium, etc., cover vast areas of this country. The soils are mainly Entisols and Inceptisols. However, alluvial parent materials include (1) floodplains, the part of the river valley that is flooded; (2) alluvial fans, sediments carried by streams in narrow valleys and falling suddenly below and deposited in lower broader valleys; and (3) delta, sediments carried by streams passing through floodplains into lakes, streams, seas, and oceans. These are land masses found in the mouths of rivers, lakes, and in estuaries. Glacial parent materials are carried by glaciers. Materials directly deposited by ice are glacial till. These parent materials contain rock debris of mixed sizes ranging from boulders to clays. Glacial till may be deposited in irregular ridges known as moraines. Eolian parent materials include (1) dune sand, (2) loess, and (3) aerosolic dust. Dune sand is formed by piling up of windblown medium and fine-grained sand into hills. These are found along beaches of large lakes, seas, oceans, and in barren deserts. Loess is formed by deposition of windblown materials largely silt sized with some fine sand and coarse clay. Aerosolic dust is composed of very fine particles (approximately 1–10 μm) and deposited generally with rainwater. Colluvial deposits are found on valleys around mountains and foothills. Like glacial parent materials, these are also composed of irregular sizes of materials. Many soils on transported parent materials are immature soils and are included in Aridisols, Entisols,

Table 3.2 Time required to form soils—some examples

Soil horizon or profile	Depth (cm)	Age (years)	Years required to form 1 cm soil	References
Entisol on volcanic ash	35	45	1.3	Mohr and van Baren (1954)
A_1 of a Hapludoll on loess in Iowa	33	400	12	Simonson (1959)
Histosol in Wisconsin	200	3,000	15	Buol et al. (1997)
Solum of a Hapludalf on weathered loess in Iowa	100	4,000	40	Arnold and Riecken (1964)
Solum of an Ultisol in Australia	300	29,000	97	Butler (1958)
Solum of an Oxisol in Africa	100	75,000	750	Aubert (1960)

Inceptisols, and Vertisols orders. Some mature soils have also developed from old sediments. For example, Beavers et al. (1955) reported Mollisols in Illinois, USA, and Mcintosh (1984) observed Spodosols in Otago, New Zealand developed from eolian parent materials.

Within the same climatic and vegetation zones, soil properties are found to vary with parent materials. Such variation in soil properties is expressed by the term lithosequence. On the other hand, different soils on slopes of the same landscape form a catena.

3.7 Relief Is the Configuration of Land Surface

Relief or topography gives a condition at the site where a soil is formed. Through its effects on infiltration and runoff, topography influences drainage and the amount of water available for soil development. For topographic conditions such as in hill slopes, some soils tend to remain dry for a considerable period of the year. In lowlands, soils tend to remain waterlogged all the time. In these two contrasting situations, oxidation and reduction, respectively, predominate in chemical transformations.

Important components of topography are altitude (height above mean sea level), degree of slope (inclination against the horizontal land axis), and aspect (face of the sloping surface). The degree of slope affects erosion, natural or accelerated, so that soil materials are continually removed from sloping surfaces and are deposited in valleys. The steeper the slope (higher degree), the greater is the loss of soil materials. Generally shallow soils above bedrocks develop on sloping surfaces. Usually a thin A horizon develops over the thin C horizon. Such thin soils were earlier known as lithosols. On gentler slopes, well-developed B horizon may be found. The depth of soil thus depends on the degree of slope. Sometimes erosion removes the entire A horizon exposing the B horizon. If the degree of slope is very high and the soils are not stabilized by vegetation or if the vegetation is removed by deforestation, such soils are completely removed exposing the underlying rock. In addition to the effective soil moisture content, altitude influences temperature. This effect almost resembles the latitudinal variation in temperature.

Another important component of topography is aspect. North- and east-facing slopes in the Southern Hemisphere receive more sunlight, and soils are warmer and drier. The reverse slopes are affected similarly in the Northern Hemisphere. Therefore, soils and natural distribution of vegetation are different in opposing aspects. A set of soils developed under similar climate, organism, parent material, and time in varying topographies is known as a toposequence.

3.8 Soil Grows and Matures with Time

Soil develops and changes with time. The time required to develop a soil varies from some centuries to some millions of years depending on other soil-forming factors. Generally profile development and horizon differentiation are faster under warm, humid, and forested conditions because of higher weathering and leaching of bases, colloids, and other materials. Table 3.2 gives some examples of time span required for soil development.

There had been considerable confusion in the past on the method of estimating the age of a soil. Parent material is often regarded as the initial state of the soil system. So, the time of soil development is the difference from the time of parent material formation to the time of studying the soil. But, it is difficult to determine when parent material has completely formed and when pedogenesis has started. Since weathering and pedogenesis occur simultaneously in nature, the age of a soil is sometimes measured from the time the rock was first exposed to the surface or from the time the alluvial material was deposited. It is, however, better to conceive that soil incepts, grows, and matures with time through several stages of development (Fig. 3.6).

Over time, vegetation at a space develops through a series of stages of succession. As a coupled system component, soil development keeps pace with the vegetation development. In other words, succession of soil and vegetation occurs simultaneously up to a steady state. Both soil and vegetation tend to retain this steady state, unless or until abrupt changes in the state factors or driving factors take place. When the

Fig. 3.6 Succession of soil horizons over time

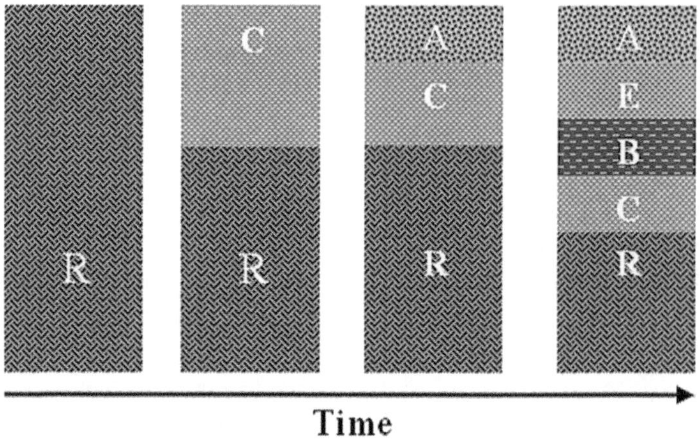

steady state is reached, further changes may occur remarkably if abrupt environmental changes occur. Otherwise, the state remains fairly stable. However, natural changes in climatic conditions often take place over time with accompanying changes in the soil–vegetation steady state.

Again, climate at most places considerably changed at different intervals over the periods of soil formation; so there were concomitant changes in the vegetation and pedogenic processes. There could be successive waves of changes in the environment and simultaneous changes in pedogenesis. It is suggested that some soils may have been polygenetic (Buol et al. 1997). Absolute dating of soil is being tried by radiocarbon (for origin of the organic components) dating method (Scharpenseel 1971), fission-track dating (Ward et al. 1993), and by principles of geomorphology and stratigraphy fortified by radiometric dating. These techniques are still advancing. However, a set of soils developed under similar all other soil-forming factors along a time gradient is known as chronosequence.

3.9 Basic Soil-Forming Processes Operate in the Formation of All Soils

There are four basic soil-forming processes, namely, addition, removal, transformation, and translocation. These processes are often simultaneous and operate usually in conjunction. Some transformations aid in translocation and removal; some others may cause chemical precipitation and addition. Addition of rainwater may cause transformation, translocation, and removal.

3.9.1 Additions Are the Inputs of Materials and Energy in Soil

Addition of materials and energy in soil generally takes place from the atmosphere, organisms, rocks and minerals, and groundwater. Atmosphere provides the inputs—solar energy, moisture, gases, dusts, and sometimes pollutants. The amount of solar energy received by a soil depends on geographical locations, topography, mineralogical composition, degree of moisture saturation, etc. Amount, intensity, and distribution of rainfall—determinants of water receipt—depend on climatic conditions. The amount of water intake by a soil may, on the other hand, depend on soil, vegetation, and slope characteristics. Addition of air is needed for the aeration and redox systems. Important gases added from atmosphere to soil to participate in various functions are O_2, CO_2, and N_2. Aerosolic dusts include fine mineral particles carried by the wind. Organisms, within and outside the soil, provide the organic inputs in various forms—plant and animal litter on the surface and decaying roots, dead-burrowing animals, soil insects, microorganisms, etc.—within the soil. Rocks and minerals provide the raw materials to form most soils. Moisture and soluble substances may also be added through capillary rise from the groundwater. Salt crusts and lime layers in some Aridisols develop in this manner. Materials added to soil by humans in managed soil systems include manures, fertilizers, and other agrochemicals. Different inorganic and organic materials are added to soil in waste disposal sites. However, maximum organic and atmospheric additions take place on the surface of the soil. Addition of organic matter and its transformation to humus (Chap. 7) make the surface mineral soil (A horizon) darker than the subsoil.

3.9.2 Soil Materials Are Removed by Physical, Chemical, and Biological processes

Materials are removed from soil by volatilization and evaporation to the atmosphere, in solution and suspension in water downward, upward, and in lateral directions, and outside the soil by biomass harvest.

Gaseous losses from soil occur due to evapotranspiration, volatilization, diffusion, etc. Moisture is continuously lost from soil through evapotranspiration and percolation. Gases, produced by biochemical or geochemical transformations—CO_2, NH_3, N_2, NO_x, SO_2, H_2S, CH_4, etc.—are always lost to variable extents depending on soil and other environmental conditions. Some of these gases have critical roles on the environment. If they are not lost from soil (as, for example, due to low diffusion in waterlogged soils), they become toxic to non-hydrophytic plants. On the other hand, most of them are well-known greenhouse gases. Soil materials are lost with the running water in erosion over the land surface. Under natural conditions, both erosion and soil formation proceed simultaneously, and erosion losses are usually compensated by gains through soil formation. At the steady state, there is a fairly stable equilibrium between losses and gains through several processes involving the interactive systems. However, accelerated erosion (Chap. 12) causes loss of the fertile surface soil and finer soil particles, and sometimes, if the slope is steep and not protected by vegetation, the A horizon or even the entire soil body is lost. Loss of soluble substances such as ions, bases, acids, salts, inorganic and organic colloids, and complexes accompanies the percolating water to the groundwater. It is called leaching which is the complete removal of soluble materials from the soil profile (Schaetzl and Anderson 2005). Leaching rate is very high in soils of the humid areas, particularly Oxisols and Ultisols and in soils coarse in texture. Sandy soils are less nutrient retentive, and whatever nutrients are added, most are leached with percolating water.

3.9.3 Transformations in Soil Are Physical, Chemical, and Biological in Nature

Complex, and often unpredictable, transformations occur in soil all the time. These transformations include physical, physicochemical, chemical, and biological processes. Transformations in soil may be reversible and irreversible. Pedochemical weathering processes operate throughout the entire period of soil formation. Soil minerals, particularly silicate minerals, are weathered into clay minerals and soluble substances which are dissolved in soil solution. The important physical transformation that occurs in soil is perhaps the aggregation of soil particles and formation of peds. Aggregation is one of the criteria that distinguish soils from other natural loose materials on the surface of the earth. Aggregation is effected by colloids, clays and humus; cementing agents, lime, Ca^{2+} ions, and microbial products; and binding agents, plant roots and fungal hyphae, etc. Flocculation of colloids which is a physicochemical transformation is the initial step of aggregation. Formation of clay–humus complex is also important for aggregation. Chemical transformations include solution, hydration, hydrolysis, oxidation, reduction, carbonation, and other acid reactions. Some of these reactions have been shown in relation to chemical weathering in Chap. 2 (Sect. 2.14). Complexation, precipitation, chelation, ion exchange, dispersion, coagulation, etc., are other important chemical and physicochemical transformations related to soil formation. Decomposition of organic matter and humification are important biological transformations related with soil formation.

3.9.4 Materials in Soil Are Translocated in All Directions

Translocation of materials occurs in all directions of the soil—lateral, upward, and downward. This may take place in solution and suspension or as mass movement along cracks and channels.

When soil materials move downward with water and are lost from the soil body, it is called leaching. If movement takes place from an upper horizon to a lower horizon, it is known as eluviation. Eluviation occurs usually from the surface horizons. However, under prolonged weathering in humid conditions, eluviation of bases, humus, clay minerals, and sesquioxides may create a bleached gray horizon enriched with silica (denoted by E) below the A horizon. E horizons are common in Alfisols and Spodosols. Translocation of clay-sized particles in suspension is called lessivage or argilluviation (L. *argilla*, white clay; *luv*, washed). Although most lessivage occurs from the upper profile to the lower (from A and E horizons to B horizon), the process does occur laterally as well. Eluvial materials are deposited in the B horizon either by coagulation, precipitation, or by mechanical impedance. This process is known as illuviation. Illuviation of humus, clay, iron oxides or hydroxides, lime, etc., is a common occurrence in soils of humid regions. Spodosols in the humid tropics are characterized by illuviation of clay, humus, and aluminum. The horizon B is supplemented with several small letters (Bw, Bt, etc.) to indicate the kind of illuviated materials. Bh denotes the illuviation of humus, Bt shows illuviation of tont (silicate clay), Bs indicates illuviation of sesquioxides (Fe and Al oxides), and so on. Significant amount of soil materials are translocated downward through the wide cracks in Vertisols. Cryoturbation translocates some soil materials in frosted soils such as Gelisols. A considerable amount of soil materials is translocated vertically and laterally by such soil animals as rodents, earthworms, and insects.

3.10 Specific Soil-Forming Processes Produce Specific Soils

There are some specific soil-forming processes for development of specific soils. Some of these processes are laterization, latosolization, ferralitization, podzolization salinization, and desalinization.

3.10.1 Laterization and Latosolization Occur Mainly in Humid Tropics

Both laterization and latosolization are processes of enrichment of iron, manganese, and aluminum oxides and hydroxides in the soil profile. When iron and aluminum come in solution to the soil profile from outside, such as saprolite below or from upslope and precipitated, the process is called laterization (Fanning and Fanning 1989). Laterization is responsible for the formation of Oxisols and related soils. It is a dominant soil-forming process of the humid tropics, usually beneath broadleaf evergreen forests. It involves long-term intensive weathering of old landforms under warm and moist environment. In these conditions, organic matter production and decomposition are both high, but potential decomposition rate is higher. So, there is very low accumulation of organic matter. Weathering dissolves most minerals including phyllosilicate clays, releasing bases which are lost by runoff and leaching. The soil is impoverished in silica, bases, and organic matter but enriched with iron, manganese, and aluminum. Soluble iron, manganese, and aluminum, residual or migrating from outside the profile, are precipitated as their oxides and hydroxides. Silica is leached, leaving mainly iron and aluminum cations which form iron- and aluminum-rich clay minerals such as hematite, boehmite, and goethite.

In latosolization, residual sesquioxides accumulate; bases and silica are leached out from the profile. Some of the bases are biocycled, leading to slightly higher concentration in the A horizon. The residual material is rich in sesquioxides of Fe and Al, Ti and Mn oxides, oxyhydroxides, and hydroxide minerals, as well as heavy elements like Ni, Zr, and Cr. Clay minerals such as kaolinite neoform from these residual ions. The Fe and Al compounds in these soils are typically highly crystalline. That is, they have neoformed into minerals with discrete-ordered structures that are not easily weatherable. This is how the oxic (Bo) horizon develops. However, Oxisols may contain both residual and migrated iron and aluminum (Schaetzl and Anderson 2005).

3.10.2 Ferralitization Is a Result of Strong Weathering in Tropical Climate

van Breemen and Buurman (2002) used the term "ferralitization" for the processes associated with strong weathering that lead to the formation of Oxisols (USDA and NRCS 2003) or Ferralsols (FAO 2006). These soils have an oxic or ferralic B horizon affected by extreme weathering of primary and secondary minerals. Almost all weatherable minerals are removed from the sand and silt fractions. In the clay fraction, weathering causes a dominance of kaolinite, gibbsite, and iron minerals, resulting in a low CEC (<16 $cmol_c\,kg^{-1}$ at pH 7). Typical of "ferralitic" weathering is the removal of silica (desilication) from primary silicates and even from quartz. Desilication leads to residual accumulation of oxides and hydroxides of Fe, Mn, and Al.

3.10.3 Podzolization Is the Accumulation of Clay and Humus in Subsoil Under Humid Climate

Podzolization is the dominant soil-forming process of the humid temperate climate under forest vegetation. Podzolization occurs under vegetation that produces acidic litter, such as coniferous forest and heather, and on coarse-textured parent materials (Schaetzl and Anderson 2005). In such situation, plenty of organic matter is produced, but the decomposition of organic matter is relatively slow. Organic residues accumulate in the surface, and their decomposition produces appreciable organic acids. The pH of the surface soil becomes very low, and at this strong acidity, most minerals are weathered. Some of these weathered minerals are lost from the profile, but the great majority are redistributed, in suspension, to the lower profile. In podzolization, humus, iron, and aluminum, in some combination, are translocated from the A and E horizons to the B horizon. A spodic horizon develops by podzolization (Lundstrom et al. 2000). The resulting soils are classified as Spodosols which are generally coarse textured containing few primary minerals and silicate clays.

3.10.4 Calcification Is the Enrichment of Lime in Soil Profiles

Calcification is the process of precipitation and accumulation of secondary calcium carbonates in soils. The accumulation of $CaCO_3$ may result in the development of a calcic horizon. It occurs mainly in the arid regions where the availability of water for leaching is low. Solubility of $CaCO_3$ is also low, and it accumulates in the profile (Southard 2000). Calcium carbonate may precipitate from soluble calcium released by weathering of minerals:

$$CO_2 + H_2O \rightarrow H_2CO_3$$
$$H_2CO_3 + Ca \rightarrow Ca(HCO_3)_2 \text{ (soluble)}$$
$$Ca(HCO_3)_2 \rightarrow CaCO_3 \text{ (precipitates)} + H_2O + CO_2$$

Calcification is also common in subhumid grassland soils (earlier Chernozem soils, Mollisols in Soil Taxonomy) and in humid climates where soil horizons are clay rich and slowly permeable (Schaetzl et al. 1996). When illuvial horizon is enriched with carbonates, it is denoted by Bk.

3.10.5 Salinization and Desalinization Are Processes of Salt Accumulation and Removal, Respectively

Salinization is the accumulation of soluble salts in soil producing saline soils. Natural salts often accumulate in surface of arid land soils as a white crust. The salts are generally chlorides and sulfates of sodium, potassium, calcium, and magnesium. Gypsum may often accumulate. In regions where the potential evaporation is greater than precipitation, some soluble salts are liberated by weathering, but due to scarcity of water, these salts cannot be leached to any considerable extent. Moreover, capillary rise of water brings some salts on surface. Water is evaporated leaving the salts in soil. A salic horizon (Soil Survey Staff 1999) may develop in soil as a result of salinization. When sodium carbonate and bicarbonate predominate among salts, exchangeable Na may be high (>15% of the cation exchange capacity). The resulting soil is then called a sodic soil. The process is alkalization. Salinization may also occur in poorly drained soils of humid areas by the capillary rise of salts from groundwater.

Desalinization is the removal of salts from the upper solum, possibly due to a change in climate, irrigation, or a change in the surface or subsurface hydrology. Desalinization may remove excess salts from the soil leaving excess sodium in the exchange complex. So, alkalization may result at the initial stage of desalinization.

3.10.6 Mottling and Gleization Occur Through Redox Transformations

Redox transformations predominate in soil formation in soils of fluctuating groundwater table and in seasonally flooded soils. In the region of alternate water saturation in the soil, Fe and Mn compounds undergo cycles of oxidation and reduction. Oxides of Fe and Mn form nodules and concretions and often in aerobic conditions precipitate as reddish brown spots in soil. These are called mottles, and the process is called mottling. Over time, continued oxidation–reduction cycles cause Fe and Mn to become concentrated in these red mottles; surrounding areas have lower chromas and grayer hues (Richardson and Hole 1979). These nodules, concretions, and mottles are called redoximorphic features (Vepraskas 1999). In the lower part of such soils and in permanently waterlogged soils, reduced iron compounds accumulate in the lower part of the B horizon giving it a bluish-green appearance. Soluble reduced iron may also form complexes with humus. These features are found in hydric soils (Chap. 13).

Study Questions

1. How would you differentiate between parent material and soil? Discuss the contrasting characteristics of an E and a B horizon.
2. What is a soil-forming factor? Explain that a soil is formed by the interaction of five soil-forming factors.
3. Do you justify the inclusion of human as a soil-forming factor?
4. Explain that soil and vegetation form a coupled system which evolves together (see also Sect. 14.6.5).
5. Compare and contrast between podzolization and laterization.

References

Amundson R, Jenny H (1991) The place of humans in the state factor theory of ecosystems and their soils. Soil Sci 151(1):99–109

Arnold RW, Riecken FF (1964) Grainy gray ped coatings in Brunizem soils. Proc Iowa Acad Sci 71:350–360

Arnold RWI, Szabolcs, Targulian VO (1990) Global soil change. Report of an IIASA-ISSS-UNEP Task Force. International Institute for Applied Systems Analysis, Laxenburg, Austria

Aubert G (1960) Influences de la vegetation sur le sol en zone tropicale humide et semi-humide. Rapp Du Sol et de la Vege Colloq Soc Bot Fr 1959:11–13

Beavers et al (1955) cited from Wilding LP, Smeck NE (1983) Pedogenesis and soil taxonomy. Elsevier, Amsterdam

Buol SW, Hole FD, McCracken RJ, Southard RJ (1997) Soil genesis and classification, 4th edn. Iowa State University Press, Ames

Butler BE (1958) Depositional systems of the riverine plain of south-eastern Australia in relation to soils. CSIRO Soil Publ 10 CSIRO Division of Soils, Canberra

Dokuchaev VV (1879) Short historical description and critical analysis of the more important soil classifications. Trav Soc Nat St Petersb 10:64–67 (In Russian)

Dokuchaev VV (1883) Russian Chernozems (Russkii Chernozems). Israel Prog Sci Trans, Jerusalem, 1967. Translated from Russian by Kaner N. Available from U.S. Department of Commerce, Springfield

Dudal R (2004) The sixth factor of soil formation. Presented at the International conference on soil classification 2004. Petrozavodsk, Russia, 3–5 Aug 2004

Fanning DS, Fanning MCB (1989) Soil morphology, genesis and classification. Wiley, New York

FAO (2006) World reference base for soil resources 2006. A framework for international classification, correlation and communication. FAO–UNESCO–ISRIC. Food and Agriculture Organization of the United Nations, Rome

Hilgard EW (1892) A report on the relations of soil to climate. U.S. Dept. of Agriculture. Weather Bull 3:1–59

Jenny H (1941) Factors of soil formation a system of quantitative pedology. McGraw-Hill, New York
Jenny H (1980) The soil resource: origin and behavior. Springer, New York
Juo ASR, Franzluebbers K (2003) Tropical soils: properties and management for sustainable agriculture. Oxford University Press, New York
Lundstrom US, van Breemen N, Bain DC et al (2000) Advances in understanding the podzolization process resulting from a multidisciplinary study of three coniferous forest soils in the Nordic Countries. Geoderma 94:335–353
Margesin R (2009) Permafrost soils. Springer, Berlin
Mcintosh PD (1984) Genesis and classification of a sequence of soils formed from aeolian parent materials in East Otago, New Zealand. Aust J Soil Res 22(3):219–242
Mohr ECJ, van Baren FA (1954) Tropical soils: a critical study of soil genesis as related to climate, rock and vegetation. Royal Tropical Institute, Amsterdam. Inter-science, New York
Nortcliff S (2009) Soils of cold and temperate region. Acco Publishing Cy, Leuven
Retallack GJ (2001) Soils of the past, 2nd edn. Blackwell Science, New York
Richardson JL, Hole FD (1979) Mottling and iron distribution in a Glossoboralf-Haplaquoll hydrosequence on a glacial moraine in northwestern Wisconsin. Soil Sci Soc Am J 43:552–558
Rowell DL (1994) Soil science: methods and applications. Longman, Harlow
Schaetzl RJ, Anderson S (2005) Soils: genesis and geomorphology. Cambridge University Press, Cambridge
Schaetzl RJ, Frederick WE, Tornes L (1996) Secondary carbonates in three fine and fine-loamy Alfisols in Michigan. Soil Sci Soc Am J 60:1862–1870
Scharpenseel HW (1971) Radio carbon dating of soils – problems, troubles, hopes. In: Yaalon DH (ed) Paleopedology- origin, nature and dating of paleosols. International Society of Soil Science/Israel University Press, Jerusalem
Simonson RW (1959) Outline of a generalized theory of soil genesis. Soil Sci Soc Am Proc 23:152–156
Soil Survey Staff (1993) Soil survey manual. Handbook 18. US Government Printing Office, Washington, DC
Soil Survey Staff (1994) Keys to soil taxonomy, 6th edn. USDA Soil Conservation Service, Washington, DC
Soil Survey Staff (1999) Soil taxonomy. US Department of Agriculture handbook no. 436. US Government Printing Office, Washington, DC
Southard RJ (2000) Aridisols. In: Sumner ME (ed) Handbook of soil science. CRC Press, London/Boca Raton
USDA, NRCS (2003) Keys to soil taxonomy, 9th edn. United States Department of Agriculture. Government Printing Office, Washington, DC
USDA-NRCS (2007) Anthropogenic soils: human-altered and -transported soils. ICOMANTH report no. 2 – version 2.0
van Breemen N, Buurman P (2002) Soil formation, 2nd edn. Kluwer, New York
Vepraskas MJ (1999) Redoximorphic features for identifying aquic conditions. N. Carolina Agricultural Research Service Technical Bulletin no. 301. North Carolina State University, Raleigh, NC
Ward PA III, Carter BJ, Weaver B (1993) Volcanic ashes: time markers in soil parent materials in the southern plains. Soil Sci Soc Am J 57:453–460

Soil Classification

4

Soil classification is the orderly arrangement of soils according to some known or inferred characteristics. Soils are classified to organize information regarding their properties and behavior so that they can be used for different purposes with the maximum benefit under sustainable management. Systematic soil classifications began in the last part of the nineteenth century and were mainly based on ideas of soil genesis and soil forming factors as influenced by the philosophy of Dokuchaiev. Dokuchaiev himself proposed a soil classification system in 1879 before presentation of his Ph.D. thesis on the Russian Chernozems in 1883 (Gimenez 2011). There are two types of soil classification: process oriented and property oriented. The process-oriented classification systems are known as genetic soil classification systems (Bokheim and Gennadiyev 2000). Soil classification systems based on observed properties are known as natural classification systems, and those on inferred properties are technical systems. There are regional and international soil classification systems. Examples of regional classification systems include Australian soil classification system (Isbell 1996, 2002), the Canadian soil classification system (Agriculture Canada Expert Committee on Soil Survey 1987), soil classification of England and Wales (Avery 1980), French soil classification system (Baize and Girard 1995), and so on. Among international systems, Soil Taxonomy (USDA 1960, 1975) and World Reference Base for Soil Resources (FAO 1998) are popularly used worldwide at present. These systems have their own terminology and diagnostic features for different taxa.

The Soil Taxonomy is a revolutionary system in that it has abandoned all the classical, indigenous, and folk soil names as laterites, podzols, groundwater podzols, prairies, chernozems, rendzinas, black cotton soils, tropical red earths, and Solonchaks used in the earlier Russian pedology literature and used in the old American system of soil classification (Baldwin et al. 1938; Thorp and Smith 1949). It has coined a new set of soil nomenclatures and managed to keep a link in the naming of higher to lower categories, so that the higher category affiliation and properties are immediately conceived from the lower category name. World Reference Base retained a few old names such as Chernozems and Solonchaks and coined some new ones such as Anthrosols, Technosols, and so on. As there are different systems of soil classification in different parts of the world, there are different nomenclatures of soil. The same soil is called in different names in different systems (such as Oxisols in Soil Taxonomy and Ferralsols in WRB). Histosols is a taxon used in both Soil Taxonomy and WRB but their diagnostic characteristics are different. This may make considerable confusion in identification and interpretation of soil (Krasilnikov et al. 2010). Therefore, correlation between Soil Taxonomy and WRB systems is often necessary.

4.1 Soil Taxonomy Is the Most Popularly Used Soil Classification System

The Soil Survey Staff of the United States Department of Agriculture (USDA 1960) published a soil classification system named "Soil Classification, A Comprehensive System." It was popularly known as "The 7th Approximation" because it was the 7th revision before its publication. Initially, there were ten soil orders—Alfisols, Aridisols, Entisols, Inceptisols, Histosols, Mollisols, Oxisols, Spodosols, Ultisols, and Vertisols. USDA (1975) published "Soil Taxonomy: A Basic System of Soil Classification for Making and Interpreting Soil Survey" and included an additional soil order Andisols. Another soil order Gelisols was further included latter (USDA and NRCS 1998), so that the total number of soil orders, the highest category in the hierarchy, is 12 at present.

Soil Taxonomy classifies a "soil individual" according to its observable and measurable properties. A soil individual is the area of soil in a landscape that has similar pedon (Chap. 1) characteristics. The properties that are observed and measured include morphological, physical, chemical, and mineralogical properties. The morphological properties are horizon differentiation, soil depth, etc.; the physical properties include color, texture, structure, compaction, etc.; and the chemical and mineralogical properties include organic matter,

K.T. Osman, *Soils: Principles, Properties and Management*,
DOI 10.1007/978-94-007-5663-2_4, © Springer Science+Business Media Dordrecht 2013

pH, base saturation percentage, clay, iron and aluminum oxides, and silicate clays of different horizons or layers. The lower boundary of the pedon for soil classification in this system is 200 cm from soil surface unless restricted by the presence of hard bed rock within this depth. If bed rock is present within this depth the upper unconsolidated part is examined for classification. This part of soil is then called the control section. There are several diagnostic criteria for identification of different categories. Certain diagnostic horizons are used to identify the higher categories. Soil moisture and temperature regimes are used for differentiation of some taxa.

The diagnostic horizons and some other diagnostic features are briefly mentioned below. The main characteristics of different orders and suborders will also be indicated, but for details of diagnostic features and identification of different taxa, the reader is referred to Soil Genesis and Classification, 4th and 5th edition (Buol et al. 1997, 2003, respectively), Soil Taxonomy, 2nd edition (Soil Survey Staff 1999), Keys to Soil Taxonomy, 9th and 11th edn. (USDA and NRCS 2003; Soil Survey Staff 2010, respectively).

4.2 There Are Some Diagnostic Horizons in Soil Taxonomy

Soil taxonomy uses some diagnostic horizons for different categories. The horizons that form in the surface are called epipedons (Gk. epi = over; pedon = soil) and those in the subsurface soil are endopedons. An epipedon is the surface soil layer where the materials are well weathered and unconsolidated and darkened by organic matter. This may include O horizon where it is present, or the A horizon and part of the eluvial E horizons, provided they are considerably darkened by organic matter. Diagnostic features of several epipedons are given below:

Epipedons	Diagnostic features
Anthropic epipedons	Surface horizons showing the signs of long-term human disturbance such as plowing and manuring. It resembles a mollic epipedon but contains high citric acid soluble P_2O_5. If the soil is not irrigated, all parts of the epipedon are dry for 9 months or more in normal years
Histic epipedons	Usually wet surface horizons rich in organic matter (at least 12% organic carbon if the soil has no clay; at least 18% organic carbon if the soil has 60% clay, and an intermediate proportional organic carbon for intermediate amounts of clay). It is generally 20–40 cm thick. When plowed, the soil contains 8% or more organic carbon if there is no clay and at least 16% organic carbon if there is 60% or more clay
Melanic epipedons	Thick (>30 cm), black horizons rich in organic matter (more than 6% organic carbon) that has developed from andic (volcanic ash) materials high in allophane
Mollic epipedons	Soft surface horizon that contains sufficient organic matter (>0.60%) to give it a dark color. Base saturation is >50%. It remains moist at least 3 months a year, when the soil is 5 °C or higher to a depth of 50 cm
Ochric epipedons	Light-colored surface mineral horizon. Color values more than 5 dry or more than 3 moist, contains less organic carbon than 0.60%, hard or very hard when dry. It is too thin, too light in color, or too low in organic carbon to be a mollic or umbric horizon
Plaggen epipedons	This is also a surface horizon influenced by long-term human activity, but it contains many artifacts such as brick chips and pottery. It is 50 cm or more thick
Umbric epipedons	It has general characteristics of mollic epipedon but has it less than 50% base saturation. It develops under comparatively higher rainfall from parent materials poor in Ca and Mg

There are also some subsurface horizons that are diagnostic of different taxa in Soil Taxonomy. They are formed below the surface of the soil through eluviation, illuviation, and other translocation and transformation processes. These horizons generally include E and B horizons. Sometimes subsurface horizons may be exposed to the surface due to removal of the surface soil by erosion or human development activities. Different diagnostic subsurface horizons and their major features are presented below:

Subsurface horizons	Diagnostic features
Agric horizon	It forms under the plow layer and contains significant amounts of silt, clay, and humus accumulated as thick, dark lamellae
Albic horizon	It is a light-colored leached horizon with color values 5 dry or more or 4 moist or more. This horizon is created by eluviation of clay and oxides of Fe and Al. It is typical of E horizon
Argillic horizon	It has accumulation of silicate clays due to illuviation or formation in place. It has 3% more clay than the eluvial layer containing <15% clay, or 8% more clay if the eluvial layer has >40% clay. The argillic horizon contains 1.2 times clay of the eluvial layers having 15–40% clay. It should be at least one-tenth thick of all overlying horizons or more than 15 cm whichever is thinner. It is observed in B horizons of humid regions
Calcic horizons	This is a thick horizon (15 cm or more) that has an accumulation of carbonates, commonly of Ca or Mg, in excess of 15% calcium carbonate equivalent. It contains 5% more carbonate than underlying horizons

(continued)

(continued)

Subsurface horizons	Diagnostic features
Cambic horizon	This horizon has a texture of very fine sand, or loamy fine sand, or finer. It may show accumulation of some clay, or sesquioxides but not too high to be an argillic or spodic horizon, respectively
Duripan	A hard pan formed by cementation with silica. Air dry fragments from more than half of the horizon do not slake in water or HCl
Fragipan	This is a horizon of high bulk density and is brittle when moist. The high bulk density is due to extreme compaction that restricts penetration of roots and water
Glossic horizon	This is a thin (5 cm or more thick) transitional horizon between E and B. Albic materials constitute 15–85%, and remaining materials are like those of the underlying horizon, usually argillic, kandic, natric or fragipan horizon
Gypsic horizon	This horizon is 15 cm or more thick and has an accumulation of calcium sulfate. It has at least 5% more calcium sulfate than the underlying material
Kandic horizon	This horizon shows an accumulation of low-activity clay (kandic group, e.g., kaolinite) and has less than 16 cmole kg^{-1} clay CEC at pH 7
Natric horizon	It has characteristics similar to an argillic horizon, except that it has prismatic or columnar structure and more than 15% exchangeable sodium percentage
Oxic horizon	This horizon is of sandy loam or finer texture and has a thickness of at least 30 cm. It contains highly weathered materials including 1:1 clays and oxides of Fe and Al and has CEC of 16 cmole kg^{-1} clay or less. It contains less than 10% weatherable minerals in the sand fraction
Placic horizon	This is an iron/manganese pan, reddish brown to black and hard, which is thin (2–10 cm) and remain within 50 cm from soil surface
Petrocalcic horizons	Horizons having similar characteristics of calcic horizons but cemented are called petrocalcic horizon
Petrogypsic horizon	Strongly cemented gypsic horizons are known as petrogypsic horizons
Salic horizon	It is at least 15-cm-thick horizon having accumulation of secondary soluble salts. The electrical conductivity of 1:1 soil water extract exceeds 30 dS m^{-1} more than 90 days in a year
Sombric horizon	This horizon resembles umbric epipedon in color and base saturation but has formed by accumulation of humus
Spodic horizon	This horizon shows an accumulation of colloidal organic matter and aluminum oxide. It may or may not accumulate iron oxide
Sulfuric horizon	This horizon is at least 15 cm thick and contains sulfuric materials that gives a pH <3.5 and shows yellow mottles of jarosite

4.3 Soil Moisture Regimes Indicate Soil Moisture Status

Some soil moisture regimes are used in distinguishing suborders in an order or other categories. The soil moisture regimes are identified on the basis of the level of moisture saturation (or cumulative and consecutive periods of dryness and moistness) of the soil moisture control section (defined latter). In determining soil moisture regimes, a soil is considered dry when the soil water potential is less than −1,500 kPa and moist when the soil water potential is greater than −1,500 kPa. Since water potential of −1,500 kPa is considered to be the wilting point for most crops, a soil is considered dry when its water potential is less than −1,500 kPa.

Soil moisture control section: Soil moisture control section is a part of the soil which lies between an upper and a lower boundary in the vertical direction.

Upper boundary: The depth to which the dry soil (soil water potential < −1,500 kPa) is wetted by 2.5 cm of water in 24 h.

Lower boundary: The depth to which the dry soil is wetted by 7.5 cm of water in 48 h.

These limits depend on soil texture, structure, pore-size distribution, presence of pans, etc. Usually the soil moisture control section varies from 10 to 30 cm below the soil surface if the soil is fine-loamy to clayey, from 20 to 60 cm if the texture is coarse loamy, and from 30 to 90 cm if the texture is sandy. The soil moisture regimes used in identifying different categories of soil taxonomy are presented below:

Soil moisture regimes	Characteristics
Aquic	For considerable periods of the year, the soils are water saturated and remain in a state of reduction due to the absence of dissolved oxygen in ground water, for example, in tidal marshes, landlocked depressions
Aridic/torric	Aridic/torric regimes are used for the same moisture conditions but in different taxa. The soil moisture control section is dry for more than half of the cumulative days per year and moist in some or all parts for <90 consecutive days, for example, in soils of arid climate and the dryness is due to soil physical conditions
Udic	The soil moisture control section is not dry in any part for as long as 90 consecutive days and dry for <45 consecutive days in the 4 months following the summer solstice, for example, in soils of humid climate

(continued)

(continued)

Soil moisture regimes	Characteristics
Ustic	This moisture regime is intermediate between the aridic regime and udic regime. Here plant growth is restricted for some part of the year by low moisture supply
Xeric	The soil moisture control section is dry in all parts for 45 or more days in the 4 months following the summer solstice and moist in all parts for 45 or more days in the 4 months following the winter solstice. This soil moisture regime is typical of Mediterranean climates

4.4 Soil Temperature Regimes Differ in Mean Annual and Mean Seasonal Soil Temperatures

Soil temperature regimes are based on mean annual, mean summer, mean winter, and the differences between mean summer and mean winter soil temperatures either at a depth of 50 cm or at a densic (relatively unaltered materials that are non-cemented and rupture resistant), lithic (a coherent underlying material), or paralithic (unaltered materials that are an extremely weakly to moderately rupture resistant) contact which one is shallower. The following soil temperature regimes are identified:

Soil temperature regimes	Characteristics
Cryic	1. In mineral soils, the mean summer soil temperature is:
	If the soil is not saturated with water in some part of the summer and has an O horizon <8 °C
	If the soil is not saturated with water in some part of the summer and has no O horizon <15 °C
	If the soil is saturated with water in some part of the summer and has an O horizon or histic epipedon <6 °C
	If the soil is saturated with water in some part of the summer and has no O horizon <13 °C
	2. In organic soils, the mean annual soil temperature is <6 °C
Frigid	A soil warmer in summer than a cryic regime, but its mean annual temperature is lower than 8 °C and the difference between mean summer and mean winter soil temperatures is more than 6 °C either at a depth of 50 cm from the soil surface or at a densic, lithic, or paralithic contact, whichever is shallower
Mesic	The mean annual soil temperature is 8 °C or higher but <15 °C, and the difference between mean summer and winter temperature is >6 °C
Thermic	The mean annual soil temperature is 15 °C or higher but <22 °C, and the difference between mean summer and winter temperature is >6 °C
Hyperthermic	The mean annual soil temperature is 22 °C or higher, and the difference between mean summer and winter temperature is >6 °C
Isofrigid	The mean annual soil temperature is >8 °C, and the difference between mean summer and winter temperature is <6 °C
Isomesic	The mean annual soil temperature is 8 °C or higher but <15 °C, and the difference between mean summer and winter temperature is <6 °C
Isothermic	The mean annual soil temperature is 15 °C or higher but <22 °C, and the difference between mean summer and winter temperature is <6 °C
Isohyperthermic	The mean annual soil temperature is 22 °C or higher, and the difference between mean summer and winter temperature is <6 °C

4.5 There Are Some Other Diagnostic Features in Soil Taxonomy

Other diagnostic features including abrupt textural changes (contrasting textures between adjacent horizons), anhydrous conditions (active layer in soils of cold deserts and dry permafrost), coefficient of linear extensibility (COLE—ratio of the difference between the moist length and dry length of a clod to its dry length), durinodes (weakly cemented to indurated nodules), fragic soil properties (properties of a fragipan, but the thickness and volume do not meet the requirements of a fragipan), n value (relation between the percentage of water in a soil under field conditions and its percentage of inorganic clay and humus), plinthite (iron-rich mixture of clay and other minerals in platy, polygonal, and reticulate patterns), resistant minerals (durable minerals in sand fraction), slickensides (polished soil surfaces created by mass slides of soil), and weatherable minerals are sometimes used for identification of some taxa.

4.6 There Are Six Categories in Soil Taxonomy

In this hierarchical system, soil orders are divided into suborders, suborders into great groups, and so on. The hierarchy and number of known soil taxa in each category are shown below:

Order	12 orders
Suborder	64 suborders
Great group	319 great groups
Subgroup	2,484 subgroups
Family	>8,000 families
Series	Unknown number of series

The highest category in Soil Taxonomy is the soil order. *Orders* are distinguished by the presence or absence of diagnostic horizons and other diagnostic features, indicating the predominant soil forming processes. An order includes soils whose properties suggest that they are similar in their genesis. Subdivisions of orders on the basis of genetic homogeneity are suborders. *Suborders* of an order are differentiated by the presence or absence of properties associated with wetness, soil moisture regimes, major parent material, and vegetation. Subdivisions of suborders are *great groups*. They are identified according to similar kind, arrangement, and degree of expression of horizons, giving emphasis on upper sequum, base status, soil temperature, and moisture regimes. Subdivisions of great groups are *subgroups*. One subgroup has the central concept of a great group; it is given the name "Typic" before the great group name. Other subgroups in a great group are differentiated on characteristics that are intergrades between those of the central concept and those of the orders, suborders, or great groups. Soils of a subgroup having similar physical and chemical properties affecting their response to management and especially to the penetration of plant roots are grouped into a *family*. Differences in texture, mineralogy, temperature, and soil depth are bases for family differentiation. *Soil series* is the lowest but most specific category of soil taxonomy. A soil series includes soil individuals that have similar (within an acceptable range) pedon characteristics. Pedon characteristics are horizon arrangement, color, texture, structure, consistence, mottling, pH, etc. A soil series is usually given a name after the place name where the polypedon is found for the first time. Similar polypedons belong to the same soil series. It is extremely difficult to know the number of soil series so far identified. New soil series are continually being added as more detailed study is being done of soils throughout the world.

Soil Taxonomy has a systematically derived nomenclature in categories from order to subgroup. Within this hierarchy, the name of a taxon indicates the major characteristics of the soil and the higher categories to which it belongs. An example is given below:

Order: Entisol (Soils with little or no horizon differentiation)
Suborder: Aquent (wet Entisol)
Great group: Sulfaquent (wet Entisol that has sulfidic materials)
Subgroup: Typic Sulfaquent (soils that meet the central concept of Sulfaquent)

Family and series names are, however, somewhat arbitrary.

4.7 Each Soil Order Has Its Own Characteristic Features

4.7.1 Alfisols Are Well-Developed Soils with High Base Status

Alfisols are fine textured soils with high content of exchangeable bases. They have mostly developed in the humid temperate and also in the humid tropical regions under deciduous forests. They have accumulation of clay in the B horizon to form argillic, kandic, or natric horizon, with BSP >35% in the lower part or below the argillic or kandic horizon. They may contain petrocalcic horizons and duripan, fragipan, and plinthite. There is enough moisture for plant growth for 3 consecutive months during the growing seasons. There is relatively little accumulation of organic matter in mineral horizons. These soils are highly fertile and are extensively cultivated with widely diverse cropping patterns as favored by climatic conditions. Some are used for hay, pasture, range, and forests. Alfisols comprise 9.6% of the ice-free land of the earth. Alfisols have five suborders. They are:

Aqualfs: Aqualfs are Alfisols that have aquic conditions (shallow groundwater table that saturates soil with water) for some time in most years within 50 cm of the mineral horizon and redoximorphic features in the upper 12.5 cm of the argillic, natric, or kandic horizon. Aqualfs are abundant in humid regions and are primarily used for rice cultivation. They are fairly fertile, and other crops including corns (maize), soybeans can be grown if artificially drained. Nearly all Aqualfs are believed to have supported forest vegetation in the past.

Cryalfs: Cryalfs are more or less freely drained Alfisols of the cold regions (cryic soil temperature regime) and occur mostly at high elevations, as in the Rocky Mountains in the Western United States. They normally have a udic moisture regime. Most of the Cryalfs are used as forest because of their short, cool growing season.

Udalfs: Udalfs are the more or less frequently drained Alfisols that have udic soil moisture regime and a frigid, mesic, isomesic, or warmer temperature regime. Udalfs are very extensive in the United States and in Western Europe. Most Udalfs with a mesic or warmer temperature regime have or had deciduous forest vegetation, and many of the frigid temperature regimes have or had mixed coniferous and deciduous trees.

Ustalfs: Ustalfs have an ustic soil moisture regime and a frigid, mesic, isomesic, or warmer temperature regime. Ustalfs are the Alfisols of subhumid to semiarid regions. They occur in the United States, Africa, India, South America, Australia, and southeastern Asia. Sorghum, wheat, and cotton are commonly cultivated with irrigation.

Xeralfs: Xeralfs have xeric soil moisture regime common of regions that have Mediterranean climate. They are found in South Africa, Chile, Western Australia, Southern Australia, and the Western United States. They are dry for extended periods in summer, but enough is available in winter. Small grains and other annuals are common crops without irrigation. Grapes and olives are also common crops where the climate is thermic. With irrigation, a wide variety of crops can be grown.

4.7.2 Andisols Are Soils with Andic (Volcanic Ash) Properties

Andisols are characterized by andic materials. Andic materials include volcanic ash, pumice, and cinders deposited during volcanic eruptions. These materials undergo transformation to amorphous or poorly crystallized silicate minerals, including allophane, imogolite, and ferrihydrite. Andisols are young soils and have not had enough time to be highly weathered. These soils are fine textured and have a high content of fresh weatherable minerals and a high cation exchange capacity. They may also contain considerable organic matter as aluminum–humus complex. They have low bulk densities. These soils are widely distributed in all geographical regions near sources of volcanoes. Andisols are generally fertile and are used for agriculture unless restricted by slope, altitude, soil moisture and temperature regimes, etc. One of the most important characteristics of Andisols is their high capacity to fix phosphorus on the surface of the amorphous minerals (Cordova et al. 1996). This is perhaps the principal chemical constraint of Andisols. Some Andisols are left under tundra and forests. Andisols cover more than 124 million ha or approximately 0.7% of the earth's surface. Major areas of Andisols include Chile, Peru, Ecuador, Columbia, Central America, the USA, Kamchatka, Japan, the Philippines, Indonesia, and New Zealand. Andisols have seven suborders. They are:

Aquands: Aquands are the Andisols with aquic conditions at or near the surface. These soils have dark-colored surface horizons that meet the requirements for a histic, umbric, or mollic epipedon. Aquic conditions result in redoximorphic features. Aquands occur locally in depressions and along floodplains where water tables are at or near the soil surface for at least part of the year.

Cryands: Cryands are defined as Andisols with cryic soil temperature regimes. These soils are the Andisols of high latitude (e.g., Alaska, Kamchatka) and high altitude (e.g., Sierra Nevada in the USA). They are usually occupied by cold tolerant forests.

Torrands: Torrands are more or less well-drained Andisols of dry regions. They have an aridic/torric soil moisture regime and a frigid or warmer soil temperature regime. Natural vegetation is mostly desert shrubs. These soils are not extensive occurring mostly in the western part of North America, Hawaii or other Pacific regions. Most of the soils formed under grass or shrub vegetation.

Udands: Udands are more or less well-drained Andisols of moist regions. They have a udic soil moisture regime. Udands are Andisols of the humid climates. They are the most extensive Andisols. These soils are moderately extensive on the Pacific Rim, including Washington, Oregon, and Hawaii in the USA. Most Udands formed under forest vegetation.

Ustands: Ustands are defined as Andisols with ustic soil moisture regimes. These soils are distributed in the intertropical regions that experience seasonal precipitation distribution. They are found mostly in Mexico, Western USA, Pacific Islands, and the eastern part of Africa. Most Ustands formed under grass, shrub, or forest vegetation.

Vitrands: Vitrands are relatively young Andisols that are coarse-textured soils and are dominated by volcanic glass. Most Vitrands are found near volcanoes. Vitrands are abundant in Oregon, Washington, and Idaho of the USA where they form mainly under coniferous forest vegetations. They are the Andisols that have a frigid or warmer soil temperature regime. They have a low water-holding capacity. Vitrands are restricted to ustic and udic soil moisture regimes.

Xerands: Xerands are more or less well-drained Andisols that have a xeric soil moisture regime and a frigid, mesic, or thermic temperature regime. They are temperate Andisols with very dry summers and moist winters. Most Xerands formed under coniferous forest vegetation and some formed under grass or shrub vegetation.

4.7.3 Aridisols Are Soils of Drylands

Aridisols are soils of the arid regions including cold polar, cool temperate, and warm deserts. Aridisols may also occur in semiarid areas outside the zones broadly classified as arid, for example, in local conditions imposing aridity such as steep, south-facing slopes in Northern Hemisphere and in soils whose physical properties limit water infiltration or favor excessive drainage. Aridisols are classified on the basis of their soil moisture regime which is dry in all parts >50% of the time in most years, and not moist for as much as 90 consecutive days when the soil is warm enough (>8 °C) for plant growth. In an aridic/torric soil moisture regime, potential evapotranspiration greatly exceeds precipitation during most of the year. In most years, little or no water percolates through the soil. This hydrologic regime has a distinctive influence on the development of such soils. However, shifting sands of deserts are not included in Aridisols.

There is low chemical weathering, low leaching, and sparse plant growth. Soil organic matter content is low so that ochric horizons abundantly develop in Aridisols. Salts (chlorides, sulfates, carbonates) released by limited chemical weathering are not usually translocated to considerable depths but are accumulated on the surface and, where there is some downward movement of water, in the B horizon. Aridisols have one or more of the following within 100 cm of surface a calcic, cambic, gypsic, natric, petrocalcic, petrogypsic, or salic horizon. An argillic horizon is found in some Aridisols. This is believed to have developed under a moister climate of the past. Aridisols are sparsely vegetated, mostly in xeric shrub lands with xerophytes, cactus, and thorns. They may be cultivated if irrigation can be given, but source of irrigation water is also scanty there. Aridisols comprise about 12% of the world's ice-free land surface. Aridisols have seven suborders. They are:

Cryids: Cryids are the Aridisols of the cold climates. These soils are characteristically developed at high elevations, dominantly in the mountain and basin areas of the USA and Asia and other parts of the world. Cryids commonly show evidence of periglacial features.

Salids: Salids are Aridisols with accumulations that are more soluble than gypsum. The most common form is sodium chloride, but sulfates and others may also occur. These soils are common in depressions in the deserts or in closed basins in wetter areas bordering deserts. Some salts may be brought to the upper horizons by capillary rise of groundwater.

Durids: Durids are the Aridisols that have an accumulation of silica. There is a duripan which is cemented partly with opal or chalcedony. The soils commonly have calcium carbonate. The duripan restricts movement of water and penetration of roots. These soils occur in the western part of the USA particularly in Nevada. They are not known to occur outside the USA.

Gypsids: Gypsids are the Aridisols that have an accumulation of gypsum. These soils occur in Iraq, Syria, Saudi Arabia, Iran, Somalia, West Asia, and in some of the most arid regions of the USA. When the gypsic horizon occurs as a cemented impermeable layer, it is recognized as the petrogypsic horizon.

Argids: Argids are the Aridisols that have accumulation of clay. These soils have an argillic or natric horizon. The presence of an argillic horizon is commonly attributed to a moister paleoclimate. Most Argids occur in North America with a few recognized in the deserts of North Africa or the Near East.

Calcids: Calcids are the Aridisols that have accumulation of residual calcium carbonate or was added as dryfall. Precipitation is inadequate to leach or move the carbonates to great depths. These soils are extensive in the Western USA and other arid regions of the world.

Cambids: These are the Aridisols with the least degree of soil development. They have a cambic horizon that has its upper boundary within 100 cm of the soil surface. These soils are the most common Aridisols in the USA and other parts of the world.

4.7.4 Entisols Are Young Soils That Lack Horizon Development

Entisols are defined as soils that have little or no sign of horizon differentiation. Most Entisols are basically unaltered from their parent materials. Actually they are affected to a limited extent by translocation processes. However, there is considerable darkening of the surface soil by organic matter. The presence of unweatherable parent materials, removal of soil materials by continuous erosion, continuous deposition of silts with floodwater in active floodplains, cold and dry climates, and insufficient time after rock exposure or sediment deposition are the causes of delayed soil development in Entisols. These soils are distributed over a wide geographic area and can be found in any climate and under any vegetation. Entisols along river floodplains are often intensively farmed and are some of the most agriculturally productive soils in the world. Most Entisols are used for pasture, rangeland, and forests. Entisols occupy about 16% of the global ice-free land surface. Entisols have four suborders. They are:

Aquents: These are the wet Entisols. They may be found in tidal marshes, on deltas, on the margins of lakes where the soils are continuously saturated with water, on floodplains along streams where the soils are saturated at some time of the year, or in areas of wet, sandy deposits. Many Aquents have gleying with bluish or grayish colors and redoximorphic features. They may have any temperature regime. Most are formed in recent sediments and support vegetation that tolerates permanent or periodic wetness. Vast areas of alluvial Aquents are used for rice cultivation in South and Southeast Asia, including Bangladesh. Some Aquents have sulfidic materials (former acid sulfate soils).

Arents: Arents are the Entisols that do not have horizons because they have been deeply mixed by plowing, spading, or other methods of moving by humans. Arents may have 3% or more, by volume, fragments of diagnostic horizons in one or more sub-horizons at a depth between 25 and 100 cm below the soil surface.

Fluvents: Fluvents are mostly brownish to reddish soils that are formed in recent alluvial sediments, mainly on floodplains, fans, and deltas of rivers and small streams but not in back swamps where drainage is poor. Strata of clayey or loamy materials commonly have more organic carbon than the overlying, more sandy strata. Fluvents are often found associated with Aquents in floodplains. Rice and jute are grown in many Fluvents.

Psamments: Psamments are Entisols that are very sandy at all layers. Some Psamments form in poorly graded but well-sorted sands on shifting or stabilized deposits, in cover sands, or in sandy parent materials that were sorted in an earlier geologic cycle. Psamments occur under any climate without permafrost within 100 cm of the soil surface. They can have any vegetation and can be cropped with irrigation. Psamments on old stable surfaces commonly consist of quartz sand. These soils are poorly fertile and dry and often show nutrient deficiencies.

4.7.5 Gelisols Are Soils of the Cold Zone

Gelisols are soils that contain gelic materials (mineral or organic soil materials that show cryoturbation, cryodesiccation, and/or ice segregation in the active layer). Gelisols are soils of very cold climates that contain permafrost within 2 m of the surface. Freezing and thawing in the active layer influence soil formation in Gelisols. Permafrost restricts the downward movement of water. Thus, there are few diagnostic horizons in Gelisols, if any. Cryoturbation results in irregular or broken horizons, organic matter accumulation on the permafrost table, oriented rock fragments, and silt caps on rock fragments. These soils are limited geographically to the high-latitude polar regions and localized areas at high mountain elevations. Gelisols are the permafrost-affected soils that occur throughout the zone of continuous permafrost in Antarctica (Bockheim 1995). Gelisols have three suborders. They are:

Histels: Histels have organic horizons similar to Histosols except that they have permafrost within 2 m below the ground. They have 80% or more organic materials from the soil surface to a depth of 50 cm or to a glacic layer or densic, lithic, or paralithic contact, whichever is shallower. These soils occur predominantly in subarctic and low arctic regions of continuous or widespread permafrost. The natural vegetation in Histels is mostly mosses, sedges, and shrubs. The soils are used as wildlife habitat.

Turbels: Turbels are Gelisols that commonly show cryoturbation and contain tongues of mineral and organic horizons, organic and mineral intrusions, and oriented rock fragments. Organic matter is accumulated on top of the permafrost, and ice wedges are common features in Turbels.

Orthels: Orthels are soils that show little or no cryoturbation (less than one-third of the pedon). These soils occur primarily within the zone of discontinuous permafrost, in alpine areas where precipitation is greater than 1,400 mm per year. The natural vegetation is mostly lichens, mosses, sedges, shrubs, black spruce, and white spruce. The soils are used mostly as wildlife habitat. They occur throughout the Gelisol area in Alaska. The vegetation is mostly mosses, sedges, shrubs, and black spruce.

4.7.6 Histosols Are Soils Developed from Organic Soil Materials

Histosols are permafrost-free soils dominated by organic soil materials. Organic soil materials consist of organic debris accumulating at the surface in which the mineral component does not significantly influence the properties of soils. Organic soil materials have either:

1. Under water-saturated conditions 18% organic carbon (30% organic matter) or more if the mineral fraction has 60% or more clay, or 12% organic carbon (20% organic matter) if the mineral fraction has no clay, or a proportional intermediate organic carbon for intermediate content of clay
2. If never saturated with water for more than a few days, 20% or more organic carbon

Histosols typically form in settings where poor drainage inhibits the decomposition of plant and animal remains, allowing these organic materials to accumulate over time. Thus, they have developed in organic parent materials, and they are mostly soils that are commonly called bogs, moors, or peats and mucks. Peat is the name given to slightly decomposed organic material in soil, while muck is used for the rotten, highly decomposed material. The peat is used for fuel, potting soil in greenhouses and for packing. Histosols can be cultivated only if artificially drained. Histosols serve as important habitats for wetland plants and animals and as carbon reservoirs. Histosols are ecologically important because of the large quantities of carbon they contain. Typically, Histosols have very low bulk density (Chap. 5) and are poorly drained because of their occurrence in low-lying areas and high organic matter content. Most Histosols are acidic, and many are deficient in plant nutrients. Many Histosols are not suitable for cultivation because of poor drainage and low chemical fertility. However, many other Histosols formed on recent glacial lands can be very productive when drained. They can sometimes be used for orchards and vines if carefully managed. However, there is a great risk of wind erosion, shrinkage, subsistence, and compaction. Histosols occupy only 1.2% of the global ice-free land surface. Histosols have four suborders mostly distinguished on the basis of the state of organic matter and drainage. They are:

Fibrists: Fibrists are the wet, slightly decomposed Histosols. The largest extent is in southern Alaska of the USA. Most of these soils support natural vegetation of widely spaced, small trees, shrubs, and grasses.

Folists: Folists are the more or less freely drained Histosols that consist primarily of horizons derived from leaf litter, twigs, and branches resting on bedrock or on fragmental materials. Most of these soils support forest vegetation. Some of the soils mainly support grass. A few of the soils are used for specialty crops or for urban or recreational development.

Hemists: Hemists are the wet Histosols in which the organic materials are moderately decomposed. They are extensive in Minnesota and Alaska. Most Hemists support natural vegetation and are used as woodland, rangeland, or wildlife habitat. Some have been cleared and drained and are used as cropland.

Saprists: Saprists are the wet Histosols in which the organic materials are well decomposed. The largest extent in the USA is in Michigan, Florida, Wisconsin, Minnesota, and Alaska. Small areas are common on the Atlantic and gulf coasts. Many Saprists support natural vegetation and are used as woodland, rangeland, or wildlife habitat. Some of the soils, mostly those with a mesic or warmer temperature regime, have been cleared and drained and are used as cropland.

4.7.7 Inceptisols Are Soils That Show Beginning of Horizon Differentiation

Inceptisols are soils that exhibit only the beginning of soil profile development. They are weakly developed soils in that they have minimal horizon differentiation. They are more developed than Entisols and lack many characteristics of mature soils. Inceptisols may have many kinds of diagnostic horizons except argillic, natric, kandic, spodic, and oxic horizons. The most common horizon sequence is an ochric epipedon over a cambic horizon, with or without an underlying fragipan. Inceptisols typically have a cambic horizon, but one is not required if the soil has a mollic, umbric, histic, or plaggen epipedon or if there is a fragipan or duripan or any placic, calcic, petrocalcic, gypsic, petrogypsic, salic, or sulfuric horizon. Inceptisols are soils of humid and subhumid regions. Inceptisols are widely distributed and occur under a wide range of environmental settings. They are often found on fairly steep slopes, young geomorphic surfaces, wet sites, and on resistant parent materials. Inceptisols occupy 9.9% of the global ice-free land surface. Inceptisols have six suborders. They are:

Anthrepts: Anthrepts are more or less freely drained Inceptisols that have either an anthropic or plaggen epipedon. Most have a cambric horizon. Anthrepts can have almost any temperature regime and almost any vegetation. Anthrepts are usually cultivated soils, but some Anthrepts have been diverted to other land uses.

Aquepts: Aquepts are the wet Inceptisols. The water table remains at or near the surface for much of the year. Most Aquepts have formed in depressions, on nearly level plains, or on floodplains. Aquepts may have almost any particle-size class except fragmental. Many Aquepts in floodplains are used for cultivation of rice.

Cryepts: Cryepts are Inceptisols of the cold regions such as high mountains or high latitudes. They do not have permafrost within 100 cm of the soil surface. They may be formed in loess, drift or alluvium (Chap. 3), or in solifluction (mass wasting of water-saturated soil material down the slope, over impermeable surface) deposits. Cryepts occur in the USA in the high mountains of the West, southern Alaska, as well as in other mountainous areas of the world. Vegetation is mostly conifers or mixed conifers and hardwood forests. Few soils are cultivated.

Udepts: Udepts are mainly the more or less freely drained Inceptisols that have a udic or perudic soil moisture regime. They are Inceptisols of humid climates. They are found on nearly level to steeply sloping surfaces. Most of the soils were originally covered with forest vegetation, with some shrubs or grasses. The Udepts of the USA are most extensive in the Appalachian Mountains, on the Allegheny Plateau, and on the west coast. Many Udepts are now under cropping.

Ustepts: Ustepts are mainly the more or less freely drained Inceptisols that have an ustic soil moisture regime. Rainfall occurs mainly during the summer. Some Ustepts are found in older deposits on steep slopes. Native vegetation is commonly grass but some supported trees. Most are used as cropland or pasture.

Xerepts: Xerepts are mainly more or less freely drained Inceptisols that have a xeric soil moisture regime. They are Inceptisols of the temperate regions with very dry summers and moist winters. Xerepts are moderately extensive in the USA and are the most common in California, Oregon, Washington, Idaho, and Utah. The vegetation commonly is coniferous forest on soils with frigid or mesic temperature regimes and shrubs, grass, and widely spaced trees on the soils with a thermic temperature regime.

4.7.8 Mollisols Are Soils of the Grasslands

Mollisols are dark-colored, base-rich, mineral soils of the grasslands. They have a mollic epipedon. They may have an argillic, natric, calcic, or an albic horizon. Some have a duripan or a petrocalcic horizon. Mollisols do not have permafrost, organic soil materials and a spodic horizon. Mollisols may have any of the defined temperature regimes. Mollisols can have any soil moisture regime, but enough available moisture to support perennial grasses seems to be essential. Mollisols are used mainly for small grain in the drier regions and corn (maize) or soybeans in the warmer, humid region. Mollisols comprise 6.9% of the ice-free land of the earth.

Albolls: Albolls are the Mollisols that have an albic horizon and fluctuating groundwater table. Most of these soils are saturated with water to or near the soil surface at some time during winter or spring in normal years. These soils developed mostly on broad, nearly level to sloping ridges, on back slopes, or in closed depressions. Most Albolls have developed under grass or shrub vegetation.

Aquolls: Aquolls are the Mollisols that are wet and that have an aquic soil moisture regime. In these soils, the water table remains at or near the surface for much of the year. They have developed under grasses, sedges, and forbs, but a few have had forest vegetation. In the USA, Aquolls are most extensive in glaciated areas of the midwestern states where the drift was calcareous.

Cryolls: Cryolls are more or less freely drained Mollisols of the cold region. They are abundant in the high mountains of the Western USA, on the plains and mountains of Eastern Europe, and in Asia. The vegetation of the Cryolls on the plains was mostly grasses. Cryolls in the mountains have either forest or grass vegetation. Cryolls in Alaska support spruce, birch, and aspen trees.

Rendolls: Rendolls are shallow Mollisols over calcareous parent materials such as limestone, chalk, drift composted of limestone, or shell bars, of humid regions. These soils are extensive in some parts of the world and formed under forest vegetation or under grass and shrubs.

Udolls: Udolls are Mollisols of humid climates mainly under tall grass prairie (an extensive, level or slightly undulating, treeless tract of land covered with coarse grasses) vegetation, but some could have supported boreal forests (Chap. 14) several thousand years ago. Most of these soils occur in the eastern part of the Great Plains.

Ustolls: Ustolls are Mollisols of semiarid and subhumid climates having an ustic soil moisture regime. Rainfall occurs mainly during a growing season, often in heavy showers, but is erratic. Drought is frequent and sometimes may be severe. Natural vegetation in Ustolls may be grass in the Great Plains and forest in the mountains of Western USA.

Xerolls: Xerolls are the temperate Mollisols with very dry summers and moist winters within a Mediterranean climate. Xerolls have a xeric soil moisture regime. Xerolls are extensive in parts of Turkey, northern Africa near the Mediterranean, and in some of the southern republics of the former USSR and in several states in the USA.

4.7.9 Oxisols Are Highly Weathered Tropical Soils with Enrichment of Kaolinite and Oxides of Fe, Al, and Mn

Oxisols develop under a climate characterized by small seasonal variation in soil temperature and no seasonal soil freezing, and high annual precipitation. They may have a wide range of soil moisture regimes from aridic to perudic. Oxisols with aridic soil moisture regimes are often considered as paleosols (Chap. 3). Usually, Oxisols develop under climatic conditions where precipitation exceeds evapotranspiration for some periods of the year to facilitate the removal of soluble weathering products and favors the residual concentration of kaolinite and sesquioxides, which are essential to form an oxic horizon. Oxisols have the upper boundary of an oxic horizon and no kandic horizon within 150 cm or 40% or more clay by weight in the fine-earth fraction. Oxisols do not have either of the following: (a) permafrost within 100 cm of the soil surface or gelic materials within 100 cm of the soil surface and (b) permafrost within 200 cm of the soil surface. Oxisols consist mainly of quartz; kaolinite; oxides of Fe, Mn, and Al; and organic matter. Oxisols are poorly fertile weathered soils occurring on gentle slopes of geologically old surfaces in tropical and subtropical regions. The natural vegetation ranges from tropical rainforests to desert savannas. Although many Oxisols are extremely infertile, some Oxisols may be made productive when cultivated with appropriate management. Oxisols comprise 7.5% of the global ice-free land surface. Oxisols have five suborders. They are:

Aquox: Aquox are the Oxisols that have a water table at or near the surface for much of the year in shallow depressions and in seepage areas at the base of slopes. There is a tendency to accumulate iron in the form of secondary nodules, concretions, and plinthite.

Perox: Perox are well-drained Oxisols with a perudic soil moisture regime. They are found in continuously humid climates, where precipitation exceeds evapotranspiration in all months.

Torrox: Torrox are the Oxisols of the arid region. They have an aridic (torric) soil moisture regime. Torrox may become productive soils for a variety of crops if water and fertilizers are applied. They occur mainly in Southern Africa, Hawaii, and some areas of Australia.

Udox: Udox are well-drained Oxisols with a udic soil moisture regime. These soils develop in humid areas. There is usually adequate rainfall in normal years to allow for continuous crop growing. Udox occur mostly in South America and in parts of Africa and Asia.

Ustox: Ustox are the Oxisols that have an ustic soil moisture regime. These soils are found in semiarid and subhumid climates. There is at least 90 consecutive dry days in normal years. Available soil moisture is then very low, and cropping is not done in that time. One crop may be grown in the season when rainfall occurs. Two crops may be grown with irrigation in some areas. Ustox occur over a large portion of the interior of South America and in extensive areas of Africa.

4.7.10 Spodosols Are Soils with Accumulation of Amorphous Mixtures of Organic Matter and Aluminum in B Horizon

Spodosols typically form in coarse-textured parent materials and have a reddish-brown spodic horizon beneath a light-colored E horizon. Sometimes there is a fragipan or another sequum (a sequum is a couplet of an eluvial horizon above an illuvial horizon, usually an E and an underlying B horizon [(Schaetzl and Anderson 2005); many soil profiles in humid regions have an E–B sequum. Those soils that have two sequa are termed bisequal soils (Schaetzl 1996)] that have an argillic horizon below the spodic horizon. Some Spodosols have a placic horizon either on or within a spodic horizon or on a fragipan. Some Spodosols have thicker layers than a placic horizon that are cemented by spodic materials (humus–aluminum–clay complex) and organic matter. The particle-size class is mostly sandy, sandy-skeletal, coarse loamy, loamy-skeletal, or coarse-silty. Spodosols are most extensive in areas of cool, humid, or perhumid climates. They may also form, however, to a limited extent, in warm, humid tropical regions, where they occur mostly in areas of quartz-rich sands with fluctuating groundwater table. Most Spodosols in cool temperate regions are covered with coniferous or, less commonly, hardwood forests. Plenty of Spodosols are found in boreal forest regions. Some have been cleared for agriculture. Spodosols are naturally infertile, but some Spodosols may be made productive by good management. Spodosols occupy 2.6% of the global ice-free land surface. Spodosols have four suborders. They are:

Aquods: Aquods are Spodosols that have an aquic soil moisture regime. They are poorly drained soils with a water table at or near the surface for much of the year. A wide variety of hydrophytic (water-loving) plants, ranging from sphagnum in cold areas to palms in the tropics, grow on these soils.

Cryods: Cryods are Spodosols that have a cryic soil temperature regime. They are found in high latitude or high elevations. They are abundant in Alaska, in the mountains of Washington and Oregon of the USA, and Canada. Natural vegetation is mostly coniferous forest or alpine tundra.

Humods: Humods are the relatively freely drained Spodosols that have a large accumulation of organic carbon in the spodic horizon. These soils have developed under coniferous forests and in Western Europe, commonly found in sandy materials where heather (a shrubland characterized by open, low growing woody vegetation) is dominant. In the tropics, most Humods have supported a rain forest.

Orthods: Orthods are the relatively freely drained Spodosols that have a moderate accumulation of organic carbon in the spodic horizon. They are most extensive in the Northeastern United States and the Great Lakes states. Most Orthods are used as forest or have been cleared and are used as cropland or pasture. Orthods are naturally infertile, but they can be highly responsive to good management.

4.7.11 Ultisols Are Low Base Status Soils with an Argillic or a Kandic Horizon

Ultisols are red to yellow soils that are quite acidic, often having a pH of less than 5 and that develop in humid tropical (some in temperate) areas under forest vegetation. They are highly weathered soils and have an argillic or a kandic horizon with low base saturation, less than 35% BSP (by summation of exchangeable bases). The low base saturation status is mainly due to formation in parent material high in silica but low in bases. In some soils, the low base status results from intense leaching of parent material, while in others, a low base status and small quantities of weatherable minerals were initial parent material characteristics. They may have any soil temperature regime and any soil moisture regime except aridic. Leaching is high, and bases released by weathering usually are removed by leaching. The red and yellow colors result from the accumulation of iron oxide which is highly insoluble in water. Kaolinite, gibbsite, and aluminum-interlayered clays are common in the clay fraction. Major nutrients, such as calcium and potassium, are typically deficient in Ultisols. They are poorly fertile soils which may not be productive for most crops without addition of lime and fertilizers. Ultisols occupy 8.5% of the global ice-free land surface. Ultisols have five suborders. They are:

Aquults: Aquults are the Ultisols that have a water table at or near the surface for much of the year. Aquults are found extensively on the coastal plains of the USA, particularly on the Atlantic and Gulf of Mexico. Most of the soils are forested.

Humults: Humults are freely drained Ultisols rich in organic matter (>0.9% or more organic carbon in the upper 15 cm of the argillic or kandic horizon) of mid or low latitudes. Rainforests are the usual natural vegetation.

Udults: These are more or less freely drained Ultisols that have a udic soil moisture regime. They develop in humid areas with well-distributed rainfall. Most of these soils have a forest vegetation, but some have a savanna.

Ustults: These are freely drained Ultisols that have an ustic soil moisture regime and a relatively low content of organic carbon. These soils are generally found in semiarid and subhumid climates. The vegetation commonly consists of forest or savanna plants.

Xerults: Xerults are freely drained Ultisols that have a xeric soil moisture regime. They are found in areas with very dry summers and moist winters typically of Mediterranean or temperate climates. Natural vegetation consisted mostly of coniferous forest plants.

4.7.12 Vertisols Are Soils That Crack Deeply and Widely Upon Drying

Vertisols are clayey soils that have deep, wide cracks for a considerable time of the year and have slickensides (a shiny surface of the cracks produced in soils containing a high proportion of swelling clays) within 100 cm of the mineral soil surface. They shrink when dry and swell when moistened. They are generally sticky in the wet season and hard in the dry season. Most Vertisols have an ustic soil moisture regime; some have an aridic and a udic regime. Vertisols generally have 50–70% clay with a relatively large proportion of fine clay in the clay fraction. The clays in Vertisols consist predominantly of 2:1 and 2:2 layer clay minerals, but some have considerable amounts of other clay minerals. The natural vegetation is predominantly grass, savanna, open forest, or desert shrub. Most Vertisols are well suited to farming if there is plenty of rainfall or irrigation water and if suitable management practices are followed. Because of the low permeability and tendency to remain waterlogged for long periods, Vertisols are often considered as problem soils (Chap. 11). Vertisols are extensive in some parts of the world. They were known as black cotton soils in India. Vertisols occupy 2.4% of the global ice-free land surface. Vertisols have six suborders. They are:

Aquerts: Aquerts are the Vertisols that have aquic soil moisture regime. They have a water table at or near the surface for much of the year but are also dry enough for periods for cracks to open. They are found in low areas such as glacial lake plains, floodplains, stream terraces, and depressions.

Cryerts: Cryerts are the Vertisols that have a cryic soil temperature regime. They are soils of the cold climate. They are fine textured soils and periodically shrink and swell forming cracks that commonly open in late summer. Cryerts occur on the cold prairies of Canada where they are commonly derived from lacustrine deposits. They also occur in the US Rocky Mountains.

Torrerts: Torrerts are the Vertisols of arid climates. Their cracks commonly stay open for most of the year but may close for at least a few days during rains. Many of these soils are found in closed depressions that may be ponded from time to time by runoff from higher areas. Some Torrerts are found in the southwest of the USA. These soils are commonly used for rangeland.

Uderts: Uderts are the Vertisols of humid areas. They have a udic soil moisture regime. The cracks may not open completely some years due to high precipitation. In the USA, the soils occur on gentle slopes and are derived dominantly from marine shales, marls, and alluvium. Many of these soils supported grass, but some support hardwood or pine forests.

Usterts: These are the Vertisols in temperate areas that do not receive high amounts of rainfall during the summer. They have an ustic soil moisture regime. Cracks open and close once or twice during the year. They are found extensively in the USA, Australia, Africa, and India. If irrigated, Usterts can be used intensively, but large areas are used for grazing due to a lack of machinery to till soils.

Xererts: Xererts are the Vertisols of Mediterranean climates, which have xeric soil moisture regime. These soils have cracks that regularly close and open each year. In the USA, most of the soils supported grasses.

4.8 FAO/UNESCO Soil Classification Is Now World Reference Base for Soil Resources

FAO/UNESCO published a soil map of the world at the scale of 1:5,000,000 (FAO 1971–1981) that needed a legend for different soil mapping units. This legend was treated as a soil classification system in different regions of the world. The legend for the map was further revised which contained 28 "major soil groupings" (FAO 1988). A new map of soil resources of the world at the scale of 1:25,000,000 was presented at the 14th International Congress of Soil Science, Kyoto, Japan in 1990. The FAO legend was later replaced by the World Reference Base for Soil Resources (FAO 1994, 1998). In these replacement and updates, some earlier units were renamed, some new units were added, and some were merged. In the updated version, there are 30 "reference groups." Some group names resemble Soil Taxonomy, some have been taken from classical Russian and European names, and others have been synthesized. This is known as FAO/

WRB soil classification system. The WRB reference groups have been adopted in 1998 by the International Union of Soil Sciences as the standard for soil correlation and nomenclature (FAO AGL 2003). The reader is referred to "World Reference Base for Soil Resources 2006," World Soil Resources Report 103 by FAO (2006) for recent updates. Now, there are 32 "reference soil groups" (RSG). The names and characteristics of the RSG are given below:

Acrisols
Acrisols are soils that have a higher clay content in the subsoil than in the topsoil due to clay migration forming an argic subhorizon with low-activity clays and low base saturation. Acrisols are found in humid tropical, humid subtropical, and warm temperate regions and are most extensive in Southeast Asia, the southern fringes of the Amazon Basin, the southeast of the United States of America, and in both East and West Africa. Sedentary farming can be done in Acrisols with complete fertilization and careful management.

Albeluvisols
Albeluvisols are soils that have an illuvial clay horizon with an irregular or broken upper boundary resulting in tonguing of bleached soil material into the illuviation horizon. Albeluvisols are distributed in Europe, North Asia, and Central Asia, with minor occurrences in North America. Albeluvisols are developed in the continental regions that had permafrost in the Pleistocene of northeast Europe, northwest Asia, and southwest Canada and also in the loess and cover sand areas and old alluvial areas in moist temperate regions, such as France, central Belgium, southeast of the Netherlands, and west of Germany. Albeluvisols are not much suitable for cropping due to their acidity, low nutrient levels, tillage and drainage problems, and for its short growing season and severe frost during the long winter. They are usually vegetated with forest.

Alisols
Alisols are soils that have a higher clay content in the subsoil than in the topsoil due to clay migration forming an argic subsoil horizon with low base saturation and high-activity clays. They occur predominantly in humid tropical, humid subtropical and warm temperate regions. Major occurrences of Alisols are found in Latin America (Ecuador, Nicaragua, Venezuela, Colombia, Peru and Brazil), West Indies (Jamaica, Martinique and Saint Lucia), West Africa, East Africa, Southeast Asia, and northern Australia. Alisols also occur in China, Japan, and the southeast of the United States of America. The undulating topography on which Alisols are usually found makes soil susceptible to erosion and truncation. Alisols are naturally poorly fertile and have toxic concentrations of Al.

Andosols
Andosols are soils that develop in volcanic ashes. Andosols may also develop in other silicate-rich materials under acid weathering in humid and perhumid climates. Andosols occur in volcanic regions all over the world. Important concentrations are found in South America, Central America, Mexico, United States of America, Japan, the Philippines, Indonesia, Papua New Guinea, New Zealand, Fiji, Vanuatu, New Caledonia, Samoa, Hawaii, Kenya, Rwanda, Ethiopia, Madagascar, Italy, France, Germany and Iceland. Andosols are generally fertile soils and have high agricultural potential. The strong phosphate fixation caused by active Al and Fe of Andosols is a problem. Ameliorative measures to reduce this effect include application of lime, silica, organic material, and phosphate fertilizer.

Anthrosols
Anthrosols are soils that have been modified profoundly through human activities, such as addition of organic materials or household wastes, irrigation, and cultivation. Anthrosols with plaggic horizons (horizon resulting from manuring) are most common in northwest Europe. Anthrosols with irragric horizons (horizon developed as a result of irrigation) are found in irrigation areas in dry regions, for example, in Mesopotamia, near oases in desert regions, and in parts of India. Anthrosols with an anthraquic horizon (puddle layer or plow pan) overlying a hydragric horizon (subsurface horizon having characteristics of wet cultivation) are characteristics of paddy soils which occupy vast areas in China and in parts of South and Southeast Asia (Bangladesh, Sri Lanka, Viet Nam, Thailand, and Indonesia). Anthrosols with hortic horizons (horizon resulting from deep cultivation and fertilizer application) are found all over the world where humans have fertilized the soil with household wastes and manure. Rye, oats, barley, potato, sugar beet, and summer wheat are common crops on European Anthrosols with a plaggic horizon. Rice is grown intensively in Anthrosols with anthraquic horizons. Puddling of wetland rice fields (involving destruction of the natural soil structure by intensive tillage when the soil is saturated with water) is done to reduce percolation.

Arenosols
Arenosols are sandy soils, including both soils developed in residual sands and soils developed in recently deposited sands such as dunes in deserts and beach lands. Arenosols are one of the most extensive RSG in the world. Vast expanses of deep Aeolian sands are found on the Central African plateau. Other areas of Arenosols occur in the Sahelian region of Africa, various parts of the Sahara, central and Western Australia, the Near East, and China. Most Arenosols occur in arid and semiarid regions, but they may be found in humid areas as well. Agricultural potential of Arenosols is limited because of their coarse texture, high percolation, and low water and nutrient retaining capacity. They are, however, easy to plow, easy of rooting and harvesting. Some Arenosols may used for cereals and vegetables with sprinkler and drip irrigation . Most Arenosols are left under natural vegetation.

Calcisols

Calcisols are soils with secondary accumulation of lime. Calcisols are common in highly calcareous parent materials and widespread in arid and semiarid environments. Many Calcisols occur together with Solonchaks that are actually salt-affected soils. Natural vegetation of Calcisols comprises of shrubs, grasses, and herbs. Drought-tolerant rainfed crops might be grown, but Calcisols may be made productive with irrigation. Extensive areas of Calcisols are used for production of irrigated winter wheat, melons, and cotton in the Mediterranean zone. Sorghum, alfalfa, and fodder crops are tolerant of high Ca levels. A high pH of Calcisols (usually >7.5) often limits micronutrient availability. Zinc deficiency is the most ubiquitous micronutrient problem affecting cereal crops, including corn (maize), rice, and wheat in Calcisols of Turkey (Yilmaz et al. 1997; Cakmak 2008), Syria, Iraq, India, and Pakistan.

Cambisols

Cambisols exhibit the beginning of soil formation. A weak horizon differentiation may be noticed in the subsoil. Transformation of parent material is evident from structure formation and mostly brownish discoloration, increasing clay percentage, and/or carbonate removal. Cambisols may occur in both dry climates and also in humid tropics and subtropics. The young alluvial plains and terraces of the Ganges–Brahmaputra system (India and Bangladesh in Southeast Asia) are probably the largest continuous surface of Cambisols in the tropics. Cambisols generally make good agricultural land and are used intensively. Cambisols with high base saturation in the temperate zone are among the most productive soils on earth. Cambisols on irrigated alluvial plains in the dry zone are used intensively for production of food and oil crops. Cambisols with groundwater influence in alluvial plains are highly productive paddy soils.

Chernozems

Chernozems are soils with a thick black surface layer that is rich in organic matter. Chernozems are distributed mainly in the middle latitude steppes (grassland plains without trees) of Eurasia and North America. Chernozems are regarded as the most productive soils and are mainly used for arable cropping. Wheat, barley, and corn (maize) are the principal crops grown, alongside other food crops and vegetables. Some Chernozems are used for livestock rearing. Maize is widely grown in the warm temperate belt.

Cryosols

Cryosols comprise mineral soils formed in a permafrost environment. Water is present primarily in the form of ice. Cryosols are widely known as permafrost soils. Cryosols are circumpolar in both the Northern and Southern Hemispheres. Major areas with Cryosols are the Russian Federation, Canada, China, Alaska, and in parts of Mongolia. Most areas of Cryosols in North America and Eurasia are in the natural state and support sufficient vegetation for grazing animals, such as caribou, reindeer, and musk oxen. Overgrazing leads rapidly to erosion and other environmental damage.

Durisols

Durisols are very shallow to moderately deep, moderately well-, to well-drained soils that contain a layer of cemented secondary silica within 100 cm of the soil surface. Extensive areas of Durisols occur in Australia, South Africa and Namibia, and the United States of America with minor occurrences in Central and South America. The agricultural use of Durisols is limited to extensive grazing (rangeland). Durisols in natural environments generally support enough vegetation to contain erosion, but elsewhere erosion of the surface soil is widespread.

Ferralsols

Ferralsols are the classical, deeply weathered, red or yellow soils of the humid tropics. These soils have a clay assemblage dominated by low-activity clays, mainly kaolinite, and a high content of sesquioxides. Ferralsols are abundant on the continental shields of South America (especially Brazil) and Africa (especially Congo, Democratic Republic of the Congo, southern Central African Republic, Angola, Guinea, and eastern Madagascar) and Southeast Asia. Most Ferralsols have great soil depth and good physical properties. They are well drained but may in times be droughty because of their low available water storage capacity. The chemical fertility of Ferralsols is poor, weatherable minerals are almost absent, and cation retention by the mineral soil fraction is weak. Under natural forest, the bulk of plant nutrients is contained in the biomass; available plant nutrients in the soil are concentrated in the soil organic matter. Productivity of these soils may be improved by manuring and mulching.

Fluvisols

Fluvisols are young soils in alluvial deposits. Fluvisols occur on all continents and in all climates. Major concentrations of Fluvisols are found along rivers and lakes, for example, in the Amazon basin; the Ganges Plain of India and Bangladesh; the plains near Lake Chad in Central Africa; the marshlands of Brazil, Paraguay, and northern Argentina; and in deltaic areas of the Ganges–Brahmaputra, Indus, Mekong, Mississippi, Nile, Niger, Orinoco, Plate, Po, Rhine and Zambezi. Some Fluvisols have a thionic horizon (sulfidic material; acid sulfate soils) are found in the coastal lowlands of Southeast Asia (Indonesia, Vietnam, and Thailand), West Africa (Senegal, Gambia, Guinea Bissau, Sierra Leone, and Liberia) and along the northeast coast of South America (French Guiana, Guyana, Suriname, and Venezuela). Paddy rice cultivation is widespread on tropical Fluvisols. Many dryland crops are grown on Fluvisols as well, normally with irrigation in the dry period.

Gleysols

Gleysols are wetland soils (Chap. 13) that are saturated with groundwater for long enough periods to develop a characteristic gleyic color pattern. The upper part of the soil profile may be reddish, brownish, or yellowish colors but essentially grayish/bluish colors beneath. Gleysols occur in nearly all climates, from perhumid to arid. The largest extent of Gleysols is in subarctic areas in the north of the Russian Federation (especially Siberia), Canada, and Alaska and in humid temperate and subtropical lowlands, for example in China and Bangladesh. Paddy rice is cultivated in many Gleysols. Adequately drained Gleysols can be used for arable cropping. The soil will be puddle if cultivated wet. Therefore, Gleysols in depression areas with unsatisfactory possibilities to lower the groundwater table are best kept under a permanent grass cover or swamp forest.

Gypsisols

Gypsisols are soils with substantial secondary accumulation of gypsum. These soils are found in the driest parts of the arid climate zone. They were earlier called desert soils. Major occurrences of Gypsisols are in Mesopotamia, near East and Central Asian republics, Libyan and Namib deserts, southeast and central Australia, and the United States of America. Gypsisols that contain only a low percentage of gypsum in the upper 30 cm can be used for the production of small grains, cotton, alfalfa, etc. Gypsisols in young alluvial and colluvial deposits have a relatively low gypsum content. Where such soils are in the vicinity of water resources, they can be very productive. Wheat, apricots, dates, corn (maize), and grapes can be grown satisfactorily if irrigated at high rates in combination with forced drainage.

Histosols

Histosols are formed in organic material. They include soils developed in predominantly moss peat in boreal, arctic and subarctic regions, moss peat, reeds/sedge peat (fen) and forest peat in temperate regions, and mangrove peat and swamp forest peat in the humid tropics. Vast majority of Histosols occurs in lowlands. Common names are peat soils, muck soils, and bog soils (Chap. 13). The majority of Histosols occur in the boreal, subarctic and low arctic regions of the Northern Hemisphere. Most of the remaining Histosols occur in temperate lowlands and cool montane areas; only one-tenth of all Histosols are found in the tropics. Extensive areas of Histosols occur in the United States of America and Canada, Western Europe, and northern Scandinavia. Tropical forest peat borders the Sunda Shelf in Southeast Asia. In the temperate zone, millions of hectares of Histosols have been opened for agriculture. In many instances, this has initiated the gradual degradation, and ultimately the loss, of the precious peat.

Kastanozems

Kastanozems are dry grassland soils of the short-grass steppe belt. They have a similar profile to that of Chernozems, but the humus-rich surface horizon is thinner and not as dark as that of the Chernozems. They show more prominent accumulation of secondary carbonates. Major areas of Kastanozems are in the Eurasian short-grass steppe belt (southern Ukraine, the south of the Russian Federation, Kazakhstan, and Mongolia); in the Great Plains of the United States of America, Canada, and Mexico; and in the Pampas and Chaco regions of northern Argentina, Paraguay, and southern Bolivia. Kastanozems are potentially rich soils but lack enough moisture for high yields. Irrigation is nearly always necessary for good crop production, but care must be taken to avoid secondary salinization of the surface soil. Small grains and irrigated food and vegetable crops are the principal crops grown. Extensive grazing is another important land use on Kastanozems.

Leptosols

Leptosols are very shallow soils over continuous rock and soils that are extremely gravelly and/or stony. Leptosols were earlier called lithosols. Some Leptosols on calcareous rocks belonged to rendzinas. Leptosols are the most extensive RSG on earth. Leptosols are found from the tropics to the cold polar tundra and from sea level to the highest mountains. Leptosols are particularly widespread in montane areas, notably in Asia and South America, in the Sahara and the Arabian deserts, the Ungava Peninsula of northern Canada, and in the Alaskan mountains. Leptosols support grasses and forests. Some Leptosols have been planted to teak and mahogany in Southeast Asia; in the temperate zone, they are mainly under deciduous mixed forests whereas acid Leptosols are commonly under coniferous forest.

Lixisols

Lixisols comprise soils that have a higher clay content in the subsoil than in the topsoil due to clay migration leading to an argic subsoil horizon. Lixisols have a high base saturation and low-activity clays at certain depths. Lixisols are found in seasonally dry tropical, subtropical, and warm temperate regions on Pleistocene and older surfaces. These soils occur in sub-Sahelian and East Africa, South and Central America, Indian subcontinent, Southeast Asia, and Australia. Many Lixisols bear natural savannah or open woodland vegetation. Tillage of wet soil or use of excessively heavy machinery compacts the soil and causes serious structure deterioration. Tillage and erosion control measures such as terracing, contour plowing, mulching, and use of cover crops help to conserve the soil.

Luvisols

Luvisols are soils that have an illuvial argic subsoil horizon with high-activity clays and a high base saturation. Luvisols extend mainly in temperate regions such as in the west and center of the

Russian Federation, the United States of America, and Central Europe but also in the Mediterranean region and southern Australia. Most Luvisols are fertile soils and suitable for a wide range of agricultural uses. Luvisols with a high silt content are susceptible to structure deterioration where tilled when wet or with heavy machinery. Luvisols on steep slopes require erosion control measures. Luvisols in the temperate zone are widely grown to small grains, sugar beet, and fodder; in sloping areas, they are used for orchards, forests, and/or grazing.

Nitisols
Nitisols are soil that are deep, red, and well drained with diffuse horizon boundaries and a subsurface horizon with more than 30% clay and moderate to strong angular blocky structure. These soils develop in tropical regions. Weathering is relatively advanced. More than half of all Nitisols are found in tropical Africa, notably in the highlands of Ethiopia, Kenya, Congo, and Cameroon. Nitisols are also found in tropical Asia, South America, Central America, Southeast Africa, and Australia. Nitisols are among the most productive soils of the humid tropics. The deep and porous solum and the stable soil structure of Nitisols permit deep rooting and make these soils quite resistant to erosion. Nitisols are planted to plantation crops such as cocoa, coffee, rubber, and pineapple and are also widely used for food crop production on small holdings. There is high P sorption in some Nitisols, and P fertilizer application is profitable there.

Phaeozems
Phaeozems are soils of relatively wet grassland and forest regions in moderately continental climates. Like Chernozems and Kastanozems, they also have dark, humus-rich surface horizons, but they are more extensively leached and are less rich in bases. Phaeozems are found in the humid and subhumid Central Lowlands and eastern most parts of the Great Plains of the United States of America. Phaeozems are also found in the subtropical pampas of Argentina and Uruguay. Some Phaeozems are present in China and the Russian Federation. Phaeozems are porous, fertile soils and make excellent farmland. In the United States of America and Argentina, Phaeozems are in use for the production of soybean and wheat and other small grains. Phaeozems on the high plains of Texas produce good yields of irrigated cotton. Phaeozems in the temperate belt are planted to wheat, barley, and vegetables.

Planosols
Planosols are soils with a light-colored, surface horizon that shows signs of periodic water stagnation and abruptly overlies a dense, slowly permeable subsoil with significantly more clay than the surface horizon. Planosols occur in subtropical and temperate regions with clear alternation of wet and dry seasons, including Brazil, Paraguay, Argentina, East and Southern Africa, United States of America, Bangladesh, Thailand, and Australia. Land use on Planosols is normally less intensive than that on most other soils under the same climate conditions. Vast areas of Planosols are used for extensive grazing. Planosols in the temperate zone are mainly in grass, or they are planted to arable crops such as wheat and sugar beet. Planosols in Southeast Asia are widely planted to a single crop of paddy rice.

Plinthosols
Plinthosols are soils with plinthite (an Fe-rich or Mn-rich mixture of kaolinitic clay, gibbsite, etc., with quartz and other constituents forming a layer of hard nodules or a hardpan or irregular aggregates), petroplinthite (a continuous, fractured, strongly cemented to indurated nodules) or pisoliths (discrete strongly cemented to indurated nodules). Plinthosols are most common in the wet tropics, notably in the eastern Amazon Basin, the central Congo Basin, and parts of Southeast Asia and northern Australia. Poor natural soil fertility caused by strong weathering, waterlogging in bottomlands and drought cause serious limitations of these soils to agriculture.

Podzols
Podzols are soils with a typically ash-gray eluvial subsurface horizon, bleached by loss of organic matter and iron oxides, on top of a dark accumulation horizon with brown to black humus. Podzols mainly occur in humid areas in the boreal and temperate zones. Podzols are extensive in Scandinavia, the northwest of the Russian Federation, and Canada. The low nutrient status, low level of available moisture, and low pH make podzols unattractive soils for arable farming. Aluminum toxicity and P deficiency are common problems. They are mainly under natural forests in the temperate and boreal regions. Some Podzols have been cleared for cropping with irrigation, liming, and fertilization.

Solonchaks
Solonchaks are soils that have a high concentration of soluble salts. They are largely confined to the arid and semiarid climate zones and to coastal regions in all climates. Solonchaks are most extensive in the Northern Hemisphere, notably in the arid and semiarid parts of northern Africa, the Near East, the former Soviet Union, and Central Asia; they are also widespread in Australia and the Americas. Saline soils are regarded as problem soils (Chap. 11). Strongly salt-affected soils have little agricultural value. They are used for extensive grazing of sheep, goats, camels, and cattle, or lie idle. When they are reclaimed, they may give good yields of cereals and vegetables. Heavy irrigation to meet crop requirement and leaching requirement are needed.

Solonetz
Solonetz are soils with a high proportion of adsorbed Na and/or Mg ions. Solonetz that contain free soda (Na_2CO_3) are

strongly alkaline (field pH>8.5). These soils were earlier called alkali soils and sodic (Regional Salinity Laboratory 1954). Solonetz occur predominantly in areas with a steppe climate in particular in flat lands with impeded vertical and lateral drainage. Major Solonetz areas are found in Ukraine, Russian Federation, Kazakhstan, Hungary, Bulgaria, Romania, China, United States of America, Canada, South Africa, Argentina, and Australia. These soils are also regarded as problem soils, and their management for agriculture needs improvement of soil structure and porosity, lowering of pH by acidifying agents and lowering the ESP (Chap. 11). Reclaiming Solonetz involves cost, labor, and time.

Stagnosols
Stagnosols are soils with a perched water table showing redoximorphic features caused by surface water. Stagnosols are periodically wet and mottled in the topsoil and subsoil, with or without concretions and/or bleaching. Stagnosols are found in West and Central Europe, North America, southeast Australia, and Argentina. The agricultural suitability of Stagnosols is limited because of their oxygen deficiency resulting from stagnating water above a dense subsoil. They have to be drained for arable agriculture.

Technosols
Technosols are soils strongly influenced by human activity and containing many artifacts. They include soils from wastes (landfills, sludge, cinders, mine spoils and ashes), pavements with their underlying unconsolidated materials, soils with geomembranes and constructed soils in human-made materials. Technosols are found worldwide around human habitation and development sites. They are more likely to be contaminated soils. Many Technosols have to be treated with care as they may contain toxic substances resulting from industrial processes. Many Technosols may be planted to forests.

Umbrisols
Umbrisols accommodate soils in which organic matter has accumulated within the mineral surface soil (in most cases with low base saturation) to the extent that it significantly affects the behavior and utilization of the soil. Umbrisols occur in cool, humid regions, mostly mountainous and with little or no soil moisture deficit. Umbrisols are common in the Andean ranges of Colombia, Ecuador, and, to a lesser extent, in Venezuela, Bolivia, and Peru. They also occur in Brazil, India, Nepal, China, and Myanmar. Many Umbrisols are under a natural or near-natural vegetation cover. Coniferous forest predominates in Brazil in the United States of America. Umbrisols in tropical mountain areas in South Asia and Oceania are under montane evergreen forest.

Vertisols
Vertisols are deeply and widely cracking soils with a high proportion of swelling clays. Most Vertisols occur in the semiarid tropics, with an average annual rainfall of 500–1,000 mm, but Vertisols are also found in the wet tropics, e.g. Trinidad (where the annual rainfall sum amounts to 3,000 mm). The largest Vertisol areas are on sediments that have a high content of smectitic clays. Vertisols are prominent in South Africa, Australia, southwest of the United States of America, Uruguay, Paraguay, and Argentina. Although these soils are also problem soils, they have considerable agricultural potential if proper management is adopted. Tillage is usually difficult in Vertisols, but they have good chemical fertility, and occurrence of many of them is on extensive level plains in humid areas. Many Vertisols are productive. The crops include wheat, barley, sorghum millet, cotton, chickpeas, flax, and sugarcane. Construction of roads and buildings and other structures on Vertisols is risky.

4.8.1 Correlation of Reference Soil Groups of WRB with Soil Taxonomy

Reference soil groups	Soil taxonomy equivalents
Acrisols	Ultisols
Albeluvisols	Alfisols
Alisols	Ultisols
Andosols	Andisols
Anthrosols	Entisols, Inceptisols
Arenosols	Entisols (Psamments)
Calcisols	Aridisols (Calcids)
Cambisols	Inceptisols
Chernozems	Mollisols
Cryosols	Gelisols
Durisols	Aridisols (Durids)
Ferralsols	Oxisols
Fluvisols	Entisols (Fluvents Fluvaquent)
Gleysols	Inceptisols, Entisols Alfisols, Mollisols
Gypsisols	Aridisols (Gypsids)
Histosols	Histosols, Gelisols (Histels)
Kastanozems	Mollisols (Ustolls, Xerolls)
Leptosols	Entisols
Lixisols	Alfisols
Luvisols	Alfisols
Nitisols	Inceptisols, Alfisols, Oxisols, Ultisols
Phaeozems	Mollisols (Udolls and Albolls)
Planosols	Alfisols, Ultisols, Mollisols
Plinthosols	Alfisols, Oxisols, Ultisols
Podzols	Spodosols
Regosols	Entisols
Solonchaks	Aridisols (Salids)
Solonetz	Aridisols (several other orders)
Stagnosols	Entisols, Inceptisols, Alfisols, Ultisols Mollisols
Technosols	Entisols
Umbrisols	Entisols, Inceptisols
Vertisols	Vertisols

Study Questions

1. What do you mean by a diagnostic horizon? Write the features of an anthropic, a plaggen, and a mollic epipedon. What are characteristics of argillic, albic, kandic, and calcic horizons? How does anthropic epipedon differ from plaggen epipedon?
2. What do you mean by soil moisture control section? How does torric regime differ from aridic regime? Which of ustic and udic soil moisture regime is more favorable for crop growth?
3. How do you distinguish between Entisols and Inceptisols? What soil orders are found in tropical areas? What soil orders are typical of humid temperate climate? What are the agricultural significance of Alfisols and Mollisols?
4. How do Chernozems, Phaeozems, and Kastanozems differ? How do you distinguish between Anthrosols and Technosols?
5. Make a list of reference soil groups suitable for cultivation of paddy rice. What are the land uses of Cryosols, Leptosols, Histosols, and Vertisols?

References

Agriculture Canada Expert Committee on Soil Survey (1987) The Canadian system of soil classification, 2nd edn. Canadian Government Publishing Centre, Ottawa

Avery BW (1980) Soil classification for England and Wales (higher categories). Soil Survey technical monograph no. 14, Harpenden

Baize D, Girard MC (1995) Referentiel pedologique 1995. Institut National de la Recherche Agronomique, Paris

Baldwin M, Kellogg CE, Thorp J (1938) Soil classification. Soils and men: yearbook of agriculture. US Government Printing Office, Washington, DC

Bockheim JG (1995) Permafrost distribution in the Southern Circumpolar region and its relation to the environment: a review and recommendations for further research. Permaf Periglac Process 6:27–45

Bokheim JG, Gennadiyev AN (2000) The role of soil-forming processes in the definition of taxa in soil taxonomy and the world soil reference base. Geoderma 95:53–72

Buol SW, Hole FD, McCracken RJ, Southard RJ (1997) Soil genesis and classification, 4th edn. Iowa State University Press, Ames

Buol SW, Southard RJ, Graham RC, McDaniel PA (2003) Soil genesis and classification, 5th edn. Iowa State Press/Blackwell, Ames

Cakmak I (2008) Zinc deficiency in wheat in Turkey. In: Alloway BJ (ed) Micronutrient deficiencies in global crop production. Springer, Dordrecht

Cordova J, Valverde F, Espinosa J (1996) Phosphorus residual effect in Andisols cultivated with potatoes. Better Crops Int 10(2)

FAO (1988) Soil map of the world. Revised legend, FAO–UNESCO–ISRIC. World Soil Resources report no. 60, Rome

FAO (1994) World reference base for soil resources. ISSS–ISRIC–FAO. Draft. Rome/Wageningen

FAO (1998) World reference base for soil resources. ISSS–ISRIC–FAO. World Soil Resources report no. 84, Rome

FAO (2006) World reference base for soil resources 2006. A framework for international classification, correlation and communication. FAO–UNESCO–ISRIC. FAO, Rome

FAO AGL (2003) http://www.fao.org/ag/agll/default.stm. Accessed 8 July 2011

Gimenez JE (2011) The world reference base for soil resources (WRB) and its application to some soils of Argentina. São Paulo, Unesp, Geociências 1:15–21

Isbell RF (1996) The Australian soil classification. CSIRO Publishing, Collingwood

Isbell RF (2002) The Australian soil classification, Revised edn. CSIRO Publishing, Collingwood

Krasilnikov P, Arnold RW, Ibanez JJ (2010) Soil classifications: their origin, the state-of-the-art and perspectives. In: 19th world congress of soil science, soil solutions for a changing world, Brisbane, Australia, 1–6 Aug 2010 (Published on DVD)

Regional Salinity Laboratory Staff (1954) Diagnosis and improvement of saline and alkali soils. USDA agricultural handbook no. 60. US Government Printing Office, Washington, DC

Schaetzl RJ (1996) Spodosol–Alfisol intergrades: bisequal soils in NE Michigan, USA. Geoderma 74:23–47

Schaetzl RJ, Anderson S (2005) Soils genesis and geomorphology. Cambridge University Press, Cambridge

Soil Survey Staff (1999) Soil taxonomy, 2nd edn. United States Department of Agriculture, Government Printing Office, Washington, DC

Soil Survey Staff (2010) Keys to soil taxonomy, 11th edn. USDA, NRCS, Washington, DC

Thorp J, Smith GD (1949) Higher categories of soil classification. Soil Sci 67:117–126

USDA (1960) Soil classification, a comprehensive system. United States Department of Agriculture, Government Printing Office, Washington, DC

USDA (1975) Soil taxonomy: a basic system of soil classification for making and interpreting soil survey. United States Department of Agriculture, Government Printing Office, Washington, DC

USDA, NRCS (1998) Keys to soil taxonomy, 8th edn. United States Department of Agriculture, Government Printing Office, Washington, DC

USDA, NRCS (2003) Keys to soil taxonomy, 9th edn. United States Department of Agriculture, Government Printing Office, Washington, DC

Yilmaz A, Ekiz H, Torun B, Gultekin I, Karanlik S, Bagci SA, Cakmak I (1997) Effect of different zinc application methods on grain yield and zinc concentration in wheat cultivars grown in zinc-deficient calcareous soils. J Plant Nutr 20(4–5):461–471

5 Physical Properties of Soil

Physical properties of soil include color, texture, structure, porosity, density, consistence, temperature, and air. Colors of soils vary widely and indicate such important properties as organic matter, water, and redox conditions. Soil texture, structure, porosity, density, and consistence are related with types of soil particles and their arrangement. There are two types of soil particles—primary and secondary. Primary particles include sand, silt, and clay, categorized on the basis of their effective diameter. There are important differences in physical, chemical, and mineralogical properties of these fractions. Their relative proportion in a soil is called soil texture. It is a fundamental property of soil. It is not easily altered. There are 12 textural classes ranging from sand to clay. Soil structure is the arrangement of soil particles into different geometric patterns. It is classified into different types on shape, classes on size, and grades on stability. Soil structure is amenable. Soil texture and structure together regulate porosity, density, compactness, retention, and movement of water and air in soil. Soil temperature is slightly higher than air temperature in a place. It influences life processes of soil biota including plants. Soil air is retained in soil pores; its composition is variable, and it contains higher carbon dioxide and moisture and lower oxygen concentration than atmospheric air. Soil air has a great role in respiration of plant roots and microorganisms and transformation of mineral and organic matter.

5.1 Soil Has Varied Colors

Soils have varied colors, ranging from gray to black, green, and yellowish to red or brown. Soil color reflects environmental conditions, soil forming processes, and other influences on the soil. It generally indicates the status of organic matter and moisture, drainage, and the state of oxidation and reduction. It may also indicate mineralogical composition of the soil such as presence of iron, manganese, and dark minerals. Humus-rich soils have dark gray to black coloration. Saturation of soil with moisture may also give darker colors. Well-drained soils have generally yellowish-red to brown color. Oxidized iron compounds give reddish color to soils (SSSA 2010). Iron compounds under impeded drainage in the subsoils give them green colors (gleying). Manganese compounds, if present in large quantities, may give a black color. Color does not affect the behavior of the soils, however.

A soil may have a ground color (matrix color) different from mottle (discontinuous spotted streaks and small spots or blotches of contrasting color, usually red or yellow) colors. In describing soil color, both matrix and mottle colors are mentioned. The pattern may relate to the aeration or drainage of the soil. Gleyed colors (greenish) are low chroma matrix colors with or without mottles. If the soil is gleyed, it is likely to be reduced (due to prolonged moisture saturation) and wet for much of the year. Mottles occur in soils that undergo seasonal oxidation and reduction due to fluctuating groundwater table (Sect. 3.10.6).

Colors vary in their purity, lightness or darkness, and intensity. Munsell color charts are used to provide standard descriptions and names of soil color. Soil color is expressed by the color notations—hue, chroma, and value.

Hue: Hue denotes the specific color or purity of color (red, orange, yellow, green, blue, indigo, violet).

Value: Value refers to the lightness or darkness of the color, ranging from 0 (most dark) to 10 (most light).

Chroma: Chroma is the light intensity which range from 0 (least intense) to 10 (most intense).

The color notations are given in the order: hue, value, and chroma. For example, 10YR 6/4 is a light yellowish-brown soil color; 10YR is the hue, 6 is the value, and 4 is the chroma. Soil color varies with soil moisture; therefore, the state of soil moisture is to be mentioned when soil color is recorded

K.T. Osman, *Soils: Principles, Properties and Management*,
DOI 10.1007/978-94-007-5663-2_5, © Springer Science+Business Media Dordrecht 2013

Table 5.1 Dimensions of sand, silt, and clay particles

ISSS system		USDA system	
Soil particles	Diameter, mm	Soil particles	Diameter, mm
Coarse sand	2.00–0.20	Very coarse sand	2.00–1.00
Fine sand	0.20–0.02	Coarse sand	1.00–0.50
Silt	0.02–0.002	Medium sand	0.50–0.25
Clay	<0.002	Fine sand	0.25–0.10
		Very fine sand	0.10–0.05
		Silt	0.05–0.002
		Clay	<0.002

(e.g., 10YR 6/4 dry, light yellowish-brown). Soil color is influenced by several factors, including mineralogical composition, organic matter content, soil moisture, soil development, and climate.

5.2 Soils Are Composed of Variously Sized Mineral Particles

In the mineral fraction of the soil, there are two types of particles—primary particles and secondary particles. Primary particles are individual discrete particles, and secondary particles are their aggregates. Two or more primary particles cling together to form a secondary particle. Soil particles are classified into three categories—sand, silt, and clay. Soil particles vary widely in their shapes; they are not all spheres. So, they are classified on their "effective diameter," the diameter of a sphere that has a velocity of fall in a liquid medium equal to the particle in question. Clay particles are smaller than 0.002 mm in diameter. The size limits of sand and silt are different in different systems of classification, although the upper limit of sand is 2 mm in all systems. Particles greater than 2 mm diameter are known as pebbles (2 mm–7.5 cm), cobbles (7.5–25 cm), stones (25–60 cm), and boulders (>60 cm). These particles are not generally considered as soil constituents because they hardly affect the fertility and productivity of soils. However, in some soils, the percentage of stones may be very high, and they can interfere with the movement of tillage implements.

5.2.1 Soil Particles Are Classified According to Size

Dimensions of sand, silt, and clay according to ISSS (International Soil Science Society) and USDA (United States Department of Agriculture) systems are given in Table 5.1. A given mass of soil may be separated into its constituent size classes or the "soil separates." Some important properties of the soil separates are given in Table 5.2.

Table 5.2 Important properties of sand, silt, and clay particles

Particles	Properties
Sand	2.00–0.02 mm in the International and 2.00–0.05 mm in USDA systems; visible to the naked eye; gritty in feeling; little or no capacity to hold water or nutrients; and bind other particles, loose when wet, very loose when dry
Silt	0.02–0.002 mm in the International and 0.05–0.002 mm in USDA systems; most silt particles are not visible to the naked eye, but can be seen through an ordinary microscope; low to medium capacity to attract water, nutrients, and other particles; feels smooth when wet; and like talcum powder when dry
Clay	<0.002 mm in both systems; can be seen only through an electron microscope; high capacity to hold water, nutrients, and other particles; sticky and malleable when wet; and forms hard clods when dry

Besides these, there are important mineralogical differences among soil particles. Sand and silt particles consist of primary minerals such as quartz, feldspars, and mica, while clay particles are mainly secondary minerals such as kaolinite, smectite, vermiculite, illite, chlorite, and hydrated oxides of Fe and Al. Clay particles have large surface area and electrical charges, both negative and positive, on their surfaces. Because of these properties, clays have high water- and nutrient-holding capacity and they participate in chemical reactions in the soil.

5.2.2 Close Packing of Soil Particles Creates Some Degree of Fineness or Coarseness Known as Soil Texture

Sand particle is coarse, clay is fine, and silt is medium. A sand particle is at least 1,000 times greater than a clay particle. Therefore, when sand particles are packed together, they leave larger gaps between them (Fig. 5.1). On the other hand, gaps between clay particles are small. These gaps produced by packing of soil particles are called "soil pores." Larger pores, the macropores, generally accommodate air, and smaller pores, the micropores, contain water. Thus, the coarseness or fineness of soil determines its air–water relationships. It also influences other soil characteristics and activity of roots and microorganisms. The coarseness or fineness of soil due to the relative proportion of sand, silt, and clay is called soil texture.

Texture is one of the most important physical properties of soil that affect its fertility and productivity. Actually, the whole soil environment is regulated by soil texture. There are field and laboratory methods for its determination. Experienced field workers have learned to determine soil texture by examining a lump of soil between fingers. Laboratory techniques have been developed based on the velocity of fall of a particle in a liquid medium, which is proportional to the square of the radius of the particle and inversely to the viscosity (a fluid's

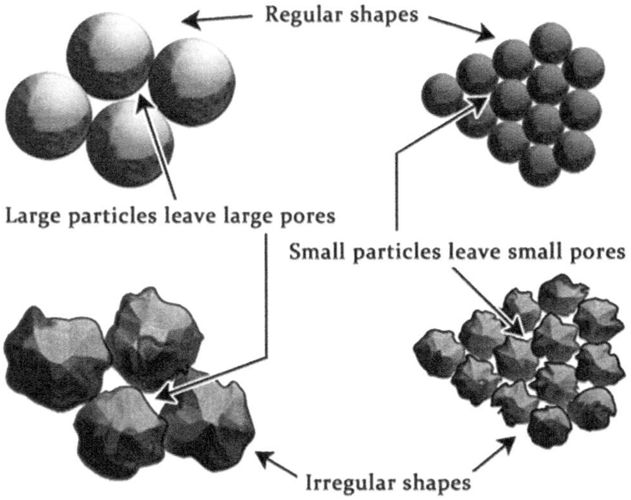

Fig. 5.1 Large particles leave large pores and small particles leave small pores when closely packed

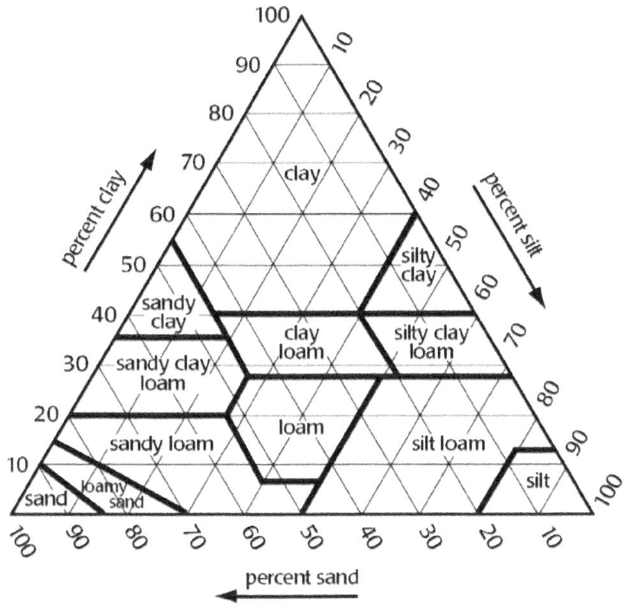

Fig. 5.2 Soil textural triangle devised by the USDA

Table 5.3 Soil textural classes with their particle size ranges

		Percentage composition		
Texture	Textural class name	Sand	Silt	Clay
Coarse	Sand	80–100	0–20	0–20
	Loamy sand	70–80	0–30	10–15
	Sandy loam	50–80	0–50	0–20
Medium	Loam	30–50	30–50	0–20
	Silt loam	0–50	50–100	0–20
	Silt	0–20	90–100	0–10
Fine	Sandy clay loam	50–80	0–30	20–30
	Clay loam	20–50	20–50	20–30
	Silty clay loam	0–30	50–80	20–30
	Sandy clay	50–70	0–20	30–50
	Silty clay	0–20	50–70	30–50
	Clay	0–50	0–50	30–100

internal resistance to flow) of the liquid (Stokes' Law, Hillel 1980). Percentages of sand, silt, and clay are determined by either "pipette method" or "hydrometer method," and a textural class name is assigned to it with the help of the "textural triangle" developed by the USDA (Fig. 5.2).

The relative proportion of sand, silt, and clay in a soil is also called "mechanical composition," and the technique of its determination is called mechanical analysis.

5.2.3 There Are 12 Soil Textural Classes

Soils are rarely composed of a single-size class of particles; they are mixtures of different size classes. But one or two size classes usually dominate the physical behavior of the soil. The soil is then named after the name of that soil separate. Thus, a sandy soil displays the properties of sand particles. When a soil equally exhibits the properties of sand, silt, and clay, then it is called loam (approximately 40% sand, 40% silt, and 20% clay, Table 5.3). A higher proportion of sand in loam produces sandy loam. In this way, textural class names are given to soils on the basis of percentage composition of sand, silt, and clay. Twelve textural classes have so far been identified.

Soil textural class names can be determined from this triangle if the percentages of any two size fractions are known. For example, lines for 40% sand and 20% clay intercept in the area demarcated as loam.

5.2.4 Soil Texture Regulates Soil Behavior

Sandy soils are loose and friable, they absorb water rapidly and drain it quickly, and they can be worked easily in both moist and dry conditions. They are called "light-textured soils." On the other hand, clay soils are generally compact and stiff, sticky when wet and hard when dry, and require much energy to work both in wet and dry conditions. They are called "heavy-textured soils." Actually, sandy soils are heavier; they have higher mass in dry condition because they have smaller total volume of pores, that is, higher volume of solids. Soil texture influences many other soil properties and growth and activity of soil organisms including plants.

Porosity and air–water relationships: A soil is said to be porous when it favors movement of air and water, elongation of roots, and drainage of excess water. Coarse-textured soils have higher number of macropores while fine-textured soils have higher number of micropores. Fine-textured soils hold more water, and coarse-textured soils drain away quickly. Texture controls air capacity, aeration, maximum water-holding capacity, field capacity, and saturated and

Table 5.4 Physical behaviors of soil with different textures

Soil textural class	Infiltration	Water-holding capacity	Nutrient-holding capacity	Aeration	Workability	Leaching
Sand	Good	Poor	Poor	Good	Good	High
Silt	Medium	Medium	Medium	Medium	Medium	Medium
Clay	Poor	Good	Good	Poor	Poor	Low
Loam	Medium	Medium	Medium	Medium	Medium	Medium

unsaturated flow of water. Coarse-textured soils are more absorptive and have higher infiltration rates. They are also leached very easily. Sandy soils tend to remain dry, and clay soils usually become waterlogged.

Soil fertility: Fine-textured soils are more fertile than coarse-textured soils. Fine-textured soils have higher contents of weatherable minerals. They contain more organic matter and exchangeable bases. Fine-textured soils have more total *N* and available nutrients. Sandy soils need more fertilizers, but nutrients are lost more easily from these soils. So, small amounts of fertilizers can be applied at a time in coarse-textured soils, and they need frequent fertilizer application. However, sandy soils make good container media used in nurseries for raising seedlings.

Table 5.5 Most favorable soil textures for some crops

Crops	Most favorable soil texture
Rice	Sandy clay loam to clay
Wheat	Loam
Corn (maize)	Sandy loam to loam
Barley	Loam
Pearl millet	Sandy loam to clay loam
Black gram	Loam to clay loam
Green gram	Loam
Pea	Loam
Soybean	Sandy clay loam to clay loam
Sunflower	Sandy loam to loam
Groundnut	Sandy loam to loam
Tea	Sandy loam to loam
Coffee	Sandy loam to clay loam

Irrigation: Sandy soils are wetted with low amount of water, but they are quickly dried up. So, frequent irrigation is needed in sandy soils. On the other hand, huge amount of water can be applied at a time in fine-textured soils.

Root growth and distribution: Profuse and deep roots develop in porous soils. Fine-textured compacted soils are not good root media.

Microbial activity: There is higher microbial population in fine-textured soils. Microbial transformations are also generally more rapid in such soils.

The physical behaviors of different soil textures are summarized in Table 5.4.

5.2.5 Crop Plants Need Favorable Soil Textures

Crop plants have adapted in different soil conditions during their evolution. There are suitable ranges of moisture, air, and nutrient requirements of specific crops. Different crops generally prefer different soil textures (Table 5.5).

5.2.6 Soil Texture Is Not Easily Altered

Soil texture is not usually changed by management practices. It inherits from the parent materials and originates through weathering and pedogenic processes such as, recrystallization, eluviation, and illuviation. It may, however, be altered by erosion, deposition, truncation, and some other human interventions. Coarse sediments may be deposited by inundation with flood water on fine-textured soils. Removal of surface soils by runoff and wind erosion may expose coarse or fine-textured subsoils. Truncation of the surface soil for brick making or other purposes may expose the compact heavy substrata. Sands are often mixed with fine-textured soils for improving drainage of lawn, playground, greenhouse, and nursery soils.

5.3 Soil Structure Is the Arrangement of Soil Particles

"Soil structure is a key factor in the functioning of soil, its ability to support plant and animal life, and moderate environmental quality. Favorable soil structure and high aggregate stability are important for improving soil fertility, increasing agronomic productivity, enhancing porosity and decreasing erodibility" (Bronick and Lal 2005). In some soils, the particles remain separated as individuals as in dunes and beeches. These soils are called single grained. In most soils, there is aggregation or structure formation; they are structured soils. Soil particles form "aggregates" or "peds" of various sizes and shapes. Peds differ from "clods" and "concretions"; clods

Table 5.6 Types of soil structure and their characteristics

Types	Characteristics
Spheroidal	The peds are roughly spherical or granular. They are formed mostly by biotic activity in the surface A horizon. Soils with granular structure are porous and absorptive. When granular peds are highly porous, the structure is called crumb
Platelike	The peds are thin with vertical axes shorter than the horizontal axes. Soils with platy structure are usually compact. They are mostly found in B horizons of clay soils
Blocklike	The vertical and horizontal axes of peds are more or less equal; they are generally found in the B horizons of the humid regions; blocks with sharp edges are angular blocky, and those with dull edges are sub-angular blocky. Soils with blocky structure may be porous
Prismlike	The horizontal axes are greater than the vertical axes of peds; they are generally found in the B horizons. They are common in soils of arid and semiarid regions; if the peds have flat angular tops the structure is called prismatic, and if the top is rounded the structure is columnar.

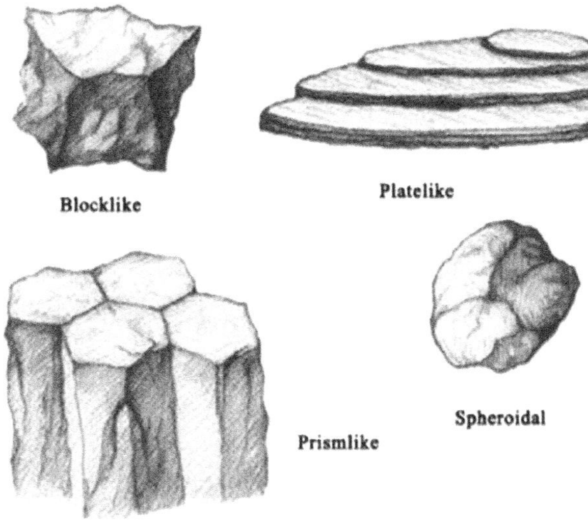

Fig. 5.3 Approximate shapes of different soil structures

or chunks are artificially formed (such as by plowing) hard soil mass. Concretions are hard lumps produced by the precipitation of dissolved substances (usually iron and manganese oxides). The arrangement and organization of primary and secondary particles in a soil mass is known as soil structure. In some soils such as heavy clays, all the particles adhere together. Structure of these soils is called massive. Soil structure is classified on the shapes of aggregates into types, on size of peds into classes, and on distinctness and stability of the aggregates into grades. There are four types of soil structure (Table 5.6). These are spheroidal (granular and crumb), blocklike (angular blocky and sub-angular blocky), platelike, and prismlike (prismatic and columnar) (Fig. 5.3).

5.3.1 Classes of Soil Structure Are Based on Size of Aggregates

On the basis of size of aggregates, soil structure is divided into five classes: very fine, fine, medium, coarse, and very coarse. Their dimensions are given in Table 5.7.

Table 5.7 Classes of soil structure

Size class	Diameter of granules, mm	Thickness of plate, mm	Diameter of blocks, mm	Diameter of columns, mm
Very fine[a]	<1	<1	<5	<10
Fine[a]	1–2	1–2	5–10	10–20
Medium[a]	2–5	2–5	10–20	20–50
Coarse[a]	5–10	5–10	20–50	50–100
Very coarse[a]	>10	>10	>50	>100

[a]Platy structure is designated as thick or thin instead of fine or coarse (Soil Survey Staff 1993)

5.3.2 Grades of Soil Structure Refer to Stability of Peds

Grades of soil structure are identified on the basis of the visibility in a horizon and stability of aggregates. Stability of aggregates refers to their resistance to destruction by water. Three structural grades are identified: weak, moderate, and strong.

Weak: Poorly formed, indistinct peds which are not durable and much unaggregated material.

Moderate: Moderately well-developed peds which are fairly durable and distinct.

Strong: Very well-formed peds which are quite durable and distinct.

To characterize a soil structure, the sequence grade, class, and type is followed, for example, strong coarse angular blocky structure, moderate thin platy structure, and weak fine prismatic structure.

5.3.3 Formation of Soil Structure Results from Complex Processes

Soil structure formation results from complex interactions of many factors including the environment, soil management, plant influences, and soil properties such as mineral composition,

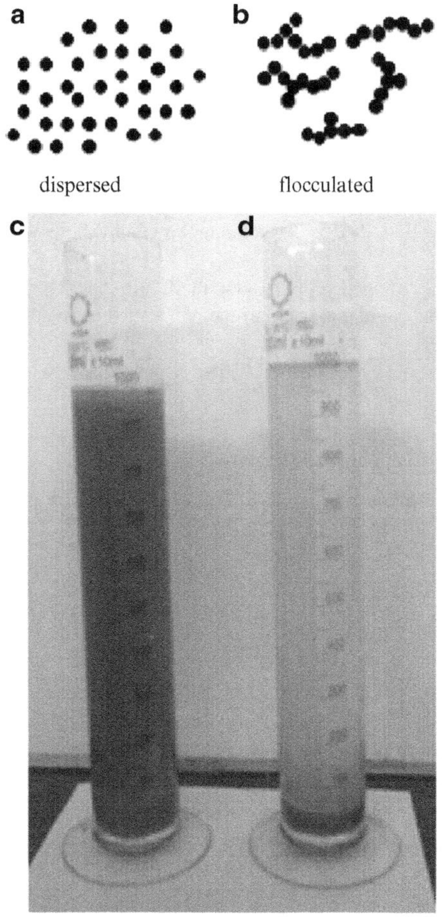

Fig. 5.4 Dispersion and flocculation; (a) dispersed and (b) flocculated particles; (c) clay particles dispersed in water and (d) flocculated and sedimented clay particles

texture, soil organic matter, pedogenic processes, microbial activities, exchangeable ions, nutrient reserves, and moisture availability (Kay 1998). There are several mechanisms of aggregation. Aggregates are formed in stages, with different bonding mechanisms dominating at each stage (Tisdall and Oades 1982). Sand and silt particles are inert materials. They can come closer but cannot hold themselves together because they do not possess the power of adhesion and cohesion. Clays form coatings on them and cement several sand and silt particles into a larger unit. Soil particles are bound mechanically by plant roots and fungal hyphae. There are also various chemical compounds which act as cements and gums in soils. These materials include flocculated clays, hydrated oxides of Fe and Al, organic compounds produced by root exudation, microbial synthesis and excretion and organic matter decomposition, and polyvalent cations.

5.3.3.1 Flocculation of Clay Particles

Clay particles are electrically charged particles. They carry both negative and positive charges, but there are, generally, higher number of negative charges on their surfaces. They attract cations. These cations are enveloped with water molecules. However, some cations are loosely held and some are strongly held depending on the type of clay minerals and ionic size and valence of the cations. When clay particles adsorb and get satisfied largely with monovalent cations like Na^+ and K^+, some negative charge deficit remains on their surfaces. Then, they behave like large anions, repulse one another, and remain dispersed in a liquid medium (Fig. 5.4). Clay particles may come closer in a suspension only when they are flocculated. Flocculation (a process wherein colloids come out of suspension in the form of floc or flakes by the addition of a clarifying agent) is effected by replacing monovalent cations on colloidal surfaces by polyvalent cations like Ca^{2+}, Mg^{2+}, Fe^{2+}, Fe^{3+}, and Al^{3+}. Flocculation of clay particles produces clay domains. These clay domains are cemented together by themselves or by other cementing agents to form microaggregates. Microaggregates are bound together to produce macroaggregates. However, flocculation alone is not enough for aggregation; it involves a combination of different processes such as hydration, pressure, and dehydration and require cementation of flocculated particles. Due to flocculating and binding effects of Ca^{2+} ions, calcareous soils are generally well structured and sodic soils with high exchangeable Na^+ are poorly structured soils. In acid soils, H^+ (monovalent, but more strongly held than Na^+ or K^+ due to smaller ionic radii), Fe^{3+}, and Al^{3+} ions give soils generally good structures. Liming with calcite or dolomite is a strategy for improving soil structure in sodic soils. All clays are flocculated if the concentration of salts in the soil solution is high enough (Wild 1988). Therefore, saline nonsodic soils with prevalence of sodium chlorides and sulfates (white alkali soils) remain usually well structured.

5.3.3.2 Mechanical Binding of Soil Particles

Soil particles are mechanically bound into larger aggregates by plant roots and fungal hyphae. When decomposable organic matter is added to soils, fungi with extensive hyphal network profusely proliferate. Their hyphae grow in pores, push soil particles apart, and surround particles into stable envelopes. However, fungal species and type of hyphae may affect the aggregation. Fast-growing hypha is more efficient in producing aggregates than slow-growing hypha. Plant roots also bind soil particles together, but there are wide differences among the plants in their ability to bind soil particles. Sod-forming plants like grasses are familiar soil binders. There are a number of ways, in addition to the mechanical binding, in which plant roots affect aggregation. They excrete mucilage that cements particles; they harbor bacteria in the rhizosphere that produce gums extracellularly and through decomposition of organic matter.

Fig. 5.5 Polyvalent cations may form bridges between clay particles

5.3.3.3 Cementation of Soil Particles

Cementing materials in soil includes organic substances, silicate clays, hydrated oxides of Fe and Al, and cations. Organic substances are humus, decomposition products of soil organic matter, root exudates, compounds of microbial synthesis and excretion, etc. Organic cementing compounds may be transient or persistent (Tisdall and Oades 1982). Transient materials include some microbial and root exudate polysaccharides and polysaccharide gums whose effects last only a few weeks. Persistent category includes aromatic organic compounds linked by polyvalent metal cations to clay surfaces and also other strongly adsorbed polymers. Larger aggregates, low in organic matter, are bound by transient agents only, probably low molecular weight polysaccharides. Chesters et al. (1957) showed that polysaccharide gums are the principal constituent of humus that stabilizes the structure in agricultural soils. They also found that crump stability was much better correlated with the polysaccharide content of the soil than with the total humus content. Many investigations on aggregate formation have indicated that well-known natural organic cements are the polysaccharides—levans and dextrans, polyuronides, etc. They stabilize the aggregates when they are dehydrated.

When dehydrated, oxides of Fe and Al bind soil particles firmly together. However, aluminum oxides are more efficient than iron oxides in this regard. Colloidal clays are strong cementing agents; their cementation depends on their mineralogical composition, specific surface, and the state of deflocculation and dehydration. Clays linked with polyvalent cations, as Ca^{2+} in most instances, form stable aggregates (Fig. 5.5).

Cations surrounded by water molecules form bridges between two or more clay particles. When they are dehydrated, they form stable aggregates.

However, aggregation is the result of complex pedogenic processes affected by environmental and management factors.

5.3.4 "Soil Structure Is the Key to Soil Fertility"

Soil structure modifies the effects of soil texture. Clay soils, which are likely to be impervious, may be porous if they are well structured. Soil structure has a tremendous environmental and agricultural significance. Soil structure regulates pore size, number of pores, and distribution of pores and total porosity of the soil. Thus, retention and movement of soil water including infiltration, permeability, percolation, drainage, and leaching all depend on soil structure. Under natural condition, rhizosphere soils are drier than surrounding soils due to uptake of water by plant roots and transpiration by plant leaves. The ease with which water will move to the root zone depends on soil structure. Nutrient uptake by plants, a function of nutrient availability in soil and nutrient absorption capacity of roots, is influenced largely by soil structure. Soil structure controls soil air, aeration, and evaporation. Root growth, rooting depth, and volume of roots differ in soils of different porosity. Soil structure influences accommodation, movement, and activity of soil organisms. Microbial processes like organic matter decomposition, mineralization, nitrification, and nitrogen fixation are influenced by soil structural conditions. So, Baver (1960) rightly states, "Soil structure is the key to soil fertility."

5.3.5 Unlike Soil Texture, Soil Structure May Readily Be Altered

Soil structure may be altered by natural forces and by soil management practices. Natural forces include wetting, drying, freezing, thawing, and raindrop impacts. Soil management practices such as tillage, irrigation, fertilizer and manure application, liming, and cropping patterns all have positive and negative impacts on soil structure. Over-tilling, over-irrigation, and mono-cropping damage soil structure. On the other hand, manuring, particularly green manuring, and fallowing improve soil structure. Cultivation of soil for prolonged period reduces soil organic matter and destroys aggregation.

5.3.6 Puddling Is the Process of Destruction of Soil Structure

Puddling is the process of destruction of aggregates by plowing soil in wet conditions. Making soil into almost massive and free flowing mud is accomplished by puddling. Puddling is done for making soil water retentive and impervious as is necessary for rice cultivation. Puddling is a consequence of tillage, desired or undesired, of soil in waterlogged condition. It makes favorable tilth for rice, but it may be undesirable for the crops in succession in the cropping patterns. After rice harvest, when the soil is dry, it becomes too hard to work with. Alternate puddling and restructuring are done in wetland–dryland successive cropping systems.

All soils can be puddled, but the clay soils are puddled more easily and strongly. Clay mineral composition has also an important influence on puddling. Soils dominated by 2:1 type clays are more easily puddled than those with 1:1 type clays. Again, Na-saturated clays are more easily puddled than Ca-saturated clays. However, sandy soils with low clay contents can also be puddled. Long cycles of puddling and restructuring make the surface soils compact and impervious.

It is found in many soils with plaggen epipedons (surface soil horizons compacted by tillage).

5.4 Density Is the Mass Per Unit Volume

Density of soil is obtained by dividing the mass (dry weight, usually at 105°C for 24 h) by its volume. If a core of soil has a total volume V_T, which is the sum of the volume of solids, V_S, and volume of pores, V_p, and the dry weight of the soil is W_S, then

$$\text{Total volume}, V_T = V_p + V_S, \text{ and}$$

$$\text{Bulk density}, D_b = \frac{\text{Mass of oven dried soil}}{\text{Total volume}}$$

$$\text{Bulk density}, D_b = \frac{W_S}{V_T} = \frac{W_S}{V_p + V_S}$$

Bulk density can easily be determined by the core method. Standard aluminum or stainless steel cores with sharp edges may be used. The cores are pushed into the soil with the help of a wooden hammer at soft strokes so that the natural structure is not disturbed. Let the internal diameter of a core be 5 cm and the length of the core be 8 cm. The weight of the core (W_1) is 165 g, and the weight of the oven-dried soil with the core (W_2) is 377 g. Then,

$$\text{Volume of the core} = \pi r^2 L,$$

$$\text{where } r = \frac{\text{diameter}}{2} = 2.5\text{ cm}$$

$$= 3.14 \times 2.5^2 \times 8 = 157\text{cm}^3$$

$$\text{Weight of core} \quad W_1 = 165\text{g}$$

$$\text{Weight of core + ovendried soil } W_2 = 377\text{g}$$

$$\text{Weight of oven dry soil} \quad W_2 - W_1 = 212\text{g}$$

$$\text{Then, Bulk Density} = \frac{W_2 - W_1}{\pi r^2 L} = \frac{212}{157} = 1.35\text{gcm}^{-3} \text{or} 1.35\text{Mgm}^{-3}$$

$$(1\text{Mgm}^{-3} = 1\text{gcm}^{-3})$$

Most mineral soils have bulk densities between 1.0 and 2.0 Mg m^{-3}. Bulk density of soil varies due to organic matter content, texture, compaction, and porosity of the soil. Histosols have very low bulk (and particle) density, as low as 0.70 Mg m^{-3} (Brady and Weil 2002). A loose soil has a lower bulk density; soils which show massive structures and less porosity will show higher bulk densities. Compacted soils have bulk density values between 1.6 and 1.7 Mg m^{-3}. Average bulk density of a loam-textured mineral soil is taken to be 1.3 Mg m^{-3}. On this assumption, the mass of soil of 1 ha (10,000 m^2) furrow slice (0–15-cm depth) may be calculated as:

Mass of 1 ha-15 cm soil = Area x depth x density = $10{,}000 \times 0.15 \times 1.3 = 1{,}950\,\text{Mg ha}^{-1} = \text{App.}\,2 \times 10^6\,\text{kg ha}^{-1} = 2$ million kg ha^{-1}. This value is often needed for the estimation of fertilizer requirements, particularly on mass basis.

Particle density is the density of the solids,

that is, particle density of soil = Mass of oven-dried soil/volume of solids.

$$\text{Particle density}, D_p = \frac{W_S}{V_S}$$

Density of different soil-forming silicate minerals may range from 2.60 to 2.75 Mg m^{-3} (1 Mg m^{-3} = 1 g cm^{-3}). Most soils have particle density values within this range. Density of quartz which prevails in the sand and silt fractions of most soils is 2.65 Mg m^{-3}. So, an average particle density of 2.65 Mg m^{-3} is often applied to mineral soils. Soils high in organic matter have lower particle density, generally between 0.9 and 1.3 Mg m^{-3} (Brady and Weil 2002). Conversely, soils that are rich in iron can have high particle densities. Ferromagnesian minerals have densities ranging from 2.9 to 3.5 Mg m^{-3}. Density of iron oxides and other heavy minerals can exceed 4 Mg m^{-3} (Smettem 2006). Particle density values obtained from different sources for some common soil constituents are given in Table 5.8.

Table 5.8 Particle density of some soil minerals and humus

Substance	Particle density, Mg m^{-3}
Quartz	2.65
Feldspar	2.55
Mica	2.81
Calcite	2.70
Hematite	5.10
Dolomite	2.85
Apatite	3.25
Kaolinite	2.40
Soil solids	2.65
Humus	1.30

5.5 Pores Are Void Spaces Between Soil Particles and Aggregates

Spaces in soil that are not occupied by solids are called pores. Actually, they are not void; they are filled with either air or water, or both, depending on the soil moisture content and condition of rainfall and irrigation. Roots and soil organisms, both macro- and microflora and fauna, occupy these pores. There are small or large pores depending on the texture and structure. Large pore spaces allow fast infiltration and percolation of water through a soil. On the

Table 5.9 Classes of soil pores

Class of pores	Equivalent diameter (μm)
Macropores	>75
Mesopores	30–75
Micropores	5–30
Ultramicropores	0.1–5
Cryptopores	<0.1

After SSSA (1997)

other hand, small pores have strong attractive forces to hold water in the pore. Large pores, also called macropores (>0.75 mm), accommodate air and conduct water into the soil where it fills the micropores (<0.75 mm). Clay soils have numerous micropores and hold large quantities of water, but since they have few macropores, they produce very slow infiltration rates. The pores in the clays may be so small and hold water so tenaciously that the water is not available to plants. Sandy soils have numerous macropores, but few micropores (Table 5.9).

Genetically, there are two types of pores in soil—biopores and aggregate pores. Biopores are created by organisms including roots, earthworms and insects, and humans. Roots exert pressures during their elongation and expansion pushing the soil particles apart and leaving long and continuous pores within the soil after their death. Continuous pores are also created by the movement of earthworms and insects, particularly the soil scramblers. Soil management practices, such as plowing and intercultural operations, also loose soils forming biopores. Aggregate pores may be intra-aggregate pores and inter-aggregate pores. Intra-aggregate pores are left by packing of the primary particles within the aggregates, and inter-aggregate pores are found between the aggregates. Inter-aggregate pores are usually larger than intra-aggregate pores. In general practice, meso-, micro-, ultramicro-, and crypto-pores are taken together as fine pores or micropores. Macropores and micropores are also called non-capillary pores and capillary pores, respectively.

5.5.1 Percentage of Soil Volume Occupied by Pores Is Known as Porosity

$$\text{Porosity,\%} = \frac{V_{\text{P}}}{V_{\text{T}}} \times 100$$

where, total volume is V_T and volume of pores is V_P.

Porosity of the soil can be obtained from bulk density and particle density according to the following formula:

$$\text{Porosity} = \left\{1 - \frac{\text{Bulk density}}{\text{Particle density}}\right\} \times 100$$

Table 5.10 Relationship of bulk density with soil porosity

Bulk density, Mg m^{-3}	Estimated total porosity, %
1.20	55
1.30	51
1.40	47
1.50	43
1.60	39
1.70	35

Total porosity of soil is of little practical significance. The number and proportion of macropores and micropores and distribution of different size classes of the pores seem to be more important.

5.5.2 Anything That Affects Bulk Density Also Affects Porosity

Size distribution of pores and porosity of soil are affected by a number soil factors such as texture, structure, organic matter, bulk density, and soil organisms and management factors such as tillage, cropping, and irrigation.

Soil texture: It has already been mentioned that sandy soils have a large number of macropores but a few micropores. They are generally well aerated, but they have low water retention capacities. On the other hand, clay soils have a large number of micropores but a few macropores. They retain high amount of water but, unless well aggregated, they are poorly drained. However, the total porosity is higher in fine-textured soils than in coarse-textured soils. Sandy, silty, and clay soils may have total porosity ranges of 25–50, 35–50, and 40–70%, respectively.

Soil structure: Soil structure influences all of number, size, and volume of pores. Number of air-filled pores increases with the increase in the relative proportion of larger aggregates. Again, crumb structure is the most porous followed by the granular structure. The platy structure has the lowest porosity, and angular and sub-angular blocky structures have intermediate porosity.

Organic matter: Organic matter produces aggregation of soil. The number of different size classes of peds and the stability of peds are correlated well with the organic matter content.

Bulk density: Porosity is inversely proportional to bulk density, that is, soils with lower bulk density have higher porosity. Estimated porosity for different bulk density values are given in Table 5.10.

Action of soil organisms: Soil organisms influence porosity in a number of ways. First, they bind soil particles into aggregates; second, they exert pressure and push soil particles apart; third, they secrete or excrete organic gums which bind particles together; fourth, they decompose organic matter and produce polysaccharide gums; and fifth, they leave pores after their death. Soil fauna that live and move within the soil produces interconnected pores. Earthworms and arthropods are particularly efficient pore formers.

Tillage: Tillage is intended for loosening the soil and breaking larger peds and clods into smaller fragments. So, tillage normally increases porosity. Tillage increases aeration, expels toxic gases from the soils, and makes the conditions favorable for seed germination and root growth. However, the effect of tillage on porosity is generally temporary. Soils usually return to their original compaction level before the next tillage. Heavy rains or irrigation following tillage may cause a collapse of the soil structure, that is, the peds may disintegrate and the porosity is reduced. Use of heavy machineries for tillage may compact the soil, particularly below the plow layer. Decline in soil aggregation and soils becoming more compact have been observed due to long-term effects of tillage. Compaction results in a decrease in the total pore space and an increase in the bulk density.

Cropping: The surface soil is exposed to rain and sun for varying periods of time during crop production. At this time, falling raindrops break up the desiccated peds at the soil surface and the primary particles are dispersed. Dispersed soil particles clog the pores. The net result is a reduction in porosity. Irrigation and drainage have both positive and negative effects on soil porosity. Surface irrigation systems including flood, channel, and basin irrigation usually break up the weak aggregates and reduce the porosity.

5.6 Consistence Is Resistance of Soil to Deformation Under Pressure

The resistance of soil to deformation or rupture under pressure is known as consistence. It is determined by the cohesive and adhesive properties of the soil mass. It is denoted by the behavior of a soil sample between fingers. This behavior may be different at different soil moisture contents. A soil may stick to the fingers or plow when it is wet; it may be hard when it is dry. Therefore, whether consistence has been determined in wet, moist or dry conditions have to be mentioned.

Wet soils may be sticky or plastic. If the soil adheres to finger and feels like paste or gum, it is said to be at sticky consistence. Clay soils generally show sticky consistence. Some soils do not adhere to the fingers under wet condition. They are non-sticky. These soils are usually coarse-textured soils. Plastic soil is capable of being molded or deformed by relatively moderate pressure into various shapes when wet. Plastic consistence is attained when soil samples can be rolled between fingers to form threads. Earthen wares are prepared from plastic soils.

Moist soils may be firm, loose, or friable. Some soils are compact masses in moist condition. They are firm soils. Loose consistence is found in soils dominated by sands. When a soil mass is ruptured by pressing lightly under moist condition between fingers into constituent peds, it is considered to have attained the friable consistence. Friable consistence is the most desired consistence for the cultivation of a soil. It is attained in well-structured granular soils with sufficient organic matter content. Friable soils readily break apart and are not sticky when moist.

Soils may be soft or hard under dry conditions. Soft consistence is exhibited by some well-granulated soils which are broken by light pressure into its aggregates. These are sandy loam to clay loam soils with plenty of organic matter. Clods of some soils cannot be broken by usual pressure by fingers. They need more pressure to break. It is called hard consistence. A hard consistence is not desirable for cultivation. Soils may also be cemented and indurated. Soil particles may be cemented by calcium carbonate, silica, and oxides of iron and aluminum into hard mass. When a cemented horizon is so hard that a sharp blow of a hammer is required to break the soil apart, the soil is considered to be indurated (Foth 1990).

5.7 Soils Get Warmth Mainly from Solar Radiation

Soils get warmth mainly from the solar radiation. A part of the solar radiation that enters the atmosphere reaches the soil surface. Some of the radiation is reflected from the surface, and a large part is absorbed by the soil. Some heat from soil surface is transmitted to depths of soil. Transfer of heat occurs by conduction (objects in contact), convection (through air currents), and radiation (heat transfer over a distance without contact or air flow).

Soil temperature is an important physical property that regulates life processes of soil biota including plants. Soil temperature may affect plant growth through its effects on many physical, chemical, and biological processes of the soil and physiological processes of plants. Soil temperature has tremendous ecological impacts, for example, through evaporation, thawing of permafrost (permanently frozen soils of the cold climates), emission of CO_2 via soil respiration, and organic matter decomposition.

Soil temperature may range from well below 0°C in the temperate (in winter) and boreal areas (all the year round in permafrost) to more than 40 °C in summer in the tropical areas. The source of heat in soil is primarily the solar radiation, which falls differently in different geographical and environmental situations. Again, the specific heat, amount of heat required to raise the temperature of unit mass of a substance by 1°C, varies with soil composition (minerals, organic matter content) and conditions (moist or dry, bare or covered).

Soil temperature has tremendous bearing on crop production. Growth and yield of crops can considerably be improved if the soil temperature can favorably be manipulated.

5.7.1 Heat Capacity of Unit Mass Is Known as Specific Heat

Specific heat is the ratio of the amount of heat needed to raise the temperature of a substance by 1 °C to the amount of heat needed to raise the temperature of water by 1 °C.

$$\text{that is specific heat} = \frac{\text{Amount of heat required to raise } 1^{\circ}\text{C temperature of 1g substance}}{\text{Amount of heat required to raise } 1^{\circ}\text{C temperature of 1g water}}$$

Specific heat is a ratio; it has no unit. On the other hand, heat capacity is the amount of heat needed to increase 1°C temperature of unit mass or volume of substance. When mass is considered, the unit is cal $g^{-1}\,°C^{-1}$ or J $g^{-1}\,°C^{-1}$. The amount of heat needed to raise 1 °C temperature of unit volume of substance is known as the volumetric heat capacity. It is expressed by cal cm^{-3} $°C^{-1}$ or J cm^{-3} $°C^{-1}$. Heat capacity per unit mass is the specific heat. Among soil constituents, specific heat of water and humus are 1.00 cal $g^{-1}\,°C^{-1}$ and 0.46 cal $g^{-1}\,°C^{-1}$, respectively. Most soil minerals have specific values between 0.19 and 0.21 cal $g^{-1}\,°C^{-1}$ (Bowers and Hanks 1962). The amount of heat needed to increase the temperature of soil is strongly related to water content. It takes only 0.2 cal of heat energy to increase the temperature of 1 g dry soil 1 °C compared with 1.0 cal per gram per degree for water. This means that it takes five times more energy to raise the water temperature 1 °C than the soil temperature.

5.7.2 Soil Components Differ in Thermal Conductivity and Diffusivity

Suppose a metal rod is heated at one end. Why does the other end get warmed? The metal has the capacity of conducting heat from one end to the other. Heat is conducted by a material from the warmer point to the cooler point. The surface soil is heated by the solar radiation during daytime when the subsoil remains cooler. So, heat is transmitted through the soil from the warmer surface to the cooler subsoil by the process conduction. Heat is conducted by molecular energy transfer from a point of higher temperature to a point of lower temperature. Thermal conductivity is the amount of heat transferred per unit area, per unit time, per unit °C. It occurs in the direction of negative temperature gradient, that is, higher temperature to lower temperature. Thermal conductivity may be expressed by Fourier's Law as

Table 5.11 Thermal conductivity of different substances

Component	Thermal conductivity, cal $cm^{-1}\,s^{-1}\,°C^{-1}$
Quartz	20×10^{-3}
Other soil minerals	7×10^{-3}
Organic matter	0.6×10^{-3}
Water	1.43×10^{-3}
Air	0.06×10^{-3}

$$Q_{\mathrm{h}} = -k\frac{\Delta T}{x},$$

where Q_{h} is the heat flux, the amount of heat transferred across a unit cross sectional area in unit time; k is the thermal conductivity; and $\Delta T / x$ is the temperature gradient over distance x. The minus sign is used to denote the direction of temperature gradient. Thermal conductivity of soils expresses the rate at which the subsoil will be warmed by conduction of heat from the surface soil.

Thermal conductivity of soil depends on heat capacity and thermal conductivity of different soil components such as mineral matter, humus, water, and air. Different soil minerals have different thermal conductivities. Some values obtained from different sources are shown in Table 5.11. Thermal conductivity of finer soils is generally higher than coarse-textured soils. It varies in the sequence: sand > loam > clay. It is due to reduction of area of contact between the particles. Thermal conductivity increases with increasing wetness of the soil up to a limit beyond which it declines. Thermal conductivity of organic matter is considerably less than common soil minerals. Organic matter also increases porosity, that is, it reduces the contact between soil particles. So, thermal conductivity is low in soils with high organic matter content. Likewise, thermal conductivity increases with compaction of the soil.

Thermal diffusivity is the ratio of thermal conductivity and volumetric heat capacity. It's unit is $cm^2\,s^{-1}$. Thermal diffusivity is defined as the change in temperature produced in a unit volume by the quantity of heat flowing through the volume in unit time under a unit temperature gradient.

5.7.3 Reflectivity of Incident Radiation Is Called Albedo

The term "albedo" was first introduced into optics by Johann Heinrich Lambert in 1760. Albedo is defined as the ratio of total reflected to incident radiation. It is the fraction of solar energy (shortwave radiation) reflected from the earth back into space. It is a measure of the reflectivity

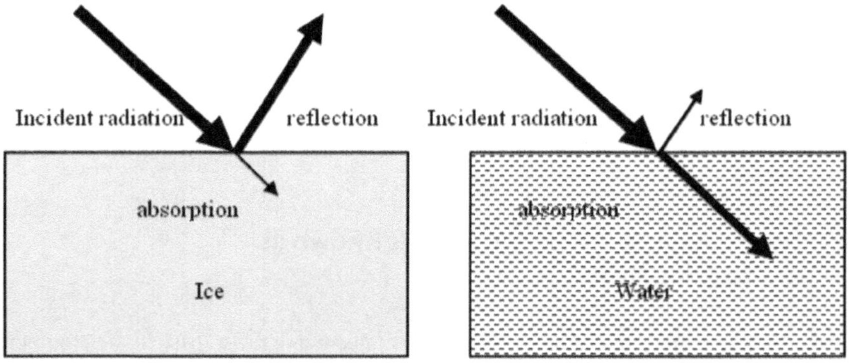

Fig. 5.6 Reflectivity of incident radiation by ice and water

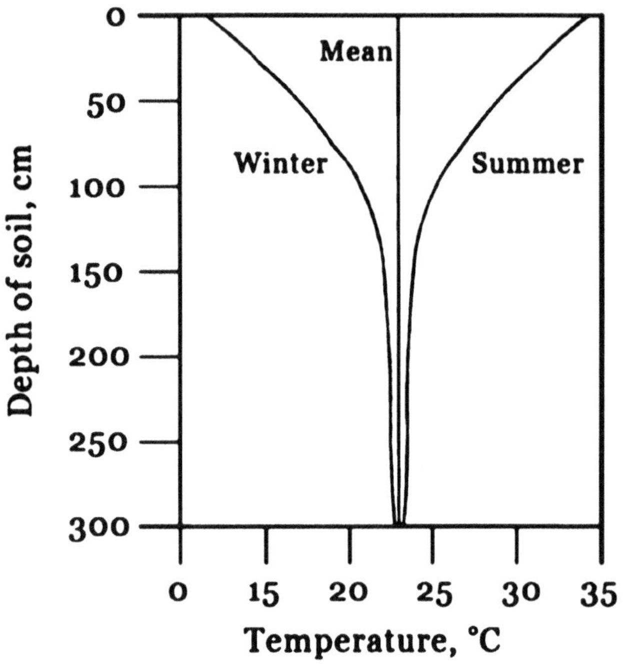

Fig. 5.7 Variation of soil temperature with depth in winter and summer

of the earth's surface. It has no unit, and it ranges from 0 to 1. Sometimes, it is expressed in percentage. Dark-colored soil surface has an albedo of 0.1–0.2 (reflectivity 10–20%), while light-colored smooth surface may have 0.5 (reflectivity 50%).

Ice, especially with snow on top of it, has a high albedo; most sunlight hitting the surface bounces back toward space. Water is much more absorbtive and less reflective (Fig. 5.6). Likewise, organic matter absorbs more solar energy and reflects less. Therefore, higher content of water and organic matter in soil and the roughness of the surface reduce its albedo. The lower the albedo, the warmer is the soil. Most soils and soil particles have albedo around 0.20.

5.7.4 Environment and Soil Conditions Affect Soil Temperature

Soils obtain heat mainly from the sun. The amount of solar radiation that a soil receives, absorbs, and transmits depends on geographic, climatic, edaphic, and topographic conditions. These factors greatly influence soil temperature.

Geographical location: The direction and intensity of solar radiation received by a soil surface depends on the geographical location, particularly latitude. Tropical soils are warmer than temperate and tundra soils. Soil temperature also varies with altitude; high altitude soils, that is, montane soils, are cooler. Temperatures of soils near large water bodies such as lakes and seas are often buffered by the heat absorbed by these water bodies. When the mean annual soil temperature is below 0°C, the depth of freezing in winter may exceed the depth of thawing in summer. As a consequence, a layer of permanently frozen soil, called permafrost, may develop. Permafrost may be found in vast areas covered with boreal and tundra vegetation. In some areas, permafrost may extend up to the soil surface. In some other areas, the surface layer thaws in summer and freezes in winter. It is known as the active layer. When the active layer thaws, it saturates the soil because the water cannot move downward for the frozen soil. The active layer is used by plants for water and nutrients in the growing period.

Soil color: Dark soils absorb more energy from the sun than light-colored soils; dark soils are generally warmer. Red or yellow soils will show a more rapid temperature rise than those that are white.

Soil texture: Sand particles are heated faster than clay particles, and clay particles hold more moisture than sandy soils. So, sandy soils are usually warmer than the clay soils.

Fig. 5.8 Angle of solar radiation influences soil temperature

Organic matter content: Organic matter is heated very slowly, and it transmits heat also very slowly. So, soils with higher organic matter content are relatively cooler.

Depth of soil: Temperature of surface soil varies greatly with time of the day and season of the year (Fig. 5.7). Generally, surface soils are warmed and cooled more rapidly than the subsoil. Heat is not conducted very fast within the soil. So, the subsoil shows little daily or weekly fluctuations in temperature. In temperate regions, surface soils in general are warmer in summer and cooler in winter than the subsoil.

Soil water: Water absorbs much of the solar radiation but warms up slower than mineral particles. So, dry soils are generally warmer than wet soils. Again, evaporation of soil water requires some energy and causes a cooling effect on the surface soil. The low temperature of a wet soil is due partially to evaporation and partially to high specific heat of water.

Soil cover: Both reduction and increase in temperature may be caused by a cover on the soil, either vegetative or organic mulch. Mulches lower soil temperature by reflecting solar radiation and increasing soil moisture. It tends to increase soil temperature by reducing loss of heat by radiation and reducing evaporation from the soil surface. So, a cover keeps the soil warmer in the cool season and cooler in warm season.

Topography: Topography determines the angle and direction of solar radiation on soil. Surfaces where solar radiation falls perpendicularly get warmer than places where it falls obliquely. South-facing slopes in the northern hemisphere are relatively warmer than the northern aspect (Fig. 5.8).

5.7.5 Soil Temperature Is Related to Air Temperature

Soil materials have lower specific heat than air. For the same amount of solar radiation, the soil is heated more and faster than the air in the soil-atmosphere interface. So, surface soil temperature is generally 1–5°C higher than the air in the immediate soil surface. This interface is in a state of continuous interaction, and variation in temperature of both soil and air is brought about by differences in solar radiation and due to cloudiness, cold waves, warm waves, rainstorms, and drought. However, the fluctuation in temperature of the soil surface is more in magnitude and frequency than the soil at depths. Wide fluctuations in surface soil temperatures occur within daytime, between day and night and among seasons. It is, however, assumed that although soil temperature varies differently at different depths in the soil, the average temperature is the same at all depths (Hillel 1980). Further, the concept of "mean annual soil temperature (MAST)" has been introduced. Once-a-month measurement of temperature at a depth of 50 cm has been recommended to obtain direct measurement of the MAST (Buol et al. 1997). By measuring the temperature at 50cm, daily temperature fluctuations are avoided (Pikul and Allmaras 1984).

Soil temperature can be estimated from air temperature with enough precision (Soil Survey Staff 1975). To get mean annual soil temperature, 1 °C is added to mean annual air temperature in the United States, and 2.5 °C is added to obtain mean annual soil temperature in the tropical latitude (Van Wambeke 1985). Murtha and Williams (1986) found that in tropical Australia, mean monthly soil temperatures are best approximated for air temperatures by adding 4 °C in ustic soil moisture regimes (dry soils; in most years, part of the soil moisture control section is dry, soil moisture at permanent wilting point, for 90 cumulative days) and 2 °C in udic soil moisture regimes (moist soils; the soils are not dry in any part of the soil moisture control section as long as 90 cumulative days).

5.7.6 Soil Temperature Regulates Soil Processes and Plant Growth

Soil temperature regulates physical, chemical, and biological properties of soils and growth of plants and microorganisms.

5.7.6.1 Plant Growth and Yield

Soil temperature has a profound role on plant growth although it is often thought to be more influenced by air temperature. Plants need a suitable temperature for root and shoot growth and absorption of water and nutrients. In the warm regions, summer soil temperature may be too high, and in the cold regions, winter soil temperature may be too low for normal growth and development of plants. Such management practices as irrigating the warm soil or drying the cold soil and mulching may be adopted to avoid adverse effects of soil temperature in such situations.

Table 5.12 Suitable temperatures for different seed germination

Crops	Suitable temperature (°C)
Wheat	20
Alfalfa	25
Barley	20
Bean	16–30
Beet	10–30
Cabbage	7–35
Carrot	7–30
Cauliflower	7–30
Corn	16–32
Oats	20–24
Red clover	25
Sweet clover	18–25
Tomato	16–30
White clovers	18–20

Source: http://www1.agric.gov.ab.ca/$department/deptdocs.nsf/all/agdex1203

All crop plants have their suitable temperature ranges for profitable growth and yield. Corn (maize) yield increases linearly as the average surface soil temperature increases from 15 to 25°C. Optimum soil temperature for potato is 17°C. Tuber formation in potato is practically absent at above 30°C soil temperature. However, different temperature ranges may be suitable for a crop plant at different stages of its life cycle. Root growth is favored by elevation of soil temperature because it increases rate of cell division and elongation. Root activity, that is, absorption of water and nutrients, also increases as the temperature rises within a range.

5.7.6.2 Microbial Growth

The temperature that allows for the most rapid growth of microorganisms during a short period, 12 or 24 h, is known as "optimum growth temperature." On the basis of optimum growth temperature, microorganisms are divided into the following categories:

- *Psychrophiles*: Organisms that have optimum growth temperature of 15°C or lower; they can grow at 0 °C and a maximum temperature of 20°C. Facultative psychrophiles are able to grow best at 20–30°C.
- *Mesophiles*: Organisms that grow best within a temperature range of approximately 25–40°C. Most soil microorganisms are mesophiles. They function best at 25–35 °C.
- *Thermophiles*: Organisms that grow best at temperature above 45°C; facultative thermophiles can grow in mesophilic range. True thermophiles cannot grow at mesophilic range.

Microbial processes: The most important microbial processes in soil which regulate soil fertility and productivity include organic matter decomposition, ammonification, nitrification, denitrification, nitrogen fixation, phosphorus, and sulfur transformation. They are all enzyme-regulated processes that occur at faster rates when the temperature is elevated within a range. For example, the most favorable limit of temperature for nitrification is 25–35 °C. Optimum temperature for other microbial transformations falls in this range. Nitrification proceeds very slowly at a temperature below 10 °C. Likewise, denitrification is enhanced by elevation of soil temperature (25–60 °C). Soil microbial activity and organic matter decomposition ceases below 5 °C, which is often called the "biological zero." Organic matter decomposition and release of bound nutrients, that is, nutrient recycling, occur very slowly if soil temperatures are not favorable. Under such conditions, plant may suffer from nutrient deficiency, particularly of nitrogen, phosphorus, and sulfur.

Seed germination: An optimum range of soil temperature is needed for germination of seeds of every plant species. Maximum number of seeds germinates at this temperature and in less time.

If the soil is too cool or too hot, germination is delayed. Seeds may be damaged, and reseeding may be necessary. Again, if the temperature requirement of the crop is known, sowing dates can be adjusted by early or late sowing. Otherwise, there will be a poor crop stand that may not produce the desired yield. However, some seeds need cold or warm stratification for germination. The suitable temperature ranges for seed germination of different crops are given in Table 5.12.

5.7.7 Soil Temperature May Be Controlled by Soil Moisture and Soil Cover

Sometimes, soil temperature needs to be controlled for good growth of plants. This can be done by two different steps. One is the regulation of soil moisture status. Since water has higher specific heat than soil particles and needs more energy to be warm, higher moisture content would decrease temperature of warm soils. On the other hand, soil particles get more easily heated than water, so draining the land would improve temperature of cold soils. The other step may be covering the soil with plant residues and other organic mulches or by black plastic cover. Organic mulches, such as straw, may increase or decrease soil temperature depending on their effects on absorption and reflection of solar radiation, infiltration, and evaporation of water. Mulches will tend to lower soil temperature because more solar radiation will be reflected and less absorbed, and the water content of the soil will tend to be greater because more water will infiltrate. Mulches will, again, tend to increase soil temperature because it reduces loss of heat by radiation and it reduces water evaporation from the soil surface. The net effect is to reduce soil temperature in the warm season and increase temperature in the cool season. Black plastic mulch increases soil temperature because it increases absorption of solar radiation, reduces heat loss by radiation, and reduces evaporation of water from the soil surface.

5.8 Soil Air Has a Composition Slightly Different from Atmospheric Air

Soils contain air in the large pores if they are not saturated with water. In a loam-textured surface soil, approximately 25% of its volume, is filled with air. Sometimes, as in dry soils, some micropores may also be filled with air. On the other hand, air in the macropores may be pushed out of the soil by water immediately after rainfall or irrigation. They are temporarily filled with water, but after some time, gravity pulls this superfluous water into the groundwater table leaving the spaces for air. However, soils get this air from the atmosphere. So, soil air has a composition closely related to the atmospheric air. The atmosphere consists of nitrogen (78.084% by volume) and oxygen (20.9476%) with smaller quantities of water vapor (variable), carbon dioxide (0.0314%), inert gases such as argon (0.934%), and some other rare gases. The concentration of oxygen in soil is less because of utilization by living organisms, and concentrations of CO_2 are higher because of biological respiration. The soil CO_2 level in active soil reaches 0.15% at 15 cm, but can reach 5% and affect the pedosphere chemistry (Adl 2003). O_2 content in well-aerated soils may be about 20%. It is reduced as the water content increases and the rate of exchange of gas between soil and atmosphere decreases. In some cases, oxygen can be as low as 5–10%. If the soil remains saturated with water or waterlogged for a long period, say several days, O_2 content may completely be depleted.

However, the composition of soil air is highly variable. It depends on the ease of exchange of air between the atmosphere and the soil. The more readily the exchange occurs, the closer is the composition of soil and atmospheric air. Renewal of air in soil is not so rapid because of the discontinuity of soil pores and the obstruction offered by soil water.

5.8.1 Soil Properties, Climate, and Management Affect Soil Air Composition

Soil texture, structure, and porosity: Soil texture, structure, and porosity affect the amount and composition of soil air and its renewal. Light-textured soil or sandy soil contains much higher air than heavy soil. Sandy soils generally have low total porosity but large individual pores. Clay soils generally have high total porosity but small individual pores. Soils with large pores promote more rapid gas exchange than the small pores. Soils with large pores generally have good drainage and aeration.

Soil water: Diffusion of air occurs slowly in a water-saturated soil. Under prolonged waterlogging, the O_2 in soil air is diminished and the CO_2 concentration highly increases. In waterlogged soils, some volatile gases like H_2S, CH_4, and C_2H_4 may accumulate and exert toxic effects on plant roots.

Soil depth: Depth of soil influences composition of soil air. The diffusion of gases occurs more readily in the surface soil. Therefore, oxygen generally decreases and CO_2 generally increases in soil air with depth in the soil profile.

Microbial activity: Soil microorganisms utilize O_2 for respiration and decomposition of organic matter and produce CO_2. Hence, soils rich in organic matter contain higher percentage of CO_2 in soil air.

Cultivation: Cultivation is intended to aid in the aeration of soil. Both mass flow and diffusion increase as a result of cultivation.

Season: Soil moisture and temperature have profound influence on diffusion and mass flow of soil air. Gaseous exchange is greater during the dry period. Therefore, relatively high O_2 and low CO_2 levels are found in the dry season than monsoon. High temperature during summer season encourages microorganism activity which results in higher production of CO_2.

5.8.2 Renewal of Soil Air Occurs by Mass Flow and Diffusion

There are two processes by which soil air is renewed: mass flow and diffusion.

Mass flow: Mass flow occurs when the entire mass of air moves from one place to another, particularly from the soil to the atmosphere and from the atmosphere to the soil, under conditions of fluctuating air pressure. Air is driven away from or into the soil pores by this pressure. Movement occurs due to atmospheric pressure changes, soil temperature changes, entry of water following rain and irrigation, evaporation of water from pores, and driving force of wind near soil surface. However, mass flow is much less important than diffusion, except perhaps in layers at or very near the soil surface.

Diffusion: Diffusion is the predominant process of soil aeration. In diffusion, individual gas constituent moves separately under the influence of partial pressure gradient. When partial pressure of CO_2 in soil air increases due to root and microbial activity, molecules of CO_2 diffuses from soil to the atmosphere. When O_2 in soil air is consumed for respiration in soil, its partial pressure is reduced and a partial pressure gradient is created between the soil and the atmosphere, then O_2 diffuses into the soil. Generally, there is net diffusion of CO_2 into soil and net diffusion of O_2 into the atmosphere. Through diffusion, each gas moves in a direction determined by its own partial pressure. The diffusion of gases in soil is regulated by numbers of air-filled pores and their continuity, both decreasing with increasing soil water content, and the tortuosity of air-filled pores which increase with soil water content.

5.8.3 Soil Air Affects Other Soil Properties and Growth of Plants and Microorganisms

Soil air is an important soil component that affects many other soil properties and processes. It influences the physiological functions of the plants and microorganisms.

Organic matter decomposition: The rate of organic matter decomposition, the products of organic matter decomposition, and the amount of organic matter in soil all are affected by soil air. Organic matter decomposition is very slow in anaerobic soils. Organic matter also undergoes partial decomposition in the absence of O_2. Partially decomposed organic matter accumulates in marshes and may form Histosols there. Under aerobic conditions, the ultimate product of organic matter decomposition is CO_2, but in an anaerobic soil, gases like CH_4 and C_2H_4, alcohols, and organic acids are produced. H_2S may also be formed and accumulated in poorly aerated soils.

Oxidation–reduction: Oxidation–reduction reactions in soil are regulated by the aeration. Some substances including many compounds of iron, manganese, sulfur, phosphorus, and nitrogen get oxidized under aerobic condition and reduced under anaerobic condition. They may undergo alternate oxidation and reduction due to waterlogging and draining by seasonal flooding and fluctuating water table. These oxidation–reduction reactions in soil determine the redox potential of soil.

$$2FeO + 2H_2O \Leftrightarrow 2FeOOH + 2H + 2e^-$$

Redox potential: Redox potential, E_h, is a measure of the state of oxidation and reduction of a substance. It is the tendency of a chemical species to receive or donate electrons and get oxidized or reduced. Each chemical species has its own intrinsic reduction potential; the more positive the potential, the greater the species' affinity for electrons and tendency to be reduced. There are many chemical substances in soil which undergo oxidation and reduction as the condition of moisture and air dictates. Redox potential is measured in volts (V) or millivolts (mV).

Table 5.13 Sensitivity of crop plants to poor aeration

Very sensitive	Sensitive	Tolerant	Very tolerant
Wheat	Corn	Sorghum	Rice
Barley	Millet	Mustard	
Oats	Cabbage	Mulberry	
Beans	Peas		
Sugar beets			

The voltage difference at an inert electrode immersed in a reversible oxidation–reduction system is measured. Redox potential measures the affinity of a substance for electrons compared with that of hydrogen, whose redox potential is set at 0. Strong oxidizing agents have a positive redox potential; strong reducing agents have a negative redox potential.

Oxidation–reduction reactions affect solubility and availability of some nutrient elements such as iron, manganese, sulfur, and phosphorus. Oxidation–reduction also affects the pH of the soil.

On being waterlogged, oxygen content in soil air is depleted. Therefore, poor aeration reduces the redox potential.

Plant growth: The critical oxygen concentration for roots is 5–10%, although some plants are able to grow at lower O_2 concentrations. For example, cotton grass (*Eriophorum angustifolium*) roots grow at 2–4% and rice (*Oryza sativa*) roots can grow at 0.8% O_2 (Schulze et al. 2005). Rice and some other hydrophytes have developed their own mechanism of gas exchange under waterlogged conditions. They have arenchyma tissues which ventilate air from the atmosphere to the soil. Soils around rice roots show oxidized red-brown mottles even under reduced conditions. On the other hand, roots of most crop plants fail to function normally in a completely anaerobic condition. Roots often decay in virtual absence of O_2 in soil air. Plants need O_2 for seed germination too. Oxidation of storage foods during germination provides the developing seedling with necessary energy. Under anaerobic conditions, toxic materials may accumulate in seeds and damage the seeds.

Plants widely differ in their ability to tolerate low oxygen levels in soils (Table 5.13). Sensitive plants are wilted or killed by saturating the soil with water for even a short period.

Nutrient uptake: Soil air affects mineralization, mineral transformation, nitrification, denitrification, and nitrogen fixation, thereby influencing the overall soil fertility. The processes of mineralization, nitrification, nitrogen fixation, etc. all are reduced by O_2 stress in the soil. Loss of nitrogen through denitrification is high in anaerobic soils. The mechanisms of nutrient uptake also need the participation of O_2.

Microbial growth: Microorganisms are classified into aerobes and anaerobes on their O_2 requirement. Therefore, the aeration of the soil not only regulates the total microbial population but also affects the proportion of different groups and their functions. For example, nitrifiers are abundant in aerobic soils, and the denitrifiers are most abundant in anaerobic soils. Most of the nitrogen fixers are aerobic; so, nitrogen fixation is reduced by poor soil aeration.

Study Questions

1. Why does soil color vary among soils? What is the significance of soil color? How is soil color determined?
2. How do distribution and arrangement of soil particles affect other soil properties?
3. What are the reasons of spatial and temporal variation in soil temperature? How does soil temperature influence soil organisms and plant roots?
4. Explain that bulk density is a function of the interaction of mineral and organic matter in soil.
5. Why does the composition of soil air differ from atmospheric air? How does soil air affect growth of soil organisms?

References

Adl SM (2003) The ecology of soil decomposition. CAB International, Oxon

Baver LD (1960) Soil physics, Modern Asia edition. Charles E. Turtle Company, Tokyo

Bowers A, Hanks RJ (1962) Specific heat capacity of soils and minerals as determined with a radiation calorimeter. Soil Sci 94(6):392–396

Brady NC, Weil RR (2002) The nature and properties of soils, 13th edn. Pearson Education Inc., New Delhi

Bronick CJ, Lal R (2005) Soil structure and management: a review. Geoderma 124:3–22

Buol SW, Hole FD, McCracken RJ, Southard RJ (1997) Soil genesis and classification, 4th edn. Iowa State University Press, Ames

Chesters G, Attoe OJ, Allen ON (1957) Soil aggregation in relation to various soil constituents. Soil Sci Soc Am Proc 21:272–277

Foth HD (1990) Fundamentals of soil science, 8th edn. Wiley, New York

Hillel D (1980) Fundamentals of soil physics. Academic, New York

http://www1.agric.gov.ab.ca/$department/deptdocsnsf/all/agdex1203. Accessed 8 July 2011

Kay BD (1998) Soil structure and organic carbon: a review. In: Lal R, Kimble JM, Follett RF, Stewart BA (eds) Soil processes and the carbon cycle. CRC Press, Boca Raton

Murtha GG, Williams J (1986) Measurement, prediction and interpretation of soil temperature for use in soil taxonomy. Tropical Australian experiences. Geoderma 37:189–206

Pikul JL Jr, Allmaras RR (1984) A field comparison of null-aligned mechanistic soil heat flux. Soil Sci Soc Am J 48:1207–1214

Schulze ED, Beck E, Hohenstein KM (2005) Plant ecology. Springer, Berlin

Smettem KRJ (2006) Particle density. Encyclopedia of soil science. Taylor & Francis, Boca Raton

Soil Survey Staff (1975) Soil taxonomy. U.S. Department of Agriculture handbook 436. U.S. Government Printing Office, Washington, DC

Soil Survey Staff (1993) Soil survey manual. U.S. Government Printing Office, Pittsburg

SSSA (1997) Glossary of soil science terms. Soil Science Society of America, Madison

SSSA (2010) Soil color lesson. Soil Science Society of America, Madison

Tisdall JM, Oades JM (1982) Organic matter and water-stable aggregates in soils. J Soil Sci 33:141–163

Van Wambeke A (1985) Calculated soil moisture and temperature regimes in Asia. SCS-USDA. SMSS technical monograph no. 9, Washington, DC

Wild A (1988) Russell's soil condition and plant growth. Longman Group Ltd., London

Soil Water, Irrigation, and Drainage

6

Water is a precious natural resource, and freshwater is scarce too. So, water—soil water or irrigation water—need cautious management. In soil, water is held in pores and on particles in various forms and under different forces. All of it is not available to plants. The amount of water held between field capacity (FC) and permanent wilting point (PWP) is the amount of plant's available water in soil. The FC and PWP correspond to −10 and −1,500 kPa soil water potential, respectively. Soil water potential is the amount of free energy that soil water possesses, and it is the driving force of water in soil. Water moves to the direction of the gradient of soil water potential, and the rate of movement is proportional to the potential difference ($\Delta\psi$) between the two points and the hydraulic conductivity of the soil. Water moves along soil-root-stem-leaf-air pathway because there is a water potential gradient along the soil–plant–atmosphere continuum (SPAC). Water is needed by plants for many physiological functions. Their growth is hampered if adequate water is not available. Yields of many crops are reduced significantly due to water stress. Therefore, irrigation should be applied well ahead of developing water stress and at growth stages when water is urgently needed. Sometimes, over irrigation adversely affects crop yield. Excess water, applied or natural, must be removed by artificial drainage for growing most crops.

6.1 Freshwater Is a Limited Resource

Water is needed for all life on earth. Water covers 70% of the earth's surface, but freshwater is scarce. Water in oceans and seas comprises about 97.2% of the total volume of water. Freshwater is only 2.8% of the total volume, of which groundwater is 0.6%, and soil water is less than 0.1% of the total. The vast majority of the freshwater is in the form of ice and permanent snow cover (75%), or fresh groundwater (24.5%). Only 0.3% of the total amount of freshwater is stored in lakes, reservoirs, and rivers (Lal and Shukla 2004). The major reservoirs of water are oceans ($1,350\times10^6$ km^3), polar ice and glaciers (29×10^6 km^3), aquifers (8×10^6 km^3), lakes (0.1×10^6 km^3), soil moisture (0.1×10^6 km^3), atmosphere (0.013×10^6 km^3), rivers (0.002×10^6 km^3), and biosphere (0.001×10^6 km^3).

6.2 Movement of Water Through Biosphere, Hydrosphere, Lithosphere, and Atmosphere Forms the Hydrological Cycle

Water continuously moves among the earth's biosphere, atmosphere, lithosphere, and hydrosphere; this cycling of water is called the hydrologic cycle. The hydrologic cycle involves several pools and fluxes. Pools include various reservoirs such as atmosphere, oceans, lakes, rivers, soils, glaciers, snowfields, and groundwater. However, the amount of water that was once created has remained constant. It has not been destroyed nor lost; it only changes state and pools. Fluxes of water in the hydrologic cycle include evaporation, transpiration, condensation, precipitation, runoff, infiltration, percolation, groundwater flow, sublimation, and melting (Fig. 6.1). The generation of precipitation is commonly considered as the beginning of the terrestrial hydrological cycle. Precipitation may be in the form of rainfall or snow. The rain or melt water may be intercepted by vegetation cover or detained by land surface depressions, may infiltrate into the soil, or may run over land surface into streams. The infiltrated water may store in the soil as soil moisture or may percolate to deeper layers and ultimately to groundwater.

6.2.1 Precipitation

The main source of the earth's water supply is the precipitation. It comes from condensation of small water vapor droplets around available nuclei commonly dust particles, or from ice crystal process in the clouds. Collision and coalescence make water droplets increase in size and when they attain about 2 mm in diameter, they begin to descend to the earth's

K.T. Osman, *Soils: Principles, Properties and Management*,
DOI 10.1007/978-94-007-5663-2_6, © Springer Science+Business Media Dordrecht 2013

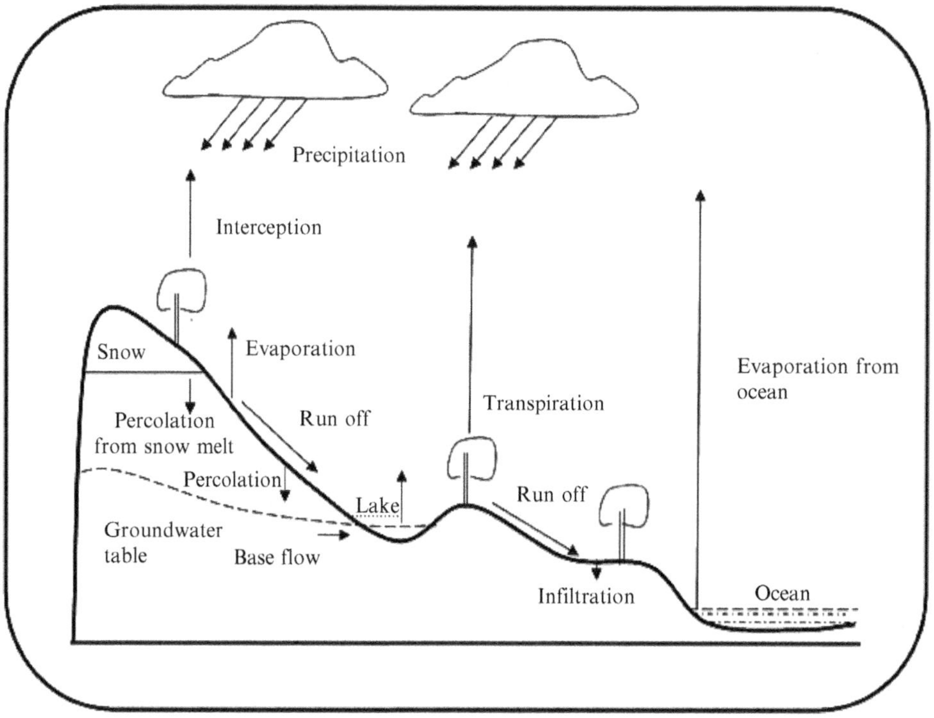

Fig. 6.1 The hydrologic cycle

surface as rain. By collision and sticking together, ice crystals may form snowflakes which can reach the ground in the form of snow or rain, depending on the temperature of the lower atmosphere. Four main types of precipitation are the following: (1) frontal precipitation, where the lifting is due to relative movement of two large air masses; (2) precipitation caused by horizontal convergence; (3) convective precipitation; and (4) orographic precipitation. These types generally do not occur separately, but some may dominate under certain conditions.

6.2.2 Interception

A part of the precipitation, rain or snow, may be intercepted by vegetation and other surface covers before reaching to the land surface. Some of the rainfall evaporates, and some may flow down on vegetation stems; the flow along stems is known as stemflow. Rainfall interception varies with species composition, age, and density of vegetation cover. Twenty-five to thirty-five percent rainfall may be intercepted by a densely growing coniferous forest. Rainfall interception by deciduous hardwood forests may be about 15% for the period with leaves and about 7% for the period without leaves. Net interception in the tropical rainforest is approximately 10% of rainfall. Interception can be significant in large urban areas; intercepting surfaces in these areas are flat rooftops, potholes, parking lots, cracks, and others.

6.2.3 Evapotranspiration

Water is transferred from the earth's surface to the atmosphere through evaporation, the process by which water changes from a liquid to a gas. Land, lakes, rivers, and oceans send up a steady stream of water vapor. On the other hand, transpiration is a biological process by which water vapor is released to the atmosphere from vegetation largely through their stomata. Approximately 80% of all vaporization occurs from the oceans, with the remaining 20% coming from inland water and vegetation. The combined loss of water to the atmosphere via the processes of evaporation and transpiration is called evapotranspiration.

Evaporation requires large amounts of energy; for example, 2.4×10^6 J (5.7×10^5 cal) of heat energy is needed to convert 1 kg liquid water to water vapor. Water vapor may transfer heat from the point of evaporation to the point of condensation. Vaporization and condensation of water constitute a medium of energy transfer. At any given temperature, the air can hold only a given amount of moisture known as the saturation humidity. The relative humidity (expressed as %) is the ratio of the measured humidity (g water m^{-3} air) to the saturation humidity. Evaporation ceases when 100% relative humidity is reached. As an air mass cools, its saturation humidity decreases, and relative humidity of the air mass increases, which leads to condensation. Precipitation occurs when 100% relative humidity is exceeded and occurs usually due to cooling of the air at a higher altitude.

6.2.4 Infiltration

When water is poured on a dry soil surface, it rapidly soaks into the soil. Water gradually penetrates the surface with the wetting front advancing gradually. This process by which water of rain, irrigation, and snowmelt on the ground surface enters the soil is called infiltration. When water infiltrates into a dry soil, a definite wetting front can be found. This is the boundary between the wetted upper part and the dry lower part of the soil. The rate of water entry in soil surface is the infiltration rate which may alternatively be defined as the meters of water entering the soil per unit time. Infiltration rate is the highest at the beginning of a rainfall event; it gradually decreases as the rainfall continues and the soil is moistened. Infiltration rate is the minimum when the soil becomes saturated with water. Infiltration can continue only if there is space for additional water. It depends on the porosity of the soil and the rate at which previously infiltrated water moves away from the surface through the soil. If the precipitation exceeds the infiltration, water will first be ponded in depressions and then start running off as overland flow or stream flow. Soil properties that affect the infiltration rate include soil texture, structure, porosity, organic matter content, water content during infiltration, compaction, and frozen surface. A sandy soil usually has a higher infiltration rate than a clayey soil. A crust on the soil surface can seal the pores and restrict the entry of water, and a compacted zone or an impervious layer close to the surface restricts the entry of water into the soil. Soils that have stable strong aggregates have a higher infiltration rate than soils that have weak, massive, or plate-like structure. The content of water in the soil affects the infiltration rate of the soil. The infiltration rate is generally higher when the soil is initially dry; it decreases as the soil becomes wet. Factors that affect the porosity, especially macroporosity, also affect infiltration rate.

6.2.5 Percolation

Percolation is the downward movement of water through the soil profile. Water enters the soil by infiltration, and the infiltrated water move down to the water table by percolation. A part of the water that infiltrates gets evaporated, some is absorbed and transpired by plants, some is retained in pores and on particles, and the remaining water percolates. The factors that affect infiltration also affect percolation. In addition, the depth of the groundwater table (GWT) also influences the percolation. The deeper the GWT, the higher is the rate of percolation. Percolation affects hydraulic conductivity, soil permeability, and drainage.

6.2.6 Runoff

Runoff is the movement of water across the surface of the land. Rainfall or snowmelt, in excess of infiltration and ponding, runs off the ground as overland flow and stream flow. There is a general relation—the greater the infiltration, the smaller is the runoff. Infiltration can be increased and runoff can be reduced by soil management practices. Excess water in saturated soils flows into channels, streams, and rivers, to the ocean, to terminal lakes, or swamps. Groundwater can meet (base flow) stream flow in rivers and lakes if the water table is near the surface. Runoff is influenced by soil physical properties such as texture, structure, porosity, compaction, and chemical properties such as organic matter content and exchangeable bases, and slope characteristics.

$$R = P + S - I - ET,$$

where
R = runoff
P = precipitation
S = storage
I = infiltration
ET = evapotranspiration

6.3 Some Forces Attract Water in Soil

There are two attractive forces between water molecules: hydrogen bonding and the van der Waals force. Hydrogen bonding takes place as a result of the electrical structure of water molecules. The negative lone-pair electrons of one water molecule are attracted to a positive partially screened proton of another water molecule (Kirkham and Powers 1972). Thus, each corner of the four corners of the water tetrahedron can be attached, by electrostatic attraction, to four other water tetrahedron molecules. Hydrogen bonds have a binding force of about 1.3–4.5 kcal per mole of water (Kramer 1983).

A van der Waals force is one that exists between neutral nonpolar molecules and, therefore, does not depend on a net electrical charge. This attractive force occurs because the electrons of one atom oscillate in such a way as to make it a rapidly fluctuating dipolar atom, which in turn polarizes an adjacent atom, making it, too, a rapidly fluctuating dipole atom such that the two atoms attract each other. van der Waals force is a very weak force amounting to about 1 kcal $mole^{-1}$ (Kramer 1983). This force contributes little to the attraction of water to itself.

6.4 Adhesion and Cohesion Are Forces of Water Retention in Soils

Soil particles attract water molecules. When some water falls onto a lump of oven dry soil, the water molecules will be strongly adsorbed and the molecules will spread themselves over the surfaces of the soil particles to form a thin film of water. A film of water on soil particle may contain 15–20 layers of water molecules. The force of attraction is called adhesion. With increasing distance from the soil particle surface toward the outer edge of the water film, the strength of adhesion decreases and the energy and mobility of the water increase. Water held by adhesion is called adhesion water. Adhesion water is always present in field soils, but it is so strongly adsorbed that it moves little. It can be removed by drying the soil in an oven. Adhesion water is the most immobile water in the soil, and is generally unavailable for use by plant roots and microorganisms.

The mutual force of attraction between water molecules is called cohesion. It has already been stated that water molecules are attracted together by hydrogen bonding and van der Waals forces. When water is added to a dry soil, water is first adsorbed on soil particle surfaces. The thickness of the water film around soil particle increases as more water is added. As the thickness increases, the force of adhesion decreases. At a point, further water is held by cohesion. The water held by cohesion is known as cohesion water. The cohesion water is slightly mobile and may be partly available to plants. There is a water-energy gradient across the water films, and the outermost water molecules have the greatest energy, the greatest tendency to move, and the greatest availability for plant roots.

Soil holds water in two ways, as a thin film on individual soil particles and as water stored in the pores of the soil. Water stored in the pores of the soil is said to be in capillary storage. Adhesion contributes mainly to the retention of water on particle surfaces, and cohesion contributes to the retention of water both on particles and in pores.

6.5 Soil Water Content May Be Expressed on Volume and Mass Basis

Water content is a measurement of the amount of water in the soil either by weight or volume and is defined as the water lost from the soil upon drying to constant mass at 105°C (SSSA 1997). It is expressed in units of either volume of water per unit bulk volume of soil ($m^3\ m^{-3}$), the volumetric water content, or mass of water per unit mass of dry soil ($kg\ kg^{-1}$), gravimetric water content (Fig. 6.3).

Volumetric water content, θ_v, is defined as

$$\theta_v = \frac{V_w}{V_T},$$

where V_w is the volume of water and V_T is the total volume= $(V_s+V_v)=(V_s+V_w+V_a)$ (soil volume+water volume+air space).

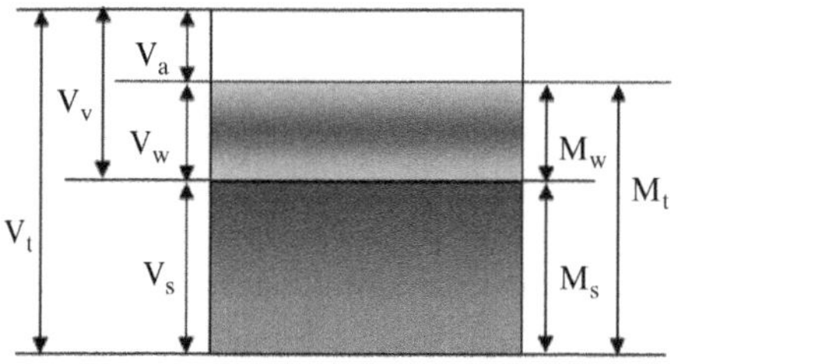

Fig. 6.2 Soil composition by phase (*s* soil (dry), *v* void (pores filled with water or air), *w* water, *a* air, *t* total, *V* volume, *M* mass)

The unit of θ_v is $cm^3\ cm^{-3}$ or $m^3\ m^{-3}$. Often, it is expressed as percent volume by volume (Fig. 6.2).

Gravimetric water content, θ_m, is given by

$$\theta_m = \frac{M_w}{M_s},$$

where θ_m is the gravimetric water content, $g\ kg^{-1}$ or $kg\ kg^{-1}$, M_w is the mass of water (mass of moist soil – mass of dry soil), and M_s is the mass of oven dry soil. Gravimetric water content is usually expressed as percent by weight.

The gravimetric method is a direct, absolute technique for estimating the total water content of soils at any given time. The method involves drying a soil sample in an oven at 105°C for 24 h to determine the soil moisture content. Water content equals the initial field soil weight minus the oven dry weight. The volumetric water content θ_v and the gravimetric water content θ_m are related to each other as

$$\theta_v = \theta_m \frac{\rho_b}{\rho_w},$$

where ρ_w is the density of water (1,000 kg m^{-3} at 20°C) and ρ_b is the bulk density.

Conveniently the quantity of soil water in a specific soil depth increment may be expressed in terms of soil water storage or equivalent depth of soil water (units of length). Equivalent depth of soil water, D_e (m), is calculated as $D_e = \theta_v \cdot D$, where D is the soil depth increment (m) having volumetric water content θ_v.

6.5.1 Degree of Saturation

Study of the saturation of soil with water is needed for tillage, irrigation, drainage, and other management purposes. The degree of saturation, S_w, is defined as

$$S_w = \frac{V_w}{V_v} = \frac{\theta_v}{\varphi},$$

where $\varphi = V_v/V_T$ is the porosity and V_v is the volume of void or pore space. Values of S_w can range from 0 (oven dry) to 1 (completely saturated).

6.6 There Are Several Forms of Soil Water

Water may be present in soils with chemical compounds, on particle surfaces, in micro- and macropores. There are three main general forms of water in soil: hygroscopic water, capillary water, and non-capillary water or gravitational water.

6.6.1 Hygroscopic Water

When an oven dry soil is exposed to a saturated air (100% relative humidity), it absorbs some water. This water is known as hygroscopic water. It is the water that an air dry soil will lose when it is heated at 105°C. Hygroscopic water remains in a vapor state on soil particle surfaces. This water is very tightly held in soil so that plant roots cannot absorb this water.

6.6.2 Gravitational Water

Water occupies all pore spaces in the soil immediately following rainfall or irrigation. The soil is then said to have reached saturation. When rain or irrigation stops, water retained temporarily in the macropores is drained away by the pull of gravity. This excess water is called gravitational water. It is held in non-capillary pores and so it is also known as non-capillary water. It is lost to the groundwater table within 1 or 2 days after cessation of rainfall. It is of little use for plant growth because it cannot stay in soil for a considerable period. However, percolation of water may take some time, and some of this extra water can sometimes be used by plants or lost to evaporation. So, gravitational water can be taken to be physiologically available but physically unavailable to plants. Some soils are underlain by compact subsoils (plow pans, crusts). These soils tend to remain saturated (i.e., waterlogged) for a longer period. Percolation in these soils is very slow, and growing plants may utilize part of this water. Since this water is very temporary, it is also called superfluous water.

6.6.3 Capillary Water

When gravitational water has drained off, soil water is comprised of capillary water, hygroscopic water, and water vapor. Capillary water is held by surface tension as a film of moisture on surface of soil particles and peds, and as minute bodies of water filling part of the pore space between particles. Water held in the wall of the micropores is tightly bound and is not available for plants. As shown in Fig. 6.4, a part of the capillary water is available to plants.

6.7 There Are Some Soil Moisture Constants

There are four soil moisture constants: hygroscopic coefficient, maximum water-holding capacity, field capacity, and permanent wilting point. Although they are called "constants", they are highly variable from soil to soil and in the same soil under different management. All these constants are actually soil water contents at specific situations. These parameters are affected by soil texture, structure, porosity, and pore size distribution, organic and inorganic colloids, slope, and management including tillage, drainage, mulching, and irrigation.

6.7.1 Hygroscopic Coefficient

The amount of water that can be absorbed by an oven dry soil at 105°C when it is exposed to an atmosphere of 100% relative humidity is called hygroscopic coefficient (HC). This water is retained on soil particle surfaces (mostly organic and inorganic colloidal particles) and remains in the vapor state. Hygroscopic coefficient is also defined as the water content in soil at 31 bar tension or −3,100 kPa soil water potential.

6.7.2 Maximum Water-Holding Capacity

The amount of water that is held in soil when all the pores, small and large, are filled with water is called the maximum water-holding capacity (MWHC). It is the amount of water at 100% saturation. Water is retained on all available spaces, including particle and ped surfaces, micropores and macropores. Therefore, the sum of hygroscopic, capillary, and gravitational water is the MWHC.

6.7.3 Field Capacity

When a saturated soil is allowed to drain freely under the force of gravity and if there is no evaporation loss, moisture content of the soil will approach an equilibrium level after a time. The amount of water retained then is called field moisture capacity or simply field capacity (FC). It is the moisture content that a given soil reaches and maintains after it has been thoroughly wetted and allowed to drain freely. At this time, soil holds water on particle surfaces and in micropores. Usually, it is the upper limit of moisture content that a soil can retain. It is the moisture content when all macropores or transmission pores have emptied and filled with air.

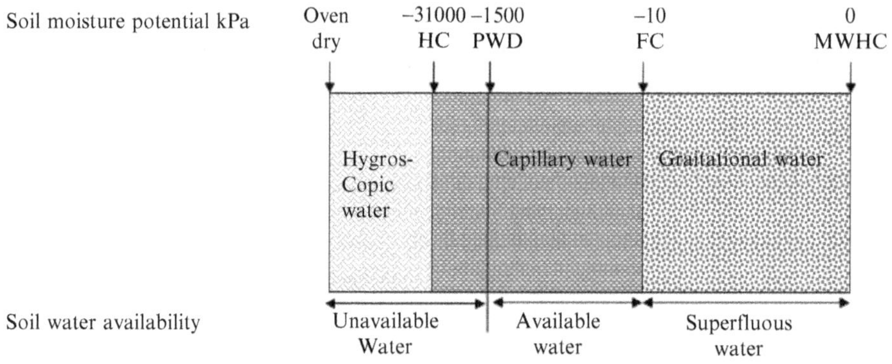

Fig. 6.3 Soil water forms and constants

Soils are heterogeneous mixtures; most natural soils do not have a well-defined field moisture capacity. Because they continue to drain for a long period of time, clayey soils rarely attain a field capacity. On the other hand, soils with impeded drainage never attain a field capacity. Field capacity is the upper limit of plant's available water in soil. At FC, the soil water potential is around −10 kPa (−10 kPa for sandy and −30 kPa in loamy and clayey soils according to some authors, Foth 1990).

6.7.4 Permanent Wilting Point

Unless water is added, the water content of a soil gradually decreases due to losses by drainage, plant uptake, and evapotranspiration. At a point, water content decreases to such an extent that the force of its retention in soil far exceeds the suction of plant roots. Then plants cannot draw water from the soil. This is the moisture content at which plant leaves wilt permanently and do not regain turgidity even they are watered again. This is called permanent wilting, and the percentage of soil moisture at this point is known as permanent wilting percentage or permanent wilting point (PWP). Moisture content at PWP differs widely among soils and plants. The PWP is higher in soils with higher clay content and with 2:1-type expanding-lattice clays (Lal 1979). Some plants are very sensitive to soil moisture stress. All plants do not wilt at the same soil water content; plants vary in their capacity to draw water from soil. Therefore, PWP often refers to soil water content at 15 bar tension or −1,500 kPa soil water potential without referring to the degree of water saturation or the uptake by plants. PWP is the lower limit of the plant available water in soil. However, some plants can absorb water from soil at potentials much lower than this; for example, creosote bush (*Larrea divaricata*) can absorb water at soil water potential −6,000 kPa (Salisbury and Ross 1978). But the amount of water actually held by the soil between −1,500 and −6,000 kPa is small.

6.8 Water That Can Be Absorbed by Plants Is Available Water

The FC is the upper and the PWP is the lower limit of plant's available water. Therefore, plant available water capacity (AWC) is the difference between FC and PWP:

$$\text{AWC} = \text{FC} - \text{PWP}.$$

AWC is the amount of water retained in soil between −10 and −1,500 kPa (−0.1 and −15 bars) soil water potential (Fig. 6.3). The available water in soil does not depend on soil properties alone; it also depends on climatic factors such as rainfall and humidity, and plant factors such as root characteristics, including rooting depth and extension, and leaf characteristics such as leaf area and stomatal density. Some factors affect the FC but not the PWP such as organic matter content, but others may affect both FC and PWP such as clay content. Increase in clay content increases both FC and PWP. However, it is generally conceived that the whole range of water between PWP and FC (i.e., available water, AW) is not equally available to plants (Taylor 1952). A portion of the available water (about 75%) is readily available to plants; it is called readily available water. On the other hand, many investigators believe that a gradual decrease in the ease of water uptake by plants takes place with the gradual decrease in available water content.

The amount of available water is related to the surfaces and pores both depending on texture and structure, or the soil matrix. Silt loams have the largest available water-holding capacity. These soils may contain more than 15% by volume of available water. Sands have the smallest available water-holding capacity. Sandy soils tend to be droughty soils due to retention of low available water and higher hydraulic conductivity. Field capacity for sand is more likely nearer −10 kPa (−0.1 bar) than −30 kPa (−0.3 bar) as in loamy and clayey soils. This means that much of the plant available water in sandy soils can move rapidly to roots when water near roots is depleted.

Jhonson (2009) suggested that sandy soils contain 25–100 mm m^{-1} (2.5–10% by volume), loams contain 100–175 mm m^{-1} (10–17.5% by volume), and clay soils contain 175–250 mm m^{-1} (17.5–25% by volume) available water.

6.9 Water Potential Expresses Energy Relationships of Water

There are two types of energy—kinetic and potential. Kinetic energy is acquired by virtue of motion and is proportional to the velocity squared. Movement of water in soils is relatively slow (usually <0.1 m h^{-1}); so its kinetic energy is negligible. It is the potential energy that characterizes soil water movement within the soil body. Soil water moves from a point of higher potential energy to a point of lower potential energy of water and always tends to reach equilibrium with its surroundings. The standard state for soil water is defined as pure and free water (no solutes and no external forces) at a reference pressure, temperature, and elevation, and is arbitrarily given the value of zero (Bolt 1976).

Water potential is the amount of work that an infinitesimal unit quantity of water at equilibrium is capable of doing when it moves (isothermally and reversibly) to a pool of water at similar standard (reference) state (similar pressure, elevation, temperature, and chemical composition). Alternatively, water potential is the chemical potential or specific free energy of water. Soil water potential refers to the potential energy contained in the soil water. It is a measure of the relative potential energy of water in the soil in comparison with that of pure free water. It is theoretically obtained by deducting the free energy of soil water from free energy of pure free water. Pure free water has the highest water potential, and its value is 0. So, all natural surface water has a negative water potential. The primary forces acting on soil water held within a rigid soil under isothermal conditions can be conveniently grouped as follows: (1) matric forces resulting from interactions of the solid phase with the liquid and gaseous phases, (2) osmotic forces owing to differences in chemical composition of the soil solution, and (3) body forces induced by gravitational and other inertial force fields (centrifugal). Consequently, the total potential, $\psi_{T,}$ can be expressed as the algebraic sum of the component potentials corresponding to the different fields acting on soil water:

$$\psi_T = \psi_m + \psi_s + \psi_p + \psi_g.$$

ψ_m is the matric potential resulting from the combined effects of capillarity and adsorptive forces within the soil matrix, ψ_s is the osmotic potential resulting from osmotic effects of dissolved salts, ψ_p is the pressure potential resulting from external pressure on water, and ψ_g is the gravitational potential.

6.9.1 Matric Potential, ψ_m

The matric potential is the portion of the water potential that can be attributed to the attraction of the soil matrix for water. The matric potential used to be called the capillary potential, because the matric potential is largely due to capillary action. However, as the water content decreases in a porous material, water that is held in pores due to capillarity becomes negligibly small when compared to the water held directly on particle surfaces. The term matric potential, therefore, covers phenomena beyond capillarity.

6.9.2 Osmotic Potential, ψ_s

The osmotic potential is the portion of the water potential that can be attributed to the attraction of solutes for water. If pure water and solution are separated by a membrane, pressure will build up on the solution side of the membrane that is equivalent to the energy difference in the water on the two sides of the membrane. This pressure, which is usually called the osmotic pressure, is numerically equivalent, but opposite in sign, to the solute potential or osmotic potential.

6.9.3 Pressure Potential, ψ_p

The pressure potential is the potential energy due to the weight of water at a point under consideration, or to gas pressure that is different from the pressure that exists at a reference position. Sometimes this pressure potential energy is divided into two separate components: the air pressure potential, which occurs under unsaturated conditions when the soil has an air phase, and the hydrostatic pressure potential, which occurs when the soil is saturated and there is a hydrostatic pressure from an overlying water phase.

6.9.4 Gravitational Potential, ψ_g

The gravitational potential is the potential associated with gravity. The reference height or datum assigned can vary according to need and is often based on utility. The reference level usually depends on the direction of water movement, infiltration or percolation. If the reference level is below the point in question, work must be done on the water and the gravity potential is positive; if the level is above the point in question, work is done by the water and the gravitational potential is negative (Baver et al. 1972). The driving force of water in soil is the difference in soil water potential. The difference in water potential per unit distance of flow is the water potential gradient which determines the rate of water movement in soil. Under nonsaline, unsaturated conditions, two

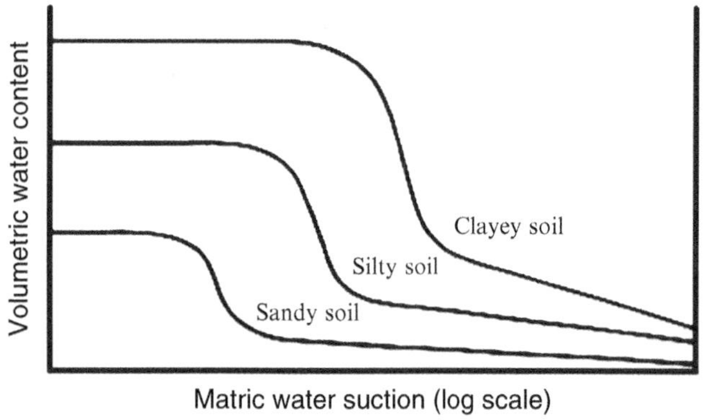

Fig. 6.4 Soil water characteristic curve

most important components of soil water potential are the matric potential and the gravitational potential; both determine the direction of flow of water. Under nonsaline, saturated conditions, important components of soil water potential are the pressure potential and the gravitational potential, and the difference in the sum of these two potentials, called the hydraulic head difference, governs the soil water flow.

6.10 Soil Water Content Can Be Expressed in Relation to Soil Moisture Suction

The soil water characteristic curve for a soil is defined as the relationship between the water content and suction (soil water suction is equal in magnitude to soil water potential; suction is positive, and potential is negative) for the soil (Williams 1982). Water content represents the volumetric water content (θ_v), and the suction is either matric suction or the total suction (matric plus osmotic suction). In place of water content, the degree of water saturation, S_w which is the percentage of soil pores filled with water, is sometimes used. At high suction (i.e., greater than about 1,500 kPa) matric suction and total suction may be considered equivalent.

Soil moisture characteristic curves (Fig. 6.4) have taken on numerous forms because of different terminologies used. Fredlund and Xing (1994) suggested that the term soil water characteristic curve be used to represent the relationship between volumetric water content and matric suction.

6.11 Hysteresis Expresses the Lag of Water Content at the Same Water Potential During Wetting and Drying

Usually, there is a difference in the content of soil moisture at the same water potential in the same soil between drying and wetting courses (Fig. 6.5). This phenomenon is called hysteresis of soil moisture. When the soil is wetted from the

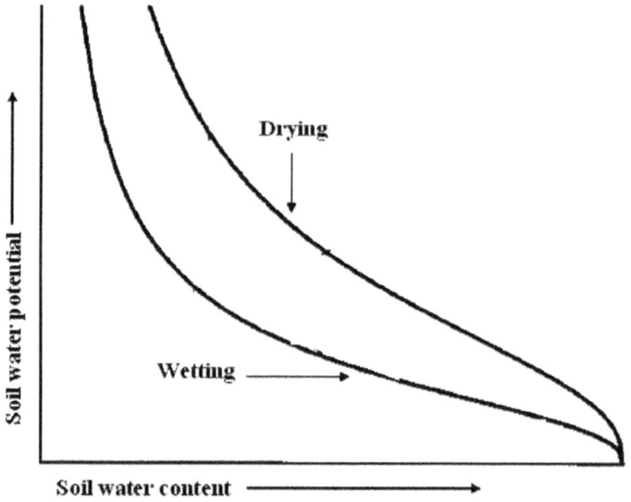

Fig. 6.5 Idealized soil water hysteric curve

residual water content or drained from the saturated water content, the main wetting or drying curves are followed, respectively (Simunek et al. 1999).

Hysteresis is a well-known phenomenon in soil water retention in unsaturated soil (Maqsoud et al. 2004). In general, unsaturated soils undergoing drying processes such as evaporation or drainage tend to retain a greater amount of water than at the same magnitude of water potential during wetting processes such as infiltration or capillary rise. It follows then that if capillary cohesion is directly dependent on the magnitude of matric potential, there is likely to be some amount of wetting–drying hysteresis in the relationship between capillary cohesion and water content. The hysteresis effect can be attributed to four main causes (Hillel 1980; O'Kane et al. 2004): (1) geometric nonuniformity of individual pores, resulting from the so-called "ink bottle" effect, (2) different spatial connectivity of pores during drying or wetting process, (3) variation in liquid–solid contact angle, and (4) air entrapment. Hysteresis can significantly influence water flow and solute transport in variably saturated porous media like soil (Vachaud and Thony 1971; Royer and Vachaud 1975; Gillham et al. 1979; Kaluarachchi and Parker 1987; Jaynes 1992).

6.12 Soil Water Moves in Saturated and Unsaturated Conditions

Soil water movement occurs in saturated and unsaturated conditions. Soils are said to be saturated when soil pores, both micro- and macro-, are entirely filled with water. In this condition, the volumetric water content (θ_v) is equal to the total porosity (φ) and the air capacity $\theta_a = 0$.

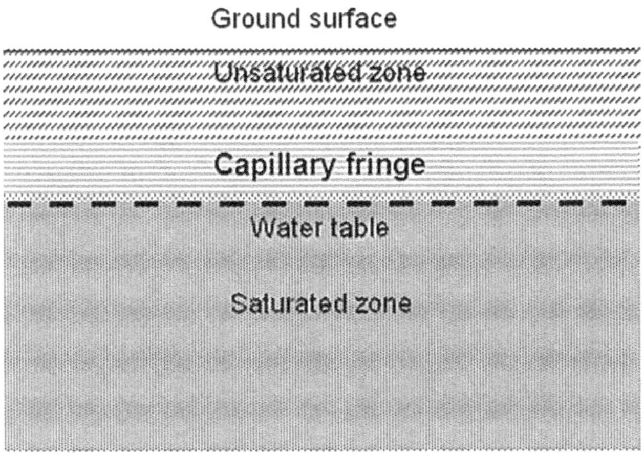

Fig. 6.6 Saturated and unsaturated soil zones

Soil pores can be assumed to be fully saturated usually below the water table. Above the water table soil may also become saturated following a rainfall event or irrigation. Saturated soil conditions occur below the water table, while unsaturated conditions generally predominate above the water table (the vadose zone) (Fig. 6.6). However, localized zones of saturation can exist especially following precipitation or irrigation. In saturated soil, water movement is predominately horizontal, with lesser components of flow in the vertical direction. As a rule, water movement in the unsaturated zone is vertical but can also have large lateral components (Radcliffe and Rasmussen 2002).

6.12.1 Poiseuille's Equation

In an approximated water-filled cylinder of given radius (r), the volumetric flow rate (Q) is described by Poiseuille's equation as

$$Q = \frac{\pi r^4 \rho g \Delta H}{8 \eta L},$$

where η is the viscosity of water, ρ is the density of water, g is the acceleration due to gravity, and ΔH is the difference in total head at a distance L. Poiseuille's equation shows that volume of water flowing per unit time varies with the fourth power of the radius, and the water flux Jw ($=Q/\pi r^2$) varies with the radius squared. Hence in a soil, other factors being equal, water would move 100 times faster through water-filled pores of 1 mm radius than those of 0.1 mm radius.

6.12.2 Darcy's Equation

Darcy (1856) found that the volumetric flow rate of water per unit cross-sectional area (A) through a sand filter of a given thickness (L) was proportional to the total head gradient across the sand ($\Delta H/L$). He termed the proportionality constant saturated hydraulic conductivity (K_s). According to Darcy,

$$J = \frac{Q}{A} = -K_s \frac{\Delta H}{L},$$

where J is the fluid flux. Using partial derivative notation, this can be written as

$$J = -K \frac{\partial H}{\partial Z}.$$

6.12.3 Saturated Hydraulic Conductivity

The saturated hydraulic conductivity (K_s) of a porous medium, such as soil, refers to its ability to conduct water when all pores are full of water. It is a compound parameter, which comprises properties of the soil and water at the specified temperature and pressure. For measurements of K_s, a saturated soil column of uniform cross-sectional area with a diameter large enough for the validity of the assumption of one-dimensional flow is subjected to a hydraulic gradient. The resulting flux of water is measured, and the proportionality constant in Darcy's law gives the value of K_s of the soil column.

$$\text{Hydraulic conductivity } K_s = \frac{k \rho g}{\eta},$$

where k is the intrinsic permeability of the soil. Soils with low porosity, few large pores, and poor interconnectivity between pores have low values of K_s. Saturated hydraulic conductivities is the highest in coarse-textured soils due to larger pores and declines in fine-textured soil. Higher K_s occurs in the coarse-textured soils in spite of the generally lower porosity. Hydraulic conductivity is then the function of macroporosity rather than total porosity.

6.12.4 Unsaturated Flow

The unsaturated zone, also known as the vadose zone, is above the water table up to the surface. Following rain or irrigation, this zone also becomes saturated, but afterward the excess gravitational water drains away. Then soil water content lies below saturation; most large pores are filled with air. There is the capillary fringe, above the water table, and water is under tension there but is near saturation. The thickness of this capillary fringe varies from soil to soil, mainly depending on texture and porosity. Sands have thinner capillary fringes than unstructured clays. Because the large pores are empty, water moves in the unsaturated zone generally through smaller pores and by film adjustment. So, the hydraulic conductivity is much smaller under unsaturated condition.

Flux density, q in saturated flow, is proportional to the driving force, the hydraulic gradient, $\Delta H/L$ as

$$q \alpha \frac{\Delta H}{L}$$

or

$$q = K(\theta)\frac{\Delta H}{L},$$

where $K(\theta)$ is the unsaturated hydraulic conductivity of the soil. This equation is equivalent to Darcy's law. Buckingham (1907) was first to describe hydraulic gradient-dependent flow through unsaturated media. So the above equation is known as the Darcy–Buckingham equation.

The unsaturated hydraulic conductivity [$K(\theta)$] is dependent on both moisture content and matric potential. The above equation can be written in terms of suction (ψ_m, or its negative suction head) and gravitational component (ψ_z) as $H = \psi_m + \psi_z$. Contrary to saturated flow, where ψ_p is a function of z only, in unsaturated flow, ψ_m is a function of both z and time (t). Therefore, the derivative ΔH is a partial derivative, and the equation can be written as

$$q = K(\theta)\frac{\partial(\psi_m + \psi_z)}{\partial z},$$

where ∂z is change in length (L).

Here, it was only intended to concisely inform the readers about water movement in soil. Readers are referred to Baver et al. (1972), Kirkham and Powers (1972), Warrick (2002), Lal and Shukla (2004), Kirkham (2005), and Miyazaki (2006) for further reading.

6.13 Plant Water Moves Along a Potential Gradient

In the soil–plant–atmosphere system, water always moves from higher water potential to lower water potential, that is, water moves along a potential gradient. Generally, soil water potential is higher than root water potential, root water potential is higher than stem water potential, which is again greater than leaf water potential; so there is a potential gradient along the root surface to the leaf surface. Consequently, there is a continuous flow of water from soil to roots and root to leaf followed by transpiration (Fig. 6.7).

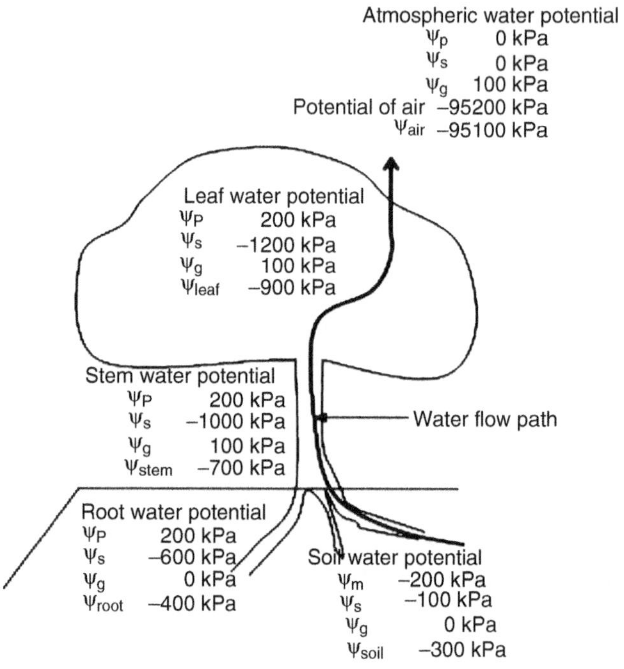

Fig. 6.7 Hypothetical distribution of water potential along the SPAC

Plant water potential is the sum of three potentials—osmotic potential (ψ_s), pressure potential (ψ_p), and gravitational potential (ψ_g). Matric potential (ψ_m) is negligible in plant tissue because little suction is applied by cell walls on water molecules. But, turgidity exerts positive pressure on water increasing its tendency to move from cell to cell; so the pressure potential is positive. Pressure potential may be negative in highly flaccid cells, but this is a rare case. Osmotic potential is created by the dissolved ions in cell sap. It is always negative:

$$\text{Plant water potential}, \psi_{plant} = \psi_s + \psi_p + \psi_g.$$

ψ_s is negative, ψ_p and ψ_g are positive, and thus, theoretically, ψ_{plant} may be negative or positive depending on the magnitude of the pressure and gravitational potential. In practicality, ψ_s dominates over $\psi_p + \psi_g$ in plant tissues, and cell water potential is usually negative.

Water moves from soil to plant roots because of the potential gradient. Likewise, there must be a potential gradient along the root-stem-leaf pathway to have an unhindered flow of water along this path.

There is also a large difference in the water potential between plant leaf and the surrounding air. This difference is necessary for transpiration to occur. Nobel (1991) shows representative values for the water potential and its components in the soil–plant–atmosphere continuum. In light of his values, Fig. 6.7 is constructed to show the gradient in water potentials along the soil–plant–atmosphere continuum (SPAC) and the direction of flow of water.

However, the water potential widely varies with weather and soil conditions. There is diurnal variation as well. In a rapidly transpiring plant tissue at midday, the pressure potential is considerably reduced. Air water potential is higher on a humid day.

6.14 Soil Water Regulates Plant Growth

Life, plant or animal, without water is impossible. Plants need water for their normal growth and development. However, plants vary widely in their water requirements. Many plant species need very large amount of water. They are hydrophytes or hygrophytes adapted to grow in waterlogged soils. Some crop plants such as rice (*Oryza sativa*) have hydrophytic characters. Some plants are adapted to soil moisture shortages; they are xerophytes which need relatively low water for growth. However, most plants, particularly crop plants, are mesophytes which are upland plants and need intermediate level of soil water. For these plants, soil moisture level should be maintained at the readily available water level, that is, above the refill point.

More than 95% water that is absorbed by plants is lost to the atmosphere through transpiration. It is an essential loss of water. The primary role of water in the plant body is the maintenance of turgidity. Water is needed by plants in almost all processes, including photosynthesis and respiration and synthesis of carbohydrates, proteins, nucleic acids, and other biomolecules. It is needed for various growth processes such as cell division, root elongation, production of leaves and branches, flowering, and fruiting. Therefore, desired growth, yield, and quality of crops can be assured only when plants are timely supplied with optimum quantity of water.

6.15 Plants Suffer from Water Stress in Some Situation

Between rains or two spells of irrigation, plants get their water from the available soil water storage at the root zone, which gradually decreases due to evapotranspiration. As the readily available water is diminished, and unless water is replenished, plants begin suffering from water stress. The most obvious sign of water stress is the wilting (temporary at the beginning and permanent at the advanced stage). Plants may suffer from water shortage even without producing any signs. Plants suffer from water stress in two situations: (1) if soil moisture level is too low, so that the matric potential is low (matric suction is high) and (2) if the osmotic potential is very low, that is, if the soil is highly saline.

Reduction in cell expansion, cell wall synthesis, chlorophyll synthesis, protein synthesis, nitrate reductase, stomatal opening, CO_2 assimilation, and xylem conductance occurs due to water stress. On the other hand, abscisic acid synthesis, respiration, proline accumulation, and sugar level increase by water stress. There are a number of modifications in plant structures and processes as a consequence of drought. These include sensitivity of stomatal response, osmotic adjustment, smaller cell volume, reduced leaf area, increased leaf thickness, hairy leaves, and increased root–shoot ratio, as well as several changes in enzyme and hormone production and activity. Dehydration usually causes severe damage and disorganization of membranes and organelles, mechanical rupture of protoplasm, degradation of cell membranes, protein denaturation, and gene mutations. Contents of proteins, glycolipids, and phospholipids in chloroplasts generally decrease due to water stress. Water stress causes reduction in seed germination; seedling emergence; seedling survival; leaf area; shoot growth; number of leaves, branches, flowers, and fruits; shedding of leaves, flowers, and fruits; and ultimately growth and yield. Plants produce a number of visual symptoms at moisture stress. The following visual moisture-stress symptoms may be used to monitor crop need of irrigation in the field. But it is not wise to wait for symptoms to appear. By the time most of these symptoms appear, it is too late and the damage has already been done. It is wise to monitor the soil moisture conditions regularly to decide time and amount of irrigation. Moisture-stress symptoms in some crops are summarized in Table 6.1.

Table 6.1 Moisture-stress symptoms in some crops

Crops	Symptoms
Alfalfa, corn (maize), potatoes, small grains, mint	Dull, bluish green color of leaves
Beans, onions, sunflower, capsicum	Wilting
Sweet corn	Leaf curling
Tomato	Grayish green color, droopy, wilting, leaf rolling
Grapes	Dark, grayish green color of leaves

6.16 Some Plants Have the Capacity to Tolerate or Avoid Drought

Plant responses to water stress may be classified as (a) physiological responses to short-term changes, (b) acclimation to a certain level of water availability, and (c) adaptations to drought. Short-term responses are linked to stomatal regulation reducing water loss by transpiration and maximizing CO_2 intake. Midterm responses (acclimation) include the adjustment of the osmotic potential by solute accumulation, changes in cell wall elasticity, and morphological changes. Long-term adaptation to drought includes genetically fixed patterns of biomass allocation, specific anatomical modifications, and sophisticated physiological mechanisms, with an overall growth reduction to balance resource acquisition.

Some plant species can cope with soil water stress. They are said to possess stress tolerance or stress avoidance. Some plant species (e.g., *Agave deserti*) store water in their buds, stems, or leaves and utilize this stored water under conditions of severe drought. There are some other plant species which avoid water stress by deep root system and mechanism of low transpiration. Some crops such as cluster bean (*Cyamopsis tetragonoloba*) and cowpea (*Vigna unguiculata*) are drought tolerant for their deep taproot system (Kumar 2005). Drought tolerance in *Brassica carinata*, *B. napus*, and *B. campestris* is related to their better-developed root system (Liang et al. 1992). Drought tolerance in plants involves plant responses at cellular and at whole-plant level such as synthesis and accumulation of organic compatible solutes, synthesis of stress proteins, upregulation of antioxidant enzymes, development of deep and dense root system, epicuticular wax, and leaf rolling (Parry et al. 2005; Reynolds et al. 2005; Neumann 2008). List of some dryland crops are given in Sect. 11.2.2.

6.17 Crop Water Requirement Varies with Crop Types

After the excess gravitational water has drained away, further removal of water from soil takes place by evaporation and transpiration. The amount of water needed by a crop, that is, crop water use is equivalent to the evapotranspiration (ET) for a given time. Evapotranspiration is influenced by prevailing weather conditions, available water in the soil, crop species, and growth stage. At full cover, a crop will have the maximum ET rate if soil water is not limited, that is, if the soil root zone is at or near field capacity. Full cover is a growth stage at which most of the soil is shaded by the crop canopy. Different crops reach full cover at different growth stages and times after planting. Crop water requirement (CWR) is not, therefore, a specific limit for any crop species. Atmospheric conditions such as relative humidity and temperature; soil conditions such as soil water content, soil texture, and hydraulic conductivity; and crop characters such as rooting depth and extension, vegetative growth, growth stage, leaf area, stomatal density, etc., all influence the crop water requirement. A crop may have short or long growing period. CWR is influenced by the length of crop duration. For example, a crop variety with maturity period of 150 days will use more water than 120 days variety. Under similar environmental condition, a plant having little leaf area and root system would require much less water than a plant having higher leaf area and dense root system. In addition, crop density may also influence water requirement. Water requirement of some important crops is given below:

Water requirement for total growing period, mm	Crops
300–500	Bean, cabbage, onion, pea
450–700	Barley, melon, millet, oats peanut, potato, rice, sorghum, soybean, sugar beet, tomato, wheat
500–900	Corn (maize), pepper, sugar beet, tomato
700–1,300	Sunflower, citrus, cotton
1,500–2,500	Banana, sugarcane

These are typical average estimates for the humid region. Crop water requirement varies with climate, being the highest in the arid and semiarid regions where the supply of water in rainfall is the least (Hillel 1990).

6.18 Water Supply Influences Crop Quality

We need to be aware that water supply influences both quantity and quality of crops. Quantity is the economic yield, while quality is characterized by some physical characteristics such as size, color, texture, and odor and chemical characteristics such as protein, sugar, and vitamin contents of vegetables, grains, and fruits. Irrigation usually increases yield, but yield is not always related with quality. Sometimes, increased yield is associated with decreased contents of sugar and proteins for the so-called dilution effect. However, adequate irrigation throughout periods of active vegetative growth generally results in an improvement in crop quality. But moderate moisture stress during the maturation has often been found to be desirable for some crops such as sugarcane and sugar beet. The withdrawal of irrigation water several weeks before harvest is a common practice in sugarcane cultivation.

6.19 Irrigation Water Requirement Includes Crop Water Requirement and Others

Irrigation water requirement is different from the crop water requirement. Crop water requirement is the amount of water used by a crop in its growing period. On the other hand, irrigation water requirement is the amount of water needed to be applied to the field. Some of CWR is satisfied by rainfall. So, irrigation water requirement is usually the amount of water between evapotranspiration and precipitation, provided that no water from precipitation is lost by runoff and percolation. IWR depends on crop, soil, weather, rainfall, method of irrigation, and depth of the water table. In areas and in seasons when rainfall is sufficient and evenly distributed, the whole CWR may be satisfied by rainfall. In many humid tropical and humid temperate regions, agriculture is mainly rainfed. No supplementary irrigation is needed in some seasons. In regions of heavy rainfall, on the other hand, the removal of excess water by artificial drainage is a problem. In areas of deficit rainfall and in periods of drought in humid areas, the shortage of water is met by irrigation. Irrigation water requirement includes water for evaporation and transpiration, for leaching of salts if the soil is saline and any system loss due to drawing and distribution of water. Tentatively, IWR should be

$$\mathrm{IWR} = (\mathrm{ET} + \mathrm{LR} + \mathrm{RP} + \mathrm{SL}) - (\mathrm{P} + \mathrm{S} + \mathrm{CR})$$

where, ET = evapotranspiration, LR = leaching requirement, SL = system loss, RP = runoff and percolation, P = precipitation, S = soil water storage during sowing/transplantation, and CR = capillary rise of water from the water table.

Table 6.2 Most critical growth period for water stress

Crop	Critical period
Alfalfa	Early spring and immediately after cutting
Corn (maize)	Tasseling, silk stage until grain is fully formed
Sorghum	Boot, bloom, and dough stages
Sugar beets	Post thinning
Beans	Bloom and fruit set
Small grains	Boot and bloom stage
Potatoes	Tuber formation to harvest
Onions	Bulb formation
Tomato	After fruit set
Cabbage	During head formation and enlargement
Lettuce	Just before harvest when the ground cover is complete

6.19.1 Crop Water Use Efficiency

Crop water use efficiency (WUE) is judged by the amount of yield obtained per unit of irrigation water applied. Sometimes, WUE is taken to be the ratio of yield (Y) to ET:

$$\mathrm{WUE} = \frac{Y}{\mathrm{ET}}.$$

WUE is not a function of the watering alone; it also depends on other soil and crop management factors. However, under similar management, some crops use water more efficiently than others, that is, they produce more yield per unit of water under standard management. Crop species and varieties differ due to differences in rooting habit, and size, number, and orientation of leaves, leaf area index, stomatal density, crop duration, crop density, season, and soil characteristics. The increases in water use efficiency result from increased transpiration as a fraction of the ET. Crop breeding can change rates of photosynthesis relative to transpiration, crop duration, vegetative growth habits, and physiological behavior related to water uptake and partitioning of assimilates to economic yield. Such changes may aid in the efficient use of water.

6.19.2 Full Irrigation

Under adequate water supply, full irrigation is practiced. The whole irrigation water requirement is fully applied, and no water stress develops under such conditions in any growth stage. Then, crop yield is expected to reach the potential yield, provided that other factors are not limiting.

6.19.3 Deficit Irrigation

In areas of water scarcity, water becomes the primary limiting factor to improve agricultural production. Accordingly, maximizing yield per unit of water, and not yield per unit of land, is a more viable objective for on-farm water management (Ali 2010). However, crops do not need equal amount of water at all growth stages. Some growth stages are not much sensitive to soil water stress, while some others are critical (Table 6.2). In many cases, water stress reduces biological yield but not economic yield. Sometimes economic yield increases at a mild water stress during a particular growth stage (Kirkham 2005). This has led to adoption of deficit irrigation which is the deliberate and systematic under-irrigation of a crop (English 1990). Deficit irrigation means less application of water than a plant would normally use. It is based on the assumption that in field crops, imposing water stress at specific growth stages may not cause significant yield reduction, and irrigation in these stages can be ignored. Deficit irrigation is practiced where essential resources such as water, capital, energy, and/or labor are limited.

Techniques of deficit irrigation include (1) increasing interval between irrigations, that is, less frequent irrigation, (2) omitting irrigation at the growth stage less sensitive to moisture stress ,(3) wetting partial root zone, (4) wetting alternate furrows, and (5) allowing root zone soil–water depletion to a particular level. Advantages of deficit irrigation include increase in water use efficiency, more profitability than full irrigation, increase in quality of produce (protein content, sugar content, grain size, etc.), and reduction in nutrient loss through leaching. Disadvantages include reduction in crop yield unless crop behavior is understood and possibility of creating soil salinity particularly in arid and semiarid regions.

6.20 Choice of Irrigation Methods Depends on Crop Types and Farm Facilities

There are four principal methods of irrigation: surface irrigation, sprinkler irrigation, drip or trickle irrigation, and subsurface irrigation. Surface irrigation can be subdivided into uncontrolled and controlled flooding systems. Controlled flooding is further subdivided into basin irrigation, contour levee irrigation, border irrigation, furrow irrigation, and corrugation irrigation. Each of these methods has advantages and disadvantages. A method is suited to a particular set of physical conditions such as crop type, soils, slope, farm facility, and water availability. Choice of irrigation methods depends on availability of funds, water, and labor; costs of labor, fuel, operation, and maintenance; area, amount, and frequency of irrigation needed; farm facilities; crop type; and additional purposes such as fertilizer application and frost control.

Fig. 6.8 Open flooding or wild flooding irrigation (Photo courtesy of USDA-NRCS)

Fig. 6.9 Border flooding irrigation (Photo courtesy of USDA-NRCS)

Fig. 6.10 Furrow irrigation in grape vines (Photo Alberto Conu, FAO. With permission)

6.20.1 Surface Irrigation

Surface irrigation methods are the most widely used methods. These methods are suitable for most crops, dense or sparse, but are not applicable to highly permeable or sloping soils. Surface irrigation methods are relatively easy to operate and maintain. Some labor cost may be involved in land leveling which is simple and straightforward. Drawing and distribution of water is also easy, and the system can run very efficiently under careful management. However, water is drawn to the fields in these methods through open earth channels. Control over water supply and distribution is low. There is greater loss of water in surface irrigation systems. Surface irrigation methods produce less water use efficiency (yield of crop per unit of irrigation) than sprinkler and drip irrigation methods. Surface irrigation could easily be managed in areas with abundant sources of water.

6.20.2 Uncontrolled Flooding

Uncontrolled flooding is the oldest and the simplest irrigation practice, with water being drawn from streams and other reservoirs and distributed over the land without any boundary (Fig. 6.8).

The method is very basic, easy to set up and operate, and low-cost. Uncontrolled flooding can be used in many crop types, but the distribution of water in field may be uneven. Loss of water may also be high.

6.20.3 Controlled Flooding

Border Flooding Irrigation: For border flooding irrigation, a land on a smooth surface is uniformly leveled. Main irrigation canals are constructed in the borders with opening for water at suitable places (Fig. 6.9). Water is let into the field so that it spreads all over the surface.

To achieve uniform irrigation application, the border should not have a cross slope. The borders can vary in length and width depending on the soil type, the slope, the available stream size, and farming practices.

Furrow Irrigation: It is the most common method for irrigating row crops such as corn (maize), sugar beet, potatoes, sunflower, vegetables, orchards, and vineyards (Fig. 6.10).

Fig. 6.11 Contour levee irrigation (Photo courtesy of Jim Hill)

Fig. 6.12 Improvised basin irrigation in a forest plantation (With permission from Photo-Library, FAO)

The method is applicable to most soil types, except those which are highly permeable, steeply sloping, or easily erodible. Slopes are limited to 2% in arid climates and only 0.3% in humid climates with intense rainfalls. Furrows are connected to the distribution channels. Seedlings may be planted on top of ridges, sides, and base of furrows. Salinity may pose a problem if planted on top of ridges where salts tend to accumulate. Positioning the seedlings on the sides of the ridges may avoid this problem. The spacing between the furrows is governed by the soil type, closer in lighter soils and wider in heavy soils and distance between crops.

Contour Levee Irrigation: This method is similar to basin irrigation but is used on hillsides and steeply sloping land (Fig. 6.11). The basins are cut into the slope and run along the contour to form a terrace. Water is applied and held in the basin until the required depth has been applied.

Basin Irrigation: The land is divided by dikes to form basins, and water is turned into the basin from a channel in basin irrigation method (Fig. 6.12). The land within each basin is then leveled to allow uniform water distribution. The method may suit well to smallholder irrigation where farmers grow a variety of crops within a small land area.

Fig. 6.13 Sprinkler irrigation in a tea garden (Photo courtesy of Ashim Saha, Bangladesh Tea Research Institute, Srimangal, Bangladesh)

This method may successfully be used in orchards where the basin is prepared around each tree base and water is ponded in each basin. The basins are linked by water distribution channels. Water loss by deep percolation is likely in basin irrigation like other flooding systems. In rice systems, the water is often turned into one basin and then flows through a series of basins before discharging into the field drain.

Corrugation Irrigation: This is similar to furrow irrigation; here the ridges are smaller and shallower and closer spaced than furrow irrigation. The method is suitable for irrigation of close-growing crops such as barley and wheat.

6.20.4 Sprinkler Irrigation

Sprinkler irrigation is applicable to almost all types of crops, including closely growing, row crops and orchards. The sprinkler system consists of sprinklers or other sprinkler devices operating under pressure, a pump unit, and a water distribution network, either surface or buried (Fig. 6.13). There are different sprinkler devices, including revolving head sprinklers (single- and multiple-nozzle sprinklers), fixed head sprinklers, nozzle lines and perforated pipes, and a number of different sprinkler systems classified according to their mobility. Sprinkler systems may be mobile, semipermanent or permanent. The method can be used in all soil types and terrains. This method achieves high crop water use efficiency compared to surface irrigation systems. This method generally requires less amount of water and involves less loss of water. But for the high primary installation costs, this method has not become popular. It does not also function well under windy conditions.

Fig. 6.14 Drip irrigation system (Photo A.K. Kimoto, FAO. With permission)

6.20.5 Drip Irrigation

Drip irrigation systems can be used for field and glasshouse crops. It usually consists of a pump, filter, flow meter and pressure gauge, fertilizer injector, valves, pipe networks, and emitters. The method involves trickling or dripping small quantities of water, usually drop by drop, from a pipe onto the soil surface near the root zone (Fig. 6.14).

Because the rate of application is low, almost all the water is absorbed into the soil; there is little or no runoff. If the application rate and frequency of irrigation can be matched with crop need, the irrigation efficiency can be very high. In addition, the system can be used to apply fertilizers direct to the root zone. In this method, the installation costs may be high, and there can be significant problems with blocking of the emitters from sand and silt, chemical precipitation from the water, and algae.

6.20.6 Subsurface Irrigation

In subsurface irrigation system, water is applied below the ground surface. The water reaches the plant either through buried pipes or drains, or through seepage from irrigation or drainage canals (Fig. 6.15). However, openings of underground pipes may be clogged, and the operation of the system may not be sustainable. Watering through deep drains may not be possible in very light-textured soils.

6.21 Irrigation Water Should Be of Proper Quality

Irrigation water quality refers to the suitability of a given water for irrigation. This is determined by the physical, chemical, and biological characteristics of water. Physical characteristics include turbidity and temperature. Some surface waters are turbid due to the presence of suspended solids, including sand, silt, and clay. Fluid sludge and other wastewaters may also be turbid due to inorganic and organic solids. These particles in water, if used for irrigation, clog the soil pores and make the soils impervious. Suspended particles may also damage turbines and clog sprinkler systems. Industrial wastewater may be unusually hot or cold that should not be immediately used for irrigation.

Chemical characteristics of water that affect suitability for irrigation include acidity, salinity, sodicity, and contamination of trace elements. Surface waters suitable for irrigation may be slightly acidic or alkaline (pH 6–8). Some waters, such as industrial wastewater, may be strongly acidic due to the presence of dissolved acids. Water used for irrigation may contain soluble salts such as chlorides and sulfates of Ca^{2+}, Mg^{2+}, Na^{+}, and K^{+}, and often their carbonates, bicarbonates, and nitrates. The total dissolved salts may be determined by evaporating a sample of water or estimated from its electrical conductivity (EC) measured by a conductivity meter according to the relationship

$$\text{TDS (mg L}^{-1}) = 640\ \text{EC (dS m}^{-1}).$$

TDS is also termed as "salinity hazard" of irrigation water. It is, however, convenient to use the EC value as the salinity index. According to Richards (1954), EC limits of <0.25, 0.25–0.75, 0.75–2.25, and >2.25 dS m^{-1} at 25°C indicate waters of low, medium, high, and very high salinity, respectively. The EC of most irrigation waters lies between 0.15 and 1.5 dS m^{-1}. In the soil, low water salinity is concentrated several times due to evapotranspiration. At high salinity level, many crops suffer from osmotic stress. The osmotic potential of water may be estimated from

$$\psi_s(\text{kPa}) = -36\ \text{EC (dS m}^{-1}).$$

The relative proportion of Na to Ca and Mg is called sodicity of water. Apart from total salinity, excess sodium in water has some detrimental effects on plants. Sodicity is denoted by the sodium adsorption ratio (SAR) calculated

Fig. 6.15 Subsurface irrigation system (image courtesy of Mr.Jafar Afsar)

Table 6.3 Permissible levels of trace elements in irrigation water

Elements	Concentration mg L^{-1}	Elements	Concentration mg L^{-1}
Aluminium[a]	1.00	Fluoride[ab]	1.00
Arsenic[a]	0.10	Iron[b]	5.00
Boron[a]	0.30	Lead[c]	2.00
Cadmium[b]	0.01	Manganese[c]	10.00
Chromium[b]	0.10	Nickel[c]	2.0
Cobalt[b]	0.05	Selenium[c]	0.05
Copper[b]	0.20	Zinc[c]	5.0

Source: [a]Awad (1984), [b]Rowe and Abdel-Magid (1995), [c]McLaughlin et al. (2000)

from values of soluble sodium, calcium, and magnesium in water according to the following relationship:

$$SAR = \frac{Na^{+}}{\sqrt{\dfrac{Ca^{2+} + Mg^{2+}}{2}}}.$$

Continued use of water of high SAR (>15) eventually leads to a greater exchangeable sodium percentage (ESP) in soil. Soil structure is destroyed, colloids are dispersed, and the soil is compacted by high sodium saturation. Hardness in water is created by dissolved carbonates and bicarbonates of Ca^{2+} and Mg^{2+}. Higher concentration of Ca^{2+} and Mg^{2+} indicates greater hardness. If hard water is used in sprinkler and drip irrigation systems, calcium and magnesium salts may form a crust inside supply and distribution pipes and nozzles (Southorn 1997).

Water may contain small quantities of trace elements, including heavy metals. Crops may absorb these elements and concentrate them in their edible parts. The well-known itai-itai disease in Japan was caused by the consumption of rice grown in soil irrigated with Cd-contaminated river water. Permissible limits of some of these elements in irrigation water are summarized in Table 6.3.

Groundwater Contamination with Arsenic and Human Health in Asia

Arsenic in groundwater is a major health concern in Asia. A large population uses shallow tube well water for domestic uses including drinking. Health problems associated with elevated levels of arsenic include thickening and discoloration of the skin, stomach pain, nausea, vomiting; diarrhea; numbness in hands and feet; partial paralysis; and blindness. Arsenic has been linked to cancer of the bladder, lungs, skin, kidney, nasal passages, liver, and prostate. Arsenicosis is the principal health problem associated with arsenic poisoning. United States Environment Protection Agency has set the arsenic standard for drinking water at 0.01 mg L^{-1} to protect people from long-term chronic exposure to arsenic. These problems are mainly related with As contaminated drinking water, but there is also the risk of contaminating the food chain. Shallow aquifer water is the major source of irrigation water in rice growing areas of Asia. Studies show that soils become contaminated with As as a result of long term irrigation with contaminated groundwater. Data indicate that rice and vegetables may contribute significantly to daily intake of As in Bangladesh (Heikens 2006).

Bangladesh

Groundwater from shallow aquifers is the main source of drinking-water in Bangladesh. Some of the shallow aquifers contain As concentrations above the national drinking-water standard of 0.050 mg L^{-1}, particularly in the south and south-western part of the country. Approximately 20% of the shallow tube wells (STW) exceed the standard and 10,000–30,000 people have

(continued)

(continued)
been diagnosed with arsenicosis. Thirty million people use water that exceeds the Bangladesh drinking-water standard for As. Approximately 95% of all groundwater drawn from shallow aquifers is used for irrigation, mainly for dry-season rice cultivation (Alam et al. (2002). The cause of groundwater contamination with As in Bangladesh is geochemical.

China
Endemic arsenicosis was found successively in many areas in mainland China during 1980s. The population exposed to As levels in drinking-water exceeding the national standard of 0.050 mg L^{-1} may be more than two million. More than 10,000 arsenicosis patients were diagnosed by 2001. The main As contaminated provinces are Xinjiang, Inner Mogolia, Shanxi, Ningxia, Jilin, and Qinghai. In some areas arsenicosis arises from burning As-rich coal indoors.
More areas with high As concentrations can be expected within these and other provinces in China because of geochemical and hydrological characteristics (Xia and Liu 2004).

India
Arsenic contaminated groundwater occurs mainly in the states West Bengal, Bihar, Uttar Pradesh, Assam, Jharkland, Chhattisgarh and Madhya Pradesh. In West Bengal, investigations suggest that eight districts show As content in well-water above 0.050 mg L^{-1}. According to UNICEF, over 13.8 million people are at risk of arsenicosis in West Bengal (Chakraborti et al. 2002).

Nepal
Arsenic contamination was detected in the Terai region of southern Nepal where nearly half of Nepal's total population is living. This region borders India and more than 90% of the people use approximately 200,000 shallow tube wells for their drinking water. Fifteen thousand STWs were tested and 23% of then exceeds the WHO drinking-water guideline of 0.010 mg L^{-1}, whereas 5% exceeds the Nepal interim As guideline of 0.050 mg L^{-1}. An estimate shows that 0.5 million people are drinking water that has As levels exceeding 0.050 mg L^{-1} (Shrestha et al. 2003).

6.22 Over Irrigation Is Harmful

By over irrigation, we mean (1) application of water surpassing field capacity, (2) application of more water than crop requirement, and (3) irrigation given more frequently than actually needed. Over irrigation involves loss of water, O_2 depletion in the root zone, enhanced leaching of nutrients, waterlogging and rise of water table, salinization, reduction in organic matter decomposition and nutrient release, and unfavorable microbial transformation in the soil. If over irrigated, crop plants may suffer from O_2 stress, and crop growth and yield may fall.

6.23 Waterlogging Is Undesirable for Most Crops

A soil is termed waterlogged when it remains saturated with water for a prolonged period. For example, a soil becomes saturated with water after a rainfall event, but the excess gravitational water in most cases will be drained away within 1 or 2 days. Some soils would remain saturated even after a week; it is then called waterlogged. Waterlogging leads to O_2 depletion in soil, rising of the groundwater table, salinization, and unfavorable microbial transformations. It affects metabolic activities of plants and microorganisms. Seeds and roots of seedlings decay due to waterlogging. Moreover, some fungal diseases occur under wet conditions.

Waterlogging creates problem for most crops because most common crops are mesophytic plants, which grow best when there is both sufficient water and air in the root zone. When plant roots respire, they absorb oxygen (O_2) from the soil atmosphere and release carbon dioxide (CO_2) back into it. In waterlogged soils, the air content of the soil is low because most pores are filled with water. Moreover, renewal of soil air is very restricted under these conditions. Consequently, root respiration is restricted by the oxygen deficiency. On the other hand, carbon dioxide accumulates to toxic levels, directly hampering the root growth and the roots ability to absorb nutrients. Anaerobic conditions in the soil may also lead to the formation of toxic concentrations of reduced iron and manganese compounds, sulfides, and organic gases.

However, the duration of waterlogging and its timing in relation to the developmental stage of the crop are also important. Waterlogging of the entire root zone for a period of 3 to 5 days can be fatal when it occurs during the seedling stage, whereas a well-developed crop is likely to suffer relatively little. Furthermore, a vigorously growing healthy crop is able to withstand waterlogging better than a poor one.

6.24 Drainage May Be Natural or Artificial

Some soils are naturally well drained, that is, excess water is quickly removed by natural processes. For example, water does not stand but runs off rapidly on sloping surfaces. Rainwater, irrigation water or snowmelt water in porous soils underlying pervious subsoils, infiltrates quickly; water in excess of the field capacity is removed to the water table by the gravity. Groundwater table lies well below the root zone for efficient base flow. Removal of excess water by runoff is known as surface drainage, and the removal of gravitational water through the profile is known as internal drainage, both taken together form the natural soil drainage. Excess water accumulates in some other soils due to their topography or physical conditions, for example, in low-lying soils, stiff clay soils, and soils underlying compact impervious layers. They are naturally poorly drained soils requiring improvement of their drainage situation by artificial meas. The systems of improvement of natural drainage are called artificial drainage systems.

Fig. 6.16 Interceptor drainage system (Photo with permission from the Department of Agricultural and Biological Engineering, Pennsylvania University)

6.25 Some Land Needs Artificial Drainage

In some periods of the year in humid regions, there may be excess water on land. This may or may not be harmful depending on the quantity of water, the periods of occurrence, and whether the excess occurs during a critical part of the growing season. Many lands are also naturally well drained, that is, the excess water is removed quickly by percolation or surface runoff. In some regions, rainfall is so distributed that there remains no excess water. But when large quantities of excess water occur for prolonged durations at critical periods, its removal by artificial drainage is needed. Land drainage is applied for the following two quite different purposes:

- Reclamation of naturally waterlogged land for agricultural use
- Improvement of the drainage conditions of existing agricultural land

Drainage for land reclamation and agricultural extension was an act of significance up to the past century. At present, it has become questionable in respect of conservation of biodiversity and ecology of wetlands. About two-thirds of the rainfed cropland is naturally sufficiently drained. Only one-third of rainfed cropland is then in need of improved drainage (Smedema et al. 2000). About 100–150 million ha of rainfed land is currently under artificial drainage. It needs extension in more 250–300 million ha. Currently, the drained area is 25–50 million ha, which is only 10–20% of the total irrigated land area (Smedema et al. 2004).

6.25.1 There Are Many Artificial Drainage Systems

In a drainage system, there is generally a network of drains, either surface or buried, with field drains connected to lateral drains and those attached to main drains to dispose of the water to streams. Drainage systems are generally classified into surface and subsurface drainage systems. Surface drainage is practiced by open ditch drains. There are three general surface drainage types:

- Interceptor drains
- Natural drains
- Regular drains
- Raised beds

6.25.1.1 Interceptor Drains

Usually deep ditch drains are dug to intercept the runoff or seepage water at points of maximum interception. Interceptor drains are generally needed in sloping lands where the groundwater table of the upper slope is elevated and seepage water ponds the depressions in the lower slopes or stagnates the level foot hills (Fig. 6.16).

6.25.1.2 Natural Drains

Surface ditch drains are so aligned that the drains collect water from the ponded depressions of the field. Generally, the arrangement of the natural drains is irregular being concentrated around depressions. Depth of ditches and their spacing depend on amount of water ponded and the slope of the land.

6.25.1.3 Regular Drains

This is a systematic open ditch drainage method where regularly spaced field drains are connected with collector drains which are again joined to main drains (Fig. 6.17). This system is practiced in more or less level lands requiring large amount of water removal. This is an easy to construct but requiring much labor and continuous maintenance system. Since, this system can operate satisfactorily over a long period of time and the skill and cost of installation or maintenance is lower than the subsurface systems, it is the most popular drainage system. Field drains are generally 0.5 m deep and similarly

Fig. 6.17 Regular surface drains (Photo courtesy of TG Drains, Lake Road, Victoria)

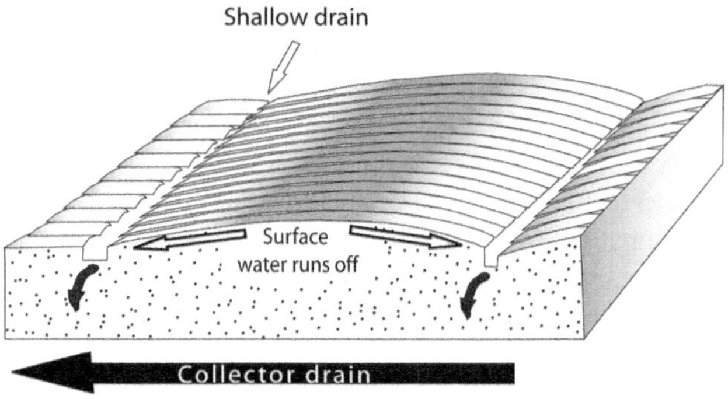

Fig. 6.18 Raised beds

wide, but specifically depth, width, and spacing depend of soil type, crop type, and rainfall characteristics.

6.25.1.4 Raised Beds

Raised beds are used to grow field crops including vegetables like cabbage and cauliflower, and small fruits such as tomato and strawberry.

Raised beds facilitate drainage. Raised beds are high ridges created by turned furrows that are flanked by ditches to carry excess water (Fig. 6.18). Beds are 15–20 cm high with a flat surface. Raised beds encourage wet soils to dry quicker and can help to prevent waterlogging of plant roots

6.25.1.5 Subsurface Drainage Systems

Subsurface drainage is designed to remove excess water from the soil profile. The water table level is controlled through a series of drainage pipes (tile or tubing) that are installed below the soil surface, usually just below the root zone. Subsurface drainage pipes are typically installed at a depth of 75–100 cm and at a spacing of 5–25 m. The subsurface drainage network generally outlets to an open ditch or stream. There are four main types of subsurface drainage systems:

(a) Tile and pipe drains

Fig. 6.19 Pipe drainage (Photo courtesy of USDA-NRCS)

(b) Mole drains
(c) Gravel mole drains
(d) Vertical drains

6.25.1.6 Tile and Pipe Drainage

Subsurface pipe drainage may be used in deep permeable soils that underlie compact impervious layers near the root zone so that the groundwater table is high. This system can also be used to drain naturally poorly drained soils, but pipes need to be installed so closely together. Subsurface pipe drainage is effected with either arranging short clay pipes known as "tiles" or plastic perforated pipes (PE or PVC; Fig. 6.19).

Tiles are expensive and difficult to lay. Tile drainage is presently little used. It has now been replaced by slotted PVC or flexible corrugated plastic pipes. Specifically designed drainage trenchers dig trenches, lay the pipes, and place permeable backfill (washed sand and small diameter gravel to facilitate water movement) into the trench on top of the laid pipe. Backfill is not needed in very permeable soils. The diameter of the pipes, depth of the pipeline, spacing of pipeline, etc., depends on the soil type, crop type, amount of rainfall, amount of water to remove, depth of groundwater table, etc. Generally pipes are placed 1–2 m below the surface at 5–25 m apart.

6.25.1.7 Mole Drains

Mole drains are used in moderately heavy to heavy soils where downward movement of water through the soil profile is slow. Mole drains are channels at suitable depth of soil created by pulling a torpedo-like mole by a mole plow (Fig. 6.20). The mole plow contains a blade to which a torpedo mole is attached to its bottom. The success and longevity of mole drains is dependent on soils. Once a mole channel is formed, it should maintain the channel for many years. Mole drains are not

Fig. 6.20 Mole drainage system (image courtesy of Mr. Jafar Afsar)

suitable in soils with slaking or dispersive clays. They are also not suited to loose soils due to collapse of the mole channels.

6.25.1.8 Gravel Mole Drains

Gravel mole drains are constructed by filling deep unlined channels with small diameter gravels. Gravel mole drains are applicable to situations where subsurface pipes are unsuitable, mole drains are likely to collapse, and in slaking soils not suitable for pipe drains or regular mole drains. Gravel mole drains are more expensive than regular moles due to the amount of backfill and the close spacing required.

6.25.1.9 Vertical Drains

A network of wells is constructed across the fields, and water is pumped out of the wells to lower the groundwater table. This method of subsurface drainage is known as vertical drainage, while the pipe drain or ditch drain is called horizontal drainage. The use of vertical drainage is mostly limited to irrigated land, while the horizontal method is by far the most common method in rainfed land.

Study Questions

1. Explain the relations among infiltration, percolation, and runoff. Why the infiltration capacity of some soils is very low? How can infiltration capacity of soil be improved?
2. What do you mean by available water in soil? What are the factors that affect content of available water in soil? Why there is so much variation in soil moisture content at permanent wilting of plants?
3. Define soil water potential. Discuss the components of soil water potential. How does soil water potential contribute to the movement of water in soil?
4. What are the effects of water stress on plants? Discuss the merits and demerits of early, late, over, and deficit irrigation. Discuss the factors that affect the need of irrigation water of crops.
5. What are the effects of prolonged waterlogging on plants? What are the merits of subsurface drainage systems? What are their demerits? Why are some soils naturally poorly drained?

References

Alam MGM, Allinson G, Stagnatti F, Tanaka A, Westbrooke M (2002) Arsenic contamination in Bangladesh groundwater: a major environmental and social disaster. Int J Environ H Res 12:236–253

Ali MH (2010) Fundamentals of irrigation and on-farm water management, vol 1. Springer, New York

Awad AS (1984) Water quality assessment for irrigation, Advisory Bulletin no. 1, Department of Agriculture NSW

Baver LD, Gardner WH, Gardner WR (1972) Soil physics, 4th edn. Wiley, New York

Bolt GH (1976) Soil physics terminology. Int Soc Soil Sci Bull 49:16–22

Buckingham E (1907) Studies on the movement of soil moisture. Bulletin no. 38. Bureau of Soils. USDA, Washington, DC

Chakraborti D, Rahman MM, Paul K, Chowdhury UK, Sengupta MK, Lodh D, Chanda CR, Saha KC, Mukherjee SC (2002) Arsenic calamity in the Indian subcontinent: what lessons have been learned? Talanta 58:3–22

Darcy H (1856) Les fontaines publiques de la ville de Dijon. Dalmont, Paris

English MJ (1990) Deficit irrigation. I. Analytical framework. J Am Soc Civil Eng 108:91–106

Foth HD (1990) Fundamentals of soil science, 8th edn. Wiley, New York

Fredlund DG, Xing A (1994) Equations for the soil-water characteristic curve. Can Geotech J 31:521–532

Gillham RW, Klute A, Heermann DF (1979) Measurement and numerical simulation of hysteretic flow in a heterogeneous porous medium. Soil Sci Soc Am J 43(6):1061–1067

Heikens A (2006) Arsenic contamination of irrigation water, soil and crops in Bangladesh: risk implications for sustainable agriculture and food safety in Asia. FAO, Bangkok

Hillel D (1980) Fundamentals of soil physics. Academic, New York

Hillel D (1990) Role of irrigation in agricultural systems. In: Stewart BA, Nielsen DR (eds) Irrigation of agricultural crops, Agronomy series no. 30. American Society of Agronomy, Madison

Jaynes DB (1992) Estimating hysteresis in the soil water retention function. In: van Genuchten MT et al (eds) Proceedings of the international workshop on indirect methods for estimating the hydraulic properties of unsaturated soils, University of California, Riverside

Jhonson C (2009) Biology of soil science. Oxford Book Company, Jaipur

Kaluarachchi JJ, Parker JC (1987) Effects of hysteresis with air entrapment on water flow in the unsaturated zone. Water Resour Res 23(10):1967–1976

Kirkham MB (2005) Principles of soil and plant water relations. Elsevier/Academic, Amsterdam

Kirkham D, Powers WL (1972) Advanced soil physics. Wiley-Interscience, New York

Kramer PJ (1983) Water relations of plants. Academic, New York

Kumar D (2005) Breeding for drought resistance. In: Ashraf M, Harris PJC (eds) Abiotic stress: plant resistance through breeding and molecular approaches. Haworth Press, New York

Lal R (1979) Physical properties and moisture retention characteristics of some Nigerian Soils. Geoderma 21:209–223

Lal R, Shukla MK (2004) Principles of soil physics. Marcel Dekker, New York

Liang ZS, Ding ZR, Wang STR (1992) Study on type of water stress adaptation in both *Brassica napus* and *B. juncea* L. species. Acta Botanika 12:38–45

Maqsoud A, Bussière B, Mbonimpa M, Aubertin M (2004) Hysteresis effects on the water retention curve: a comparison between laboratory results and predictive models. In: 57th Canadian geotechnical conference, Geo, Quebec

McLaughlin MJ, Hamon RE, McLaren RG, Speir TW, Rogers SL (2000) Review: a bioavailability-based rationale for controlling metal and metalloid contamination of agricultural land in Australia and New Zealand. Aust J Soil Res 38:1037–1086

Miyazaki T (2006) Water flow in soils, 2nd edn. CRC Press/Taylor & Francis, Boca Raton

Neumann PM (2008) Coping mechanisms for crop plants in drought-prone environments. Ann Bot 101:901–907

Nobel PS (1991) Physicochemical and environmental plant physiology. Academic, San Diego

O'Kane JP, Pokrovskii A, Flynn D (2004) The fest model for testing the importance of hysteresis in hydrology. In: Proceedings of EGU conference, vol 6, 07303 Nice, France

Parry MAJ, Flexas J, Medrano H (2005) Prospects for crop production under drought: research priorities and future directions. Ann Appl Biol 147:211–226

Radcliffe DE, Rasmussen TC (2002) Soil water movement. In: Warrick AW (ed) Soil physics companion. CRC Press, Boca Raton

Reynolds MP, Mujeeb-Kazi A, Sawkins M (2005) Prospects for utilizing plant-adaptive mechanisms to improve wheat and other crops in drought and salinity-prone environments. Ann Appl Biol 146:239–259

Richards LA (1954) Diagnosis and improvement of saline and alkali soils. United States Department of Agriculture handbook no. 60. USDA, Washington, DC

Rowe DR, Abdel-Magid IM (1995) Handbook of wastewater reclamation and reuse. CRC Press, Boca Raton

Royer JM, Vachaud G (1975) Field determination of hysteresis in soil-water characteristics. Soil Sci Soc Am J 39(2):221–223

Salisbury FB, Ross CW (1978) Plant physiology, 2nd edn. Wadsworth Pub, Belmont

Simunek J, Kodesova R, Gribb MM, van Genuchten MT (1999) Estimating hysteresis in the soil retention function from cone permeameter experiments. Water Resour Res 35(5):1329–1345

Smedema LK, Abdel-Dayem S, Ochs WJ (2000) Drainage and agricultural development. Irr Drain Syst 14:223–235

Smedema LK, Vlotman WF, Rycroft DW (2004) Modern land drainage: planning, design and management of agricultural drainage systems, 2nd edn. A A Balkema Publishers, London

Soil Science Society of America (1997) Glossary of soil science terms. Soil Science Society of America, Madison

Southorn N (1997) Farm irrigation: planning and management. Inkata Press, Melbourne

Shrestha RR, Shrestha MP, Upadhyay NP, Pradhan R, Khadka R, Maskey A, Maharjan M, Tuladhar S, Dahal BM, Shrestha K (2003) Groundwater arsenic contamination, its health impact and mitigation program in Nepal. J Environ Sci Health A38:185–200

Taylor SA (1952) Use of mean soil moisture tension to evaluate the effect of soil moisture on crop yields. Soil Sci 74:217–226

Vachaud G, Thony JL (1971) Hysteresis during infiltration and redistribution in a soil column at different initial water contents. Water Resour Res 7(1):111–127

Warrick AW (2002) Soil physics companion. CRC Press, Boca Raton

Williams PJ (1982) The surface of the earth, an introduction to geotechnical science. Longman Inc., New York

Xia Y, Liu J (2004) An overview on chronic arsenism via drinking water in PR China. Toxicology 198:25–29

Soil Organic Matter

7

Soil organic matter includes plant and animal residues at various stages of decomposition, ranging from fresh undecomposed materials through partially decomposed and short-lived products of decomposition to well-decomposed humus. Plant and animal residues contain sugars, celluloses, hemicelluloses, proteins, lignins, waxes, and lipids. When these materials are incorporated into soils, sugars, celluloses, and hemicelluloses, most proteins undergo rapid decomposition. With decomposition, the proportion of resistant compounds such as lignin gradually increases. Resistant decomposition products and newly synthesized substances accumulate in soil and form complexes with its mineral components. Humus is formed in the latter stage as a result of such continued decomposition, resynthesis, and complexation. Humus is a fairly stable substance and requires very long time, for example, centuries to millennia, to be completely decomposed. Most mineral soils contain less than 5% by weight of soil organic matter (SOM), but this little quantity of SOM has tremendous biological significance. It provides food for microorganisms, stores nutrients, retains water, acts as mulch, and performs as a soil conditioner and aggregating agent. It makes the soil friable and fertile. Soil organic matter contributes significantly to the sequestration of carbon.

7.1 Soil Organic Matter Is a Vital Component of Soil

Soil organic matter is an essential component of the soil; it affects the state and functioning of other soil components too. Loose materials on earth that do not contain organic matter are not soils. Some soils contain high organic matter, even more than 80% by weight; they are organic soils—the Histosols (Chap. 4). Most mineral soils contain <5% by weight SOM (Stanley 2000). Tropical soils (Oxisols and Ultisols) are known to have low organic matter contents, but Juo and Franzluebbers (2003) reported an average SOM content of about 2%.

By the phrase "soil organic matter," we may perceive the whole range of living and dead organic materials present in the soil. Living organic materials are the plant roots and other soil biota—flora and fauna. Dead materials include plant and animal residues and their intermediate decomposition products. As a component of the soil and as chemical reactants, these dead materials are generally considered as soil organic matter. Surface litter (unless incorporated within the soil) is generally not included as part of the soil organic matter (Juma 1999).

7.2 There Are Three Categories of Soil Organic Matter

Three categories of soil organic matter are recognized on the basis of the stages of decomposition. The categories are fresh or undecomposed, partially decomposed, and fully decomposed organic matter.

Fresh or Undecomposed Organic Matter: Fresh soil organic matter includes freshly fallen leaves, twigs, branches, flowers, fruits, and dead animal tissue if they have been incorporated into the soil. These materials begin to decompose immediately after their incorporation. It becomes often very difficult to separate decomposed and undecomposed organic matter. For all practical purposes, fresh organic matter includes materials of which the origin and structure can be identified. It comprises about 1–10% of total SOM (Lickacz and Penny 2001).

Partially Decomposed Organic Matter: It includes organic residues in soil that have undergone decomposition to a considerable extent. The structure and origin of the materials may or may not be identified. It consists of gradually decomposing detritus and partially broken down plant and animal tissues. It comprises about 10–40% of total soil organic matter.

Fresh, undecomposed and partially decomposed organic matter together constitute the "active soil organic matter." They are so called because they are easily attacked by soil

K.T. Osman, *Soils: Principles, Properties and Management*,
DOI 10.1007/978-94-007-5663-2_7, © Springer Science+Business Media Dordrecht 2013

organisms, they are decomposed very fast, and they provide food and energy to heterotrophic soil organisms.

Fully Decomposed Organic Matter or Humus: Humus is said to be "completely decomposed organic matter." It is not actually "completely" decomposed in the sense that it is still organic, and complete decomposition of organic matter produces CO_2 and water with some bases. It is "fully" decomposed in the sense that it has undergone decomposition for a considerable time, say some hundred or thousand years, and that further decomposition proceeds slow. Humus may be defined as a brown to black, amorphous, colloidal organic matter that has undergone decomposition to such an extent that further decomposition is very slow. Humus is composed of stable natural organic complexes formed by the products of biological decomposition and resynthesis. Humus is fairly resistant to decomposition due to its physical state and chemical nature. Humus is present as coating on soil particles, as gum within aggregates, and as organomineral complexes, which make them inaccessible to decomposing organisms. Again, humus contains refractory substances such as lignin and polyphenols. Because of these refractory substances, humus is called passive soil organic matter. It comprises about 40–60% of total SOM.

7.3 Composition of Soil Organic Matter Is Variable

Soil organic matter is a highly variable entity because different organic compounds are found in different plant and animal residues and different organic compounds predominate at different stages of decomposition. SOM includes materials that are added in plant and animal residues, those that are formed during decomposition, substances synthesized by microorganisms, and compounds in root exudates and microbial secretions.

Plant residues contain 15–40% cellulose, 10–43% hemicellulose, 25–40% lignin, 1–10% waxes and lipids, and 1–15% proteins (Schulze et al. 2005). After addition, microbial decomposition proceeds very fast in the initial stage and very slowly in the latter stages. Susceptible or "labile" substances such as sugars (Cheshire 1979), amino acids (Stevenson 1982), other organic acids (Stevenson 1967), and celluloses are quickly lost, and "stable" or resistant substances such as lignin increase in proportion. Some carbohydrates and proteins are newly synthesized by microorganisms for their body materials. These substances are also added to soils after their death. Microbial polysaccharides and proteins form complexes with lignin and other compounds rich in polyphenols and accumulate in humus (Coleman et al. 2004).

Some carbohydrates may be detrital, in that they originally entered the soil in plant debris, but most are thought to be products of microbial synthesis in situ. Hydrolyzed products of soil polysaccharides are pentoses, hexoses, deoxyhexoses, hexosamines, and uronic acids (Jenkinson 1988). There is a considerable amount of amino acid polymers, such as peptides, and a small quantity of amino sugars and free amino acids. Peptides and polypeptides generally remain linked with other organic fractions and with clay and polyvalent cations. They are stable substances which predominate with lignins in humus. Humic substances are defined as condensed polymers of aromatic and aliphatic compounds produced by decomposition of plant and animal residues and by microbial synthesis. They are amorphous, dark-colored, and hydrophilic, with a wide range in molecular weight from a few hundreds to several thousands (Evangelou 1998). Three fractions of materials may be obtained during acid–alkali fractionation of humus. They are humic acid, fulvic acid, and humin (Kononova 1966). But these fractions do not represent any discrete group of compounds; they are actually products obtained in different steps of the fractionation procedure. These fractions, however, provide a convenient way of describing the composition of humus. Chemical characteristics of these fractions are mentioned below.

1. Humic acid—soluble in dilute alkali, but precipitates in acid solution
2. Fulvic acid—soluble both in alkali and acid solutions
3. Humin—insoluble both in alkali and acid solutions

Generally, humic and fulvic acids have a similar structure, but they differ in molecular weight and elemental and functional group contents. Fulvic acid is lower in molecular weight and contains more oxygen-containing functional groups and elemental oxygen, but less nitrogen and carbon per unit weight than humic acid (Schnitzer and Khan 1972; Sposito et al. 1976; Schnitzer 1991). According to Aswathanarayana (1999), humic acid contains 560, 47, and 355 g kg^{-1} C, H, and O, respectively, while fulvic acid contains 457, 54, and 448 g kg^{-1}, respectively.

7.4 Many Factors Affect Soil Organic Matter Content

Many environmental, edaphic, and management factors affect soil organic matter content. Climate affects soil organic matter content on a global scale. Organic matter content increases with increasing rainfall up to a certain level beyond which it decreases. This may be attributed to the higher biomass accretion due to higher moisture supply. After a point, eluviation of organic matter increases. On the other hand, decomposition of organic matter increases as the temperature increases. The rate of decomposition doubles for every 8 or 9°C increase in mean annual temperature (Anonymous 2010). So, soil organic matter content decreases from lower to higher temperature regions. Soil texture, structure, water, air, etc., affect the organic matter content. Higher amount of

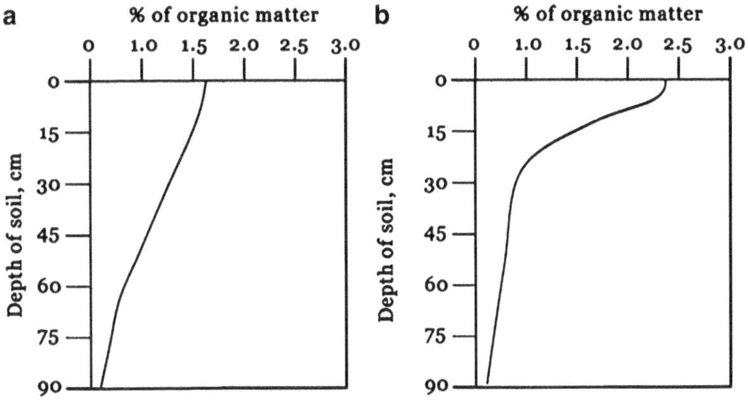

Fig. 7.1 Variation in SOM with depth in an agricultural (**a**) and in a forest soil (**b**)

biomass is produced in fine-textured fertile soils. Higher biomass adds larger amounts of organic matter. Decomposition of organic matter is faster in moist arable soils. Under saturated conditions, there is a deficiency of O_2 which reduces organic matter decomposition with a net result of accumulation of soil organic matter.

Soil organic matter content is the highest in the surface soil because organic inputs are generally the highest in the surface. It gradually decreases with depth in agricultural and grassland soils and abruptly, with some exceptions, in forest soils (Fig. 7.1). In some "buried" soils, subsoil may contain a higher organic matter content. Some soils as Spodosols contain illuvial humus accumulation in the B horizon.

Vegetation—natural or managed—has a profound bearing on the SOM level because different vegetation produces different amount of biomass and of varying decomposability. Cultivation reduces soil organic matter content. Forest and grassland soils are particularly rich in organic matter. When they are brought under agriculture, their organic matter is gradually reduced by enhanced decomposition. Tillage generally favors organic matter breakdown. However, manures and composts, fallowing, crop rotations, green manure, etc., may improve SOM levels, at least temporarily. Soils at the bottom of slopes generally have higher organic matter because these areas are generally wetter and better in fertility. Some organic matter is lost by runoff from soils of upper slopes.

7.5 Soil Organic Matter Performs Many Physical, Chemical, and Biological Functions

Organic matter performs many physical, chemical, and biological functions in soil. Physical functions mainly include enhancement of aggregation, aeration, water movement, and reduction of evaporation and thermal conductivity. Higher organic matter content gives a darker color. Organic residues accumulated on surface soil may act as mulch (Chap. 11). Organic matter improves soil structure and porosity (Chap. 5), infiltration, water-holding capacity, and drainage (Chap. 6). Organic matter reduces runoff and erosion (Chap. 12). Humus can hold the equivalent of 80–90% of its weight in moisture and therefore increases the soil's capacity to withstand drought conditions (Olness and Archer 2005).

Chemical functions include enhancement of ion exchange, buffering, pedochemical weathering, chelation, and translocation of substances within the soil. Humus colloids have dissociable $-NH_2$, –OH, and –COOH groups. Therefore, they are electrically charged particles with high surface area. Among various soil colloids, humus has the highest number of negative charges on its surface. Humus has, therefore, very high cation exchange capacity (Chap. 8). Humus colloids also contain some positive charges against which anions are adsorbed and exchanged. Several organic compounds in humus, particularly those having active groups like $-NH_2$, –OH, and –COOH, act in buffering which is an important property of the soil. Because of this property, soils do not suffer from sudden change in pH. Chelates are formed by complexation of humic substances with metals like Al, Fe, Mn, Zn, Cu, and some other heavy metals like Cd and Pb. The availability, absorption, redistribution, and leaching of these metals are regulated by chelation. Organic matter also forms complexes with pesticide residues and controls their mobility to water bodies, both surface and groundwater.

Biological functions include food for soil organisms and a storehouse of nutrients. When fresh organic residues are added to soils, they are attacked by the detritivores and the heterotrophic soil microorganisms. They utilize organic substances as their carbon and energy sources. However, the food quality of the residues depends on their chemical makeup. Some residues are high in carbohydrates and proteins; they are very rapidly consumed. Some others are rich in lignins; they are slowly degraded. Much of the plant nutrients are returned to the soil in the form of organic residues. Upon decomposition, these nutrients are released again to the available nutrient pool of the soil. Nutrients are recycled in this way through the soil organic matter (Chap. 14). Organic matter provides more than 90% nitrogen and about

50–60% phosphorus and sulfur. It releases nutrients at a slow pace to the growing roots, and it holds nutrients on their surfaces so that loss of nutrients from a soil rich in organic matter is minimum.

7.6 Humification Is a Complex Process of Decomposition and Resynthesis

Detritivorous soil animal and heterotrophic microorganisms such as fungi and many bacteria invade the plant and animal residues at the initial stage of soil organic matter breakdown. They fragment these materials and make them accessible to saprophytes. Organic compounds are released at different steps by breaking the tissue structures. The conversion of complex organic compounds into simpler organic compounds by biological processes is known as decomposition. Organic compounds added to soils undergo a series of decomposition reactions. For example, polysaccharides are first decomposed to their monomers (sugars) and then to CO_2 and water by aerobic decomposition. Complete decomposition of all organic compounds leads to the production of inorganic substances such as carbon dioxide, water, phosphates, sulfates, and bases. It is known as mineralization. Conversion of nitrogenous materials such as amino acids and proteins into ammonia, carbon dioxide, and water is also a mineralization process, but it is popularly known as ammonification. As earlier mentioned, plant and animal residues contain many different classes of organic compounds which undergo decomposition at variable rates and produce a variety of products depending on the stage of decomposition and condition of decomposition (Coleman et al. 2004). Decomposability of a particular compound may vary considerably depending on soil moisture and nutrients, microbial population, and climate. Generally, decomposability follows the following order:

Sugars	Very rapidly decomposed
Starches	↑
Simple proteins	
Crude proteins	
Hemicelluloses	
Cellulose	
Fats	
Waxes	
Resins	↓
Lignins	Very slowly decomposed

Carbohydrates such as sugars, starches, celluloses, and hemicelluloses produce CO_2 and water when they are decomposed in the presence of O_2. Under anaerobic conditions, intermediate products like CH_4, alcohols, and organic acids are produced. Under further oxidation, they also produce CO_2 and water. Cellulose is decomposed by many species of fungi and bacteria. The enzyme cellulase splits this long chain polymer of glucose to cellobiose which in turn is split to glucose by the enzyme β-glucosidase. Hemicelluloses are polymers of simple sugars such as pentoses, hexoses, and uronic acid. Their complete oxidation also yields CO_2 and H_2O.

$$\text{Cellulose} \rightarrow \text{Cellobiose} \rightarrow \text{Glucose} \rightarrow CO_2, H_2O$$

Degradation of hemicellulose, lignin, and pectin follows similar pathways.

Bacteria of the genera *Achromobacter, Bacillus, Cellulomonas, Cellvibrio, Clostridium, Cytophaga, Vibrio, Pseudomonas, Sporocytophaga, Erwinia, Vibrio, and Lactobacillus*; fungi of the genera *Aspergillus, Chaeto mium, Fusarium, Penicillium, Rhizoctonia, Rhizopus, Trichoderma, and Verticillium*; and actinomycetes of the genera *Micromonospora, Nocardia, Streptomyces, and Thermomonospora* are responsible for the decomposition of celluloses and hemicelluloses.

Protein is first degraded to peptides by the enzymes proteases and then into amino acids by the enzymes peptidases. These enzymes are produced by many species of *Clostridium, Proteus, Pseudomonas*, and *Bacillus*. Further decomposition yields NH_3, NO_3^-, CO_2, and water. This process, ammonification as mentioned earlier, occurs as a result of hydrolytic and oxidative enzymatic reaction under aerobic conditions by heterotrophic microbes.

$$\text{Protein} \xrightarrow[\text{Proteases}]{} \text{peptides} \xrightarrow[\text{peptidases}]{} \text{amino acids}$$
$$\rightarrow NH_3, NO_3^-, CO_2, H_2O, SO_4^{2-}, \text{etc.}$$

The final stage of decomposition is characterized by gradual decomposition of more resistant organic compounds such as lignin for which actinomycetes and fungi play a major role (Stevenson 1985). Lignin is converted to phenylpropanes, polyphenols, and quinines. Some of the original lignin is left in soil. Bacteria of the genera *Flavobacterium, Pseudomonas, and Micrococcus*; fungi of the genera *Humicola, Fusarium Fames, Penicillium, Aspergillus, and Ganoderma*; and actinomycetes of the genera *Streptomyces, Nocardia*, etc., decompose lignin.

During decomposition of organic compounds, many intermediate products of decomposition may be encountered in soil at any given time. Soil organisms resynthesize some cellulose, amino acids, and proteins which are also added to soils after their death. Microbial cellulose and protein are believed to be more resistant to decomposition than those present in the original plant residues. Residual cellulose, lignin, and their derivatives, together with resynthesized compounds, form complexes with inorganic materials of the soil such as bases, metals, nutrients, and colloids to produce humus (Fig. 7.2). Formation of humus from organic

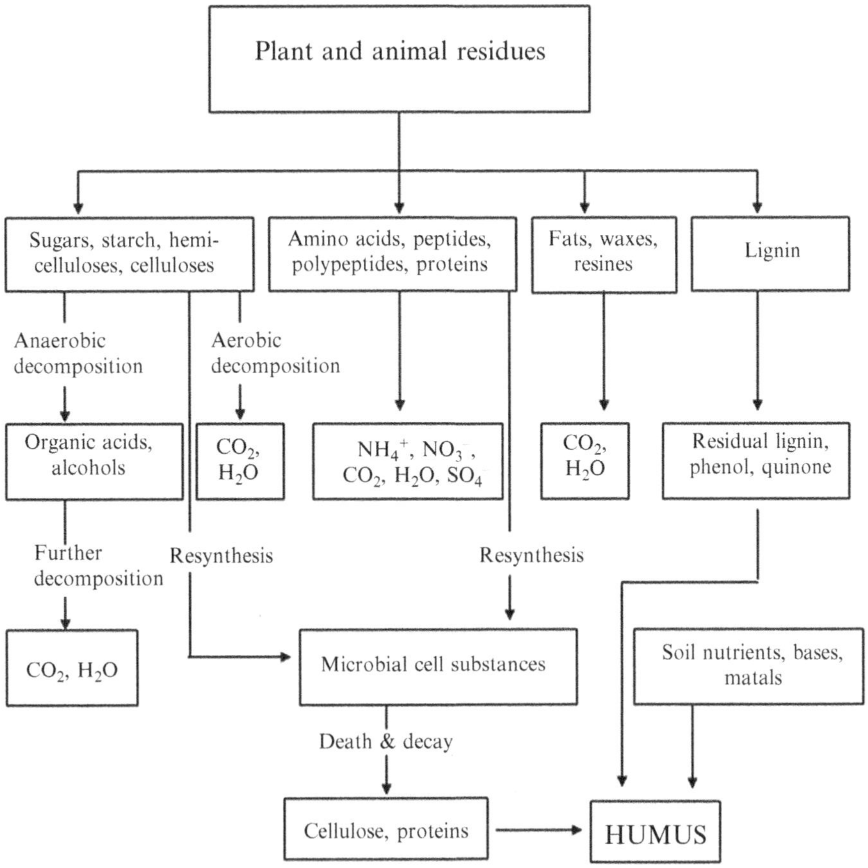

Fig. 7.2 Pathway of humification

residues in soil is called humification. Humus is more complex than original plant and animal residues, and humification is the combination of decomposition, resynthesis, and complexation. Humus contains approximately 45% lignin; 35% amino acids; 4% cellulose; 7% hemicellulose; 3% fats, waxes, and resins; and 6% other miscellaneous substances.

7.7 Managing Soil Organic Matter Is Necessary for Sustainable Soil Fertility

Decline in soil organic matter causes degradation of soil structure, reduction in porosity, water-holding capacity, and aeration and makes the soil compact and erodible. So, soil organic matter content should be maintained at a level necessary for favorable tilth and fertility. Organic matter status of most virgin soils (undisturbed, uncultivated) usually remains at a steady level for a long period. Cultivation opens the soils and oxidizes organic matter. Continuous cropping for a long time seriously declines SOM. Only a small fraction of crop residues added to soil remains as soil organic matter. So, addition of plenty of organic residues is needed for sustainable soil fertility.

As a general rule, mixing of residues and aeration of the soil associated with tillage increase organic matter decomposition and decrease soil organic matter content. However, conservation tillage methods (zero tillage, minimum tillage, mulch tillage) can be practiced for gradual improvement of SOM. Forage crops in rotations with cereals and oilseeds are beneficial for conserving SOM. Several long-term crop rotation studies have shown that crop rotations involving perennial forages tend to stabilize SOM. Inclusion of legumes further improves SOM. Greater proportion of the aerial biomass of the current crop should be left in field and incorporated within the soil. Huge amount of straw and stover are removed after harvest from corn, soybean, and other crop fields for use as fuel or livestock food. These materials may alternatively be composted and applied to fields as fertilizer and as conditioners. Manures, composts, and other organic supplements such as plant and animal residues, straw, sawdust, and oil cakes may be added to soil to supply plant nutrients at a slow rate to growing crops and improve SOM status. However, it may be difficult to collect huge amount of residues needed (about 10–20 t ha^{-1}). On farm composting from current crop residues, weeds, prunings and farm animal excreta may be a good practice. Inclusion of a legume like sun hemp, black gram, and cowpea in the cropping sequence for a short period and plowing it down at before maturity is a good practice to supply N to the successive crop and add organic matter to the soil.

7.8 Carbon–Nitrogen Ratio Is an Important Index of SOM Decomposition

Carbon–nitrogen ratio is obtained by dividing the mass of carbon by the mass of nitrogen in plant tissues, plant and animal residues, or in soil. C/N ratio in soil is the ratio of its organic carbon to total nitrogen contents. This ratio has a significant bearing on organic matter decomposition, mineralization, ammonification, and nitrification in soil.

Most fresh plant materials contain about 40% carbon. If it has 1% nitrogen, the C/N ratio becomes 40:1. But, nitrogen content in plant residues varies from 0.5 to 1.5%. So, the C/N ratio varies among different plant materials.

The rate of decomposition of organic matter and the amount of humus formed are related to the C/N ratio of the residue. Organic residues of narrower C/N ratio are decomposed at a faster rate because nitrogen is utilized by decomposing microorganisms. At the initial stage of decomposition, there is a rapid disappearance of carbon in comparison to nitrogen (Olson 1963). Therefore, the C/N ratio starts declining with the advancement of decomposition. At a point when the later slow stage of decomposition is reached, the C/N ratio is stabilized around 10:1 to 15:1. Further decomposition of organic residue and further decrease in C/N ratio are extremely slow. Therefore, C/N in arable soil varies within a very narrow range, usually between 10:1 and 12:1. A higher C/N ratio in soil denotes the presence of a greater amount of undecomposed to partially decomposed organic matter. On the other hand, a smaller C/N ratio than 10:1 usually indicates greater nitrification in soil. The C/N ratio values of different organic residues obtained from different sources are given in Table 7.1.

Table 7.1 Average C/N ratio values of some organic residues

Residues	C/N ratio
Corn silage	35–45
Alfalfa	15
Cottonseed meal	7
Soybean meal	4–6
Food wastes	14–16
Legume hay	15–19
Pig manure	10–20
Poultry manure	5–15
Plant leaves	40–80
Wheat straw	100–150
Compost	15–20
Sawdust	170
Sewage sludge	5–16
Cow manure	10–30
Vegetable wastes	11–19

Source: http://www.norganics.com/applications/cnratio.pdf

7.9 SOM Fractions Represent Different Ages and Rates of Turnover

Soil organic matter is a heterogeneous mixture of diverse organic materials in various stages of decomposition. It can be separated into different fractions by physical and chemical fractionation procedures. These fractions are, however, not actual pure organic compounds but are often arbitrary mixtures separated by the procedures adopted. There are light fractions which are free and not complexed with mineral matter. These fractions currently undergo decomposition. There are fractions of soil organic matter stabilized in clay or silt sized organomineral complexes. There are metal chelates and recalcitrant organic compounds (Post and Kwon 2000). These fractions need much time to decompose.

The heterogeneity of different organic components in soil is reflected by their highly variable radiocarbon ages. A bulk SOM ^{14}C analysis only represents the mean ^{14}C concentration of the soil sample and does not depict the ages of the various fractions of soil organic matter. A reliable estimate of the age of soil carbon may be obtained if purified fractionation can be achieved. Rethemeyer et al. (2004) observed from a ^{14}C analysis that the black particles isolated from the plow horizon of an agricultural field in Germany represented from 12,350 to 21,360 years BP (before present). The oldest and most recalcitrant C pools dominate estimates by radiocarbon dating because of the long-term time frame (200–40,000 year) that this method measures (Goh 1991). For some sites under boreal, temperate, and tropical forests, the average age of bulk carbon in detrital and Oh/A-horizon organic carbon ranges from 200 to 1,200 years. In each case, this mass-weighted average includes components such as relatively undecomposed leaf, root, and moss litter with much shorter turnover times, and humified or mineral-associated organic matter with much longer turnover times. Recent analytical and experimental advances have demonstrated that molecular structure alone does not control soil organic matter stability: In fact, environmental and biological controls predominate (Schmidt et al. 2011).

The turnover rate of the different soil organic carbon compounds varies due to the complex interactions between biological, chemical, and physical processes in soil. However, there may be a continuum of soil organic carbon compounds in terms of their decomposability and turnover time (Post and Kwon 2000). The rate of turnover of an element in a pool may be predicted from the balance between its inputs (I) and outputs (O). It is generally indicated by the mean residence time (MRT) and the half-life $(T_{1/2})$. The MRT is defined as the average time the element resides in the pool at steady state. It may again be taken as the average time required to completely renew the content of the pool at steady state. The half-life of SOM is the time required for half of the currently existing stock to decompose. The different C pools existing in the soil have different mean residence times, ranging from 1 year to a few years depending on the biochemical composition (lignin, e.g., is more stable than cellulose) to decades or more than 1,000 years (stable fraction). Some of the stable carbon fraction is physically protected by encapsulation of OM fragments by clay particles or soil macro- or microaggregates (Balesdent et al. 2000). Chemical protection refers to specific bonds of OM with other soil constituents (colloids or clays), but most often this concerns very stable organic compounds (FAO 2001).

The turnover of SOM may be expressed by the first-order model assuming constant zero-order input with constant proportional mass loss per unit time (Jenny 1980) as

$$\frac{\partial S}{\partial t} = I - kS,$$

where S is the organic matter stock in soil, t is the time, k is the decomposition rate, and kS is equivalent to output O. Assuming equilibrium $(I = 0)$, the MRT can then be calculated as $\mathrm{MRT} = I / k$ and $T_{1/2}$ as $\mathrm{MRT} = T_{1/2} / \ln 2$ (Six and Jastrow 2002).

Actually, soil organic matter represents heterogeneous mixtures containing organic materials in all stages of decay and a variety of decomposition products of different ages and levels of complexity. Thus, the turnover of these components varies continuously, and any estimate of MRT for SOM as a whole merely represents an overall average value (Goh 1991).

Pools of SOM have characteristic turnover times of less than a year, years to decades, and over several hundred years (Parton et al. 1987). Most of the roughly 80–160 Pg C in surface detritus (Matthews 1997) and an estimated 200–300 Pg C in SOM are in forms that accumulate and decompose on time scales of a century or less (Schimel 1995; Potter and Klooster 1997). The remainder, constituting the majority of carbon stored in mineral soils, is stable on time scales of centuries to millennia (Trumbore 2000). Imagine, plants and animals that contributed materials for humus of our soils lived millions of years ago.

7.10 Soil Organic Matter Contributes to Carbon Sequestration

Soil carbon sequestration is the process of transferring carbon dioxide from the atmosphere into the soil through crop residues and other organic solids and in a form that is not immediately reemitted. It represents long-term storage of carbon in soil. This transfer or sequestering of carbon helps offset emissions from fossil fuel combustion and other carbon-emitting activities while enhancing soil quality and long-term agronomic productivity. Soil carbon sequestration can be an effective option of mitigating CO_2 emission that combines with environmental conservation and soil fertility improvement (Smith et al. 2007).

Globally, soil stores approximately 1,500 Pg organic C and an additional 900–1,700 Pg inorganic C in the surface of 1 m soil. The atmosphere contains ~750 Pg C as carbon dioxide (Eswaran et al. 1993). Soil has a carbon stock three times that of the atmosphere. Thus, soil organic carbon (SOC) can play a significant role in mitigating greenhouse gas emissions. Soil C sequestration implies enhancing the concentration/pools of carbon through land-use conversion and adoption of recommended management practices in agricultural, pastoral, and forestry ecosystems and restoration of degraded and drastically disturbed soils. The SOC sequestration involves putting C into the surface 0.5–1 m depth through the natural processes of humification (Lal 2004). A wide range of soil and water management practices can be adopted to sequester atmospheric CO_2 in terrestrial ecosystems. The technical potential of C sequestration in terrestrial ecosystems is estimated at 5.7–10.1 Pg C year^{-1}. This includes carbon sequestration by vegetation and soils. Restoration of degraded and desertified soils is an important mitigation strategy because of its large technical potential for sequestering 1–2 Pg C year^{-1} (Lal 2006).

Study Questions

1. Mention the chemical composition of soil humus. What do you mean by labile and stable soil organic matter? Why humus is stable in soil?
2. How do you distinguish between humification and mineralization? What are the characteristics of different fractions of humus? These are actually arbitrary fractions—why? Mention chemical composition of humic and fulvic acid.
3. Discuss factors that affect SOM content. What is the significance of SOM? Write sources of organic matter in soil.
4. Explain that humification is a microbial process. Mention some microorganisms involved in humification.
5. How older is the humus in a soil? Why soil is an important C sequester? What is the significance of C/N ratio of an organic amendment?

References

Anonymous (2010) http://www.dpipwe.tas.gov.au/inter.nsf/WebPages/TPRY-YW6YZ?open. Accessed 12 Sept 2011

Aswathanarayana U (1999) Soil resources and the environment. Oxford/IBH Publishing Co Pvt Ltd., New Delhi

Balesdent J, Chenu C, Balabane M (2000) Relationship of soil organic matter dynamics to physical protection and tillage. Soil Tillage Res 53:215–230
Cheshire MW (1979) Nature and origin of carbohydrates in soil. Elsevier, London
Coleman DC, Crossley DA Jr, Hendrix PF (2004) Fundamentals of soil ecology, 2nd edn. Elsevier-Academic, San Diego
Eswaran H, Vandenberg E, Reich P (1993) Organic carbon in soils of the world. Soil Sci Soc Am J 57:192–194
Evangelou VP (1998) Environmental soil and water chemistry: principles and applications. Wiley, New York
FAO (2001) Soil carbon sequestration for improved land management. Food and Agriculture Organization of the United Nations, Rome
Goh KM (1991) Carbon dating. In: Coleman DC, Fry B (eds) Carbon isotope techniques. Academic, San Diego http://www.norganics.com/applications/cnratio.pdf. Accessed 5 Oct 2011
Jenkinson DS (1988) Soil organic matter and its dynamics. In: Wild A (ed) Russel's soil condition and plant growth, 11th edn. English Language Book Society, London
Jenny H (1980) The soil resource—origin and behavior. Springer, New York
Juma NG (1999) Introduction to soil science and soil resources, vol I, The pedosphere and its dynamics: a systems approach to soil science. Salman Productions, Sherwood Park
Juo ASR, Franzluebbers K (2003) Tropical soils: properties and management for sustainable agriculture. Oxford University Press, New York
Kononova MM (1966) Soil organic matter. Pergamon, Oxford
Lal R (2004) Soil carbon sequestration to mitigate climate change. Geoderma 123:1–22
Lal R (2006) Enhancing crop yield in developing countries through restoration of soil organic carbon pool in agricultural lands. Land Degrad Develop 17:197–209
Lickacz J, Penny D (2001) Soil organic matter. Alberta Agriculture and Food Agdex 536-1 (Online). http://www1.agric.gov.ab.ca/$department/deptdocs.nsf/all/agdex890?opendocument. Accessed 3 Sept 2011
Matthews E (1997) Global litter production, pools, and turnover times: estimates from measurement data and regression models. J Geophys Res 102:18771–18800
Olness A, Archer D (2005) Effect of organic carbon on available water in soil. Soil Sci 170:90–101
Olson JS (1963) Energy storage and the balance of producers and decomposers in ecological systems. Ecology 44:322–331
Parton WJ, Schimel DS, Cole CV, Ojima DS (1987) Analysis of factors controlling soil organic matter levels in Great Plain grasslands. Soil Sci Soc Am J 51:1173–1179
Post WM, Kwon KC (2000) Soil carbon sequestration and land-use change: processes and potential. Glob Chang Biol 6:317–328
Potter CS, Klooster SA (1997) Global model estimates of carbon and nitrogen storage in litter and soil pools—response to changes in vegetation quality and biomass allocation. Tellus B Chem Phys Meteorol 49B:1–17
Rethemeyer J, Grootes PM, Bruhn F, Andersen N, Nadeau MJ, Kramer C, Gleixner G (2004) Age heterogeneity of soil organic matter. Nuc Instrum Meth Phys Res B 223–224:521–527
Schimel DS (1995) Terrestrial ecosystems and the carbon cycle. Glob Chang Biol 1:77–91
Schmidt MWI, Torn MS, Abiven S, Dittmar T, Guggenberger G, Janssens IA, Kleber M, Knabner IK, Lehmann J, Manning DAC, Nannipieri P, Rasse DP, Weiner S, Trumbore SE (2011) Persistence of soil organic matter as an ecosystem property. Nature 478:49–56
Schnitzer M (1991) Soil organic matter-the next 75 years. Soil Sci 151:41
Schnitzer M, Khan SU (1972) Humic substances in the environment. Marcel Dekker, New York, 327 pp
Schulze WX, Gleixner G, Kaiser K, Guggenberger G, Mann M, Schulze E-D (2005) A proteomic fingerprint of dissolved organic carbon and of soil particles. Oecologia 142:335–343
Six J, Jastrow JD (2002) Organic matter turnover. Encyclopedia of soil science. Marcel Dekker, New York
Smith P, Martino D, Cai Z, Gwary D, Janzen H, Kumar P, McCarl B, Ogle S, O'Mara F, Rice C, Scholes B, Sirotenko O (2007) Agriculture. In: Metz B, Davidson OR, Bosch PR, Dave R, Meyer LA (eds) Climate change 2007 mitigation. Contribution of working group III to the fourth assessment report of the intergovernmental panel on climate change. Cambridge University Press, Cambridge
Sposito GK, Holtzclaw M, Baham J (1976) Analytical properties of the soluble, metal complexing fractions in sludge-soil mixtures. II. Comparative structural chemistry of the fulvic acid. Soil Sci Soc Am 40:691
Stanley ME (2000) Soil environmental chemistry. In: Environmental chemistry. CRC Press LLC, Boca Raton
Stevenson FJ (1967) In: McLaren AD, Peterson GH (eds) Soil biochemistry, vol 1. Marcel Dekker, New York
Stevenson FJ (1982) Nitrogen in agricultural soils. Agron Monogr no 22:67
Stevenson FJ (1985) Cycles of soil: carbon, nitrogen, phosphorus, sulfur, micronutrients. Wiley Inter Science Publications, New York
Trumbore S (2000) Age of soil organic matter and soil respiration: radiocarbon constraints on belowground C dynamics. Ecol Appl 10(2):399–411

8 Chemical Properties of Soil

The soil is a chemical entity. All the materials there are chemical substances. Soils are composed of solid, liquid, and gas; soluble and insoluble; and organic as well as inorganic substances. There are ions and compounds, salts, acids, bases, minerals, and rock fragments. There are also colloids which are very active chemically. They are very fine soil particles, including humus, fine silicate clays, and oxides and hydroxides of iron and aluminum. Colloids carry electrochemical charges, both positive and negative, and these charges develop by substitution in the lattice structure and dissociation of active groups. These charges hold oppositely charged ions which are replaceable by ions of similar charges in the soil solution. Along with ion exchange properties, two other important indices of soil chemical environment are pH and E_h. Soil pH is the negative logarithm of hydrogen ion concentration in soil suspensions. It denotes the degree of acidity and alkalinity and influences solubility of chemical substances, availability and uptake of nutrients, and growth and activity of soil microorganisms. Some nutrients become unavailable if the soil pH remains at extremely acid or extremely alkaline conditions. The E_h represents the redox potential which indicates the tendency of a soil to be reduced or oxidized. Redox reactions are very important in soil genesis. There are a variety of other chemical reactions occurring all the time in the soil.

8.1 Soils Are Composed of Chemical Elements

Soils contain more than 100 chemical elements, but only a few make their bulk mass. The ranges and median values of different elements in soils are given in Sparks (2003). Median values of oxygen (O), silicon (Si), aluminum (Al), iron (Fe), calcium (Ca), magnesium (Mg), and potassium (K) are 49.0, 33.0, 7.1, 4.0, 5.0, 0.5, and 1.4%, respectively. There are small quantities of other elements. Plant nutrients (Chap. 10) such as calcium, magnesium, potassium, phosphorus (P), sulfur (S), iron (Fe), manganese (Mn), zinc (Zn), copper (Cu), molybdenum (Mo), boron (B), and nickel (Ni) are present in the minerals and in the soil solution. Oxygen, silicon, and aluminum occur as constituents of minerals and as oxides. Iron occurs mainly in the form of oxides and hydroxides. Phosphorus is present in soil mainly in the mineral apatite; it also occurs as aluminum and calcium phosphates and as organic compounds such as phospholipids, inositol, and nucleic acids. Nitrogen occurs in soil mainly in the organic form as proteins, peptides, and amino acids. All the micronutrients Mo, Fe, Mn, Zn, Cu, B, and Ni occur in the inorganic form. There are many different soils in nature, and their chemical compositions are also varied.

8.2 Water in Soil with Dissolved Substances Forms Soil Solution

Soil solution refers to the aqueous liquid phase of the soil and its solutes. Solutes include free hydrated ions, ion pairs, ligands, and complexes. One will note that Ca^{2+} is the most prevalent metal cation in the soil solution, which is typical for most soils. Nitrate, chloride, and sulfate are the common anions in more acidic soils, while carbonate can be important in some basic soils. Plants and soil organisms take up the nutrients and elements they need to grow from the soil solution. A plant absorbs ions from the soil solution at the contact of roots. Thus, ions are continuously removed from soil solution by plant roots. Two processes restore the supply of nutrients at this situation. They are the following: (a) movement of more solution from other portions of soil toward the plant root (mass movement) and (b) more ions move through the soil solution toward the root (diffusion). The soil solution is a dynamic entity. Plant roots take up ions from soil solution and release exudates into it. Ions in the soil solution can be sorbed on organic and inorganic components of the soil. Ions can again be desorbed into the soil solution. Some substances in soil solution can precipitate. Weathering of minerals releases soluble ions. Ions are transported through the soil into groundwater or removed through surface runoff

K.T. Osman, *Soils: Principles, Properties and Management*,
DOI 10.1007/978-94-007-5663-2_8,

processes. Microorganisms can remove ions from the soil solution, and when they die and are decomposed, ions are released to the soil solution (Sparks 2003). The concentration of a particular ion in the soil solution is thus changed from time to time. Soils have the natural but variable capacity to replenish the solutes in soil solution. However, the actual concentration of the ion species in the soil solution changes with changes in soil moisture content (Wolt 1994).

8.3 Soil Colloids Are Very Fine Soil Particles

A colloid is an ultramicroscopic substance dispersed evenly throughout another substance; together, both are known as a colloidal system. A colloidal system consists of two separate phases: a dispersed phase (colloidal particle) and a dispersion medium. A colloidal system may be solid, liquid, or gaseous.

The solid phase of the soil comprises of differently sized particles including the colloids. They are formed by weathering minerals as finely divided secondary clay minerals. Many new clay minerals are recrystallized from soluble substances as well. Feldspar, mica, and other primary minerals produce vermiculite, smectite, kaolinite, and other clay minerals. These extremely small mineral particles are visible only by an electron microscope and pass through ultra filters. They are inorganic soil colloids of size <0.002 mm or 2 μm diameter. There are organic colloids as well (humus). Due to their minute size and chemical nature, colloids acquire some properties entirely different from the larger fraction of the soil. Colloids have large surface area and electrical charges on their surfaces. Soil colloids have high nutrient- and water-holding capacities. They are all chemically reactive.

8.3.1 Inorganic Soil Colloids Mainly Include Aluminosilicate Clay Minerals

Clay minerals are commonly hydrous aluminosilicates, usually with appreciable amounts of iron, magnesium, calcium, sodium, and potassium. Clay minerals are generally phyllosilicates (from Gk. *phyllon* meaning leaf) because they are foliated. Clay minerals are basically constituted by organization of silica tetrahedral and alumina octahedral sheets. The basic unit of the silica tetrahedral sheet is the silica tetrahedron composed of one silicon atom surrounded by four oxygen atoms (Fig. 8.1). A tetrahedral sheet is built by sharing basal oxygen of one tetrahedron by its neighboring tetrahedron. On the other hand, the basic unit of the octahedral sheet is the alumina octahedron composed of one aluminum atom surrounded by six oxygen/hydroxyls, giving an eight-sided geometric shape.

In the tetrahedral sheet, some Si may be substituted by Al, and in the octahedral sheet, Al may be substituted by Mg or Fe.

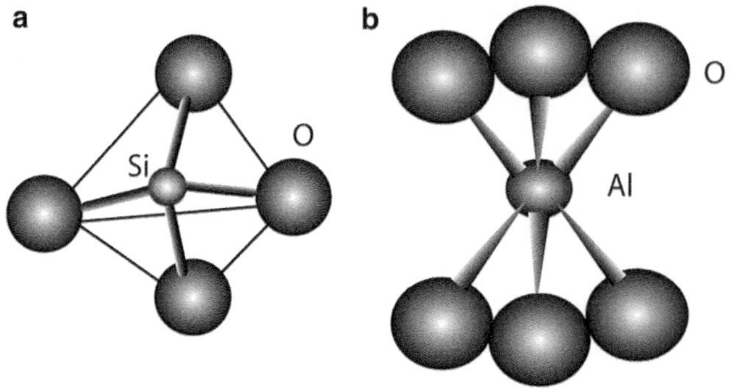

Fig. 8.1 (**a**) Silica tetrahedron and (**b**) aluminum octahedron

In some clay minerals, most Al atoms in the octahedral sheets are substituted by Mg. Al octahedral sheets are known as dioctahedral, and Mg octahedral sheets are known as trioctahedral sheets. Clay mineral crystals are formed either by 1 tetrahedral sheet bound to 1 octahedral sheet (1:1 type clay minerals) or 2 tetrahedral sheets sandwiching 1 octahedral sheet (2:1 type clay minerals). In some clays (such as chlorite), an additional magnesium hydroxide [$Mg_6(OH)_{12}$] sheet is bound to the 2:1 structure. It is, therefore, regarded as 2:1:1 type clay mineral.

8.3.1.1 Major Types of Clay Minerals in Soil

1:1 Type

A layer of 1:1 clay crystal consists of one tetrahedral and one octahedral sheet held tightly together by oxygen atoms mutually shared by silicon and aluminum cations in their respective sheets. These layers, in turn, are held together by hydrogen bonding. Consequently, the structure is fixed, and no expansion ordinarily occurs between layers when the clay is wetted. This is, therefore, known as non-expanding clay. The 1:1 type clay includes the kandite group [$Al_4Si_4O_{10}(OH)_8$], the chief members of which are kaolinite, dickite, nacrite, and halloysite, formed mainly by the decomposition of orthoclase feldspars.

2:1 Type

The 2:1 clay crystal units are characterized by an octahedral sheet sandwiched between two tetrahedral sheets. Three general groups have this basic crystal structure: smectite, vermiculite, and illite. Smectites and vermiculites are expanding type of clay, while illite is non-expanding.

Smectite Group: The smectite group has the general formula $(Na,Ca)_{0.33}$ $(Al,Mg)_2Si_4O_{10}$ $(OH)_2 \cdot nH_2O$. It includes montmorillonite, bentonite, nontronite, hectorite, saponite, and sauconite, formed by the alteration of mafic igneous rocks rich in Ca and Mg. These minerals have high swelling/shrinking potential. Wide cracks commonly form when smectite dominated soils (e.g., Vertisols; Chap. 4) are dried. The dry clods are very hard, making such soils difficult to till.

Vermiculite Group: Vermiculite group has the general formula $(Mg,Fe,Al)_3$ $(Al,Si)_4O_{10}(OH)_2{\cdot}4H_2O$. In the tetrahedral sheet of most vermiculite, considerable substitution of aluminum for silicon has taken place. For this reason, a high proportion of net negative charge develops in vermiculite. There are many strongly adsorbed cations including Ca^{2+} and Mg^{2+} and water molecules in the interlayer space of vermiculites. They bind the units tightly together. The degree of swelling and shrinking is, therefore, considerably less in vermiculites than in smectites. Vermiculites are considered limited-expanding type of clay minerals, expanding more than kaolinite but much less than the smectites.

Illite Group: Illite has the general formula (KH_3O) $(Al,Mg,Fe)_2(Si,Al)_4$ $O_{10}(OH)_2{\cdot}H_2O$. It includes hydrous micas, phengite, brammalite, celadonite, and glauconite, formed mainly by the decomposition of some micas and feldspars. They are predominant in marine clays and shales. The K^+, Ca^{2+}, and Mg^{2+} cations in interlayers of illite prevent the entrance of H_2O into the structure. So, the illite clays are non-expanding clays.

2:1:1 Type

This silicate group is represented by chlorites having the general formula $(Mg,Fe)_3(Si,Al)_4O_{10}(OH)_2(Mg,Fe)_3(OH)_6$. In chlorite clay crystals, the 2:1 layers alternate with a magnesium-dominated trioctahedral sheet. Magnesium replaces a considerable amount of aluminum in the octahedral sheet of the 2:1 layer. Thus, the crystal unit contains two silica tetrahedral sheets and two magnesium-dominated trioctahedral sheets, giving rise to the term 2:1:1 type structure. Important members of the group are clinochlore, chamosite, nimite, pennantite, and cookeite.

Other Clays

Sesquioxide Clays: Sesquioxides are predominant in tropical soils. They are formed by the intensive weathering in humid warm climates. Most of the silica and much of the alumina in primary minerals are dissolved and slowly leached away. The remnant materials, which have lower solubility, are the sesquioxides. Sesquioxides are mixtures of aluminum hydroxide $[Al(OH)_3]$ and iron oxide (Fe_2O_3) or iron hydroxide $[Fe\ (OH)_3]$. Examples of iron and aluminum oxides common in soils are gibbsite $(Al_2O_3{\cdot}3H_2O)$ and goethite $(Fe_2O_3{\cdot}H_2O)$. These substances form amorphous to crystalline clay-sized materials.

Allophanes: These clays are common in soils forming from volcanic ash. Almost all of their charge is from accessible hydroxyl ions (OH^-), which can attract a positive ion or lose the H^+ attached. These clays have a variable charge that depends on H^+ in solution.

Distribution of Clay Minerals in Different Soil Orders

Order	Dominant Clay Minerals
Andisols	Allophane, imogolite, hydrated halloysite, and halloysite
Aridisols	Smectites and vermiculites
Vertisols	Smectites
Mollisols	More smectites than kaolinites, some illites
Alfisols	Kaolinites and smectites
Ultisols	Kaolinites, sesquioxides, some smectites
Spodosols	Kaolinites, sesquioxides
Oxisols	More sesquioxides than kaolinites

8.3.2 Organic Colloids Are Derivatives of Humus

Humus or organic colloidal fraction is composed of highly decomposed residues of plant and animal remains (Chap. 7). Organic colloids are amorphous. The structure of soil organic colloid is extremely complex, with many different types of functional groups which can contribute negative and positive charges. Humus colloids consist of modified lignins, polyuronides, and other constituents composed of carbon, hydrogen, and oxygen, along with minor amounts of nitrogen, phosphorous, sulfur, and other elements.

The charge and surface area of humus far exceed that of the crystalline colloids. Humus may contribute from 30 to 90% of the total charge present in the soil even though present in relatively low amounts.

8.3.3 Soil Colloids Have Unique Properties

Because of their tiny size, all soil colloids have a large external surface area per unit mass. Certain silicate clay colloids have extensive internal surfaces as well, which occur between platelike crystal units that make up each particle. The total surface area of soil colloids ranges from 10 $m^2\ g^{-1}$ for clays with only external surfaces to more than 800 $m^2\ g^{-1}$ for clays with extensive internal surfaces.

Soil colloids carry negative and/or positive charges on their surfaces with more negative charges than positive. When an electric current is passed through a colloidal suspension, soil colloids migrate to anode, indicating that they carry a net negative charge. The magnitude of the charge is known as zeta potential. The presence and intensity of charge influence the attraction and repulsion of the particles toward each other. Due to the presence of charges, colloids attract ions of an opposite charge to their surfaces. They attract hundreds of positively charged ions or cations such as H^+, Al^{3+}, K^+, Na^+ Ca^{2+}, and Mg^{2+} to satisfy negative charges and SO_4^{2-}, NO_3^-, and $H_2PO_4^-$ to satisfy positive charges. Thus, a colloidal particle is surrounded by a swarm of adsorbed cations. These ions can be replaced by other ions, say K^+ by H^+, in solution. In addition to the adsorbed cations, a large number of water molecules encircle soil colloidal particles. Some are attracted to the adsorbed cations, each of which is hydrated; others are held in the internal surfaces of the colloidal particles.

Cohesion and adhesion are two important physicochemical properties of colloids. Colloids of similar nature tend to stick together. This tendency is primarily due to the attraction of the clay particles for the water molecules held between them. When colloidal substances are wetted, water first adheres to the particles and then brings about cohesion between two or more adjacent colloidal particles. On the other hand, adhesion refers to the phenomenon of colloidal particles sticking to other substances.

When a suspension of colloidal particles is examined under a microscope, the particles seem to oscillate. The oscillation is due to the collision of colloidal particles or molecules with those of the liquid in which they are suspended. Soil colloidal particles with those of water in which they are suspended are always in a constant state of motion. The smaller the particle, the more rapid is its movement. This is known as Brownian movement.

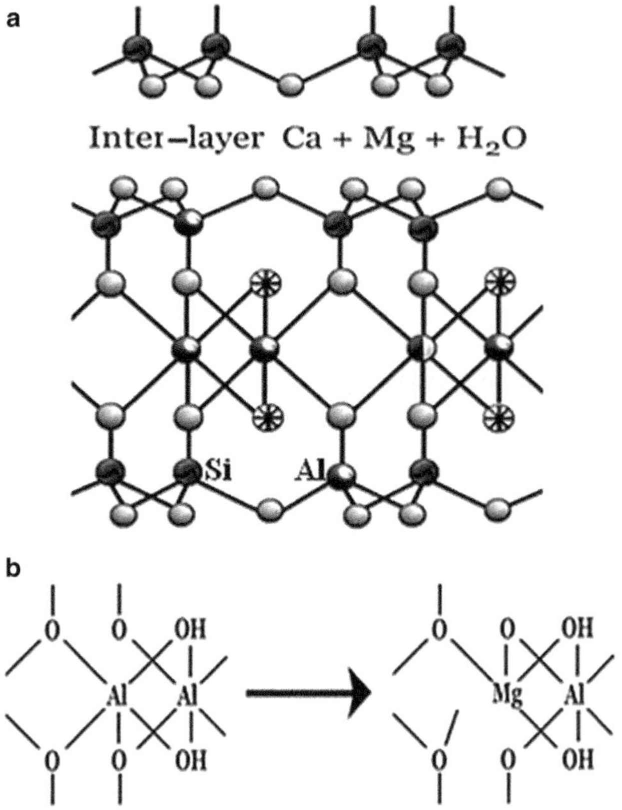

Fig. 8.2 Isomorphous substitution in the silica tetrahedral (**a**) and alumina octahedral sheet (**b**)

8.3.4 There Are Electric Charges on Soil Colloids

Charges on soil colloids are of two types: (1) permanent charges or charges independent of pH and (2) pH dependent or variable charges. The permanent charges develop on layer silicate clays (except kaolinite) by isomorphous substitution. On the other hand, pH-dependent charges develop on all types of colloids, on layer silicate colloids from broken crystal edges, and on amorphous inorganic colloids, oxides, and hydroxide colloids and humus by dissociation of active groups and association of hydrogen ions (protonation). There may be both negative and positive pH-dependent and ph-independent charges.

8.3.4.1 Permanent Charges

Isomorphous Substitution

During crystallization, some of Si^{4+} atoms in the tetrahedral sheet of crystalline colloids may be substituted by Al^{3+}. One negative charge develops due to this substitution. Si^{4+} and Al^{3+} have comparable ionic radii. This substitution of one atom by another of similar size in a crystal lattice without disrupting or changing the crystal structure of the mineral is called isomorphous (iso—similar, morphous—size and shape) substitution. Al^{3+} in the octahedral sheet may also be substituted by Mg^{2+}, Fe^{2+}, and Zn^{2+}. Thus, substitution of some atoms in crystals by other atoms of similar size but different valence creates positive and negative charges on colloidal surfaces (Fig. 8.2). These charges are not dependent on soil pH and are called permanent charges. Permanent charges develop in such clay colloids as smectite, vermiculite, mica, and chlorite (Sparks 2003).

Variable Charges

The variable charges in soils change with pH due to protonation and deprotonation of functional groups on inorganic soil minerals such as kaolinite, amorphous materials, metal oxides, oxyhydroxides, and hydroxides and layer silicates coated with metal oxides and soil organic matter (Sparks 2003). In the 1:1 type clay, the negative charges are due to deprotonation of OH at broken edges (Fig. 8.3).

The pH-dependent charges may either be positive or negative depending on the pH of the soil. In acid soils, positive charges develop because of the protonation of the OH group on the surfaces. In alkaline medium, negative charges develop due to deprotonation of the surface O and OH groups (Fig. 8.4).

Humus colloids contain phenolic (–OH), carboxylic (–COOH), amino (–NH_2), and some other active groups which undergo protonation and deprotonation depending on the pH of the soil. Positive and negative charges develop due to protonation (in acidic soil) and deprotonation (in alkaline soil), respectively (Fig. 8.5).

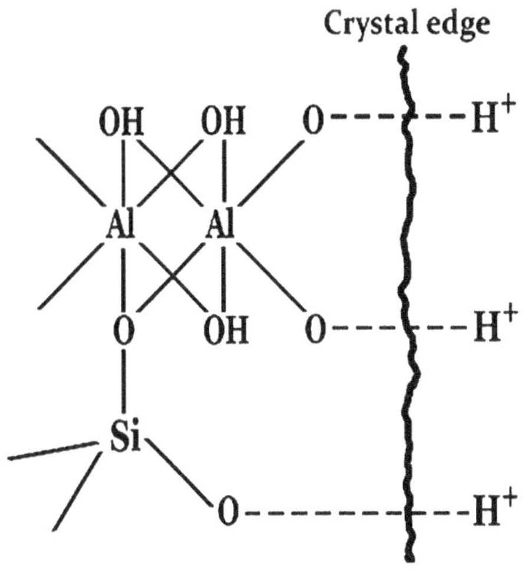

Fig. 8.3 pH-dependent charge at the broken edge of 1:1 clay

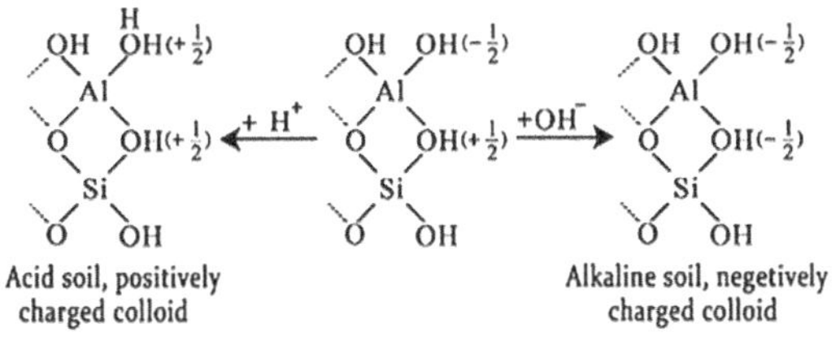

Fig. 8.4 Protonation and deprotonation of clay colloids and associated charge development

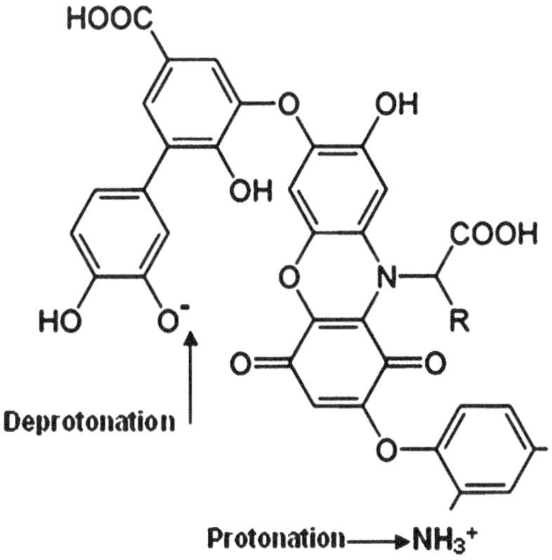

Fig. 8.5 Part of a humus molecule with protonation and deprotonation

8.4 The Soil Is a Seat of Diverse Chemical Reactions

Many diverse chemical reactions occur simultaneously in the soil. These reactions govern the solubility, availability, and mobility of important soil constituents and those introduced into the soil and groundwater system (Gee et al. 1999). Soil is a very complex chemical system, and it is difficult to predict the nature of reactions that may occur in soil at any given time. It depends on the soil environment (water, oxygen, pH, etc.) and biological activity. Many soil reactions are biologically mediated (Mulder and Cresser 1994; Gee et al. 1999). The most important chemical reactions in soils include oxidation, reduction, hydrolysis, hydration, carbonation, dissolution, chelation, precipitation, adsorption, and ion exchange. Examples of oxidation, reduction, hydrolysis, hydration, and carbonation were given in connection with biogeochemical weathering in Sect. 2.14.3. Examples of oxidation–reduction reactions are also given in connection with properties of submerged soils in Chap. 13. Oxidation and reduction of N, P, S, Fe, and Mn and hydrolysis of minerals leading to the release of nutrients are important aspects of plant nutrition (Chap. 10). Here dissolution, chelation, precipitation, adsorption, and ion exchange reactions are explained.

8.4.1 The Equilibrium Constant Represents the Solubility Product in Dissolution Reactions

Many chemical compounds are said to be insoluble, but they are actually slightly soluble in water. Moreover, the soil solution in most aerobic soils is acidic in nature; so, some minerals are slowly dissolved, although in a very low amount, in soil water. Dissolution of calcium carbonates, gypsum, oxides of Fe and Mn, and minerals is very important in relation to soil genesis, soil chemistry, and plant nutrition. During dissolution of primary minerals and formation of clay minerals, and other dissolution reactions, a solid exists in equilibrium with its solution. Thus, decomposition of kaolinite and formation of gibbsite and dissolution of lime in soil, etc., obey solubility laws (Tan 2011).

If we consider that the chemical compound to be dissolved in soil is BA and during dissolution some B^+ and A^- ions are produced, then at equilibrium, the equilibrium constant according to the law of mass action, K, is called the solubility product K_{sp}. The solubility product is defined as the product of ion concentrations in a saturated solution of a difficultly soluble solid. The dissolution reaction can be written as

$$BA \leftrightarrow B^+ + A^-,$$

Since the activity of a pure solid is unity at equilibrium, $K_{sp} = (B^+)(A^-)$

For dissolution of iron oxides such as goethite, hematite, and ferrihydrite in soils, K_{sp} values are usually extremely low $((Fe^{3+})(OH^-)^3 = 10^{-37}–10^{-44}$. However, for each mineral type, K_{sp} may increase by several orders of magnitude with decreasing crystal size (Schwertmann 1991). Because the K_{sp} values are generally very small, they are often expressed in terms of ten raised to certain negative powers. For convenience, the solubility product (K_{sp}) can be expressed as pK_{sp} defined by $pK_{sp} = -\log K_{sp}$. Visconti et al. (2010) reported that reliable values for gypsum and calcite solubility products (pKs) in a water-saturated salt-affected soil were 4.62 and 8.43, respectively.

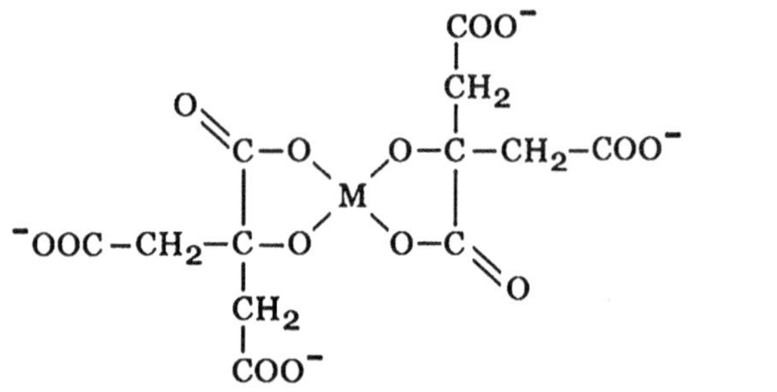

Fig. 8.6 Metal citrate chelate

8.4.2 Chelation Is the Complexation of Metals with Organic Substances

Chelation is the process by which metals are bound to ligands or functional chelate groups. Chelates are organic molecules that can form inner-sphere complexes with divalent hard metals and heavy metals (Evangelou 1998). Some well-known synthetic chelating agents are EDTA (ethylenediaminetetraacetic acid), DTPA (diethylenetriamine pentaacetic acid), CyDTA (cyclohexanediaminetetraacetic acid), EDDA [ethylenediamine-di(O-hydroxyphenyl)acetic acid], or Chel-138. These chelating agents are widely used in chemical industries and in remediation of metal contamination of the environment. There are also many natural chelating agents. All chelating agents have active groups capable of associating with the metal ion on adjacent carbon atoms. Such groups include amine, carboxyl, $R_2CH–OH$, $R_2C{=}O$, and $RCH_2–SH$. So, organic acids, amino acids, amines, and sugars can all be chelating agents if they meet the steric requirements (rearrangement of electrons in atoms or molecules affecting reactivity) for ring formation (Evangelou 1998). Figure 8.6 shows chelation of metal (M) with citrate. The common metals that form chelates in soils are Fe^{3+}, Cu^{2+}, Zn^{2+}, Mn^{2+}, Fe^{2+}, Ca^{2+}, and Mg^{2+}. Heavy metals such as Cd also frequently form chelate complexes.

8.4.3 Precipitation Is the Separation of Substances from Solution

Precipitation in soil is the formation and separation of an insoluble solid from soil solution due to a chemical change. Three types of precipitation reactions are usually found to take place in the soil system: (1) precipitation of an element due to its supersaturation in the soil solution, (2) surface precipitation, and (3) coprecipitation of elements. These reactions can potentially control the mobility of metals in soil.

A solution containing more solute than allowed at equilibrium is a supersaturated solution. Such solutions are unstable, and any further addition of solute will cause its precipitation. Precipitation from soil solution is common for Al, Fe, Si, Mn, Ca and Mg, and trace elements. Surface precipitation is the three dimensional growth of a solid on the surface of a soil particle. It occurs when the soil particle becomes increasingly covered by solid precipitate from the soil solution. Often trace elements, such as Pb, Cr, and As, will precipitate on a soil particle and become immobilized. The thermodynamic reasons for surface precipitate formation are the following: (1) the solid surface may lower the energy of nucleation by providing sterically similar sites (McBride 1991), (2) the activity of the surface precipitate is <1 (Sposito 1986), and (3) the solubility of the surface precipitate is lowered because the dielectric constant of the solution near the surface is less than that of the bulk solution (O'Day et al. 1994). The surface precipitates can form (1) through polymeric metal complexes that form on mineral surfaces and through the sorption of aqueous polymers (Chisholm-Brause et al. 1990) and (2) on a surface when the solution becomes saturated and the surface acts as a nucleation site. Coprecipitation is the incorporation of trace element into mineral structure during solid solution formation and recrystallization of minerals. The opportunity for coprecipitation is higher with Fe and Mn oxides than Al oxides or aluminosilicates. The reason for this is the higher solubility of the former two minerals under anaerobic conditions. Soil reduction generates the soluble ions, Fe^{2+} and Mn^{2+}, which then reoxidize to Fe^{3+} and Mn^{4+}, again form the insoluble oxides once the soil is aerated. Coprecipitation of trace metals may occur during these cycles of alternating reduction and oxidation (McBride 1994).

8.4.4 Adsorption Is the Attraction of Gas, Liquid, or Solid on Surfaces of Colloids

Adsorption is the adhesion of atoms, ions, or molecules of gas, liquid, or dissolved solids to a surface (Brandt et al. 1993). The substance that is adsorbed is called adsorbate, and the surface on which adsorption takes place is called the adsorbent. This process creates a film of the adsorbate on the

Table 8.1 Common adsorbents and adsorbates in soil

Common adsorbents in soil	Common adsorbates in soil
Silicate clays, hydrated oxides of iron and aluminum, humic substances, coated surface, closed or open pore	Cations: Ca^{2+}, Mg^{2+}, K^{+}, Fe^{2+}, Fe^{3+}, Al^{3+}, Cu^{2+}, Cd^{2+}, H^{+}
	Anions: $H_2PO_4^-$, HPO_4^{2-}, PO_4^{3-}, AsO_4^{3-}, CrO_4^{2-}, HCO_3^-
	Dissolved organics

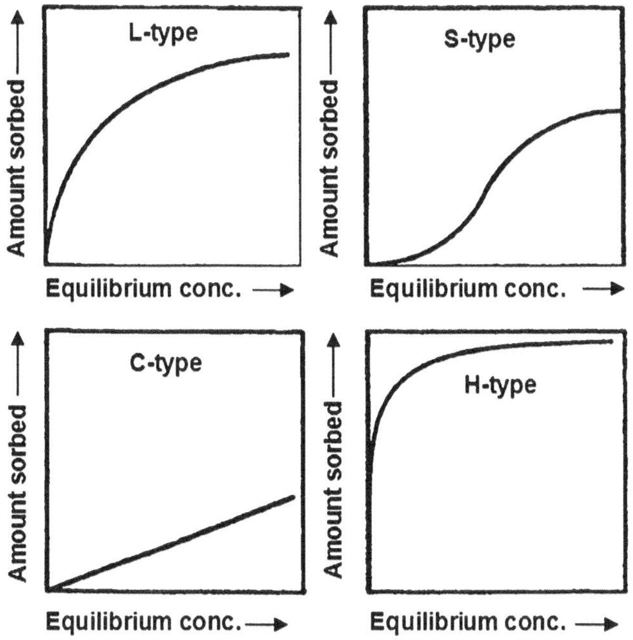

Fig. 8.7 General types of adsorption isotherm

surface of the adsorbent. Soil solids particularly the colloids act as adsorbents, and ions, polar molecules, and colloids themselves may be adsorbate. Clay and humus colloids are electrically charged particles, and against these charges, water, ions, and other colloids may be adsorbed. A list of common adsorbents and adsorbates in soil is provided in Table 8.1.

8.4.4.1 Adsorption Models

Adsorption is a reversible reaction. At equilibrium, the rate of the forward reaction equals the rate of the reverse reaction. Two major techniques commonly used to model adsorption equilibrium processes are (1) the Freundlich technique and (2) the Langmuir technique. Both of these techniques involve adsorption isotherms. An adsorption isotherm is a curve that describes the relation between the activity or equilibrium concentration of the adsorptive and the quantity of adsorbate on the surface at constant temperature. Adsorption isotherms are classified into four types, L-, S-, C-, and H-type, depending on their general shape (McBride 1994). These types are shown in Fig. 8.7. The L-type shows high-affinity adsorption between the adsorbate and adsorbent; S-type describes the adsorbate–adsorbate interactions on the adsorbent, often referred to as clustering of the adsorbate with solution ligands; C-type indicates partitioning, which suggests interaction between a generally hydrophobic adsorbate with a hydrophobic adsorbent; and the H-type is basically an extreme case of the L-type isotherms.

8.4.4.2 Freundlich Adsorption Isotherm

Freundlich equation is an empirical model used frequently in adsorption studies in soils. Freundlich developed this model in 1909 to describe adsorption of gas molecules on solid surfaces. This model shows the relationship between the concentration of a solute on the surface of an adsorbent and the concentration of the solute in the liquid. The equation can be expressed as

$$q = K_d C^{\frac{1}{n}},$$

where q is the amount of adsorption (adsorbate per unit mass of adsorbent), C is the equilibrium concentration of the adsorptive, K_d is the distribution coefficient, and n is a correction factor. The equation produces an L-type curve. By plotting the linear form of the above equation, $\log q = 1/n \log C + \log K_d$, the slope is the value of $1/n$ and the intercept is equal to $\log K_d$. The main disadvantage of this model is that it does not predict an adsorption maximum. The single K_d term in the Freundlich equation implies that the energy of adsorption on a homogeneous surface is independent of surface coverage (Sparks 2003).

8.4.4.3 Langmuir Adsorption Isotherm

Langmuir equation is another model used commonly for explaining relationships between amount of adsorption and the equilibrium concentration of adsorbates in soil. Langmuir (1918) developed this model to describe the adsorption of gas molecules on a planar surface. Fried and Shapiro (1956) and Olsen and Watanabe (1957) applied this model first to soils in relation to phosphate adsorption. Like Freundlich equation, it is best applicable at low concentrations. The equation can be written as

$$q = \frac{kCb}{1 + kC},$$

where q and C have the same meanings as mentioned earlier, k is a constant related to the binding strength, and b is the maximum amount of adsorptive that can be adsorbed. This equation also produces an L-type curve. The Langmuir model is criticized because most of the original assumptions that Langmuir had made for developing this model are not valid in a heterogeneous system like soil. The assumptions were the following: (1) adsorption occurs on planar surfaces that have a fixed number of identical sites that can hold only one molecule (monolayer coverage), (2) adsorption is reversible, (3) there is no lateral movement of molecules on the surface, and (4) the adsorption energy is the same for all sites and independent of surface coverage (i.e., the surface is homogeneous) (Harter and Smith 1981).

8.4.4.4 Surface Complexation Models

Surface complexation models based on a molecular description of the electric double layer are being used at present to characterize adsorption. The electric double-layer theory is also a century older. According to this theory, a diffuse layer of adsorbate ions is formed adjacent to the charged surface (adsorbent). Gouy (1910) and Chapman (1913) derived an equation describing the ionic distribution in the diffuse layer. The countercharge (charge of opposite sign to the surface charge) can be a diffuse atmosphere of charge or a compact layer of bound charge together with a diffuse atmosphere of charge. The surface charge and the sublayers of compact and diffuse counterions constitute the double layer. Schofield (1947) applied the diffuse double-layer theory in soil science to study the thickness of water films on mica surfaces (Evangelou 1998). Further discussions on this theory are not within the scope of this book. Readers may get details of the electric double-layer theory in Bolt (1982). Surface complexation models include (1) the constant-capacitance model (CCM), (2) the triple-layer model (TLM), (3) the Stern variable surface charge–variable surface potential model (VSC–VSP), (4) the generalized two-layer model, and (5) one *pKa* model (Goldberg 1992).

8.4.4.5 Free Ion Activity Model

The free ion activity model is used generally to describe bioadsorption of metal ions by plant cell membranes. The free ion activity model assumes that (1) the cell membrane is the primary site for metal interactions, (2) interaction of metal ions with the plasma membrane results from surface complexation reaction which occurs rapidly leading to the establishment of pseudo-equilibrium between metal species in the bulk solution and at the membrane surface, (3) the concentration of free sites remains virtually constant, and (4) the nature of the membrane remains constant during exposure to the metal ion. The model is consistent with experimental results for systems involving divalent trace metals, a fixed pH, and a known ligand concentration. Several formulations of the free ion activity model have been tried in connection with metal ion uptake by plants but often satisfactory results are not found (Hough et al. 2005).

8.4.5 Ion Exchange Is the Replacement of One Ion from Colloidal Surface by Another

Adsorbed cations on colloids can be replaced by other cations in solution (Fig. 8.8). Similarly, one anion can replace another anion. This phenomenon is known as ion exchange.

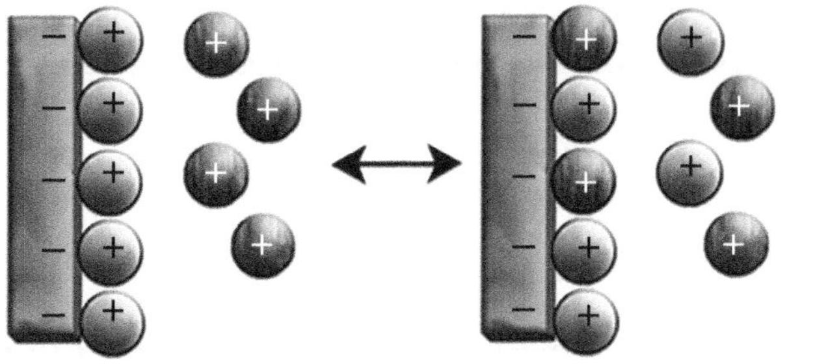

Fig. 8.8 Exchange of ions between colloidal phase and solution phase

Ion exchange mechanism is used in many ways to remove ions from solution and to separate ions of various kinds from one another. It is widely utilized in the scientific laboratory for purifications and for analysis of unknown mixtures. Ion exchange materials such as zeolites are also employed commercially to purify water and medically to serve as artificial kidneys, and for other purposes.

There are cation exchange and anion exchange. The most commonly adsorbed cations on soil colloids are H^+, Al^{3+}, Fe^{3+}, Fe^{2+}, Ca^{2+}, Mg^{2+}, K^+, and Na^+, while the most common adsorbed anions are Cl^-, SO_4^{2-}, NO_3^-, $H_2PO_4^-$, and HCO_3^-. These adsorbed ions are loosely held and are exchangeable. This phenomenon of exchange of ions between the colloidal surfaces and solution is known as ion exchange. Ion exchange is of two types: cation exchange and anion exchange.

Cation exchange takes place when one of the cations in the soil solution replaces one of the cations on the soil colloid. Cation exchange capacity (CEC) is the capacity of unit mass of soil to hold and exchange cations at pH 7.0.

8.4.5.1 Cation Exchange Capacity

CEC is the sum total of the exchangeable cations that a soil can adsorb. The unit of CEC is $cmol_c\ kg^{-1}$. CEC of a soil is the function of the amount and type of colloids. Different types of clay minerals contain different number of charges and cation exchange capacity. Among soil colloids, humus has the greatest CEC (100–300 $cmol_c\ kg^{-1}$). Among inorganic colloids, allophanes, vermiculites, and montmorillonite have CEC values higher than 100 $cmol_c\ kg^{-1}$. However, different types of soil colloids have different cation exchange capacities (Table 8.2).

Exchangeable cations remain in equilibrium with cations in solution. When concentration of a cation increases in soil solution, for example, by fertilizer addition, exchange of ions between colloid surfaces and solution takes place to reach a new equilibrium. The replacing power of cations varies with the type of ion, its size and degree of hydration, valence, and concentration. It is controlled by a number of factors; so no single order of replacement can be given. Usually the replacing power of monovalent cations are in the following

Table 8.2 Cation exchange capacity of different soil colloids

Type of colloids	CEC, $cmol_c\ kg^{-1}$
Kaolinite	1–10
Montmorillonite	80–120
Vermiculite	120–150
Chlorite	20–40
Gibbsite, goethite	1–10
Humus	100–300

http://www.plantstress.com/Articles/min_deficiency_i/soil_fertility.pdf

order: $H^+ > Cs^+ > Rb^+ > K^+ > Na^+ > Li^+$ and for divalent cations: $Ba^{2+} > Sr^{2+} > Ca^{2+} > Mg^{2+}$. In case of mixture of cations as they exist in normal soils, the replaci ng power remains in the following order: $Al^{3+} > H^+ > Ca^{2+} > Mg^{2+} > NH_4^+ > K^+ > Na^+$.

The factors affecting cation exchange capacity include soil texture, organic matter, nature of clay, and pH. CEC increases as the soil becomes heavier in texture (more clayey). CEC also increases with organic matter content. Soils with montmorillonite and vermiculite clay have much higher CEC than soils with kaolinite clay. As the number of negative and positive charges on most soil colloids varies with pH, it is customary to measure CEC at pH 7.0.

An Approximate Estimate of CEC of Tropical Soils

A fair estimate of CEC of tropical soils (Oxisols, Ultisols, some Inceptisols, and Entisols) may be obtained from their organic matter and clay contents by the thumb rule:

Add 1 $cmol_c\ kg^{-1}$ for each 10% clay (tropical soils are dominated by kaolinite and oxides of Fe and Al; they may have a maximum of 10 $cmol_c\ kg^{-1}$ CEC). Add 3 $cmol_c\ kg^{-1}$ for each 1% organic matter (humus has a maximum of 300 $cmol_c\ kg^{-1}$ CEC).

If a tropical soil, for example, contains 30% clay and 2% organic matter, its CEC may be around -

$$30 \times 1/10 = 3\ cmol_c kg^{-1} \text{for } 30\% \text{ clay}$$

$$3 \times 2 = 6\ cmol_c kg^{-1} \text{for } 2\% \text{ organic matter}$$

$$(3+6) = 9 cmol_c kg^{-1} \text{CEC for the soil}$$

The soil has an approximate CEC of 9 $cmol_c\ kg^{-1}$.

8.4.5.2 Total Exchangeable Bases

Total exchangeable bases (TEB) are the sum of exchangeable cations excluding Al^{3+} and H^+. TEB may be obtained by subtracting exchangeable H^+ and Al^{3+} from CEC or by summing up exchangeable Ca^{2+}, Mg^{2+}, K^+, and Na^+. In calcareous soils, the sum of the cations ($\sum Ca^{2+}$, Mg^{2+}, K^+, Na^+) is invariably equal to the CEC because any deficit of cations on the exchange surfaces can be made up by Ca^{2+} ions from the dissolution of $CaCO_3$ (White 2006). In noncalcareous soils, however, $\sum Ca^{2+}$, Mg^{2+}, K^+, Na^+ is less than CEC. Often, TEB is taken to be a more meaningful index than CEC of soil's ability to provide plant nutrients. Usually, the higher the TEB is, the more fertile is the soil. Most base cations are essential nutrients for plant growth. Their predominance as exchangeable cations follow the order: $Ca^{2+} > Mg^{2+} > K^+$.

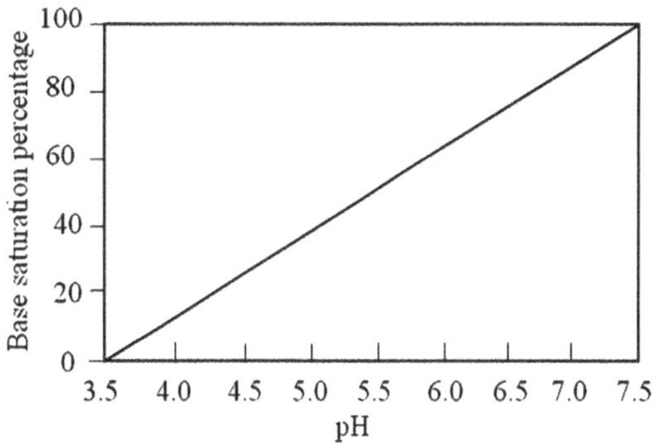

Fig. 8.9 Relationship of base saturation percentage with pH of soil

8.4.5.3 Base Saturation Percentage

Base saturation is the proportion of CEC occupied by bases. It is usually expressed as percentage.

$$\text{BSP} = \frac{\text{TEB}}{\text{CEC}} \times 100$$

$$\text{Again, BSP} = \frac{\text{CEC} - \text{exchangeable acidity}}{\text{CEC}} \times 100$$

If CEC of a soil is 10 $cmol_c\ kg^{-1}$ and the exchangeable acidity is 2 $cmol_c\ kg^{-1}$, then base saturation percentage, BSP, is 80. Such a soil must have a pH value above 7.0 because bases prevail in neutral to alkaline solution. Base saturation is roughly correlated with soil pH in many acidic to neutral soils, ranging from 20 to >60% as the pH increases from 5 to >7 (Fig. 8.9).

A high CEC does not necessarily mean high nutrient-supplying capacity, because most exchange sites remain satisfied with H and Al in acid soils. CEC and BSP together indicate better the fertility of soils. Soils with base saturation percentage lower than 50 are called dystrophic soils. They are medium to low fertility soils. Soils with base saturation percentages higher than 50 are eutrophic soils. They are high fertility soils. Alfisols have higher and Ultisols have lower exchangeable base status.

8.4.5.4 Exchangeable Sodium Percentage and Sodium Adsorption Ratio

Excessive exchangeable sodium deteriorates soil physical conditions. Soil particles, particularly colloids, remain dispersed when saturated with exchangeable Na^+. Soil aggregates are dispersed into their constituent individual soil particles. Sodium-saturated soils are impervious and have high pH (as high as 10).

$$ESP = \frac{\text{Exchangeable Na}^+}{\text{CEC}} \times 100$$

A soil is said to be sodic when it contains ESP > 15. Sodic soils are often compacted, macroporosity is greatly reduced, and water infiltration slows to the minimum. Wet sodic soil has poor infiltration and drainage, and when it dries, it becomes quite hard. Surface soil crust and hardsetting are common. In the subsoil, soil sodicity leads to decreased permeability to water and air and poor soil drainage over time. Another important index of sodicity in soil is the sodium adsorption ratio or SAR. A soil is said to be sodic if it contains SAR greater than 13. SAR is obtained by

$$SAR = \frac{Na^+}{\sqrt{\frac{Ca^{2+} + Mg^{2}}{2}}}.$$

Na^+, Ca^{2+}, and Mg^{2+} are exchangeable sodium, calcium, and magnesium, $cmol_c\ kg^{-1}$, respectively. Seilsepour et al. (2009) observed a relationship between SAR and ESP as

$$ESP = 1.95 + 1.03\ SAR.$$

Sodic soils are considered as problem soils (Chap. 11).

8.4.5.5 Anion Exchange

Colloids contain some positive charges against which anions such as SO_4^{2-}, NO_3^-, Cl^-, HCO_3^-, and $H_2PO_4^-$ can be adsorbed. Any one of these anions or all can be replaced by a suitable anion. In other words, like cation exchange capacity, soils have anion exchange capacity (AEC) as well. But adsorbed nitrate is very readily replaced by other anions so that little nitrate is found adsorbed in any given time. Nitrate leaching from arable soils has become a serious problem of groundwater pollution and health hazard. Phosphate sorption is also a mechanism of reducing leaching of phosphate. Movement of phosphate from agricultural fields has eutrophicated many adjacent surface water reservoirs. Issues related to nitrate and phosphate leaching are discussed in Chap. 10.

8.5 Soil Reaction Is the State of Acidity or Alkalinity in a Soil

Soil reaction is denoted by soil pH. The term pH was taken from French "pouvoir hydrogene" meaning "power of hydrogen." Sorenson (1909) defined pH as the negative logarithm of hydrogen ion activity which is equal to the hydrogen ion concentration in very dilute solution. The H^+ ion concentration is expressed in moles per liter:

$$pH = -\log[H^+]$$

The pH scale ranges from 0 to 14, with a value of 7 at neutrality. A pH value less than 7 indicates acidity, while a value greater than 7 indicates alkalinity. The lower the pH, the higher is the acidity. Similarly, alkalinity increases as the pH rises. Soil pH generally lies within 4–10; however, acid sulfate soils (Chap. 11) may have pH values as low as 2. Good quality agricultural soils have pH values around 6.0–7.0. Since the pH scale is logarithmic, a difference of 1 unit pH is actually a ten-fold difference. For example, a soil with pH 5 is ten times more acid than a soil with pH 6.

8.5.1 Soil pH Is Measured from Soil–Water Suspensions

For soil pH measurement, a suspension is made with soil and distilled water at a ratio of 1:1 or 1:2 or 1:2.5 (20 g air dry soil with 20 or 40 or 50 ml distilled water, respectively). The 1:2 ratio is more frequently used. Usually, air dry soil (field moist soil will also do) is mixed with distilled water by occasional stirring for equilibration for half an hour, and the pH reading is recorded by a pH meter. Soil pH value increases by about 0.5 units as the soil–water ratio increases from 1:1 to 1:2. Therefore, it is customary to mention the ratio of measurement along with the pH value.

Soil pH is sometimes measured in a 0.01 M calcium chloride suspension of soil in order to counteracting calcium release from the soil exchange complex. Soil pH value thus obtained is generally lower than that recorded in a suspension made up with distilled water. Since pH of distilled water is itself acidic, soil pH is sometimes measured in suspension made with neutral 1 N KCl solution.

8.5.2 The pH Value May Be Estimated from H^+ and OH^- Ion Concentrations

In any dilute solution, the product of H^+ and OH^- ion concentrations is 10^{-14}. Therefore, pH may be calculated if concentration of any one of H^+ or OH^- ion is known. For example, if the concentration of OH^- ions in a dilute aqueous solution is $10^{-9}\ g\ L^{-1}$, its pH may be calculated as follows:

$$\begin{aligned}
[H^+] \times [OH^-] &= 10^{-14} \\
[OH^-] &= 10^{-9}\,\text{moles L}^{-1} \\
\text{So,} [H^+] &= 10^{-5}\,\text{moles L}^{-1} \\
pH &= -\log[H^+] \\
&= -\log 10^{-5} \\
&= 5 \log 10 \\
&= 5 \times 1 = 5.
\end{aligned}$$

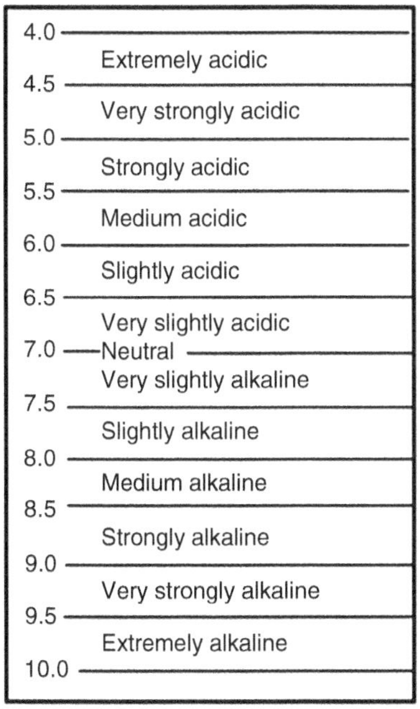

Fig. 8.10 Categories of acidity and alkalinity in soil

The pH of the soil is 5.

In any solution, the H^+ ion concentration decreases as the OH^- ion concentration increases.

8.5.3 Soils Are Classified into Different Categories of Acidity and Alkalinity on the Basis of pH

Soils are classified into several categories of acidity and alkalinity on the ranges of soil pH (Fig. 8.10).

However, for practical purposes, a soil pH range of 6.5–7.5 may be considered as neutral.

8.5.3.1 Sources of Acidity in Soil

(a) Precipitation reacts with carbon dioxide of the atmosphere; the carbonic acid thus formed is a weak acid which gives natural rainwater a pH of about 5–6. Precipitation adds acidity continuously to soils.

(b) Respiration by roots and other organisms produces carbon dioxide that also forms carbonic acid (H_2CO_3). It contributes H^+ to the soil solution.

(c) CO_2 is also produced by decomposition of organic matter in soil.

(d) Organic acids are formed by partial decomposition of organic matter.

(e) Oxidation of mineralized nitrogen and sulfur produces nitric and sulfuric acid, respectively.

(f) Soils may be acidified by fertilizers, particularly nitrogen, when used for long time in excess of crop needs.

(g) Greater precipitation results in more plant growth, causing more respiration and organic matter mineralization. Rainwater leaches soluble bases and increases soil acidity. So, soils of humid regions are more acidic than arid or semiarid soils.

(h) Soils may inherit acidity from the parent materials.

(i) Some soils may contain acid-forming compounds such as pyrite (FeS_2) which is oxidized in aerobic conditions to form H_2SO_4. These soils are called acid sulfate soils which may have pH as low as 2.0. Acid sulfate soils are formed when seawater or sulfate-rich water mixes, in the absence of oxygen, with land sediments containing iron oxide and organic matter. Mangroves, salt marshes, floodplains, swamps, wetlands, estuaries, and brackish or tidal lakes are ideal areas for acid sulfate soil formation (Lovell 2006).

(j) Aluminum ions generate soil acidity, particularly in soils that are already acidic. Al^{3+} ions cause hydrolysis during which H^+ ions are released into the soil solution.

$$Al^{3} + H_2O = Al(OH)_2^{+} + H^{+}$$

(hydroxy aluminium)

(k) Hydroxy-aluminum may occupy exchange sites releasing H^+ from clay surfaces or may further be hydrolyzed to produce more hydrogen ions.

$$Al(OH)_2^{+} + 2H_2O = Al(OH)_3 + 2H^{+}$$

(Gibbsite)

(l) Crop harvest results in base export along with harvested parts and may be responsible for some soil acidity in continuously cropped areas.

(m) Acid rains may bring some acid to soils.

8.5.3.2 Sources of Alkalinity in Soil

(a) Soil parent materials may be calcareous for the presence of free $CaCO_3$. Presence of lime or calcium carbonate in soil may be tested with dilute HCI which evolves carbon dioxide gas. Calcium carbonate produces OH^- ions on hydrolysis and contributes to alkalinity in soils. In calcareous soil, carbonate hydrolysis controls soil pH. If the soil remains calcareous, carbonate hydrolysis maintains a pH that ranges from 7.5 to 8.5 or more.

$$CaCO_3 + H_2O = Ca^{2+} + HCO_3^{-} + OH^{-}$$

(b) Soils that contain Na_2CO_3 may have soil pH as high as 10, which is caused by the greater solubility of Na_2CO_3 and greater production of OH^- by hydrolysis.

(c) The weathering of many primary minerals contributes to soil alkalinity. For example, the hydrolysis of anorthite produces a moderately strong base $Ca(OH)_2$.

$$3CaAl_2Si_2O_8 + 6H_2O = 2HAl_4Si_6O_{10} + 3Ca(OH)_2$$

(anorthite) (aluminosilicate)

(d) Calcium and magnesium are alkaline earth metals, and potassium and sodium are alkali metals. These cations are called basic cations, and their predominance makes soils neutral to alkaline.

(e) Soils that contain high exchangeable Na (>15% exchangeable sodium percentage ESP, or >13 sodium adsorption ratio, SAR) may have soil pH values above 10.

Table 8.3 Aluminum sensitivity/tolerance of some plants

Categories	Plants
Highly sensitive	Durham wheat, barley, lentils, chickpeas, lucerne, strawberry, Berseem, buffelgrass, tall wheatgrass
Sensitive	Canola, red clover, Balansa clover, white clover
Tolerant	Whistler, diamond bird wheat, ryegrass, tall fescue, Subterranean clover, chicory
Highly tolerant	Narrow-leaf lupins, oats, triticale, cereal rye, cocksfoot, Paspalum, yellow and slender serradella, consol lovegrass

8.5.4 Plant Growth Depends on Soil pH

Soil pH affects plant growth mainly by their roles on other soil properties including nutrient availability, elemental toxicity, and microbial activity. Although H^+ ions may exert some direct corrosive effect on root membranes at very low pH (Islam et al. 1980; Foy 1984), effects of soil pH on plants are mainly indirect. Plants suffer in acid soils from deficiency of Ca, Mg, P, and Mo and toxicity of Al, Fe, and Mn.

8.5.4.1 Nutrient Availability

Plant roots absorb soluble and exchangeable nutrient ions from soil. Soluble and exchangeable ions in soil remain in equilibrium at any given time. Since soil pH regulates the solubility of elements and compounds in soil, it also governs the availability of plant nutrients in soil. When soil pH decreases from 6.5, that is, as the soil becomes more acidic, the solubility of some elements including Fe, Al, Mn, Cu, and Zn increases and of some others, for example, Ca, Mg, and Mo decreases. On the other hand, when soil pH rises from 7.0, solubility of Ca, Mg, and Mo increases and that of Fe, Al, Mn, Cu, and Zn decreases. Aluminum, although highly predominant in soil, is not an essential element for plants. In strongly acid soils (pH below 5), solubility of Al along with Fe and Mn increases to such an extent that they become toxic to many plants. Moreover, plants may suffer from Ca and Mg deficiency in acid soils.

At low pH (<5.5), phosphorus is precipitated with Al, Fe, and Mn as their polyphosphates. At high pH (>8.0), P is precipitated with Ca. Both at soil acidity and alkalinity, P availability is reduced to deficiency levels. The predominance of the three ionic forms of P ($H_2PO_4^-$, HPO_4^{2-}, and PO_4^{3-}) is regulated by pH. Phosphorus as $H_2PO_4^-$ is more available than other forms, and it predominates at pH near 6.5. Availability of P is usually higher in the pH range of 6.5 and 7.0. Boron deficiency may also occur in strongly acidic and strongly alkaline soils. Availability of macronutrients (Ca, Mg, K, P, N, S) and Mo and B is restricted at low pH. On the other hand, availability of most micronutrients (Fe, Mn, Zn, Cu, Co) increases at low pH. The most satisfactory plant nutrient levels occur at a pH range of 5.5–6.5. Alkaline soils may have problems with deficiencies of such nutrients as zinc, copper, boron, and manganese. Soils with an extremely alkaline pH (greater than 9) are likely to have high levels of sodium which may deteriorate soil physical properties and exert toxicity on plants.

8.5.4.2 Toxicity of Al, Fe, and Mn

Toxicity of Al occurs due to predominance of Al^{3+} and $Al(OH)^{2+}$ ions in soils at pH below 5.2. $Al(OH)^{2+}$ is more toxic than Al^{3+}, and it predominates in soil at pH 4.5. Absorbed Al is accumulated in plant roots, impairing their normal functioning. Aluminum may block the uptake of Ca and may precipitate P in plant body. Aluminum may interfere with P metabolism including ATP and DNA functions. Plant roots become stunted with little branching; root tip and lateral roots become brown. Plants suffering from aluminum toxicity produce symptoms in leaves that resemble P deficiency.

Iron and manganese are essential nutrients, but their concentrations in very acidic soils may create toxicity to plant roots. Reduced forms of Fe and Mn which are predominant in O_2-depleted or waterlogged soils are more toxic than their oxidized forms. Toxicity by oxidized Fe frequently occurs in soils of pH below 4.0. Mn toxicity is likely in soils having 200–5,000 mg kg^{-1} Mn. Manganese concentrations in the range of 0.2–12 mM have been reported to produce severe growth limitations in solution culture studies of species such as cotton (*Gossypium hirsutum* L.) [Kennedy and Jones 1991], sweet potato (*Ipomoea batatas* L.) [Mortley 1993], sorghum (*Sorghum bicolor*) [Mgema and Clark 1995], and wheat (*Triticum aestivum* L.) [Taylor et al. 1991]. Aluminum sensitivity of some plants is shown in Table 8.3.

8.5.4.3 Microbial Population and Activity

Soil pH influences the microbial population, thereby regulating the biochemical transformations of mineral and organic materials in soils. Among bacteria, actinomycetes, and fungi, only fungi can thrive well in acid soils, and they outnumber bacteria and actinomycetes in such soils. Many fungal diseases of plant roots appear in acid soils. Organic matter decomposition, nitrogen mineralization, nitrification, and nitrogen fixation are reduced by soil acidity.

From the above discussion, it appears that growth and yield of crops are hampered by soil acidity and alkalinity. Some plants can tolerate acidity; they are tolerant to excess Al, Fe, and Mn or need less Ca, Mg, and P for normal growth. Management of acid and alkaline soils is discussed in Chap. 11. The following steps may, however, be taken if the soil is acidic or alkaline:

- Choose a set of crops suitable for the soil pH level.
- Adjust pH of the soil to the desired level by liming (if the soil is acidic) or acidifying (if the soil is alkaline).
- Ensure adequate irrigation and drainage to leach acids or bases.

8.5.5 Plants Have Preferences for pH

Plant species differ in their optimal soil pH requirement. Some crops may thrive best in very acid or alkaline soils, but most agriculturally important crops do well on soils of neutral pH (6.5–7.5). Legumes such as alfalfa, sweet clover, and soybeans are among these crops because the bacteria *Rhizobium* that fix nitrogen in their nodules proliferate best on neutral soils. Crops such as corn, sorghum, wheat, and cool-season grasses tolerate a wider pH range, including moderate acidity as well as neutral pH conditions. Soil pH preference of some plants is given below (Jett 2005):

pH range	Suitable crops
4.5–5.5	Azalea, camellia, rhododendron
5.5–6.0	Blackberry, chicory, gooseberry, magnolia, pineapple, potato, oats, rosemary
5.5–6.5	Apple, avocado, barley, cranberry, capsicum, melon, turnip, strawberry, wheat
6.0–6.5	Broad bean, bean, carrots, lemon, lentil, olives, parsley, pear, pumpkin, raspberry, soybean, squash, watermelon, white clover
6.0–7.5	Asparagus, broad bean, broccoli, cabbage, cauliflower, celery, cherry, chrysanthemum, cucumber, dahlia, French bean, garlic, grape, horse radish, lettuce, mulberry, mustard, onions, peach, pear, peas, peppers, plum, radish, spinach, sweet corn, tomato, turnips
6.5–7.5	Alfalfa, asparagus, avocado, barley, sugar beet, beetroot, grapevine, mushroom, spinach

8.5.6 Acidity May Be Active or Reserve

There are two types of soil acidity: active acidity and potential or reserve acidity. Active acidity of soil is based on the concentration of hydrogen ions in the soil solution, while reserve acidity is due to the amount of hydrogen and aluminum ions attached to clay and organic colloids in the soil. Reserve acidity is also known as exchangeable acidity. By soil pH, we measure the active acidity. Reserve acidity can be determined by replacing the adsorbed hydrogen and aluminum.

8.5.7 Soils Have Some Capacity to Resist Change in pH

Buffering capacity is the ability of a solution to resist changes in its pH on addition of small amount of acid or alkali. Some soils have high buffering capacity, that is, their pH values tend to remain unchanged upon addition of a considerable amount of acid or alkali. This quality of soil can be easily demonstrated by making a soil–water suspension and adding dilute acid slowly from a burette in it under a pH electrode. The pH reading will not change after addition of some acid in the beginning, pH will very slowly go down after some time, and when the buffering capacity is diminished, the pH reading will sharply fall. This resistance in the beginning is due to buffering capacity of the soil. Buffering substances in soil include weak acids and weak bases in soil solution; soluble acidic anions such as NO_3^-, HCO_3^-, and SO_4^{2-}; basic cations such as Ca^{2+} and Mg^{2+}; and exchangeable cations and anions and active groups on humus. Soils with high clay and organic matter content (i.e., higher CEC) have high buffering capacity. Calcareous soils often have high buffering capacities because free $CaCO_3$ effectively neutralizes acid.

Soils with a high buffering capacity need a great deal of liming or acidifying effort to alter pH. This is good if the soil has a desirable pH, but it can be a problem if the soil needs pH modification.

8.6 Redox Potential Is the Tendency of a Substance to Accept Electrons

Soils contain a variety of electron acceptors and electron donors. An electron acceptor is a chemical entity that accepts electrons transferred to it from another chemical entity. An electron acceptor is an oxidizing agent that, by virtue of its accepting electrons, is itself reduced in the process. Prominent oxidizing agents in soil are O_2, NO_3^-, Fe (III), Mn (IV), SO_4^{2-}, and CO_2. An electron donor donates electron during an oxidation–reduction reaction, and by donating electrons, it becomes oxidized. Electron donors are reducing agents, and prominent ones in soil are H_2, S, S^{2-}, Fe (II), Mn (II), CH_4, and organic compounds. However, oxidation is the loss of electrons or an increase in oxidation state by a molecule, atom, or ion.

Reduction is the gain of electrons or a decrease in oxidation state. A common redox reaction in soil is given below:

$$Fe(OH)_3 + e^- + 3H^+ \Leftrightarrow Fe^{2+} + 3H_2O$$

This reaction proceeds to the left (oxidation) if the soil is aerobic, and it proceeds to the right (reduction) if the O_2 supply in soil is low. This reaction or any oxidation–reduction reaction can be divided into two half reactions: one in which a chemical species undergoes oxidation and one in which another chemical species undergoes reduction. The driving force is the reduction potential if a half reaction is written as a reduction. If the half reaction is written as oxidation, the driving force is the oxidation potential related to the reduction potential by a sign change.

The redox potential is a measure of the affinity of a substance for electrons. It indicates the electronegativity of substances. Substances less electronegative than hydrogen have negative redox potentials, and substances more electronegative than hydrogen have positive redox potentials. The redox potential or oxidation–reduction potential (ORP) of a compound is measured under standard conditions against a standard reference half-cell. In biological systems, the standard redox potential is defined at pH=7.0 versus the hydrogen electrode and partial pressure of hydrogen=1 bar (IUPAC 2003). Redox potential is denoted by E_h and expressed in volts (V) or millivolts (mV).

8.6.1 Aeration Governs Redox Potential in Soil

Redox potential indicates the oxidation status of the soil. It varies within a wide range depending on the aeration in the soil. It may range from 100 to −300 mV in O_2-depleted soils and may be as high as 800 mV in very well aerated soils. According to Jackson (2005), well-oxidized soils, moderately well-oxidized soils, poorly oxidized soils, much-reduced soils, and extremely reduced soils may have redox potential values around 500, 300, 100, −200, and −300 mV, respectively. Thus, redox potential decreases with the depletion of oxygen level in soils. The sequence of reactions and stepwise lowering of E_h are shown in Chap. 13 on Wetland Soils.

8.6.2 There Is a Relationship Between pH and E_h of Soils

From the generalized redox reaction,

OX. (oxidizer) + ne^- + mH^+ = Red. (reducer) + m/n H_2O, the Nernst reaction can be written as

$$E_h = E_h^0 - \frac{RT}{nF}\ln\frac{(\text{Red})}{\text{Ox}}\left(H^+\right)^m,$$

where E_h^0 is the standard electrode potential, R is the gas constant, T is absolute temperature (Kelvin), and F is Faraday constant.

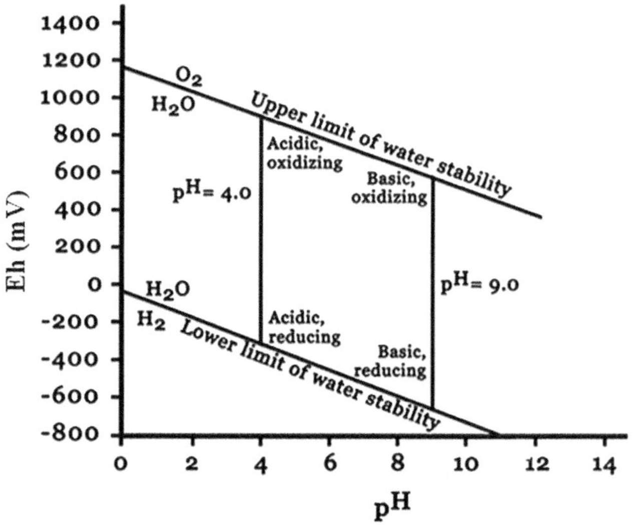

Fig. 8.11 The generalized pH–Eh relationship curve

$$\text{At } 25^\circ\text{C}, \left(\frac{RT}{F}\times 2.303\right) = 0.059 \text{ V, then}$$

$$E_h = E_h^0 - \frac{0.059}{n}\log\frac{(\text{Red})}{(\text{Ox})} - 0.059\frac{m}{n}\text{pH}.$$

Thus, E_h increases with the increase in activity of the oxidized component, decreases with increasing activity of the reduced component, and increases with an increase in H^+ ion activity or reduction in pH (Fig. 8.11). If the ratio of proton to electron (*m*/*n*) is equal, there is a predicted 0.059 V change in E_h per unit change in pH. Since the pressure of hydrogen in surface environments could not exceed 101 kPa, the maximum possible reducing potential in the presence of water would be

E_h = −0.059 pH − 0.03 log (1) = −0.059 pH (DeLaune and Reddy 2005).

Study Questions

1. What are the types of chemical reactions that occur in soils? Differentiate between precipitation and adsorption. How adsorption is related with ion exchange?
2. What are the types of colloids in soils? Distinguish between 1:1 and 2:1 clay colloids in respect of structure, swelling, and surface charge. Show the distribution of clay minerals in different soil orders.
3. What is the significance of electric charges on colloid surfaces? What is the difference between permanent and variable charges? How do charges develop on smectites and vermiculites?
4. Explain CEC, AEC, BSP, and SAR. Indicate the problems of low base status and sodicity. Why BSP in calcar-

eous soils is high? What will be the dominant cations in a soil with very low BSP? How will be its effect on soil reaction?

5. Define pH. What are the sources of H^+ ions in soil? How does soil pH affect nutrient availability and toxicity? Explain Eh and discuss its relationship with pH.

References

Bolt GH (1982) Soil chemistry. B: Physico-chemical models, 2nd rev edn. Elsevier, Amsterdam

Brandt RK, Hughes MR, Bourget LP, Truszkowska K, Greenler RG (1993) The interpretation of CO adsorbed on Pt/SiO_2 of two different particle-size distributions. Surf Sci 286:15–25

Chapman DL (1913) A contribution to the theory of electrocapillarity. Philos Mag 25(6):475–481

Chisholm-Brause CJ, O'Day PA, Brown GE Jr, Parks GA (1990) Evidence for multinuclear metal-ion complexes at solid/water interfaces from X-ray absorption spectroscopy. Nature 348:528–530

Delaune RD, Reddy KR (2005) Redox potential. In: Hillel D (ed) Encyclopedia of soils in the environment. Academic, New York

Evangelou VP (1998) Environmental soil and water chemistry principles and applications. Wiley, New York

Foy CD (1984) Physiological effects of hydrogen, aluminum, and manganese toxicities in acid soil. In: Adams F (ed) Soil acidity and liming. American Society of Agronomy, Madison

Fried M, Shapiro G (1956) Phosphate supply pattern of various soils. Soil Sci Soc Am Proc 20:471–475

Gee W, Shober A, Anderson B (1999) Chemical reactions in soils. http://www.cee.vt.edu/ewr/environmental/teach/gwprimer/group01/intro.html. Accessed 7 Oct 2011

Goldberg S (1992) Use of surface complexation models in soil chemical systems. Adv Agron 47:233–329

Gouy G (1910) Sur la constitution de la charge électrique à la surface d'un electrolyte. Ann Phys (Paris) (IV) 9:457–468

Harter RD, Smith G (1981) Langmuir equation and alternate methods of studying "adsorption" reactions in soils. In: Dowdy RH, Ryan JA, Volk VV, Baker DE (eds) Chemistry in the soil environment, Special publication no 40. American Society of Agronomy, Madison

Hough RL, Tye A, Crout N, McGrath SP, Zhang H, Young S (2005) Evaluating a free ion activity model applied to metal uptake by Lolium perenne L. grown in contaminated soils. Plant Soil 270(1):1–12

http://www.plantstress.com/Articles/min_deficiency_i/soil_fertility.pdf. Accessed 8 Nov 2011

Islam AKMS, Edwards DG, Asher CJ (1980) pH optima for crop growth. Results of a flowing solution culture experiment with six species. Plant Soil 54:339–357

IUPAC (2003) IUPAC compendium of chemical terminology. International Union of Pure and Applied Chemistry, Research Triangle Park

Jackson ML (2005) Soil chemical analysis – advanced course, Rev 2nd edn. Parallel Press, University of Wisconsin, Madison

Jett JW (2005) pH preference of plants. http://www.wvu.edu/~agexten/hortcult/homegarden/pHpref.pdf. Accessed 9 Oct 2011

Kennedy CW, Jones JE (1991) Evaluating quantitative screening methods for manganese toxicity in cotton genotypes. J Plant Nutr 14: 1331–1339

Langmuir I (1918) The adsorption of gases on plane surface of glass, mica, and platinum. J Am Chem Soc 40:1361–1382

Lovell J (2006) Guidelines for environmental assurance in Australian horticulture. Horticulture Australia Ltd., Sydney

McBride MB (1991) Processes of heavy and transition metal sorption by soil minerals. In: Bolt GH, Boodt MFD, Hayes MHB, McBride MB (eds) Interactions at the soil colloid-soil solution interface 190:149–176. Kluwer, Dordrecht

McBride MB (1994) Environmental chemistry of soils. Oxford University Press, New York

Mgema WG, Clark RB (1995) Sorghum genotypic differences in tolerance to excess manganese. J Plant Nutr 18:983–993

Mortley DG (1993) Manganese toxicity and tolerance in sweet potato. Hortic Sci 28:812–813

Mulder J, Cresser MS (1994) Soil and solution chemistry. In: Moldan B, Cerny J (eds) Biogeochemistry of small catchments, a tool for environmental research. Wiley, Chichester

O'Day PA, Brown GE Jr, Parks GA (1994) X-ray absorption spectroscopy of cobalt (II) multinuclear surface complexes and surface precipitates on kaolinite. J Colloid Interf Sci 165:269–289

Olsen SR, Watanabe FS (1957) A method to determine a phosphorus adsorption maximum of soils as measured by the Langmuir isotherm. Soil Sci Soc Am Proc 21:144–149

Schofield RK (1947) Calculation of surface areas from measurements of negative adsorption. Nature 160:408

Schwertmann U (1991) Solubility and dissolution of iron oxides. Plant Soil 130(1&2):1–25

Seilsepour M, Rashidi M, Khabbaz BG (2009) Prediction of soil exchangeable sodium percentage based on soil sodium adsorption ratio. Am-Eur J Agric Environ Sci 5(1):01–04

Sorenson SPL (1909) Enzyme studies II. The measurement and the importance of the hydrogen ion concentration in enzyme reaction. Coml Rend Tran Lab (Carlsberg) 8:1–168

Sparks DL (2003) Environmental soil chemistry, 2nd edn. Academic, Amsterdam

Sposito G (1986) Distinguishing adsorption from surface precipitation. In: Davis JA, Hayes KF (eds) Geochemical processes at mineral surfaces. Am Chem Soc Symp Ser 323:217–229

Tan KH (2011) Principles of soil chemistry, 4th edn. CRC Press, Boca Raton

Taylor GJ, Stadt KJ, Dale MRT (1991) Modelling the phytotoxicity of aluminum, cadmium, copper, manganese, nickel, and zinc stress using the Weibull frequency distribution. Can J Bot 69:359–367

Visconti F, De Paz JM, Rubio JL (2010) Calcite and gypsum solubility products in water-saturated salt-affected soil samples at 25 °C and at least up to 14 dS m^{-1}. Eur J Soil Sci 61:255–270

White RE (2006) Principles and practice of soil science: the soil as a natural resource, 4th edn. Blackwell Publishing, Malden

Wolt J (1994) Soil solution chemistry. Wiley, New York

9 Biological Properties of Soils

A large number of organisms live in the soil. They perform a variety of functions for their growth and reproduction. For these functions of soil organisms, soils behave like a living entity. Soil components photosynthesize, respire, and reproduce. In addition, they produce organic matter, consume organic matter, and decompose them. Some of them burrow in the soil, make spaces for their accommodation and movement, and mix surface and subsoil materials together. Soil becomes a dynamic body for the activity of soil organisms. The changes that are caused by soil organisms have their impact on soil fertility and productivity. A sterile soil is not a soil in the real sense. Although soil biota, which includes living roots and soil organisms, occupies a very small fraction of the total soil volume (<0.5%), it has tremendous influences on soil properties and soil processes. However, soil organisms are usually the most active in the surface soil zone of 0–15 cm, because this zone has accumulation of organic residues and available nutrients. Soil depth, organic matter and nutrients, microclimate, and physical and chemical soil environment influence the structure and function of soil biota.

9.1 Plant Roots, Rhizoplane, and Rhizosphere Are Unique Ecological Niches

Roots are the adsorbing and anchoring device of the vascular plants. Roots have some other functions as well. For example, there are food-storage roots, water-storage roots, propagative roots, pneumatophores, aerial roots, buttress roots, parasitic roots, and mycorrhizas. Roots of some plant species are effective soil binders. Dead roots leave channels that bring water and oxygen down to the organisms living in the soil. Roots bring nutrients from deeper layers to the surface. A plant with an expansive root system has a greater ability to adsorb water, oxygen, and nutrients from the soil. A well-developed root system keeps the plants healthy. The depth and volume of soil occupied by roots depend on plant species and soil conditions. Roots of most arable crop plants are limited to about 0–15 cm of the soil depth. Tree roots may, however, penetrate to greater depths. Probably, the most important ecological function, the roots perform, is the modulation of biological activity in soils of their vicinity. Fine and small roots (<5 mm) and coarse roots (>5 mm) are two major components of belowground biomass. Roots respire, and they are the major sources of carbon dioxide within the soil.

A part of the carbon fixed in the leaves during photosynthesis is carried to the roots via phloem and is translocated to the soil as a mixture of soluble and insoluble substances, together with sloughed cells. This process is known as rhizodeposition. Rhizodeposition brings a huge quantity of carbon and nutrients (up to 10–25% of carbon fixed by photosynthesis and 30–40% of the photosynthates translocated to the roots). Rhizodeposition releases several substances in the root zone, including exudates of water-soluble, low molecular weight compounds leached from the roots without metabolic control by the plant; secretions of low molecular weight compounds released by metabolic processes; and mucilages secreted by Golgi organelles in the root cap, hydrolysates of the polysaccharides of the primary cell wall and sloughed root cap cells, mucilage secreted by epidermal cells and root hairs, and those produced by bacterial degradation of dead epidermal cells. Rhizodeposition also includes mucigel, a gelatinous material at the surface of roots. There are lysates and sloughed cells from the epidermis and cortex of roots (Lavelle and Spain 2003).

Plant roots have enormous effects on population and functions of the soil organisms. Activities of soil flora and fauna on the rhizoplane (surface of the root) and in the rhizosphere (around the root) are different from other zones of the soil. Clark (1949) proposed the term "rhizoplane" to refer to the immediate surface of plant roots together with any closely adhering particles of soil or debris. Many microorganisms colonize the rhizoplane for the utilization of metabolites secreted by the roots. Rhizoplane microorganisms can influence plant growth and development. There are some bacteria which inhabit the rhizoplane and promote growth of

K.T. Osman, *Soils: Principles, Properties and Management*,
DOI 10.1007/978-94-007-5663-2_9, © Springer Science+Business Media Dordrecht 2013

plants. They are called plant growth-promoting bacteria (PGPB). Plant growth-promoting activities have been reported for several bacterial species, including *Pseudomonas, Azospirillum, Azotobacter, Klebsiella, Enterobacter, Alcaligenes, Arthrobacter, Burkholderia, Bacillus*, and *Serratia* (Han and Lee 2005). Some fungi inhabit the root surface in a mycelial state. They belong to the genera *Mortierella, Cephalosporium, Trichoderma, Penicillium, Gliocladium, Gliomastix, Fusarium, Cylindrocarpon, Botrytis, Coniothyrium, Mucor, Phoma, Pythium*, and *Aspergillus*.

The German scientist Hiltner introduced the term "rhizosphere" in 1904 to denote that region of the soil which is influenced by plant roots. Rhizosphere is characterized by greater microbiological and faunal activity than the soil away from plant roots. The rhizosphere soil differs in physical and chemical properties from the bulk soil (Whalley et al. 2005). Not only the population of microorganism is higher in the rhizosphere, but also the kinds of organisms and their requirements of metabolites are different. The rates of metabolic activity of the rhizosphere microorganisms are higher than those of the non-rhizosphere soil. It is, however, difficult to make a sharp demarcation in field between rhizosphere and non-rhizosphere zones of soil. The rhizosphere zone may be some millimeters wide, but it has no distinct boundary. It is an area of intense biological and chemical activity influenced by compounds exuded by the root and by microorganisms feeding on these compounds.

Plant roots exude a great variety of biochemical compounds such as amino acids, other organic acids, carbohydrates, sugars, vitamins, mucilage (polysaccharides), proteins, flavones, enzymes, hydrocyanic acid, glycosides, auxins, and saponins (Gupta and Mukerji 2002). Some of these substances supply food for the microorganisms; some others have growth-promoting or growth-inhibiting activity. However, the microorganisms mineralize organic matter and bring about other transformations that enhance nutrient availability for the plants. Root exudates stimulate the growth of many bacteria, including free-living nitrogen fixers of the genera *Azotobacter, Azospirillum*, and *Azoarcus*, the symbiotic nitrogen fixers of the genus *Rhizobium*, and several other bacteria and fungi, including the mycorrhizal fungi. All these activities make the rhizosphere the most dynamic environment in the soil. According to several estimates, the population of bacteria, fungi, and actinomycetes in the rhizosphere is five to ten times higher than the non-rhizosphere soil. The population of algae may be higher in the non-rhizosphere zone.

Very complex chemical, physical, and biological interactions occur between roots and their surrounding environment of soil. These interactions include root–root, root–insect, and root–microbe interactions, and they may be positive or negative. Positive interactions include symbiotic associations with bacteria and mycorrhizal fungi and root colonization by plant growth-promoting bacteria (PGPB). Plant growth-promoting bacteria were isolated from roots of a number of plants such as barley, bean, cotton, corn, groundnut, rice, various vegetables, wheat, and wood species (Manoharachary and Mukerji 2006). Negative interactions include competition or parasitism among plants, pathogenesis by bacteria or fungi, and invertebrate herbivory (Bais et al. 2006). Chemical interaction for plant–plant interference or allelopathy is one mechanism of suppression of growth of one plant by another. In addition, a number of phytotoxic compounds in plant root exudates have been identified, including 7,8-benzoflavone, catechin, juglone, 8- hydroxyquinoline, sorgoleone, and 5,7,4'-trihydroxy-3',5'-dimethoxyflavone (Bais et al. 2006). Spores or other propagules of many pathogenic fungi such as *Rhizoctonia, Fusarium, Sclerotium, Aphanomyces, Pythium, Colletotrichum, Verticillium*, and *Phytophthora* are shown to germinate as a result of stimulation and/or food sources provided by root exudates of susceptible cultivars of host plants.

9.2 Mycorrhizas Are Fungal Roots That Extend Enormously the Adsorbing Surface

The term mycorrhiza (*fungus–root*) was first applied to fungus–tree associations in 1885 by the German forest pathologist A.B. Frank (Jhonson 2009). A mycorrhiza is a symbiotic association between a fungus and a root of a vascular plant. It is a mutualistic association which provides the fungus with carbohydrates such as glucose and sucrose synthesized by the plant, and the plant, in return, gets the benefits of the huge mycelial network that adsorbs water and nutrients from a larger volume of soil. Mycorrhizal mycelia are much finer in diameter than the smallest root and can explore a large volume of soil for absorption of water and nutrients. Mycorrhizas are especially beneficial for the plant partner in nutrient-poor soils.

Seven types of mycorrhizas (arbuscular, ecto-, ectendo-, arbutoid, monotropoid, ericoid, and orchidaceous mycorrhizas) are generally encountered. Among these types, the arbuscular and ectomycorrhizas are the most abundant and widespread (Siddiqui and Pichtel 2008). Except ectomycorrhizas, the others were earlier taken together as endomycorrhiza. The hyphae of ectomycorrhizal fungi do not penetrate individual cells within the root but constitute a hyphal sheath, or mantle, covering the root tip and a Hartig net of hyphae surrounding the plant cells within the root cortex. Outside the root, the fungal mycelium forms an extensive network. On the other hand, the hyphae of endomycorrhizal fungi penetrate the cell wall and invaginate the cell membrane. Ectomycorrhizas are found to form association between the roots of woody plants such as birch, dipterocarp, eucalyptus, oak, pine, and rose families and fungi belonging to the

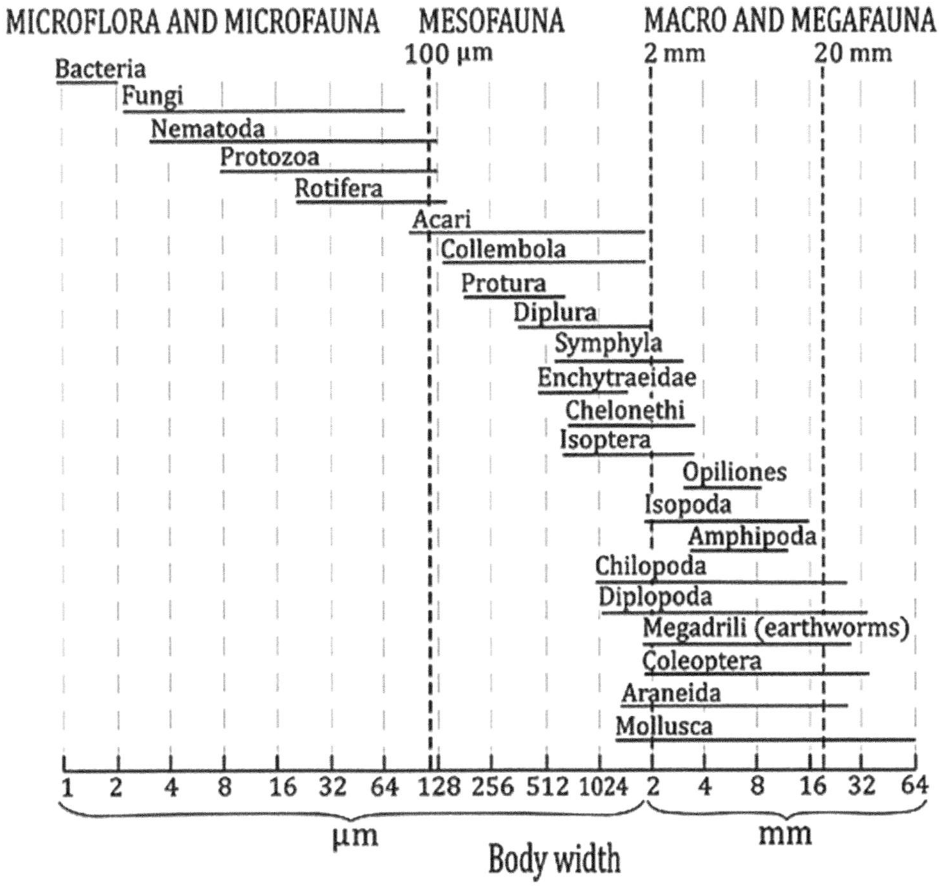

Fig. 9.1 Classification of soil fauna on the basis of body width

Basidiomycota, Ascomycota, and Zygomycota. Arbuscular mycorrhizas, or AM (formerly known as vesicular-arbuscular mycorrhizas, or VAM), are mycorrhizas whose hyphae enter the plant cells, producing either balloon-like structures (vesicles) or dichotomously branching invaginations (arbuscules). The fungal hyphae do not penetrate the protoplast, but invaginate the cell membrane. The structure of the arbuscules greatly increases the contact surface area between the hypha and the cell cytoplasm to facilitate the transfer of nutrients between them. Mycorrhizas have synergistic relations with some soil microorganisms (e.g., between *Glomus fasciculatum* and *Frankia*) and antagonistic relations with some others (such as between *G. fasciculatum* and *Pythium ultimum*).

9.3 Soil Organisms Include Macro- and Microflora and Fauna

Soil organisms are classified in many different ways. Conventionally, they are divided into flora and fauna. Besides plant roots, the other important component of soil flora is the microflora. Soil fauna is further divided into microfauna (body width <0.1 mm), mesofauna (body width 0.1–2.0 mm), and macrofauna (body width >2 mm) on the basis of their sizes. Examples of these categories are microflora—bacteria, algae, fungi; microfauna—protozoa and some nematodes; mesofauna—microarthropods and enchytraeids; and macrofauna—earthworms, termites, and millipedes.

9.3.1 Soil Fauna Are Diverse in Population and Function

Soil fauna include the major heterotrophs in soil systems; some of them are herbivores that feed on roots of living plants, but most live on dead plant matter, microbes, or both. Others are carnivores, parasites, or predators. They include life forms that live, for a part or whole of their life time, in the soil. Ladybird beetle is a transient species that hibernates in the soil but lives in the plant stratum; gnats live aboveground but they lay eggs in the soil and their larvae feed on decomposing organic residues. Cutworms are temporary soil residents; their larvae feed on plant seedlings. There are permanent soil residents such as collembolans. Swift et al. (1979) gave a classification scheme of soil fauna according to body width, which is shown in Fig. 9.1.

Body width of the fauna is related to their microhabitats. The microfauna (protozoa, small nematodes) inhabit water films. The mesofauna inhabit air-filled pore spaces; they cannot create spaces for themselves. The macrofauna make their own spaces through their burrowing activities, and like the megafauna, they can have large influences on gross soil structure (van Vliet and Hendrix 2003). Earthworms, termites, and ants alter the physical structure of the soil, influencing rates of nutrient and energy flow; they are "ecosystem engineers." Microarthropods fragment decomposing litter and improve its availability to microbes; they are litter transformers. Coleman et al. (2004) gave an account of soil fauna which is summarized below.

9.3.1.1 The Microfauna

Protozoa represent the microfauna of the soil and litter and belong to four ecological groups: flagellates, naked amoebae, testate amoebae, and ciliates (Lousier and Bamforth 1990). Protozoa are mainly distributed in the upper few centimeters of a soil profile:

1. Flagellates have one or more whiplike propulsive organs known as flagella. They are more numerous and active than other protozoa. They play a significant role in nutrient turnover by their feeding activities, with bacteria as their principal prey items.
2. Naked amoebae are very numerous and active in a wide range of agricultural, grassland, and forested soils. The principal feeding mode for the amoebae is phagotrophic (engulfing), with bacteria, fungi, algae, and other fine particulate organic matter. They inhabit very small pores in soil aggregates.
3. Testate amoebae are less numerous than naked amoebae, except in moist, forested systems where they thrive. They perform functions similar to naked amoebae.
4. Ciliates tend to be restricted to very moist or seasonally moist habitats. They are less numerous than others, with a general range of 10–500 g^{-1} of litter or soil. Like other protozoa, ciliates have resistant or encysted forms from which they can emerge when conditions become favorable for growth and reproduction.

Probably the most important ecological function the microfauna perform in soil is the regulation of microbial population and activity in the rhizosphere. Microfauna also play important roles in release of available nutrients, accretion and loss of soil organic carbon, and bioremediation of contaminated soils. Microfauna mineralize and immobilize N, P, and S and maintain nutrient balance in the rhizosphere. They suppress bacterial and fungal pathogens.

9.3.1.2 The Mesofauna

Rotifera

Rotifera are considered to be aquatic organisms and are found in water films and water-filled pores. Rotifers may also be found in tens of thousands per square meter in unsaturated soils. Rotifers have been found in leaf litter on forest floors.

Nematoda

Nematodes are also called roundworms. They are among the most numerous of the multicellular organisms in soil. They are also inhabitants of water films or water-filled pore spaces in soils. The body is cylindrical, tapering at the ends. Nematodes may be concentrated in the rhizosphere. Ingham et al. (1985) found up to 70% of the bacterial- and fungal-feeding nematodes in the rhizosphere. Nematodes seek areas of concentrated organic matter.

Tardigrada

These are micrometazoans and are called "water bears" because of their microursine appearance. Their slow movement resembles a tortoise. Tardigrades have four pairs of legs, equipped with claws on the distal end, of various sizes and forms. Tardigrades survive in extreme environmental stresses. Five types of latency (virtual cessation of metabolism) have been described: encystment, anoxybiosis, cryobiosis, osmobiosis, and anhydrobiosis. They feed on algal cells and debris and probably have a rather broad diet of various microbial-rich bits of soil organic matter. Tardigrades have also been observed to feed voraciously on nematodes.

Microarthropods

Microarthropods mainly include mites and collembolans and are found in most types of soils. Hundreds of thousands of individuals belonging to thousands of different species may be found within a square meter forest floor. Microarthropods have a significant impact on the decomposition processes of organic detritus. Many microarthropods feed on fungi and nematodes. Microarthropods in turn are prey for macroarthropods such as spiders, beetles, ants, and centipedes. Some smaller megafauna (toads, salamanders) may also feed upon microarthropods. Soil mites usually outnumber collembolans but may become more abundant in some situations. Among the mites, the oribatids usually dominate but the delicate Prostigmata may develop large populations in cultivated soils with a surface crust of algae.

Collembolans are also called "springtails" because many of the species are able to jump with a lever attached to the bottom of the abdomen. They also have a unique ventral tube (collophore), which may function in osmoregulation. They occur throughout the upper soil layer, where their major food is fungi associated with decaying vegetation. They are often the most numerous of the microarthropods in the rhizosphere.

The soil mites (Acari) are chelicerate arthropods related to the spiders and are the most abundant microarthropods in many types of soils. Four suborders of mites occur frequently in soils: the Oribatei, the Prostigmata, the Mesostigmata, and the Astigmata. Among them, the oribatids are the characteristic mites of the soil and are usually fungivores, detritivores, or both. Mesostigmatid mites are nearly all predators on other small fauna, although some few species are fungivores. Acarid mites are found associated with rich, decomposing nitrogen sources.

There is a diverse group of other small arthropods among mesofauna. These microarthropods have relatively small biomasses and probably have comparatively less impact on soil ecology. They include small spiders and centipedes, occasionally small millipedes, insect larvae, and adult insects.

Protura: Proturans are small, wingless, primitive insects readily recognized by their lack of antennae. Proturans occur in a variety of soils worldwide, often associated with plant roots and litter.

Diplura: Diplurans are small, elongate, delicate, primitive insects. Most diplurans are euedaphic, but some are nocturnal cryptozoans, hiding under stones or under bark during the day. They occur in tropical and temperate soils in low densities. They are predators on mites and other small arthropods but also ingest fungal mycelia and detritus.

Pseudoscorpionida: Pseudoscorpions resemble the scorpions, except that they lack tails and stingers. They are found in almost all types of soil. Pseudoscorpions are small cryptozoans, hiding under rocks and bark of trees, but they are found occasionally in leaf litters in the forest floor.

Symphyla: Symphylids are small, white, eyeless, elongate, many legged invertebrates that resemble tiny centipedes. They differ from centipedes in several characteristics, but superficially symphylids have 12 body segments and 12 pairs of legs, whereas centipedes have at least 15 pairs of legs, the first pair modified as fangs. Some species are pests feeding on roots of seedlings.

Pauropoda: Pauropods are myriapods with 8–11 pairs of legs and a branched antenna. They are white to colorless and blind; these characteristics make them members of the true euedaphic fauna. Pauropods occur in soils worldwide.

Enchytraeidae

Enchytraeidae is an important family of terrestrial Oligochaeta. This group of small unpigmented worms is also known as "pot worms." It consists of about 600 species in 28 genera. Species from 19 of these genera are found in soil. Members of the family are typically 10–20 mm in length and they are anatomically similar to the earthworms, except for the miniaturization and rearrangement of features overall. They possess setae (except in one genus), and a clitellum in segments XII and XIII, which contains both male and female pores.

Mesofauna live in existing pore spaces within and between the aggregates in the soil. They are small and like macrofauna they cannot make their own spaces themselves. Most of them chew and fragment organic debris and aid in their decomposition. They turn over soil organic residues and participate in nutrient recycling.

9.3.1.3 The Macrofauna

Macroarthropods

Macroarthropods are a group of larger insects, spiders, myriapods, and others. Their typical body lengths range from 10 mm to 15 cm (Shelley 2002). Many macroarthropods are cryptozoans; they dwell beneath stones, logs, under bark, or in cracks and crevices. Macroarthropods have direct effects on soil structure, porosity, aeration, and water movement. Termites and ants are important movers and mixers of soil; they bring deeper soil to the surface and on top of the litter layer. Emerging nymphal stages of cicadas may disturb soil structure. Scarabaeid beetles' larvae sometimes churn the soil in grasslands. Many macroarthropods are temporary soil residents. Macroarthropods may have a major influence on the microarthropod community. For example, Collembola are important food items for spiders.

Isopoda

Soil isopods are crustaceans; they occur under rocks and in similar habitats. They are generally saprovores. They can also feed upon roots or foliage of seedlings. Isopods possess heavy, sclerotized mandibles and are capable of fragmentation of decaying plant residues. In the laboratory, terrestrial isopods feed upon fecal pellets dropped by themselves or by any other isopods.

Diplopoda

Diplopoda (millipedes) are widely distributed saprophages. They are major consumers of organic debris in temperate and tropical hardwood forests. Millipedes become abundant in moist calcium-rich areas. They have a calcareous exoskeleton, and because of their high densities, they can be a significant sink for calcium. They can be important in calcium cycling. Millipedes are selective feeders; they avoid leaf litter high in polyphenols and favor those with high calcium content.

Chilopoda

Chilopoda (centipedes) are common predators in soil and litter. They occur in biomes ranging from forest to desert. Lithobius are the common brown, flat centipedes of litter in hardwood forests. The elongate, slim geophilomorph centipedes are euedaphic in forest habitats, where they prey on earthworms, enchytraeids, and Diptera larvae. All centipedes are predators.

Scorpionida

The scorpions (arachnid) have long legs, segmented, stinger-bearing abdomen, and chelate palpi. They are common creatures. They are inhabitants of warm, dry, tropical, and temperate regions but they are the most abundant in deserts. Scorpions are typical cryptozoans that hide under rocks or logs, or in crevices, during the day and emerge at night to feed.

Araneae

Araneae (spiders) are another familiar group of carnivores. They are found in all terrestrial environments except polar regions. Many species are found in aboveground habitats, but some are cryptozoans in litter and on the soil surface. Some small spiders are euedaphic. Some of the small litter-inhabiting spiders could be considered microarthropods.

Opiliones

Opiliones are delicate, shy, and the largest arachnids in woodlands. Their bodies are small but their legs are unusually long which suggest that their habitat is litter surface or exposed areas. There are smaller shorter-legged forms that inhabit loose leaf litter or small spaces. Some species occur high in foliage, others in subcanopy, some on soil surface, and some in litter layers.

Uropygi

Uropygi contain large species, up to 10 cm in length. It has a distinctive, long, whiplike tail but no stinger. The arachnid emits acetic acid, when disturbed, from a gland at the base of the tail. Uropygids are nocturnal predators.

The Pterygote Insects

Many winged insects (Pterygota) are residents of soils. Some are permanent soil inhabitants spending whole of its life in or on soil. Immature stages of other species are true soil dwellers but their adults are flying insects. All major winged insect orders—the Coleoptera (beetles), Lepidoptera (butterflies and moths), Hymenoptera (bees, wasps and ants), and Diptera (flies)—include soil-dwelling species. Termites belong to Isoptera and they are saprophages. The Homoptera (aphids, cicadas), Orthoptera (grasshoppers and crickets), and minor orders such as the Dermaptera (earwigs) contain soil-dwelling species. Of 26 pterygote insect orders, all but seven contain at least some soil-dwelling species.

Coleoptera

Beetles are the largest order of insects. They have soil species that are predatory, phytophagous, or saprovores. Most of them are transient members. The ground beetles (Carabidae) are among the more familiar insects active on soil surface of agroecosystems.

Hymenoptera

One of the largest orders of insects is the Hymenoptera. They have two groups of great importance: the ants and the ground-dwelling wasps. The ants are the most significant family of soil insects for their large influence on soil structure. Some bees and wasps nest in soil and also have some impacts on soils. Ants are widely distributed, numerous, and diverse.

Diptera

Many of the true flies are soil dwellers in some stages of their life cycles. Many species that live in aboveground habitats pupate in the soil. Many species of fly larvae are important saprovores in soils. They are restricted to moist soils rich in organic matter. Fly larvae have a major impact on decomposition rates of carrion. Maggots of various types hasten the decomposition rate significantly.

Isoptera

The Isoptera (termites) are among the most important of soil fauna for their impact on soil structure and on decomposition of detritus. Some termites possess a gut flora of protozoans, which enable them to digest cellulose. Their normal food is wood that has come into contact with soil. Most species of termites construct spectacular mounds. Termitidae do not have protozoan symbionts, but possess an array of microbial symbionts (bacteria and fungi) that enable them to digest the humified organic matter in tropical soils. There are wood-feeding, plant-feeding, and humus-feeding termites.

Other Pterygota

The Orthoptera, grasshoppers and crickets, lay eggs in soils and some are active on the soil surface. The Psocoptera, psocids, are a small order of insects that occasionally become abundant in leaf litter. They feed on organic detritus, algae, lichens, and fungus. The order Homoptera, cicadas, aphids, and others, has members important as belowground herbivores and as soil movers.

Gastropoda

Soil gastropods (snails and slugs) are major herbivores and detritivores in agroecosystems. They favor moist conditions and the presence of significant amounts of calcium for their metabolic needs, but some gastropods exist successfully in low pH and low calcium environments.

Oligochaeta: Earthworms

The most familiar and often the most important of the soil fauna with respect to soil processes are the earthworms. They fragment plant residues, bury them in, and mix them well with the soil. More than 3,500 earthworm species have been identified so far. Earthworms are classified within the phylum Annelida, class Oligochaeta, and order Opisthopora. Ten of the 16 families consist of the terrestrial forms commonly known as earthworms. Species within the families Lumbricidae and Megascolecidae are ecologically important.

Earthworms are soft-bodied, segmented animals. In length they range from a few millimeters to more than a meter. They consist of a simple, tube-within-a-tube body plan, the outer tube constituting the body proper and the internal tube comprising the alimentary canal. Soil material is ingested and drawn through the mouth into a muscular buccal cavity and then through the pharynx into the esophagus. Many species have a muscular esophageal gizzard that grinds and mixes food material as it passes through. Earthworms are called ecosystem engineers because they have pronounced effects on soil structure, aeration, and water movement by their burrowing activities as well as their ingestion of soil and production of castings (van Vliet and Hendrix 2003). Casts are produced after earthworms ingest mineral soil and particulate organic matter, mix them together and enrich them with organic secretions in the gut, and then egest the material as a slurry or as discrete fecal pellets within or upon the soil. Turnover rates of soil through earthworm casting range from 40 to 70 t $ha^{-1}year^{-1}$ in temperate grasslands to 500–1,000 t $ha^{-1}year^{-1}$ in tropical savannas. Earthworm burrowing in soil creates macropores of various sizes, depths, lengths, and orientations, depending on their species and soil type. Continuous macropores resulting from earthworm burrowing may enhance water infiltration by functioning as bypass flow pathways through saturated soils.

Earthworms may have some detrimental effects. These include removing and burying of surface residues that would otherwise protect soil surfaces from erosion; increasing erosion and surface sealing by free casts; creating nuisance by castings on the surface of lawns and golf greens; dispersing weed seeds in gardens and agricultural fields; transmitting plant or animal pathogens; increasing losses of soil nitrogen through leaching and denitrification; and increasing soil carbon loss through enhanced microbial respiration.

9.3.2 Soil Microflora Include Bacteria, Fungi, and Algae

9.3.2.1 Bacteria

Bacteria are unicellular prokaryotes; they do not have any distinct nuclear zone in the cytoplasm. Their sizes range from 0.3 to more than 3 μm. Bacteria may be classified on the basis of cell morphology (cocci, bacilli, spirilla), cell wall structure (Gram-positive and Gram-negative through specific staining), presence of endospores, mobility of cells, and shape and position of flagella, if any.

Bacteria live in soil as cocci (sphere, 0.5 μm), bacilli (rod, 0.5–0.3 μm), or spirilla, the bacilli being more common and spirilla being rare in soil (Baudoin et al. 2002). There are again autochthonous and zymogenous bacteria. The autochthonous bacteria get their nutrition from native soil organic or mineral matter (*Arthrobacteria* and *Nocardia*). On the other hand, the zymogenous bacteria require additional external substrate (*Pseudomonas* and *Bacillus*). The number of zymogenous bacteria increases when they get appropriate substrate and gradually declines when the added substrate is exhausted (cellulose decomposers, nitrogen transformers) (Giri et al. 2005).

On the basis of the energy source that they use (light or energy from redox reactions) and the nature of the electron donor (organic or mineral), bacteria are divided into four categories: (1) photolithotrophic bacteria using energy from light and inorganic reduced substance (essentially sulfides) as electron donors; (2) photoorganotrophic bacteria which are photosynthetic organisms using oxidizable organic substrates as electron donors; (3) chemolithotrophic bacteria using the energy produced by redox reactions and four kinds of mineral substrates as electron donors (reduced nitrogen, sulfur or iron compounds and hydrogen); (4) chemoorganotrophic bacteria are typical heterotrophic organisms taking their energy from redox reactions and using organic compounds as electron donors. There are also obligate chemoautotrophs such as *Nitrobacter* utilizing nitrite and *Nitrosomonas* utilizing ammonium, while *Thiobacillus* converts inorganic sulfur compounds to sulfate and *Ferrobacillus* converts ferrous ions to ferric ions (Baudoin et al. 2002).

Functional categories of bacteria are defined by the chemical transformations they perform. Some of the most common examples are cellulolytic, chitinolytic, nitrifiers, N-fixers, denitrifiers, etc. The most common soil bacteria belong to the genera *Pseudomonas, Arthrobacter, Clostridium, Achromobacter, Bacillus, Micrococcus, Flavobacterium, Corynebacterium, Sarcina, Azosprillium*, and *Mycobacteria*. Another group of bacteria common in soil is the myxobacteria belonging to the genera *Myxococcus, Chondrococcus, Archangium, Polyangium, Cytophaga*, and *Sporocytophaga*. The latter two genera are cellulolytic and, hence, are dominant in cellulose-rich environments. Bacteria can withstand extreme climates, although temperature and moisture influence their population. Some bacteria can thrive in arctic zones where the temperature is below freezing point and some others in arid desert soils where temperatures are very high. Based on the temperature tolerance, bacteria are grouped as psychrophilous (below 20°C), mesophyllous (15–45°C), and thermophilous (45–65°C). However, mesophyllous bacteria constitute the bulk of soil bacteria. Soil pH, farm practices, fertilizers and pesticide applications, and organic matter amendments affect population and activity of bacteria in soil.

The cyanobacteria are Gram-negative eubacteria characterized by their ability to perform oxygenic photosynthesis. They are true prokaryotic microorganisms. They have some characteristics common to algae for which they were earlier named "blue-green algae." Cyanobacteria contain a pigment known as phycocyanin, in addition to chlorophyll, which gives a special blue-green color to these organisms. The dominant cyanobacteria belong to the genera *Chroococcus, Aphanocapsa, Lyngbya, Oscillatoria, Phormidium, Microcoleus, Cylindrospermum, Anabaena, Nostoc, Scytonema*, and *Fischerella* (Benizri et al. 2002). Some cyanobacteria also possess heterocysts, which are involved in nitrogen fixation.

9.3.2.2 Actinobacteria

The Actinobacteria (formerly called Actinomycetes) are filamentous bacteria. They are mostly Gram-positive. They possess a ramified pseudomycelium whose diameter (0.5–1 μm) is much smaller than that of fungi. The Actinobacteria resemble bacteria in that they have a very simple cell structure and are about the same size in cross section. They resemble filamentous fungi in that they produce a branched filamentous network. The network compared to fungi, however, is usually less extensive. They are poorly tolerant of soil acidity and most are unable to grow in soils with acidity more than pH 5. The most conducive range of pH for Actinobacteria is between 6.5 and 8.0. Waterlogging of soil is unfavorable for the growth of Actinobacteria, whereas desert soils of arid and semiarid zones sustain sizeable populations, probably due to the resistance of spores to desiccation. The commonest genera of Actinobacteria are *Streptomyces*. In contrast, *Nocardia, Micromonospora, Actinomyces, Actinoplanes*, and *Streptosporangium* are only encountered occasionally. Although there is evidence that Actinobacteria are abundant in soils, it is generally concluded that they are not as important as bacteria and fungi as decomposers (Foth 1990).

9.3.2.3 Fungi

Fungi are heterotrophic organisms. A large number of fungi live on dead organic matter; they are saprophytes. Others live on living plant or animal tissue; they are parasites. Some fungi are deadly pathogens of plants and animals. Saprophytes depend on organic detritus for carbon and decompose them in course of their growth and nutrition. The quality and quantity of organic matter affect the population and activity of saprophytic fungi in soils. Fungi have filamentous mycelium composed of individual hyphae. The hyphae may be uni-, bi-, or multinucleate and septate or nonseptate (Hawksworth 1991).

In acid soils fungi dominate the microbial community because acidic soils are not favorable for most of bacteria and Actinobacteria. Fungi are also present in neutral or alkaline soils, and some can tolerate a pH over 9.0. Since fungi are strictly aerobic, they are generally abundant in arable soils. Their numbers decrease in waterlogged soils. However, moist organic detritus such as leaf litter or wood are ideal habitat for fungi. Some fungi dwell in deep layers of soils. They are not generally found in the surface soil. Probably, the type of organic substrate is behind this selectivity.

Fungi are classified into Phycomycetes, Ascomycetes, Basidiomycetes, and Fungi Imperfecti (Alexander 1977). Phycomycetes constitute an important class of parasitic or saprophytic fungi. The fungal body may be an undifferentiated mass of protoplasm to a well-developed and much-branched mycelium. Reproduction is mainly sexual, by the formation of conidia or sporangia. Ascomycetes include fungi which produce spores inside a sac called an ascus. Each ascus usually contains 4–8 spores depending on the species. The Basidiomycetes fungi constitute the most conspicuous group of fungi in the environment and include mushrooms, puffballs, and bracket fungi. Basidiomycetes produce a highly specialized sporangium, the basidia. Many fungi, commonly isolated from soils, come under the class Fungi Imperfecti because they produce abundant asexual spores but lack sexual stages (Lynch 1987). Many members of Asco- and Basidiomycetes can degrade very complex organic compounds such as cellulose or lignin. Many others live as mycorrhizas on roots of higher plants and obtain simple sugars from their plant partners.

The following genera of fungi are most commonly found in soils: *Acrostalagmus, Aspergillus, Botrytis, Cephalosporium, Gliocladium, Monilia, Penicillium, Scopulariopsis, Spicaria, Trichoderma, Trichothecium, Verticillium, Alternaria, Cladosporium, Pilularia, Cylindrocarpon* and *Fusarium, Absidia, Cunninghamella, Mortierella, Mucor, Rhizopus, Zygorhynchus, Pythium, Chaetomium*, and *Rhizoctonia* (Hawksworth 1991). Many yeasts belonging to true Ascomycetes such as *Saccharomyces* and those belonging to Fungi Imperfecti such as *Candida* have been isolated from soils. However, their number in soil is relatively low. Certain fungi like *Alternaria, Aspergillus, Cladosporium, Dematium, Gliocladium, Helminthosporium, Humicola*, and *Metarhizium* produce substances similar to humic substance in soil and, hence, may be important in the maintenance of soil organic matter (Hawksworth 1991).

9.3.2.4 Algae

Algae need sunlight and moisture for proliferation. They are numerous in soils, particularly in the soil surface. Soil algae mainly belong to the class Chlorophyceae. There are also diatoms in soils. These microorganisms form green scum on the surface of soils and are visible to the naked eye, although most algae are microscopic. They are, however, less numerous than fungi in soil. Algae may be unicellular (*Chlamydomonas*) or filamentous (*Spirogyra, Ulothrix*). Algae contain chlorophyll and they are photoautotrophic organisms. They use CO_2 from the atmosphere and produce O_2. Some algae are found in deeper soil layers beyond the reach of sunlight. These forms obtain their energy largely from organic matter. Algae benefit the soil by contributing organic matter to the soil. Some of the common green algae occurring in most soils belong to the genera *Chlorella, Chlamydomonas, Chlorococcum, Oedogonium,*

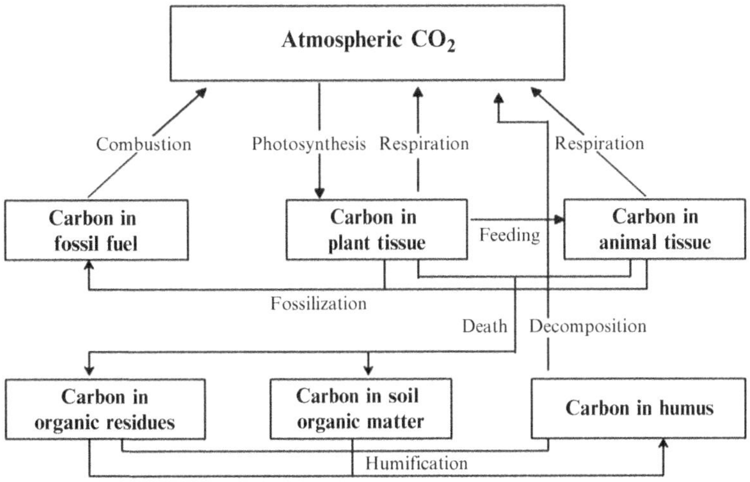

Fig. 9.2 The carbon cycle

Chlorochytrium, and *Protosiphon* (Lynch 1990). The earlier known blue-green algae are now cyanobacteria; they are included in bacteria.

9.4 Carbon, Nitrogen, Phosphorus, and Sulfur Cycles Are Biogeochemical Cycles

Soils are a sink of carbon, nitrogen, phosphorus, and sulfur. Carbon and nitrogen are added to soils from the atmosphere through physical and biological fixation mechanisms. Although sulfur may be found in the atmosphere as SO_2 and H_2S gases and particulates which fall to the ground as acid rain, phosphorus does not have an atmospheric component. In the terrestrial ecosystem, carbon and nitrogen are circulated through the atmosphere, pedosphere, and biosphere system. Huge quantities of C and N are fixed; huge quantities of C, N, P, and S are added to soils as organic residues. They are chemically and biologically transformed in soil, a large proportion are adsorbed by plants and again added to soils after their death and decay. A large proportion of these elements are exported with crop harvest and imported as fertilizers. In this way, these elements form ecological cycles. A considerable part of them is transferred to other ecosystems, such as lakes, streams, and oceans where they become a part of the greater global cycles.

9.4.1 Carbon Cycle Involves Release and Fixation of CO_2 to and from the Atmosphere

Biological energy transfer within the biosphere at landscape and ecosystem scales and within organisms goes hand in hand with the carbon cycle. Large carbon stores occur in the lithosphere, the biosphere, the atmosphere, and the hydrosphere. Mass transfers of carbon occur between the lithosphere, the biosphere, the seas and other water bodies, and the atmosphere as part of the global carbon cycle (Schlesinger 1997).

Carbon is stored as CO_2 in the atmosphere (2%), as biomass in land plants and soils (5%), as fossil fuels in a variety of geologic reservoirs (8%), and as a collection of ions in the ocean (85%). In the terrestrial ecosystem, the major carbon stores are the aboveground biomass, the litter layer, and the soil including plant roots. Within the soil, there are the living roots: the microbial biomass and the meso- and macrofauna. The soil organic matter component consists of dead residues of variable degrees of decomposition ranging from the recognizable plant and animal remains to the highly stable humic compounds. Carbon is distributed in different soil horizons through litterfall, root death and decomposition, leaching, and exudation. Carbon in soils occurs in a variety of compounds ranging in complexity from CO_2 and carbonates to large structured macromolecules.

Primary producers including plants and autotrophic microorganisms fix atmospheric CO_2 by the process of photosynthesis. Animals consume plants and other animals for their carbon. Huge carbon materials are added to soils with dead parts of organisms (Chap. 7), their excreta and exudates. These residues are deposited on the surface by litterfall and as crop residues and at variable depths by root death and exudation. Flora and fauna mix these materials with the soil. The dead organic matter is colonized by a variety of soil heterotrophic microorganisms, which derive energy for growth from the decomposition of complex organic molecules. During decomposition, essential elements are converted from organic combination to simple inorganic forms; this process is called mineralization. For example, organically combined N, P, and S appear as NH_4^+, $H_2PO_4^-$, and SO_4^{2-} ions, and a considerable amount of C is released as CO_2. The remainder of the substrate C used by the microorganisms is incorporated into microbial cell substance or microbial biomass, together with a variable proportion of other essential elements such as N, P, and S. This incorporation makes these elements unavailable for plant growth until the organisms die and decay; so the process is called immobilization. The carbon cycle is shown in Fig. 9.2.

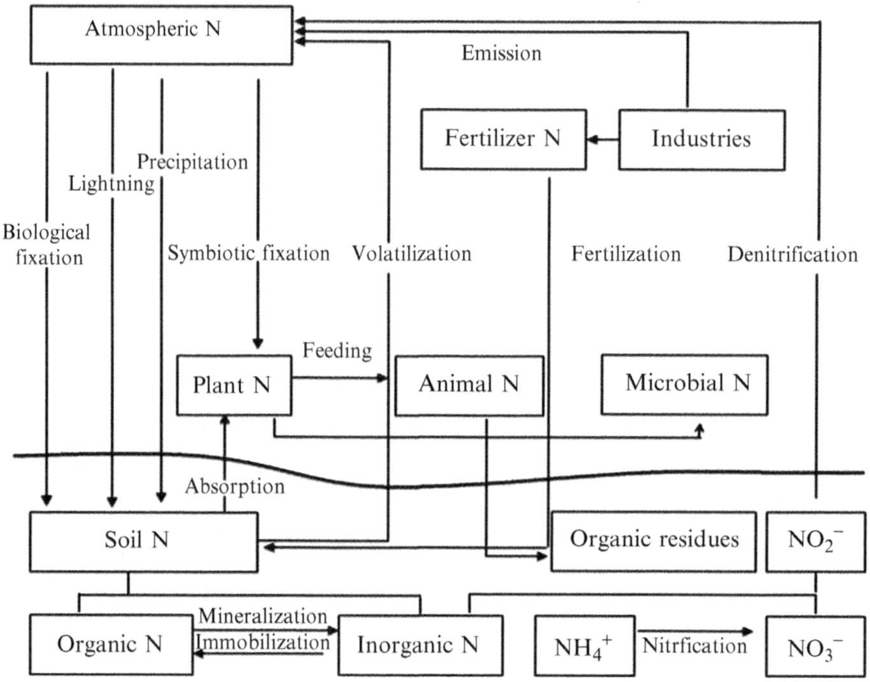

Fig. 9.3 The nitrogen cycle

During the course of decomposition of organic residues, recalcitrant substances accumulate and some new complex organic compounds are synthesized by microorganisms. Complexation of these substances with soil mineral matter produces humus. Thus, the important processes in the soil carbon cycle are immobilization, mineralization, and decomposition. In addition, humification is a concomitant process of decomposition and resynthesis by which stabilized humic materials are produced in soil.

9.4.2 Nitrogen Moves Through Soil–Organism–Atmosphere Pathway in a Cycle

The ultimate source of nitrogen is the atmosphere. It contains about 78% nitrogen in the mixture of gases called air. From atmosphere, some nitrogen is brought to the soil by atmospheric fixation and deposition with precipitation (Fig. 9.3).

A huge amount of nitrogen is chemically fixed in industries, a major part of which is used for manufacturing fertilizers. Large amount of nitrogen fertilizers is added to soils throughout the world for increased crop production. Added and native soil nitrogen is taken up by plants to make their body materials, namely, amino acids, proteins, nucleic acids, and nucleoproteins. Animals obtain their proteins from plants and other animals. Some microorganisms, with or without association with plants, utilize dinitrogen from the atmosphere. This is called biological nitrogen fixation. Plant and animal residues (excreta, anthropogenic wastes, fallen litter, crop residues, and dead organisms) return a significant proportion of nitrogen to the soil as organic nitrogen. Organic nitrogen is mineralized into inorganic nitrogen, NH_3, by soil microorganisms. Ammonia is converted into NO_3^- by the process of nitrification. Some ammonia is volatilized and some nitrate is denitrified and released again into the atmosphere. Thus, nitrogen moves in a cycle among the soil–organism–atmosphere components. This is known as the nitrogen cycle. The chief biological processes of this cycle are mineralization, immobilization, nitrification, nitrogen fixation, and denitrification.

9.4.2.1 Nitrogen Mineralization and Immobilization

During the course of decomposition of organic matter, organic nitrogenous compounds such as amino acids, peptides, proteins, and nucleoproteins that are present in soil organic matter are converted into inorganic nitrogen such as ammonium and nitrate (NH_4^+, NO_3^-). This process of conversion of organic nitrogen to inorganic nitrogen is called mineralization. It is carried out by a range of heterotrophic microorganisms, mainly fungi and bacteria. Through this process, organically bound unavailable nitrogen is made available to the plants. The rate of nitrogen mineralization is governed by the relative content of carbon and nitrogen in the organic residues. The higher the nitrogen contents of the organic matter, that is, the narrower the C/N ratio, the faster is the nitrogen mineralization. On the other hand, when low nitrogen residues are allowed to decompose, the decomposing microorganisms consume some available nitrogen (NH_4^+, NO_3^-) from the soil. In this process inorganic nitrogen is converted into organic nitrogen. The process is opposite to the mineralization and is called immobilization. Mineralization and immobilization can take place simultaneously. The dynamic nature of these competing processes causes either an increase in soil inorganic N or, conversely, a decrease when inorganic N is immobilized into organic forms.

$$\text{Organic nitrogen} \longrightarrow \text{inorganic nitrogen [mineralization]}$$
$$\text{Organic nitrogen} \longleftarrow \text{inorganic nitrogen [immobilization]}$$

Environmental factors affect the rate of N mineralization in the soil and thus the amount mineralized over time. Soil temperature and moisture content have a strong effect on N mineralization reactions. Microbial activity is limited at soil temperature near freezing and increases with rising soil temperature. Maximum N mineralization occurs when the soil temperature reaches 30–35°C. In dry soils, N mineralization is low because soil microorganism activity is limited by water availability. In saturated soils, lack of oxygen limits N mineralization because only soil microorganisms that can survive under anaerobic conditions are active. Soil management practices have a strong effect on the N mineralization potential of a soil. Intensive farming practices that rely heavily on tillage and synthetic fertilizers tend to decrease the soil's mineralizable N pool.

9.4.2.2 Nitrification

Nitrification is the biological oxidation of ammonium to nitrate. Nitrification occurs in two steps. In the first step, ammonia is oxidized to nitrite.

$$NH_3 + O_2 \rightarrow NO_2^- + 3H + +2e^-.$$

Nitrosomonas is the most frequently identified genus associated with this step, although other genera, including *Nitrosococcus* and *Nitrosospira*, may be involved. Some subgenera, *Nitrosolobus and Nitrosovibrio*, can also autotrophically oxidize ammonia. In the second step of the process, nitrite-oxidizing bacteria *Nitrobacter* oxidize nitrite to nitrate according to equation

$$NO_2^- + H_2O \rightarrow NO_3^- + 2H + +2e^-.$$

The nitrifying bacteria are Gram-negative chemoautotrophic organisms; they usually derive their energy for growth by oxidizing these inorganic nitrogen compounds. Nitrifying bacteria are most active at pH near neutrality and slightly alkaline conditions. Nitrification is low in acid soils. Acid-tolerant strains of nitrifying bacteria may carry out some nitrification in acid soils. Nitrifying bacteria are strict aerobes; they must have free oxygen to perform their work.

Nitrification is a biological process of tremendous environmental importance. Whatever nitrogen fertilizers are applied in soil, they are rapidly converted to nitrate by soil microorganisms. Adsorbed nitrate on soil colloids is quickly replaced by other anions, and very little nitrate can be retained in soil. Nitrate is transported with water and pollutes the groundwater. Excess nitrate in drinking water is involved in severe health hazards (Chap. 10).

9.4.2.3 Denitrification

Denitrification is the biological reduction of nitrate to nitrogen gas (N_2). It occurs in soils, particularly wet soils, and it causes loss of available nitrogen. Some bacteria known as denitrifying bacteria bring about this conversion. Denitrifying bacteria include representatives of *Pseudomonas, Achromobacter, Bacillus*, and *Micrococcus*. All denitrifying bacteria are aerobes, and they can use oxygen as electron acceptor for the oxidation of organic matter. When anaerobic conditions prevail (in soils with high moisture saturation, i.e., soils with low free O_2), they use the nitrates as electron acceptors and reduce them. Most denitrifiers produce nitrous oxide (N_2O) instead of dinitrogen (N_2) under aerobic conditions (Takaya et al. 2003). Denitrification involves a series of biological conversions.

$$NO_3^- \rightarrow NO_2^- \rightarrow NO \rightarrow N_2O \rightarrow N_2.$$

The net reaction may be written as

$$2\,NO_3^- + 10\,e^- + 12\,H^+ \rightarrow N_2 + 6\,H_2O.$$

9.4.2.4 Biological Nitrogen Fixation

Atmosphere contains thousands of tons of nitrogen over head, but even a single molecule of it cannot be used by plants and animals without the intervention of a group of microorganisms called diazotrophs. Diazotrophs fix atmospheric nitrogen either as free-living organisms or in association with a higher plant. The process can be represented by the following equation in which 2 mol of ammonia are produced from 1 mol of nitrogen gas:

$$N_2 + 8H^+ + 8e^- + 16ATP \rightarrow 2NH_3 + H_2 + 16ADP + 16Pi$$

Nitrogen fixers combine atmospheric nitrogen with other elements to form the organic compounds that make up the protoplasm of their cells. With the death of the nitrogen fixers, the nitrogen is liberated as ammonia and nitrates, forms available to plants.

9.4.2.5 Nonsymbiotic Nitrogen Fixation

There are many free-living bacteria capable of fixing atmospheric nitrogen in soil. They are all prokaryotes. However, there is a great diversity of metabolic types of free-living nitrogen-fixing bacteria. They may be obligate anaerobes,

Table 9.1 Important symbiotic nitrogen-fixing bacteria and their host plants

Species	Host plants
Rhizobium leguminosarum bv. *phaseoli*	Common bean
Rhizobium leguminosarum bv. *trifoli*	Clover
Rhizobium leguminosarum bv. *viceae*	Pea, vetch
Rhizobium tropici	Common bean
Rhizobium etli	Common bean
Mesorhizobium loti	*Lotus japonicus*
Azorhizobium caulinodans	Sesbania
Sinorhizobium meliloti	Alfalfa
Sinorhizobium fredii	Soybean
Bradyrhizobium japonicum	Soybean, cowpea, mung bean
Bradyrhizobium elkanii	Soybean, cowpea, mung bean

facultative aerobes, or aerobes. The dominant free-living nitrogen-fixing bacteria belong to the genera *Azotobacter, Azospirillum, Beijerinckia, Chromatium, Clostridium, Desulfovibrio, Klebsiella, Paenibacillus, Pseudomonas, Rhodopseudomonas, Rhodospirillum,* and *Thiobacillus* (Philippot and Germon 2005). Free-living nitrogen-fixing bacteria are widespread but the rates of nitrogen fixation by them are relatively low (<3 kg $ha^{-1}year^{-1}$). Free-living nitrogen fixers represent a range of bacteria including saprophytes living on plant residues, bacteria living in close association with the rhizosphere of plant roots, and bacteria which live entirely within plants (endophytes). There are both aerobes and anaerobes among nitrogen fixers. For example, *Azotobacter* and *Azospirillum* live in aerobic conditions, whereas others such as *Clostridium pasteurianum* live in oxygen-free (anaerobic) environment. The amount of nitrogen fixation by nonsymbiotic bacteria depends upon environmental conditions. These organisms are extremely sensitive to acidity. Nitrogen fixed by this means in soils with acidity more than pH 6 is negligible. Free-living nitrogen fixers need ample supply of calcium, potassium, phosphorus, and traces of iron, molybdenum, and manganese.

9.4.2.6 Symbiotic Nitrogen Fixation

Some plants, usually legumes, form symbiotic association with microorganisms, a group of bacteria known as *Rhizobium*. Most rhizobia form nodules in roots and stems of leguminous plants and can convert atmospheric nitrogen N_2 into NH_3 and then into amino acids in their tissue. This is known as symbiotic nitrogen fixation. Table 9.1 gives a list of rhizobia species and their hosts from an account by Stacey (2007).

Some Actinobacteria can also form symbiosis but on other plants than legumes. Some algae such as *Anabaena* can be associated symbiotically with the aquatic pteridophyte *Azolla*. These associations also have nitrogen-fixing capacity. But rhizobia are the most significant nitrogen fixers. However, rhizobia represent only a small fraction of the soil microflora with densities varying widely in a range of 10^2–10^4 bacteria g^{-1} soil. Rhizobia can live freely, although in small numbers, in soil, but they can fix nitrogen only when they are in nodules of plants.

Taxonomy of nodule bacteria has been subject to rapid changes. At present 12 genera and some 50 species of root- and stem-nodule bacteria are recognized. Graham (2008) gives a list of 11 genera with their host plants. These genera are *Allorhizobium, Azorhizobium, Blastobacter, Bradyrhizobium, Burkholderia, Devosia, Ensifer, Mesorhizobium, Ralstonia, Rhizobium,* and *Sinorhizobium*. Only a few of these organisms fix significant amounts of N_2 outside their host (Elliott et al. 2006).

Some shrub and tree species other than legumes form a symbiosis with a nitrogen-fixing Actinobacteria called *Frankia*. The genera that have been reported so far to fix nitrogen symbiotically are *Alnus, Ceanothus, Cerocarpus, Chamaebatia, Comptonia, Coraria, Cowania, Datisca, Dryas, Elaeagnus, Myrica, Purshia, Shepherdia, Colletia, Discaria, Kentrothamnus, Retanilla, Talguena, Trevoa, Hippophae, Allocasuarina, Casuarina, Ceuthostoma, Coriaria, and Gymnostoma* (Paul and Clark 1996). In addition, some lichenous fungi, liverworts, pteridophytes, gymnosperms, and angiosperms are able to establish symbioses with nitrogen-fixing cyanobacteria *Nostoc* and *Anabaena*. Symbiotic associations between rhizobia and leguminous plants may contribute from several tens to 350 kg N ha^{-1} $year^{-1}$ depending on environmental conditions. BNF obtained with nonleguminous angiosperms producing symbiosis with *Frankia* is between 15 and 77 kg N ha^{-1} $year^{-1}$ for *Casuarina equisetifolia* (Dommergues 1997), 29 and 117 kg N ha^{-1} $year^{-1}$ for *Alnus nepalensis* (Shrama 1993), and 18 kg Nha^{-1} $year^{-1}$ for *Myrica faya* (Vitousek and Walker 1989).

9.4.2.7 Nodulation

Roots of legumes release flavonoid compounds which trigger the production of nod factors by the bacteria. Due to this nod factor, a number of biochemical and morphological changes occur in roots; cell division is triggered to create the nodule. The bacteria attach to a point of the root and the root hair repeatedly winds around the bacteria until it fully encapsulates the bacteria. The bacteria encapsulated divide a number of times and form a microcolony. From this microcolony, the bacteria enter the developing nodule through a structure called an infection thread, which grows through the root hair into the basal part of the epidermis cell, and onward into the root cortex. They are then surrounded by a plant-derived membrane and differentiate into bacteroids that fix nitrogen.

9.4.2.8 Nitrogenase

An enzyme called nitrogenase produced by nitrogen-fixing organisms converts the gaseous nitrogen into the more available nitrogen form ammonia, which can later be assimilated

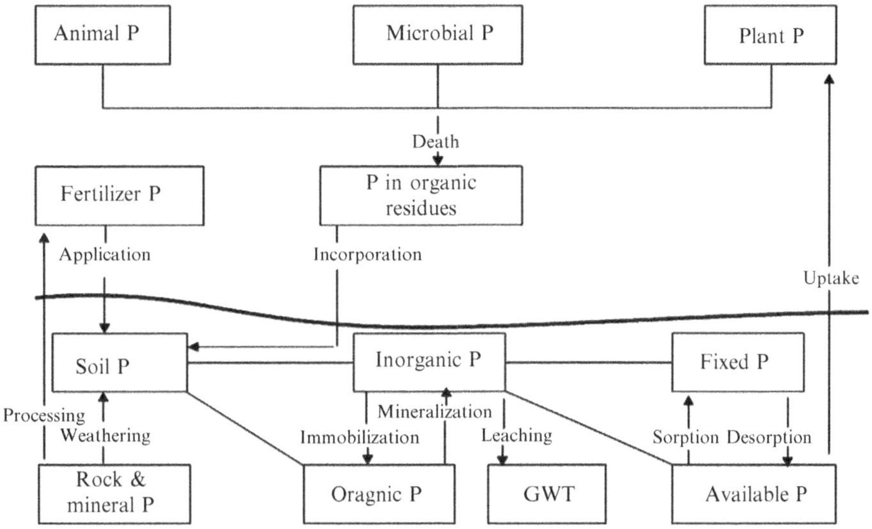

Fig. 9.4 The phosphorus cycle

into amino acids and proteins. A series of complex reactions are involved in nitrogen reduction. Nitrogenase activity requires large amounts of energy. Symbiotic nitrogen-fixing bacteria obtain required energy from the photosynthates of the host legume, but free-living bacteria must find their own source of energy, organic matter, within the soil. Nitrogenase requires the participation of about 20 genes for its synthesis and activity. Activities of these genes are suppressed at higher levels of available nitrogen (ammonium or nitrate) present in the environment. Nitrogenase is deactivated in the presence of oxygen, and all nitrogen-fixing bacteria (free-living and symbiotic) must therefore operate within oxygen-free (anaerobic) conditions. Free-living nitrogen fixers that exist only in aerobic conditions have evolved a specialized biochemical pathway to keep oxygen at very low levels within their cells.

9.4.3 Phosphorus Cycle Involves Transformations of Organic and Inorganic Phosphorus Substances

The phosphorus cycle in soil is shown in Fig. 9.4. During cycling in soil, phosphorus undergoes transformations involving interconversions of the three inorganic phosphate forms, available P, adsorbed P on organic as well as inorganic colloids, or precipitations with other ions or compounds, and primary mineral P. Also, organic P is mineralized to inorganic P and inorganic P is immobilized to organic P.

The primary source of phosphorus in soil is the mineral apatite ($Ca_5(PO_4)_3(OH, F, Cl)$). Apatite is weathered, both chemically and biologically, into available phosphates. Available P is adsorbed from soil by plants and microorganisms for their physiological activity. In this way, soluble inorganic P is converted into organic phosphate compounds in the tissue of plants and microorganisms. This process is known as phosphorus immobilization. Animals get phosphorus through the food chain. After death of all these organisms, organic phosphate is added to soils in the form of their residues. Through decomposition of organic matter, P is released to the soil in the inorganic forms. This is called phosphorus mineralization. Mineralized P may again be adsorbed or precipitated, adsorbed by plants and microorganisms, or a part may be leached to groundwater. Therefore, the general transformation processes of the P cycle are weathering, precipitation, mineralization and immobilization, and adsorption and desorption. Weathering, mineralization, and desorption increase plant available P, while immobilization, precipitation, and adsorption decrease it. Among these processes, P solubilization, mineralization, and immobilization are biologically mediated.

9.4.3.1 Mineralization and Immobilization

Mineralization is the microbial conversion of organic P to inorganic P. By their action, fungi and bacteria make organically bound phosphorus into inorganic phosphate, usually $H_2PO_4^-$, a form available to plants. On the other hand, certain microorganisms especially bacteria assimilate soluble phosphate and use for cell synthesis; this process is known as immobilization. Mineralization of phosphate is generally rapid and more in virgin soils than cultivated land. Mineralization is favored by high temperatures and more in slightly acidic to neutral soils with high organic phosphorus content. The enzymes involved in mineralization of phosphate from organic phosphorus compound are collectively called phosphatases.

9.4.3.2 Solubilization

Many microorganisms can bring insoluble inorganic phosphate into solution. Plant roots and microorganisms solubilize phosphates through secretion of organic acids (e.g., lactic, acetic,

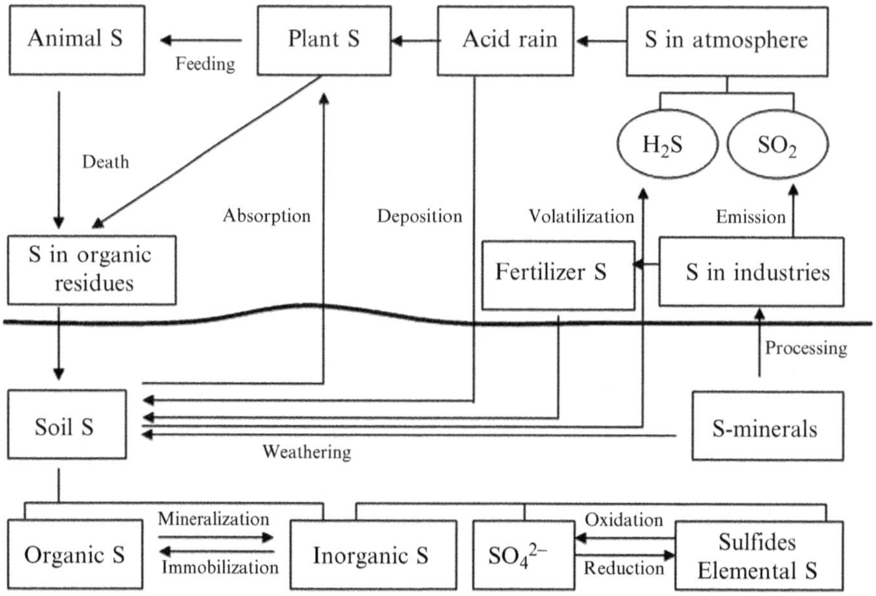

Fig. 9.5 The sulfur cycle

formic, fumaric, and succinic acids). Predominant phosphate dissolving/solubilizing microorganisms are bacteria of the genera *Pseudomonas, Bacillus, Micrococcus, Mycobacterium, Flavobacterium*, and fungi *Penicillium, Aspergillus, Fusarium*, etc. The commercially used species of phosphate solubilizing bacteria and fungi are *Bacillus polymyxa, Bacillus megaterium, Pseudomonas striata, Aspergillus* sp., and *Penicillium avamori*. Solubilization of phosphate by plant roots and soil microorganisms is substantially influenced by various soil factors, such as pH, moisture, and aeration. Phosphate solubilizing microorganisms are found concentrated in the rhizosphere. Their activity is reduced by acidity of soil. Considerable amounts of phosphate ions added in fertilizers may be transported to water reservoirs around agricultural lands. It causes eutrophication of water bodies (Chap. 10).

9.4.4 Oxidation and Reduction Are the Main Processes of Sulfur Cycling

Soils contain several valence states of sulfur, ranging from −2 (as in sulfide and reduced organic sulfur) to +6 (as in sulfate). Some primary and secondary minerals such as pyrite (FeS_2) and gypsum ($CaSO_4{\cdot}2H_2O$) are the original sources of sulfur in soil. Mineral sulfate or sulfide is chemically and biologically transformed into ionic sulfate SO_4^{2-} in soil. This SO_4^{2-} ion is the available form for plants. Sulfate is also adsorbed by microorganisms to make their body material, the process being known as sulfur immobilization. These organisms reduce SO_4^{2-} to −SH groups through the assimilatory process. Microorganisms and plants produce some amino acids and proteins containing sulfur. With plant and animal residues and after death of microorganisms, organic sulfur is added to soils. Some microorganisms decompose and mineralize organic sulfur into H_2S, which in turn is biologically oxidized into sulfate. Some organisms also have the ability to oxidize the mineral pyrite producing sulfur oxides, sulfates, and sulfuric acid. Some microorganisms can reduce sulfate into elemental S, H_2S, and FeS_2. Some sulfur is liberated as sulfur oxides to the atmosphere from industries and vehicles through burning of fossil fuel. These oxides may react with water molecules to produce sulfuric acid which falls to the ground as acid rain. On the other hand, H_2S can volatilize from wetlands to the atmosphere. Thus, the important processes of the sulfur cycle are immobilization, mineralization, oxidation, and reduction (Fig. 9.5).

9.4.4.1 Biological Oxidation of Sulfur

Biological oxidation of sulfide to sulfate is one of the major reactions of the sulfur cycle. Prokaryotes exclusively bring about the oxidation of reduced inorganic sulfur compounds. There is a phylogenetically diverse group of sulfur-oxidizing prokaryotes. Aerobic sulfur oxidation is restricted to members of the order *Sulfolobales* in the domain *Archaea*. In the Domain *Bacteria* sulfur is oxidized by aerobic lithotrophs or by anaerobic phototrophs. The nonphototrophic obligate anaerobe *Wolinella succinogenes* oxidizes hydrogen sulfide to polysulfide during fumarate respiration (Friedrich et al. 2001).

Prokaryotes oxidize hydrogen sulfide, sulfur, sulfite, thiosulfate, and various polythionates in soil at varying soil reactions. Sulfur-oxidizing aerobic prokaryotes belong to genera like *Acidianus, Acidithiobacillus, Aquaspirillum, Aquifex,*

Bacillus, Beggiatoa, Methylobacterium, Paracoccus, Pseudomonas, Starkeya, Sulfolobus, Thermithiobacillus, Thiobacillus, and *Xanthobacter* and are mainly mesophilic. Phototrophic anaerobic sulfur-oxidizing bacteria are mainly neutrophilic and mesophilic and belong to genera like *Allochromatium* (formerly *Chromatium*), *Chlorobium, Rhodobacter, Rhodopseudomonas, Rhodovulum*, and *Thiocapsa*. Lithoautotrophic growth in the dark has been described for *Thiocapsa roseopersicina, Allochromatium vinosum*, and other purple sulfur bacteria, as well as for purple nonsulfur bacteria like *Rhodovulum sulfidophilum* (formerly *Rhodobacter sulfidophilus*), *Rhodocyclus gelatinosus*, and *Rhodopseudomonas acidophila* (Friedrich et al. 2001).

9.4.4.2 Biological Reduction of Sulfur

Some bacteria and Archaea are capable of reducing sulfates to sulfides, especially to hydrogen sulfide. The sulfate-reducing bacteria, which reduce sulfate to obtain energy, are anaerobes and use sulfates as the terminal electron acceptors. Most sulfate-reducing bacteria can also reduce other oxidized inorganic sulfur compounds, such as sulfite, thiosulfate, or elemental sulfur. However, sulfur reduction involves the participation of very complex enzyme systems.

Sulfur-reducing organisms may be characterized by three groupings:

Group I. Sulfate reducers: They use lactate, pyruvate, many alcohols, and some fatty acids as electron donors when converting SO_4^{2-} to H_2S and produce acetate as an end product of metabolism.

Group II. Sulfate reducers: Use fatty acids (especially acetate) and oxidize substrate completely to CO_2 while converting SO_4^{2-} to H_2S; some may grow chemoautotrophically using H_2 as the electron donor (acetyl CoA pathway).

Group III. Dissimilatory sulfur reducers: Organisms that reduce elemental sulfur to sulfide, but cannot reduce sulfate to sulfide: $SO_4 \rightarrow S \rightarrow H_2S$; use acetate and ethanol as common electron donors.

Common sulfur-reducing bacteria belong to the genera *Desulfitobacterium, Desulfotomaculum, Desulfovibrio* (*Desulfovibrio africanus, Desulfovibrio desulfuricans, Desulfovibrio gigas, Desulfovibrio vulgaris*), and *Desulfuromonas*.

9.5 Management and Properties of Soils Affect Population and Function of Soil Microorganisms

Soil and crop management practices include plowing, sowing of seeds and transplanting of seedlings, application of manures, fertilizers, lime and pesticides, irrigation, etc. These practices affect the dynamics of soil organisms. Plowing and tillage operations facilitate aeration and expose the soil to light and heat. Fertilizers, manures, and lime provide nutrients. Cropping patterns have profound influences on soil organisms; some crops favor growth of certain groups of organisms by their excretion and exudation of sugars, enzymes, hormones, and vitamins. Liming of acid soils increases activity of bacteria and actinomycetes and lowers the population of fungi.

Pesticides (insecticides, fungicides, herbicides) often inhibit the growth and activity of nontarget organisms. Toxic pesticide residues may cause profound reduction in the normal microbial activity in the soil. Many soils have become contaminated with heavy metals and persistent organic pollutants by the use of sewage sludge as soil amendments and disposal or hazardous organic wastes on agricultural lands (Sect. 12.5). Šmejkalová et al. (2003) observed toxic effects of Cd, Pb, and Zn (although it is an essential element) on several parameters of microbial activity in soil. Cadmium and zinc exert toxic effects on nitrogen-fixing rhizobia (Leung and Chant 1990). Microorganisms are, however, sometimes employed to remediate soil pollution.

Soil texture, structure, and porosity affect the supply of moisture, oxygen, and warmth to organisms. Microbial population and activity proliferate best in the moisture range of 20–60%. Waterlogged conditions create anoxic situation (lack of oxygen) where anaerobic microflora become active and the aerobes get suppressed. Activities of soil microbes are often measured in terms of the amount of CO_2 evolved. The rate of CO_2 evolution is regulated by the oxygen supply in soil. Soil temperature influences all physical, chemical, and biological processes in soil. Extreme soil temperatures (hot and cold) are not conducive to good microbial activity. There are several groups of microorganisms based on their temperature requirement (Sect. 5.7.6). Most soil microorganisms like to grow within the temperature range of 20–45°C. Moisture and temperature in a geographic region depend also on the prevailing climate. There is seasonal variation within a climatic zone as well. For example, temperature remains below 5°C in winter in the temperate zone and the number and activity of microorganisms are reduced.

Among the chemical properties, mainly soil pH, organic matter, and nutrients influence the activity of soil organisms. Most of the soil bacteria, blue-green algae, diatoms, and protozoa prefer a neutral or slightly alkaline reaction between pH 6.5 and 7.5, and fungi grow in acidic reaction between pH 4.5 and 6.5, while actinomycetes prefer slightly alkaline soil reactions. Some organisms such as the nitrifying bacteria (*Nitrosomonas and Nitrobacter*) and the nitrogen-fixing *Azotobacter* cannot tolerate soil acidity, but nitrogen-fixing *Beijerinckia, Derxia*, and sulfur-oxidizing bacteria like *Thiobacillus thiooxidans* are active in acidic soils. Organic matter is the food of heterotrophic bacteria. So, when fresh organic residues are added to soils, their number and activity abruptly increase. As organic carbon is reduced due to decomposition, the number of heterotrophic organisms also decreases. Microbial population and activity are greatly influenced by the soil fertility in general and individual nutrient supply in particular. For example, activity and presence of nitrogen-fixing bacteria depend profoundly on the supply of available molybdenum and phosphorus.

Study Questions

1. How do you distinguish between rhizoplane and rhizosphere soils? What are the reasons of high biological activity and microbial diversity around roots?
2. What are the major groups of soil flora and fauna? Discuss the functions that micro- and macrofauna can perform in a soil.
3. Explain the significance of carbon cycling in nature. Discuss humification, mineralization, and immobilization.
4. Mention the types of interactions between roots and microorganisms. Narrate major biological transformations of the nitrogen cycle in soil. Discuss nitrification and denitrification processes and indicate their environmental impacts.
5. Explain important biological transformations that are performed by bacteria in soil. Discuss the factors that affect population and functions of bacteria in soil.

References

Alexander M (1977) Introduction to soil microbiology, 2nd edn. Academic, New York

Bais HP, Weir TL, Perry LG, Gilroy S, Vivanco JM (2006) The role of root exudates in rhizosphere: interactions with plants and other organisms. Ann Rev Plant Biol 57:233–266

Baudoin E, Benizri E, Guckert A (2002) Impact of growth stages on bacterial community structure along corn (maize) roots by metabolic and genetic fingerprinting. Appl Soil Ecol 19:135–145

Benizri E, Dedourge O, Di Battista-Leboeuf C, Nguyen CS, Piutti GA (2002) Effect of maize rhizodeposits on soil microbial community structure. Appl Soil Ecol 21:261–265

Clark FE (1949) Soil microorganisms and plant roots. Adv Agron 1:241–288

Coleman DC, Crossley DA Jr, Hendrix PF (2004) Fundamentals of soil ecology, 2nd edn. Elsevier-Academic, Amsterdam

Dommergues Y (1997) Contribution of actinorhizal plants to tropical soil productivity and rehabilitation. Soil Biol Biochem 6:931–941

Elliott GN, Chen WM, Chou J-H, Wang H-C, Sheu SY, Perin L (2006) *Burkholderia phymatum* is a highly effective nitrogen-fixing symbiont of *Mimosa* spp. and fixes nitrogen *ex planta*. New Phytol 173:168–180

Foth HD (1990) Fundamentals of soil science, 8th edn. Wiley, New York

Friedrich CG, Rother D, Bardischewsky F, Quentmeier A, Fischer J (2001) Oxidation of reduced inorganic sulfur compounds by bacteria: emergence of a common mechanism? Appl Environ Microbiol 67(7):2873–2882

Giri B, Giang PH, Kumari R, Sachdev RM, Garg AP, Oelmuller R, Varma A (2005) Microorganisms in soils: roles in genesis and functions. In: Buscot F, Varma A (eds) Soil biology, vol. 3. Microorganisms in soils: roles in genesis and functions. Springer, Berlin

Graham PH (2008) Ecology of the root-nodule bacteria of legumes. In: Dilworth MJ, James EK, Sprent JI (eds) Nitrogen-fixing leguminous symbioses. Springer, Dordrecht

Gupta R, Mukerji KG (2002) Root exudate biology. In: Mukerji KG et al (eds) Techniques in mycorrhizal studies. Kluwer, Dordrecht

Han HS, Lee KD (2005) Plant growth promoting rhizobacteria effect on antioxidant status, photosynthesis, mineral uptake and growth of lettuce under soil salinity. Res J Agric Biol Sci 1(3):210–215

Hawksworth DL (1991) The fungal dimension of diversity: magnitude, significance, and conservation. Mycol Res 95:641–655

Ingham RE, Trofymow JA, Ingham ER, Coleman DC (1985) Interactions of bacteria, fungi, and their nematode grazers: effects on nutrient cycling and plant growth. Ecol Monogr 55:119–140

Jhonson C (2009) Biology of soil science. Oxford, Jaipur

Lavelle P, Spain AV (2003) Soil ecology. Kluwer, New York

Leung D, Chant SR (1990) Effects of sewage sludge treatment of soils on nodulation and leg haemoglobin content of clover. Microbios 64:85–92

Lousier JD, Bamforth SS (1990) Soil protozoa. In: Dindal DL (ed) Soil biology guide. Wiley, New York

Lynch JM (1987) Microbial interactions in the rhizosphere. Soil Microorgan 30:33–41

Lynch JM (1990) The rhizosphere. Wiley, New York

Manoharachary C, Mukerji KG (2006) Rhizosphere biology – an overview. In: Mukerji KG, Manoharachary C, Singh J (eds) Microbial activity in the rhizosphere. Springer, Berlin

Paul EA, Clark FE (1996) Soil microbiology and biochemistry. Academic, San Diego

Philippot L, Germon JC (2005) Contribution of bacteria to initial input and cycling of nitrogen in soils. In: Buscot F, Varma A (eds) Soil biology, vol. 3. Microorganisms in soils: roles in genesis and functions. Springer, Berlin

Schlesinger WH (1997) Biogeochemistry, 2nd edn. Academic, San Diego

Shelley RM (2002) A synopsis of the North American centipedes of the order Scolopendromorpha (Chilopoda). Memoir 5, Virginia Museum of Natural History, Martinsville

Shrama E (1993) Nutrient dynamics in Himalayan alder plantations. Ann Bot 72:329–336

Siddiqui ZA, Pichtel J (2008) Mycorrhizae: an overview. In: Siddiqui ZA, Akhtar MS, Futai K (eds) Mycorrhizae: sustainable agriculture and forestry. Springer, Berlin

Smejkalova M, Smejkalova M, Mikanova O, Boruvka L (2003) Effects of heavy metal concentrations on biological activity of soil microorganisms. Plant Soil Environ 49(7):321–326

Stacey G (2007) The *Rhizobium*-legume nitrogen-fixing symbiosis. In: Bothe H, Ferguson SJ, Newton WE (eds) Biology of the nitrogen cycle. Elsevier, Amsterdam

Swift MJ, Heal OW, Anderson JM (1979) Decomposition in terrestrial ecosystems, vol 5, Studies in ecology. Blackwell, Oxford

Takaya N, Antonina M, Sakairi BC, Sakaguchi Y, Kato I, Zhou Z, Shoun H (2003) Aerobic denitrifying bacteria that produce low levels of nitrous oxide. Appl Environ Microbiol 69(6): 3152–3157

van Vliet PCJ, Hendrix PF (2003) Role of fauna in soil physical processes. In: Abbott LK, Murphy DV (eds) Soil biological fertility—a key to sustainable land use in agriculture. Kluwer, Dordrecht

Vitousek P, Walker L (1989) Biological invasion by *Myrica faya* in Hawaii: plant demography, nitrogen fixation, ecosystems effects. Ecol Monogr 59:247–265

Whalley WR, Riseley B, Leeds-Harrison PB, Bird NRA, Leech PK, Adderley WP (2005) Structural differences between bulk and rhizosphere soil. Eur J Soil Sci 56:353–360

10 Plant Nutrients and Soil Fertility Management

Seventeen chemical elements such as C, H, O, N, P, S, K, Ca, Mg, Fe, Mn, Cu, Mo, B, Zn, Cl, and Ni have so far been recognized as essential for plants. Plants cannot complete their life cycles and accomplish normal physiological functions in the absence of these nutrients. Growth and yield of crops are reduced by their deficiencies. There are some other elements, namely, sodium (Na), silicon (Si), vanadium (V), iodine (I), and cobalt (Co), reckoned to be beneficial for growth of certain plants and microorganisms. Plants often suffer from inadequate supply of nutrients by the soil. These inadequacies are met by the application of fertilizers. Fertilizers are any materials added to soils or plant leaves to supply nutrients. There are various natural and synthesized materials used as fertilizers. Composts, farmyard manures, poultry manure, oilcakes, guano, etc. are very good organic fertilizers obtained from natural materials. These substances contain low concentrations of nutrients; so they are needed in huge amounts. Moreover, their composition is not fixed; and estimating their required amounts is difficult. If added in adequate amounts and well ahead of time, they give good results. Industrial fertilizers are soluble, fast acting, and high analysis materials. They contain nutrients in available forms, and therefore, they are very efficient in correcting current deficiencies. Nutrients may be lost from applied fertilizers, especially nitrogenous fertilizers. Some nitrates and phosphates are transported from agricultural lands to surface and groundwater reservoirs. These contaminants have tremendous environmental impacts. Slow-release N fertilizers are being used to minimize loss of nitrogen from crop fields. There are some methods of fertilizer application that might reduce nutrient losses.

10.1 Soil Fertility Is the Capacity of Soils to Supply Plant Nutrients

Soil fertility may be defined as the ability of a soil to supply plant nutrients in available forms, in adequate amounts required for normal growth and reproduction of plants, and the absence of any toxicity. A soil having these qualities is a fertile soil. A soil lacking these qualities is a non-fertile (or poorly fertile) soil. Soil fertility is a state of the soil that is created by the complex interactions of physical, chemical, and biological properties. Sandy soils are poorly fertile; sands hold fewer and leach out more nutrients. Well-structured soils are more fertile; plant roots can extend to, and draw nutrients from, larger volume of soil here. Soils containing high amount of easily weatherable minerals, dominance of 2:1 type of clay minerals, a pH around 6.5, high cation exchange capacity, high base saturation percentage, high organic matter content, and absence of acidity, alkalinity, salinity, sodicity, and pollution are usually fertile. A fertile soil has a high biological activity.

However, nutrient is one of the factors of plant growth. Plants have some other demands on soil (such as water, oxygen, and warmth) as well. If the soil has the capacity to meet all these demands of plants satisfactorily, it is a productive soil. A fertile soil is, therefore, not always productive. Soil fertility is different from soil productivity. Soil productivity is the ability of a soil to produce desired yield of crops under optimum management. Soil fertility is one of the elements of soil productivity.

10.2 A Nutrient Is a Chemical Element Needed for Normal Growth and Reproduction of Plants

A plant nutrient (also known as an essential element or a nutrient element) is a chemical element, without which normal growth and reproduction of plants are not possible. Arnon and Stout (1939) proposed three criteria for an element to be essential for plants:

1. A given plant must not be able to complete its life cycle without the element.
2. The function of the element must not be replaceable by another element.
3. The element must be directly involved in plant metabolism—for example, as a component of an essential plant

K.T. Osman, *Soils: Principles, Properties and Management*,
DOI 10.1007/978-94-007-5663-2_10, © Springer Science+Business Media Dordrecht 2013

constituent such as an enzyme—or it must be required for a distinct metabolic step such as an enzyme reaction.

Although plant tissues may contain more than 90 chemical elements, only 17 elements strictly meet these criteria and are established as plant nutrients. These elements are grouped into major elements or macronutrients and minor elements or micronutrients, depending on the relative amounts of requirement. Macronutrients are needed in large amounts (>1,000 mg kg^{-1} shoot dry matter) and micronutrients (<100 mg kg^{-1} shoot dry matter) are needed in relatively small amounts (Marschner 1993). Otherwise, macro- and micronutrients are equally important in plant physiology.

Macronutrients: C, H, O, N, P, S, K, Ca, and Mg

Micronutrients: Fe, Mn, Cu, Mo, B, Zn, CI, and Ni

Besides these, Si, I, V, Co, and Na have been established as micronutrients and beneficial elements for some plants (Marschner 1993; Brady and Weil 2002). Beneficial elements are not actually essential for plants but their presence enhances growth.

10.2.1 Nitrogen Constitutes Chlorophyll, Proteins, Enzymes, Nucleic Acids, and Many Other Compounds in Plant

Plants can take up nitrogen both as NH_4^+ and NO_3^- by roots from soil. Ammonium is readily assimilated into amino acids. Nitrate is mobile; it is translocated via the xylem to the shoots. Whether in root or shoot, it must be reduced to ammonia to be suitable for assimilation. Reduction of nitrate to ammonia by plants occurs in two steps; in the first step, nitrate is reduced to nitrite by the enzyme nitrate reductase, and in the second step, the nitrite is reduced to ammonia by the enzyme nitrite reductase. It is referred to as assimilative nitrate reduction which leads, through amination and transamination, to the formation of organic compounds, including amino acids, proteins, nucleic acids, alkaloids, growth regulators, etc. The key reactions of assimilation include formation of the two amino acids glutamate and aspartate and their amides, glutamine and asparagine.

Some organisms, mostly bacteria and actinomycetes, are capable, either freely or symbiotically with higher plants, of utilizing gaseous nitrogen. N_2 is reduced to NH_3 in cells of the organisms; this NH_3 is eventually transformed into amino acids. Nitrogen, thus fixed, finds its way to the soil after death of the organisms. This is known as biological nitrogen fixation.

Nitrogen is the element needed by plants usually in the largest amount (1 to >3% of plant on dry weight basis) after C, H, and O. It is a constituent of the chlorophyll molecule, which gives plants the green color and an ability to convert solar energy into chemical energy through the process of photosynthesis. Using solar energy, plants fix atmospheric CO_2 and produce the primary food material glucose. Nitrogen is a constituent of other important biomolecules such as amino acids, proteins, nucleic acids (RNA and DNA), nucleoproteins, etc. Nitrogen is a necessary component of several vitamins, for example, biotin, thiamine, niacin, and riboflavin. It is a constituent of all enzymes. Therefore, nitrogen plays a key role in all metabolic activities of plants. Protein is a constituent of protoplasm, and nucleic acids build up the genetic materials. So, nitrogen is linked with both vegetative growth and reproduction of plants. Nitrogen enhances vegetative growth by increasing cell division leading to increased leaf numbers, stem elongation, branching, and flowering. Plants with adequate nitrogen look healthy and deep green.

10.2.1.1 Nitrogen Deficiency

Due to reduced synthesis of chlorophyll, nitrogen deficient plants show chlorosis (yellowing) of the leaves. Nitrogen is very mobile within the plant, and so yellowing starts first on older leaves, with progressively younger leaves yellowing as the deficiency becomes more severe. Nitrogen deficient plants become stunted, produce fewer branches and leaves. N-deficient crops may reach maturity earlier (early flowering and fruiting) than plants with adequate N. N-deficient crops give lower yields.

10.2.1.2 Excess Nitrogen

Excess N may cause plants to grow profusely, but at the same time may delay initiation of flowering or fruiting, which results in lower yields. Excess N can also encourage tender and succulent plant growth that may be more susceptible to certain plant diseases and insects. Plants with excess N may also be more susceptible to lodging and breakage than plants with normal N supply. There may be increased sensitivity of N-deficient plants to low temperature. Therefore, application of nitrogen in excess of normal requirement may decrease yield. Quality of crops is also affected by excess nitrogen. For example, Takano (2003) observed reduction of sugar content in fruits of peach due to excess nitrogen application.

10.2.2 Phosphorus Constitutes Nucleic Acids, Phospholipids, and Phosphoproteins

Phosphorus is required in lower amounts than other macronutrients; it ranges between 0.1 and 0.5% in plant leaf dry weight. It is an essential constituent of several compounds, including nucleic acids (DNA and RNA), phosphoproteins, phospholipids, sugar phosphates, enzymes, and energy-rich phosphate compounds. These substances control photosynthesis, respiration, cell division, and many other plant growth processes. Primarily P is involved in energy transfer through high-energy phosphate compounds ADP and ATP during metabolic reactions. Phosphorus is a vital component of the building blocks of genes and chromosomes that carry hereditary traits from one generation to the next. Phosphorus is also a component of phytin, a major storage form of P in seeds. About 50% of the total P in legume seeds and 60–70% in cereal grains is stored as phytin or closely related compounds. Phosphorus is critical in the early developmental stages of growth. It stimulates seed germination, young root formation, seedling growth, flowering, fruiting, and seed development. It enhances maturity.

10.2.2.1 Phosphorus Deficiency

Phosphorus deficiency reduces leaf expansion, leaf surface area, and number of leaves. Phosphorus deficiency reduces shoot and root growth. Generally, inadequate P slows the processes of carbohydrate metabolism including photosynthesis and respiration, which results in a buildup of carbohydrates and the development of a dark green leaf color. Other effects of P deficiency on plant growth include delayed maturity; reduced quality of forage, fruit, vegetable, and grain crops; and decreased disease resistance. Phosphorus-deficient tomato leaves develop a purple color. Leaf edges often turn scorched brown in deficiency of phosphorus. An inadequate supply of P can reduce seed size, seed number, and viability.

10.2.2.2 Excess Phosphorus

Excess P may induce K and micronutrient deficiency, particularly of iron and zinc.

10.2.3 Sulfur Constitutes Some Proteins and Activates Many Enzymes

Sulfur ranges from 0.1 to 0.5% of the plant leaf dry weight. It is a component of the amino acids cysteine, cystine, and methionine. They are highly significant in the structure and function of proteins. Sulfur activates some enzyme systems and it is a component of some vitamins. Plants contain a large variety of other organic sulfur compounds, namely, thiols and sulfolipids, and secondary sulfur compounds, namely, allins, glucosinolates, and phytochelatins. These compounds play an important role in physiology and protection against environmental stress and pests. Sulfur promotes nodulation in legumes. It catalyzes chlorophyll production. Sulfur compounds are also of great importance for food quality and for the production of phyto-pharmaceutics. Sulfur is contained in some oil glycosides, which impart characteristic odor and taste in such plants as mustard, onions, and garlic. Sulfur increases oil contents in seeds of soybean and flax. Canola seed contains up to five times the sulfur concentration of cereals. Sulfur can, in some crops, effectively reduce copper toxicity by creating Cu–S complexes.

10.2.3.1 Sulfur Deficiency

The classic symptom of S deficiency is a paleness of the younger foliage. However, often all of the foliage has a pale green color. Leaf yellowing is also caused by nitrogen deficiency, but in nitrogen deficiency, older leaves are yellowed first. In some cases, the leaf veins may be lighter in color than the surrounding tissue. Sulfur deficiency symptoms in canola include poor seed development. Sulfur deficiency delays maturity and produces mature pods on green stems, with poor seed development within the pod.

10.2.3.2 Excess Sulfur

Sulfur toxicity rarely occurs. Excess sulfate-S (SO_4^-) can reduce the uptake of some anions such as nitrate (NO_3^-) and molybdate (MoO_4^-).

10.2.4 Potassium Activates Many Enzymes, Enhances Carbohydrate and Protein Synthesis, and Acts as an Osmoregulator

Potassium is not found in any organic or structural compounds of the plant body; it is still involved in the synthesis of hundreds of organic compounds in plant cell. Potassium acts chiefly in enzyme activation, gas exchange, and osmotic regulation. It is associated with the activation of more than 60 different enzymes related to carbon, nitrogen, and phosphorus metabolism. Potassium enhances carbohydrate and protein synthesis. Potassium ions in vacuoles of plant cells regulate the osmotic potential of the cells and tissues and control the movement of water and solutes. Potassium ions in the guard cells regulate opening and closing of stomata, thereby controlling CO_2, O_2 and H_2O vapor exchange with the atmosphere during photosynthesis, respiration, and transpiration, respectively.

The benefits of proper K nutrition include improved disease resistance, vigorous vegetative growth, increased drought tolerance, improved winter hardiness, and decreased lodging.

Potassium enhances disease resistance in plants by strengthening stalks and stems, contributes to a thicker cuticle in leaf which guards against disease and water loss, controls the turgidity within plants to prevent wilting, and enhances fruit size, flavor, texture, and development.

10.2.4.1 Potassium Deficiency

Some crops exhibit characteristic deficiency symptoms when adequate amounts of K are not available. Potassium deficiency symptoms differ considerably among plants. For example, in cotton, yellowish or bronze mottling of leaves in the beginning and yellowish-green, brown specks at tip around margin and between veins afterward occurs due to K deficiency. As deficiency continues, whole leaf becomes reddish brown, dies, and sheds prematurely.

Plants become short with fewer, smaller bolls or short, weak fibers. In soybean, firing of tips and margins of lower leaves (yellowing, browning, and chlorosis along leaf margins) is a common symptom. Generally, potassium deficiency causes stunted growth with smaller dense leaves. Starting with the older leaves, the leaf tips and edges turn scorched brown and leaf edges roll. In corn, margins of the lower leaves turn brown. In some other plants, white spots and yellowing of edges of lower leaves occur. Leaf tips may be scorched.

Potassium-deficient crops are slow growing and possess poorly developed root systems. Lodging of cereal crops such as corn and small grain is common. Seeds are small and shriveled due to K deficiency. Fruits are often lacking in normal coloration and are low in sugar content.

Symptoms of K deficiency may sometimes be easily confused with deficiency symptoms for sulfur, magnesium, and zinc.

10.2.4.2 Excess Potassium

Excess potassium may cause calcium and magnesium deficiencies.

10.2.5 Calcium Is a Constituent of Cell Wall and Activator of Many Enzymes

Calcium is an intracellular regulator of plants. It regulates plant structure and functions as a constituent of cell wall and cell membrane, as activator of enzymes, and as modulator of plant growth hormones (Hepler 2005). Calcium is a constituent of plant cell wall; as calcium pectate in the middle lamella, it increases cell wall strength and thickness. Calcium produces strong structural rigidity by forming cross-links within the pectin polysaccharide matrix. It controls permeability of cell membranes and regulates movement of ions and metabolites in plants. It helps in translocations of carbohydrates and amino acids. It increases cell division and elongation; it is needed for normal mitosis, and it is involved in chromatin or mitotic spindle organization. It is essential for continued growth of apical meristem. It is an activator of several enzymes, including phospholipase, arginine kinase, adenosine triphosphatase, adenyl kinase, and amylase. It influences activities of plant growth hormones; it retards the activity of auxins and enhances cytokinin. Calcium enhances nitrate uptake.

Calcium retards senescence of leaf and abscission of flowers and fruits. It promotes cotyledon expansion.

10.2.5.1 Calcium Deficiency

Calcium is not usually a limiting nutrient. However, there are several defects that can be associated with low levels of Ca, including poor root development, leaf necrosis and curling. In extremely acidic soils, calcium deficiency may occur in some agronomic crops. The new leaves become often chlorotic in corn. Many fruits and vegetables demonstrate dramatic symptoms such as black heart in celery and broccoli, tip-burn in lettuce and cabbage, white heart or hollow heart in cucurbits, blossom end rot in tomatoes. Tree fruit with low calcium will exhibit increased storage problems such as bitter pit in apples, cork spot in apples and pears, and cracking in cherries.

10.2.5.2 Excess Calcium

Calcium has little toxic effects on plants. Most of the problems caused by excess soil Ca are the result of secondary effects of high soil pH. Excess Ca may cause reduced uptake of other nutrients such as P, K, Mg, B, Cu, Fe, and Zn.

10.2.6 Magnesium Is a Constituent of Chlorophyll and Activator of Many Enzymes

Magnesium is the central element of the chlorophyll molecule which is responsible for the fixation of solar energy into chemical energy and the production of food materials such as $C_6H_{12}O_6$ from CO_2 and H_2O. Magnesium is both an enzyme activator and a constituent of many enzymes, including those required in carbohydrate, protein, nucleic acid, and lipid metabolism.

Magnesium is needed for storage and translocation of such important plant constituents as carbohydrates, proteins, and fats. It regulates the energy balance of plants; it facilitates reactions between enzymes and ATP. It is a component of pectin, important for stability of cells, and phytin, an energy-rich phosphate store important for seed germination. It is an integral part of ribosomes and the cell matrix. It aids in stabilization of cell membranes. It is required for cell wall synthesis. It enhances P translocation, nitrogen fixation, and iron utilization. Magnesium increases the concentration of vitamin C in fruits. It increases the nutritional quality of frozen vegetables.

10.2.6.1 Magnesium Deficiency

Interveinal chlorosis of the older leaves is the classic symptom of Mg deficiency in plants. In the beginning, a pale green color appears, becoming more pronounced in the older leaves. In some plants, the leaves will curl upward and may also turn red-brown to purple in color. Chlorotic spots between leaf veins, necrosis (dead spots), and red discoloration of stems occur during prolonged periods of deficiency. Magnesium deficiency may be confused with zinc or chlorine deficiencies, viruses, or natural aging since all have similar symptoms.

10.2.6.2 Excess Magnesium

Magnesium toxicity is not generally seen. Crops grown on heavy montmorillonite clay soils that have been poorly fertilized with potassium may exhibit excesses of magnesium in their tissue. Higher tissue levels of magnesium are usually found in the older leaves. High Mg levels may induce K deficiency.

10.2.7 Iron Acts in Electron Transfer in Plant Body

Some plants need more Fe (>0.1%) than others; it is a macronutrient for these plants. Otherwise plants need <0.1% Fe. Iron is, therefore, generally placed in the middle of the macro- and micronutrients. It has many important physiological functions, the most important of which appears to be the exchange of electrons during biochemical reactions.

Two major groups of proteins—the heme proteins and the Fe–S proteins—contain iron. The heme proteins contain a heme iron–porphyrin complex, which acts as a prosthetic group of the cytochromes. Cytochromes are electron acceptors–donors in respiratory reactions. There are some other heme proteins such as catalase, peroxidase, and leghemoglobin. Catalase catalyzes the conversion of hydrogen peroxide into water and O_2, whereas peroxidases catalyze the conversion of hydrogen peroxide to water during photorespiration reactions, and in the glycolate pathway. Catalase also protects chloroplasts from free radicals produced during the water-splitting reaction of photosynthesis. Leghemoglobin acts in nitrogen fixation by nodule bacteria. Peroxidases catalyze the polymerization of phenols to form lignin.

10.2.7.1 Iron Deficiency

Iron is needed to form chlorophyll in plants; its deficiency causes chlorosis, usually first in younger leaves. Early symptoms include green leaf veins with yellowish or whitish color between veins. At advanced stage, the newest leaves may become nearly white because leaf veins and areas between veins lack any green or yellowish-green color. Brown, necrotic areas may develop along leaf margins and between veins in severe cases. Affected leaves are generally smaller than normal.

Some woody plants, including azalea, birches, blueberry, dogwood, magnolia, oaks, white pine, rhododendron, and sweet gum, are particularly sensitive to iron deficiency. The most severely affected crops suffering from Fe deficiency are sorghum, corn, and potatoes.

10.2.7.2 Excess Iron

Iron is toxic when it accumulates to high levels usually in strongly acid and waterlogged soils. It can act catalytically via the Fenton reaction to generate hydroxyl radicals, which can damage lipids, proteins, and DNA.

10.2.8 Manganese Acts in Nitrate Assimilation, Hill Reaction, and Electron Transfer

Physiological functions of manganese in plants include synthesis of chlorophyll, maintenance of lamellar structure of the chloroplast, assimilation of carbon dioxide during photosynthesis, assimilation of nitrate, electron transport, mobility of iron, and production of riboflavin, ascorbic acid and carotene. It participates in Hill reaction, where water is split during photosynthesis (Amesz 1993). It is an activator of many enzymes, including oxidases, peroxidases, dehydrogenase, kinases, and decarboxylases. As a constituent of enzymes, it helps in respiration and protein synthesis. It catalyzes oxidation and reduction reactions within plant tissue. Manganese accelerates germination and maturity.

10.2.8.1 Manganese Deficiency

Yellowing of leaves between the veins is the characteristic symptom of Mn deficiency in plants. The veins remain dark green; the color persists even when the chlorotic parts die. Most crops deficient in manganese become yellowish to olive green. Sometimes a series of brown-black specks appear. In small grains, grayish areas appear near the base of younger leaves. In cereals and grasses, grayish or brownish spots and streaks occur in the middle or basal parts of younger leaves. Manganese deficiency symptoms in dicotyledons occur as small yellow spots on the younger leaves, which turn brown or black. The abscission of developing leaves commonly occurs and flower formation is reduced. Stem and root become stunted. Their apices may die back. Flowers are often sterile.

10.2.8.2 Excess Manganese

Manganese toxicity is characterized by raised interveinal areas giving a puckered appearance; red, brown, or black spotting of the older leaves; and an uneven distribution of chlorophyll. If the toxicity continues, the plants will wilt and die prematurely. Plants particularly susceptible to manganese toxicity are lucerne, cabbage, cauliflower, cereals, clover, pineapple, potato, and tomato.

10.2.9 Copper Is a Structural Element in Regulatory Proteins and a Cofactor of Many Enzymes

Copper acts as a structural element in regulatory proteins and participates in photosynthetic electron transport, mitochondrial respiration, oxidative stress responses, cell wall metabolism, and hormone signaling (Marschner 1993). Cu ions act as cofactors in many enzymes such as Cu/Zn superoxide dismutase, cytochrome oxidase, amino oxidase, laccase, plastocyanin, and polyphenol oxidase. At the cellular level,

Cu also plays an essential role in signaling of transcription and protein trafficking machinery, oxidative phosphorylation, and iron mobilization (Yruela 2005). Copper mediated enzyme systems are involved in building and converting amino acids to proteins. Copper is important to the formation of lignin in plant cell walls, which contributes to the structural strength of the cells and the plant.

10.2.9.1 Copper Deficiency

Copper deficiency reduces photosystem I (PSI) electron transport due to decreased formation of plastocyanin (Shikanai et al. 2003). Decrease in photosystem II (PSII) activity was also observed in Cu-deficient chloroplasts. Severe Cu deficiency results in changes in the thylakoid membranes, which can influence the conformation and function of the photosystems.

Typical signs of Cu deficiency appear first at the tips of young leaves and then extend downward along the leaf margins. The leaves may also be twisted or malformed and show chlorosis or even necrosis (Marschner 1993). The death of the growing points often leads to excessive tillering in cereal crops and excessive branching in dicots. Excessive wilting, lodging, and reduced disease resistance result from the weak cell walls caused by Cu deficiency. Reduced seed and fruit yield is caused mainly by male sterility.

10.2.9.2 Excess Copper

Redox cycling between Cu^{2+} and Cu^{+} can catalyze the production of highly toxic hydroxyl radicals, with subsequent damage to DNA, lipids, proteins, and other biomolecules. Thus, at high concentrations, Cu can become extremely toxic causing symptoms such as chlorosis and necrosis, stunting, leaf discoloration, and inhibition of root growth (Marschner 1993). Toxicity of Cu may result from (1) binding to sulfhydryl groups in proteins, thereby inhibiting enzyme activity or protein function; (2) deficiency of other essential ions; (3) impaired cell transport processes; and (4) oxidative damage.

Typical toxicity symptoms of copper appear in young tissue and include dark green leaves followed by induced Fe chlorosis in which the leaves may appear nearly white and thick, short, or barbed-wire-looking roots which can be mistaken for chemical damage and depressed tillering.

10.2.10 Molybdenum Regulates Nitrogen Metabolism in Plants

Because of its involvement in the processes of N_2 fixation, nitrate reduction, and the transport of nitrogen compounds, molybdenum plays a crucial role in nitrogen metabolism of plants (Srivastava 1997).

Molybdenum is a constituent of the following enzymes: (1) nitrate reductase which catalyzes the key step in inorganic nitrogen assimilation; (2) aldehyde oxidase(s) that have been shown to catalyze the last step in the biosynthesis of the phytohormone abscisic acid; (3) xanthine dehydrogenase, which is involved in purine catabolism and stress reactions; and (4) sulfite oxidase, which is probably involved in detoxifying excess sulfite (Mendel and Hansch 2002). According to Mishra et al. (1999), nearly all nitrogenases contain the same two proteins, both of which are inactivated irreversibly in the presence of oxygen: an Mo–Fe protein (MW 200,000) and an Fe protein (MW 50,000–65,000). The Mo–Fe protein contains two atoms of molybdenum and has oxidation–reduction centers of two distinct types: two iron–molybdenum cofactors called FeMoco and four Fe–S (4Fe–4S) centers. The Fe–Mo cofactor (FeMoco) of nitrogenase constitutes the active site of the molybdenum-containing nitrogenase protein in N_2-fixing organisms.

10.2.10.1 Molybdenum Deficiency

Nitrates accumulate in Mo-deficient plant leaves and the plant cannot use them to produce proteins. As a result, the plant becomes stunted, with signs similar to those of nitrogen deficiency. The edges of the leaves may become scorched by the accumulation of unused nitrates. The main symptoms of molybdenum deficiency in nonlegumes are stunting and failure of leaves to develop a healthy dark green color. The leaves of affected plants show a pale green or yellowish green color between the veins and along the edges. In advanced stages, the leaf tissue at the margins of the leaves dies. The older leaves are the more severely affected.

In cauliflowers, the yellowing of the tissue on the outer leaves is followed by the death of the edges of the small heart leaves. Narrow, distorted leaves develop; the symptom is known as "whiptail of cauliflower."

10.2.10.2 Excess Molybdenum

Molybdenum toxicity in plants seldom occurs under field conditions. However, marginal leaf scorch and abscission may occur due to molybdenum toxicity. Excess Mo causes yellowing or browning of leaves and depressed tillering. Excess Mo levels in plants are more of a concern to animal life. Overconsumption of plant tissue high in molybdenum can lead to a condition called molybdenosis.

10.2.11 Boron Contributes to Cell Wall Development and Cell Division

Boron functions in cell wall development, cell division, fruit and seed development, sugar transport, and hormone development. Boron is involved in the functioning of meristematic tissues in plants or the cells which are rapidly multiplying. Plants can be separated into four groups based on their B requirement for growth and development: (1) Lactifers (latex-forming species) have by far the highest B requirement, (2) followed by leguminous plants, (3) then the remaining dicots and the lily family of monocots, and (4) finally the graminaceous plants (Bonilla et al. 2009).

Boron forms boron–polysaccharide complex in cell walls indicating boron cross-linking of pectin polymers. Boron is involved in membrane processes. Rapid boron-induced changes in membrane function could be attributed to boron-complexing membrane constituents. Boron may affect metabolic pathways by binding apoplastic proteins to *cis*-hydroxyl groups of cell walls and membranes and by interfering with manganese-dependent enzymatic reactions (Blevins and Lukaszewski 1998). In addition, boron counteracts toxic effects of aluminum on root growth of dicotyledonous plants.

10.2.11.1 Boron Deficiency

Boron deficiency is a widespread problem for field crop production where large losses of yield occur annually both quantitatively as well as qualitatively. In China 40% yield losses are due to boron deficiency (Goldbatch et al. 2007). Significant losses of yield or quality resulting from boron deficiency may occur as well in vegetable crops.

10.2.11.2 Excess Boron

Physiological effects of boron toxicity in plants include reduced root cell division, decreased root and shoot growth, reduced leaf chlorophyll, inhibition of photosynthesis, lower stomatal conductance, deposition of lignin and suberin, reduced proton extrusion from roots, increased membrane leakiness, peroxidation of lipids, and altered activities of antioxidation pathways (Reid 2007).

Boron toxicity is characterized by yellowing of the leaf tips, interveinal chlorosis, and progressive scorching of the leaf margins. In soybeans, the leaves may have a rust-like appearance.

10.2.12 Zinc Is a Constituent of All Six Classes of Enzymes

All six enzyme classes of enzymes (oxidoreductases, transferases, hydrolases, lyases, isomerases, and ligases) contain zinc. There are structural, catalytic, and cocatalytic Zn-binding motifs in plants. Cysteine is the preferred ligand in structural Zn sites. In catalytic sites, Zn forms complexes with histidine and water. Cocatalytic sites have aspartic acid and histidine as preferred ligands. The fourth type of Zn-binding site (protein interface) suggests a role of Zn in the quaternary protein structure (Auld 2001). The enzyme carbonic anhydrase is found in cytosol and chloroplasts; it catalyzes the reversible hydration of CO_2. It can have 2, 4, 6, or 8 units, each with one Zn atom coordinated to histidine residues at the active site. Zn metalloproteins are regulators of gene expression. A Zn-finger lateral shoot-inducing factor increases lateral shoot growth by altering cytokinin concentrations in petunia, tobacco, and Arabidopsis (Nakagawa et al. 2005). Zinc is needed for growth hormone, seed and grain production, protein synthesis, maturation of seed, and growth of stalks and height of plants.

10.2.12.1 Zinc Deficiency

Plants become stunted and leaves become dusty brown in Zn deficiency. White area develops between the veins. Leaves become characteristically undersized, sharp, small, and pointed in Zn deficiency. Decreased tillering, spike or spikelet sterility and interveinal chlorosis develop on leaves. White bud forms in corn due to Zn deficiency. Loss of leaf luster and death of shoots often occur in dicots.

10.2.12.2 Excess Zinc

Wang et al. (2009) conducted an experiment on the effect of excess Zn on mineral nutrition and antioxidative response in rapeseed (*Brassica napus*) seedlings. They observed significantly lower concentrations of P, Cu, Fe, Mn, and Mg and increased lipid peroxidation in excess supply of Zn. NADH oxidase and peroxidase activity increased, but superoxide dismutase, catalase, and ascorbate peroxidase activities decreased in leaves and roots of plants under high Zn. In their opinion, excess Zn exerts its toxicity partially through disturbing nutrient balance and inducing oxidative stress in plants. Cui and Zhao (2011) also observed similar changes in enzyme activity in corn (maize, *Zea mays*) growing in Zn contaminated soils.

10.2.13 Chlorine Controls Stomatal Opening and Internal Water Balance

Chlorine is essential for the functioning of the stomatal openings, which control internal water balance. It functions in photosynthesis, specifically during the Hill reaction and in maintenance of cation balance and transport within the plant. Chlorine aids in sugar translocation in plants. It controls water use efficiency in plants. It enhances crop maturity. It helps in disease control (Chen et al. 2010). Several enzymes such as ATPase, alpha amylase, and asparagine synthetase require Cl for stimulation or activation (Power and Prasad 1997).

10.2.13.1 Chlorine Deficiency

Chlorine deficiency symptoms include wilting due to restricted and highly branched root system often with stubby tips. Chlorine deficiency is characterized by leaflet wilting, followed progressively by chlorosis, bronzing, and finally, necrosis. Chlorine deficiency in cabbage is marked by an absence of the cabbage odor from the plant.

10.2.13.2 Excess Chlorine

Leaf/leaflet size is reduced and may appear to be thickened. Leaf margins are scorched and abscission is excessive. Overall plant growth is reduced. Chloride accumulation is higher in older tissue than in newly matured leaves.

10.2.14 Nickel Is a Component of a Number of Enzymes

Nickel may stimulate germination and growth of various crop species. Nickel functions in urea metabolism, and it may also be required in low concentrations by legumes dependent on symbiotically fixed nitrogen either for the growth of *Rhizobium* or for the utilization of fixed nitrogen or both.

Ni has been identified as a component of a number of enzymes, including glyoxalases, peptide deformylases, methyl-CoM reductase, and ureases, and a few superoxide dismutases and hydrogenases. Therefore, Ni plays a role in various important metabolic processes, including ureolysis, hydrogen metabolism, methane biogenesis, and acetogenesis (Collard et al. 1994).

10.2.14.1 Nickel Deficiency

Nickel deficiency is becoming increasingly apparent in crops, especially for ureide-transporting woody perennials. In an experiment on pecan (*Carya illinoinensis*), Bai et al. (2006) observed that foliage of Ni-deficient pecan seedlings exhibited disruption of nitrogen metabolism via ureide catabolism, amino acid metabolism, and ornithine cycle intermediates. Disruption of ureide catabolism in Ni-deficient foliage resulted in accumulation of xanthine, allantoic acid, ureidoglycolate, and citrulline, but total ureides, urea concentration, and urease activity were reduced. Disruption of amino acid metabolism in deficient foliage resulted in accumulation of glycine, valine, isoleucine, tyrosine, tryptophan, arginine, and total free amino acids and lower concentrations of histidine and glutamic acid. Ni deficiency also disrupted the citric acid cycle. Disruption of carbon metabolism was also via accumulation of lactic and oxalic acids. It appears that mouse-ear, a key morphological symptom, is likely linked to the toxic accumulation of oxalic and lactic acids in the rapidly growing tips and margins of leaflets.

10.2.14.2 Excess Nickel

Ni-toxicity symptoms include the formation of Ni^{2+}-organic acid and Ni^{2+}-NA complexes, the overproduction of NA, and high levels of free histidine. Other responses include the induction of MTs and thiol glutathione and high concentrations of glutathione, cysteine, and O-acetyl-L-serine. In addition, some enzyme activities may be enhanced such as serine acetyltransferase and glutathione reductase.

At high concentrations, nickel inhibits shoot and root growth for which plants become stunted. Excess Ni has been reported to cause leaf necrosis and chlorosis of plants. Chlorosis and along-vein necrosis appeared in newly developed leaves of water spinach by the application of excess Ni. Some plants accumulate very high concentration of Ni in their shoots without showing any toxicity symptoms. They are Ni-hyperaccumulators.

More than 310 Ni-hyperaccumulator plants have been identified, including members of the families Acanthaceae, Asteraceae, Brassicaceae, Caryophyllaceae, Fabaceae, Flacourtiaceae, Meliaceae, Myristicaceae, Ochnaceae, Poaceae, Rubiaceae, Sapotaceae, and Stackhousiaceae; Brassicaceae contains 60 Ni-hyperaccumulators capable of accumulating as high as 3% nickel (Chen et al. 2009).

10.3 There Are Positive and Negative Interactions of Nutrients in Plant

Two or more nutrients may interact in plants. Nutrient interactions are generally measured in terms of growth response and change in concentration of nutrients. When an increase in crop yields by the addition of two nutrients is more than adding only one, the interaction is positive (synergistic). If adding the two nutrients together produces fewer yields than the individual ones, the interaction is negative (antagonistic). When there is no change, there is no interaction. A nutrient may interact simultaneously with more than one nutrient. This may induce deficiencies, toxicities, modified growth responses, and/or modified nutrient composition. However, most interactions are complex (Fageria 2001). Nutrients interact for many different reasons. Some nutrients follow the same absorption pathway, some may compete for the same nutrient carrier, some nutrients may activate the biochemical reactions of others, and some nutrients may enhance the permeability of membranes that increase the movement of other nutrients. For example, N enhances photosynthesis that increases assimilation of P and S. Interactions of phosphorus with nitrogen, potassium, zinc, and copper frequently occur in crops. In these cases, the addition of one nutrient without the second may not improve growth. There are nonspecific and specific interactions between mineral nutrients in plants. A typical example of a nonspecific interaction is between nitrogen and phosphorus. Two examples of specific interaction are provided by Marschner (1993) as (a) competition between potassium and magnesium at the cellular level, which usually involves the risk of potassium-induced magnesium deficiency and (b) replacement of potassium by sodium in natrophilic species. Interactions vbetween two mineral nutrients are important when the contents of both are near the deficiency range. Increasing the supply of only one mineral nutrient stimulates growth, which in turn can induce a deficiency of the other by a dilution effect. In principle, these unspecific interactions hold true for any mineral nutrients with contents at or near the critical deficiency contents (Marschner 1993). Optimal ratios between nutrients in plants are therefore often as important as absolute contents. For example, a N/S ratio around 17 is taken to be an index of optimum nutrition of both the nutrients. Critical nutrient ratios have been proposed for some nutrients, particularly N/P, N/K, N/S, Ca/Mg, etc. However, using nutrient ratios has also been criticized because the same nutrient ratio may be obtained when both nutrients are sufficient, when both nutrients are deficient, and when one of them is deficient (Sect. 10.21.2.2).

10.4 Plants Absorb 14 Nutrients from the Soil

Plants get C from CO_2, O from O_2 and H_2O, and H from H_2O. They absorb other nutrients in ionic forms from soil. Plants can absorb nutrients as ions present in soil solution and as exchangeable ions on soil colloids. These absorbable forms are called available nutrients in soil. The available forms of different nutrients are given in Table 10.1.

Of the 14 nutrients absorbed by plants from soil, three (N, P, and K) are commonly deficient because they are removed from soil by crops in greater amounts. Some other nutrients (Mg, S, Zn, B, and Mn) may as well sometimes be deficient. Nutrient deficiencies are more common in regions with highly weathered soils (Oxisols and Ultisols), and in high rainfall areas (humid tropical). Iron, zinc, copper, and manganese deficiencies often occur in calcareous (Calcids, Calcisols) and limed soils. In acidic soils, Ca, Mg, P, and Mo and, in alkaline soils, Fe, Mn, and Zn may be deficient.

Table 10.1 Available forms of plant nutrients

Elements	Available forms
N	NH_4^+, NO_3^-, N_2[a]
P	$H_2PO_4^-$, HPO_4^{2-}, PO_4^{3-}
S	SO_4^{2-}
K	K^+
Ca	Ca^{2+}
Mg	Mg^{2+}
Fe	Fe^{2+}, Fe^{3+}
Mn	Mn^{2+}
Cu	Cu^{2+}
Mo	MoO_4^{2-}
B	H_3BO_3, BO_3^-, $B_4O_7^{2-}$
Zn	Zn^{2+}
Cl	Cl^-
Ni	Ni^{2+}

[a]Some microorganisms can utilize N_2 as a nutrient

10.4.1 Soils Should Supply Adequate Nutrients to Plants for Optimum Growth

Soils must supply plants with nutrients adequate for their normal growth and yield. But the concept of "adequate nutrient supply" is itself arbitrary. All plants do not need the same amount of any nutrient; plants widely differ in their requirements of nutrients; and even the same plant needs different amounts of nutrients at different stages of growth. Moreover, two different soils with the same nutrient level would supply different amounts of nutrients to plants if there are differences in physical and chemical conditions of the soils that affect root growth and nutrient uptake. So, a generalization of "adequate nutrient level" is difficult. However, 1.0 mg kg^{-1} extractable Cu, Mn, and Zn and 4.5 mg kg^{-1} Fe in soil were considered adequate by McKenzie (1992). Tamimi et al. (1997) proposed adequate levels of available P, K, Ca, and Mg for soils of Hawaii (Table 10.2).

10.4.2 Nutrients in Soil May Be Available or Unavailable to Plants

A nutrient can be present in soil in the following forms: dissolved in soil solution (soil water with dissolved ions, salts, acids, and bases), exchangeable cations and anions on soil colloidal surfaces, insoluble compounds, soil organic matter, and soil minerals. Sum of all these forms is the total nutrient content. As pointed out earlier, only the soluble and exchangeable forms constitute the available nutrient. Available nutrients in soil are extracted with various extractants; common extractants are dilute solutions of salts, acids, alkalis, or their mixtures. A small proportion of total nutrient is available to plants. For example, total P content in soils may lie between <400 and >1,000 mg kg^{-1}, but available P usually remains below 25 mg kg^{-1} (only 2.5–5% of total). Plants obtain nutrients readily from the soil solution. The concentration of available nutrients in soil solution and the replenishing of the soil solution in the rhizosphere are of tremendous importance to plant nutrition. Wild (1996) gives concentrations of some nutrients in soil solution as 5–200 mg L^{-1} NO_3–N, 10–100 mg L^{-1} SO_4–S, 0.01–0.60 mg L^{-1} H_2PO_4–P, 10–200 mg L^{-1} Ca^{2+}, 5–100 mg L^{-1} Mg^{2+} and 1–40 mg L^{-1} K^+.

Table 10.2 Adequate levels of available nutrients (mg kg^{-1}) in soils of Hawaii

Nutrient	Heavy soils	Light soils
P	25–35	50–85
K	200–300	200–400
Ca	1,500–2,000	3,000–4,000
Mg	300–400	600–800

10.5 Nitrogen Remains Mainly in the Organic Form in Soil

Sources of nitrogen in soil include air, soil solution, soil organic matter, crop residues, animal manure, and fertilizers.

Although air contains about 78% nitrogen, most plants cannot use this element from air. Precipitation brings some nitrogen (5–10 kg $ha^{-1}year^{-1}$) to the soil as dissolved nitrate and ammonium in rainwater. Some free living bacteria and cyanobacteria have the ability to assimilate atmospheric N_2. They utilize nitrogen for their biochemical needs and add nitrogen to the soil after their death. About 1–5 kg $ha^{-1}year^{-1}$ nitrogen may be added to soils in this way. *Rhizobium* can fix about 100–200 kg N $ha^{-1}year^{-1}$ in association with host legumes (Chap. 9).

The most immediate source of nitrogen for plants is the soil solution. It contains very low concentrations of NO_3^- (5–200 mg L^{-1}) and NH_4^+. Ammonium ion may also be held on exchange complexes. NO_3^- is very mobile and may easily be leached to the groundwater. The soil solution is continually replenished with NH_4^+ and NO_3^- by the mineralization (coupled with ammonification) of organic matter and nitrification of NH_3.

More than 90% nitrogen in soil is in the organic form. Proteins, amino acids, and nucleic acids associated with soil organic matter contain nitrogen. Organic nitrogen is slowly mineralized to supply plants with available nitrogen. Often the total N content of soil is a better index of nitrogen supply than available nitrogen. On an average, total N content of soils may range from <0.1 to >0.5%.

Crop residues and animal manures are important sources of nitrogen in soil. The amount of N in these sources depends on their amounts and concentration of N in the residues. Generally, crop residues and animal manures contain 0.5–1.5% N. Organic nitrogen is continually converted into inorganic nitrogen through mineralization. Inorganic nitrogen is also being converted into organic nitrogen through immobilization. Ammonia gets converted into nitrate by nitrification; nitrate is biologically reduced to nitrogen gas through denitrification. Nitrogen gas is bound as organic compounds by biological nitrogen fixation. These processes with the reactions involved have been discussed in Chap. 9 in connection with nitrogen cycling. Volatilization is a chemical transformation of nitrogen which involves the loss of mainly ammonia in soil to the atmosphere. Volatilization usually occurs in soils with pH above 7.5 and in calcareous, saline, and sodic soils. It may occur substantially in wet soils (Chap. 13). These transformations have profound environmental impacts.

10.5.1 Nitrate Leaching from Soils Causes Groundwater Contamination

Nitrate, whatever its sources in soil may be, undergoes the following changes:

1. It may be taken up by plants.
2. It may become incorporated in the soil organic matter, where it will remain unless it is remobilized by soil organisms.
3. It may be leached out of the soil.
4. It may be denitrified.

There is little difference between nitrate from fertilizers and that produced by microbes from organic matter. Very little nitrate is retained in soil, and much added nitrogen fertilizer is lost through leaching and denitrification in some situation. Nitrate leaching results in groundwater contamination, and it has now become a major environmental problem. Concentrations of nitrate increased in many natural waters to alarming levels around major agricultural areas. Health problems related with water contamination of nitrate include the "blue-baby syndrome" of infants and stomach cancers of human (Ewing and White 1951; Deak 1985). Infants and pregnant women are especially vulnerable to health problems from drinking water with nitrate levels above 10 mg L^{-1}. Nitrate levels of up to 3 mg L^{-1} in well water is generally taken to be naturally occurring and safe for drinking. Therefore, some soil management practices to reduce nitrate leaching are essential. The United States Environmental Protection Agency has set the primary drinking water standard for nitrate at 10 mg L^{-1} (USEPA 2007).

10.5.2 Soil Is a Source of Atmospheric Nitrogen Oxides

The term NOx represents the more reactive nitrogen oxides, including nitric oxide (NO) and nitrogen dioxide (NO_2). These oxides play a crucial role in the chemistry of the atmosphere. NO and NO_2 react with sunlight and unburned gasoline and contribute to the development of photochemical smog. Atmospheric NOx also reacts to produce nitric acid which is very soluble and, along with sulfuric acid, significantly contributes to acid rain. There is another environmentally important oxide of nitrogen, the nitrous oxide (N_2O). An increase in N_2O in the atmosphere would result in a decrease in the ozone layer. Because it can absorb infrared radiation, the excess production of N_2O can contribute to global warming. Concentrations of N_2O in the atmosphere are rising exponentially at 0.3%, driven mostly by microbial activity in nitrogen-rich soils associated with agriculture (Hall et al. 2008).

Soils are a major source of atmospheric NOx (Yienger and Levy 1995). A major part of the global emission of NOx, N_2O, and NH_3 occurs from the soil. Including emissions from animal waste management, about three-quarters of the global emission of NH_3 and N_2O take place from soil. Soils contribute about one-third to the global emission of NOx (and NH_3) and more than 50% to the global atmospheric N_2O budget. The contribution to the global budgets of agricultural soil and land-use-related emissions is 36% for NOx, 31% for N_2O, and 70% to NH_3. In the IPCC (1992) assessment, the global N_2O emissions from natural soils were estimated to be in the range 3.2–7.7 Tg N year^{-1}. IPCC (1995) estimated the direct contribution from agricultural soils receiving mineral nitrogen fertilizers, manure, and N_2-fixing legumes at 3.5 (1.8–5.3) Tg N year^{-1}.

NOx and N_2O are emitted from soils by the microbial processes of nitrification and denitrification. Some N_2O is produced in soils by a nonbiological process known as chemodenitrification. These processes depend on biogeochemical and physical properties of the soil, including microbial species, soil texture, soil water, pH, redox potential, and nutrient status (Conrad 1996). Soil emission fluxes are also tightly linked to land-use management including application of natural and synthetic fertilizers, tillage, irrigation, compaction, planting, and harvesting (Frolking et al. 1998).

10.6 Phosphorus Is an Element of Agronomic and Environmental Significance

Sources of phosphorus in soil include soil solution, minerals, organic matter, crop residues, manures, and fertilizers. Many factors influence the content of phosphorus in soil. Among these are (1) the type of parent material from which the soil is derived, (2) the degree of weathering, and (3) the climatic conditions. In addition, soil phosphorus levels are affected by erosion, crop removal, and phosphorus fertilization. There are different pools and fractions of phosphorus in soil. These pools and fractions are discussed in the following sections.

10.6.1 Three Major Phosphorus Pools in Soil Include Solution and Active and Fixed Phosphorus

Among the three pools of phosphorus in soil, the solution P pool is relatively small containing only a fraction of a kilogram of P in a hectare of land. Here it mainly remains in the primary orthophosphate form ($H_2PO_4^-$) if the soil is not alkaline. In weakly alkaline conditions, the HPO_4^{2-} ion is dominant, and in strongly alkaline conditions, the PO_4^{3-} ion predominates. Plants absorb P mainly as the primary orthophosphate ion. A vigorously growing crop may quickly deplete the P in the solution pool if it is not being continuously replenished. The active P pool contains P in the solid phase and remains loosely held on particles so that it may be easily released to the soil solution when solution P pool is impoverished. It includes weakly adsorbed P, some calcium and aluminum phosphate compounds, and organic P that is easily mineralized. Because the solution P pool is very small and because P may be frequently released from the active P pool, the active P pool becomes the main source of available P for crops. The ability of the active P pool to replenish the soil solution P pool in a soil is what makes a soil fertile with respect to phosphate. The fixed P pool contains inorganic phosphate compounds that are very insoluble and organic compounds that are resistant to mineralization. Phosphate may remain fixed in soil for a very long time so that it hardly affects the P availability in soil. The inorganic phosphate compounds in this fixed P pool are more crystalline in their structure and less soluble than those compounds in the active P pool. Some fixed P may be converted to the active P pool. There may be an equilibrium between soluble P and active P and between active P and fixed P $(\text{soluble P} \Leftrightarrow \text{active P} \Leftrightarrow \text{fixed P})$.

10.6.2 Inorganic Phosphorus Is One Major Fraction of Soil Phosphorus

Inorganic phosphorus in soil consists of P in solution and as insoluble compounds and minerals. The concentration and ionic forms of phosphorus in soil solution that were mentioned above are governed by two types of reactions: precipitation–dissolution (Sect. 8.4.3) and adsorption–desorption processes (Sect. 8.4.4). Some adsorbed phosphate may become trapped on the surface of soil minerals if any Fe or Al oxide coating is precipitated on the mineral. The trapped phosphate is then termed as occluded phosphate (Solis and Torrent 1989). Most inorganic phosphorus compounds in soils fall into one of two groups: (1) those containing calcium and (2) those containing iron and aluminum (and less frequently, manganese) (Brady and Weil 2002). Common iron phosphate compounds are strengite ($FePO_4{\cdot}2H_2O$), vivianite [$Fe_3(PO_4)_2{\cdot}8H_2O$], and tinticite [$Fe_6(PO_4)_4(OH)_6{\cdot}7H_2O$]; aluminum compounds are variscite ($AlPO_4{\cdot}2H_2O$), berlinite ($AlPO_4$), and tarakanite [$H_6K_3Al_5(PO_4)_2{\cdot}8H_2O$]; and calcium compounds are fluorapatite [$3Ca_3(PO_4)_2{\cdot}CaF_2$], oxyapatite [$3Ca_3(PO_4)_2{\cdot}CaO$], carbonate apatite [$3Ca_3(PO_4)_2{\cdot}CaCO]_3$, and hydroxyapatite [$3Ca_3(PO_4)_2{\cdot}CaCO_3$]. Inorganic P represents the dominant component of the soil P in calcareous soils accounting for about 75–85% of total soil P (Jiang and Gu 1989). Calcareous soils and those developed from phosphate-containing rocks are particularly rich in apatites—the tricalcium phosphates. Apatites are slowly weathered and it very slowly releases phosphorus in soil solution. In highly weathered well-developed soils, most apatites have been exhausted due to weathering. Long-term fertilizer P addition increases the inorganic P contents significantly in many agricultural soils. The contents of inorganic P fractions in a calcareous soil of China receiving annual fertilizer P application were in the order of Ca_{10}–P > Ca_8–P > Fe–P > Al–P > occl uded P > Ca_2–P (Wang et al. 2010).

10.6.3 Organic Phosphorus Is Another Major Fraction

The organic phosphate fraction may account for 50–80% of the total phosphorus in some soils. This fraction represents only a few individual organic phosphorus compounds; the majority is bound to the complex of humus. During fractionation in alkaline solution, some major organic P fractions are identified. These fractions include sugar phosphates (inositol phosphate), phospholipids, and nucleic acids (Tisdale et al. 1997). Inositol phosphate may comprise more than 50% of total organic P in some soils. Phospholipids may range from 1 to 10%. Phosphoglycerides are probably the dominant fraction, but little is known about the other phospholipids in soils. Orthophosphate monoesters are the major functional class of organic P in most soils, of which the higher inositol phosphates are usually dominant (Turner et al. 2002). Orthophosphate diesters dominate organic P inputs to the soil from plants and microbes, but are rapidly degraded upon release and typically represent only a small proportion of the soil organic P (Anderson 1980). Nucleic acids and their derivatives are also present in soils in similar extents to the phospholipids. Both immobilization of inorganic P and mineralization of organic P can occur concurrently in the soil, making the organic P fraction a dynamic entity.

10.6.4 Availability of Phosphorus in Soil Is Governed by Its Chemical Environment

Phosphorus availability depends on several soil conditions. Some of these are the amount of clay, the type of clay (kaolinite and iron oxides retain more P than 2:1 clays), the pH, the soil P, the soil temperature, the compaction, the aeration, the moisture content, the kind and amount of fertilizer, the time and method of application, the granule size of fertilizer, and placement. Only a small fraction of total P becomes available in soil. For example, Olsen P content may be only 10 mg kg^{-1} in a soil containing 400–1,000 mg kg^{-1} total P.

Availability of phosphorus in soil is indicated by the abundance of orthophosphate ions in soil solution. It is a pH-dependent phenomenon (Hopkins and Ellsworth 2005) as soil pH determines the ionic forms of phosphate and directs the reactions in which these ions participate. At low pH, iron, aluminum, and manganese become highly soluble in soil. They react with soluble phosphate ions to form insoluble phosphates. At high pH in sodic and calcareous soils, soluble calcium reacts with soluble phosphate ions to form insoluble calcium phosphates. Therefore, phosphorus deficiency is likely in acid, alkaline, sodic, and calcareous soils. Liming in acid soils will improve P availability, but over-liming will again induce P deficiency. Soluble $H_2PO_4^-$ ions may further be adsorbed by soil colloids, including clay and hydrated oxides of Fe and Al. Adsorbed phosphate is not readily available to plants. It becomes available when it is again desorbed.

10.6.5 Phosphorus Availability Is Judged from Its Chemical Extractability

There are several methods of determining "available" phosphorus content of soils. These methods employ a variety of extractants, which are dilute solutions of salts, acids, and alkalis with or without buffering agents. These methods actually extract the phosphorus which is "extractable" by the particular extracting solution. To have an idea of the phosphorus-supplying capacity of a soil to a crop, the amount of extracted phosphorus is correlated with the amount of phosphorus taken up. Extractable P is often called "soil test P." The fundamental goal of soil P testing has always been to identify the optimum soil test P concentration required for plant growth. Other objectives of soil P testing include (1) to "index" the P-supplying capacity of soils, thus estimating the time before fertilization would again be required; (2) to group soils, in terms of the likelihood of an economic response to P, based on their physical and chemical properties; and (3) to identify when soils are sufficiently excessive in P to contribute to nonpoint source pollution of surface waters (Sims et al. 1998).

Popular methods of determining available or extractable P include Bray and Kutrz 1 (0.025 M HCl in 0.03 M NH_4F), Bray and Kurtz 2 (0.1 M HCl in 0.03 M NH_4F) (Bray and Kurtz 1945), Mehlich 1 (0.0125 M H_2SO_4 + 0.05 M HCl) (Mehlich 1953), Mehlich 3 (0.2 M CH_3COOH, 0.25 M NH_4NO_3, 0.015 M NH_4F, 0.013 M HNO_3, 0.001 M EDTA) (Mehlich 1984), and Olsen (0.5 M $NaHCO_3$, pH 8.5) (Olsen et al. 1954) methods.

Phosphorus extracted by the Bray and Kurtz 1 method has been shown to be well correlated with crop yield response on most acid and neutral soils. For acid soils, the fluoride in the Bray and Kurtz extractant enhances P release from aluminum phosphates by decreasing Al activity in solution through the formation of various Al–F complexes. A Bray and Kurtz 1 P value of 25–30 mg kg^{-1} is often considered optimum for plant growth. Mehlich 1 method is also known as the dilute double acid method. It is used as a multielement extractant for P, K, Ca, Mg, Cu, Fe, Mn, and Zn. Mehlich 1 method extracts P from aluminum, iron, and calcium phosphates and is best suited to acid soils (pH < 6.5) with low cation exchange capacities (<10 cmol/kg) and organic matter contents (<5%). Kuo (1996) reported that Mehlich 1 method was unreliable for calcareous or alkaline soils because it extracts large amounts of non-labile P in soils with pH > 6.5, soils that have been recently amended with rock phosphate, and soils with high cation exchange capacity (CEC) or high base saturation. Mehlich 3 method is well suited to a wide range of soils, both acidic and basic in reaction. Mehlich 3 method is similar in principle to the Bray and Kurtz 1 because it is an acidic solution that contains ammonium fluoride. Acetic acid in the extractant also contributes to the release of available P in most soils. It is more effective than Mehlich 1 method for predicting crop response to P on neutral and alkaline soils because the acidity of the extractant is neutralized less by soil carbonates (Tran and Simard 1993). Olsen method was developed to predict crop response to fertilizer P inputs on calcareous soils. The Olsen P method is best suited for soils with >2% calcium carbonate, but has been shown in some research to be reasonably effective for acidic soils (Fixen and Grove 1990).

10.6.6 Inorganic Phosphorus in Soil Undergoes Precipitation–Dissolution Reactions

Precipitation is the process of formation of discrete and insoluble compounds from soluble substances in solution (Sect. 8.4.2). In soil, soluble inorganic P, phosphate ions, participate frequent precipitation reactions mainly with iron, aluminum, and calcium depending on the pH of the system and on the abundance of both phosphate and cations. In neutral or calcareous soils, Ca^{2+} is the dominant cation. With soluble phosphate ions (fertilizers or products of mineralization) it initially forms dicalcium phosphate dihydrate ($CaHPO_4{\cdot}2H_2O$), which slowly reverts to other Ca phosphates such as octacalcium phosphate ($Ca_8H(PO_4)_6{\cdot}5H_2O$), and ultimately apatite ($Ca_5(PO_4)_3OH$). In acidic soils, Al^{3+} and Fe^{3+} are the major soluble cations. They precipitate the phosphate ions of the soil solution. Iron and aluminum phosphates are the dominant precipitates in acid soils.

$$\underset{\text{very soluble}}{Ca(H_2PO_4)_2} + CaCO_3 + H_2O = \underset{\text{less soluble}}{2CaHPO_4{\cdot}2H_2O} + CO_2$$

$$6CaHPO_4{\cdot}2H_2O + 3CaCO_3 = \underset{\text{less soluble}}{3Ca_3(PO_4)_2} + 3CO_2 + 5H_2O$$

$$\underset{\text{very insoluble}}{3Ca_3(PO_4)_2} + CaCO_3 = \underset{\text{insoluble}}{3Ca_3(PO_4)_2{\cdot}CaCO_3}$$

$$\underset{\text{(soluble)}}{Al^{3+}} + H_2PO_4^- + 2H_2O \Leftrightarrow 2H^+ + \underset{\text{(insoluble)}}{Al(OH)_2H_2PO_4}$$

The presence of amorphous or partially crystalline Fe and Al oxides that can occlude P as they crystallize in acid soils makes identification of specific Fe–P and Al–P minerals difficult (Pierzynski et al. 2005). In soils that have not been extensively weathered, P mineralogy is dominated by sparingly soluble apatites and other more soluble calcium phosphates—octacalcium phosphate, monetite ($CaHPO_4$), brushite ($CaHPO_4{\cdot}2H_2O$), etc. As soils weather, apatites dissolve, Ca^{2+} ions are released in solution and taken up by plants, and precipitated or sorbed or are lost by leaching. Iron and aluminum phosphates may release phosphate ions in soil solution upon dissolution. In general, the solubility of P-bearing minerals and the availability of P to plants is the greatest under slightly acidic conditions in the pH range of 6.0–6.5 (Cowan et al. 1991).

$$\text{Hydroxyapatite}: Ca_5(PO_4)_3OH + 7H^+ \Leftrightarrow 5Ca^{2+} + 3H_2PO_4^- + H_2O$$
$$\text{Variscite}: AlPO_4{\cdot}2H_2O + 2H^+ \Leftrightarrow Al^+ + H_2PO_4^- + 2H_2O$$
$$\text{Strengite}: FePO_4{\cdot}2H_2O + 2H^+ \Leftrightarrow Fe^{3+} + H_2PO_4^- + 2H_2O$$

10.6.7 Phosphate Sorption–Desorption Regulates Phosphorus Availability in Soil

Sorption is the general term used to describe several processes collectively (surface precipitation, partitioning, polymerization, adsorption, and formation of secondary solid phases). McBride (2000) defined sorption as the transfer of ions and molecules from the solution phase to the solid phase. Sorption in soils results in the retention of a substance (sorbate) by organic and inorganic solid phases (sorbent). Sorption can occur due to physical processes involving van der Waals forces or electrostatic outer sphere complexes, for example, anion exchange, which is referred to as nonspecific adsorption, to chemical processes such as inner-sphere complexes, ligand exchange, and chemisorption, many of which are referred to as specific adsorption (Pierzynski et al. 2005). In addition, chemical precipitation may participate in the removal of ions from water and sometimes is treated as a part of sorption (McBride 2000).

Insolubilization of phosphates by precipitation with Fe, Al, Mn, and Ca and adsorption of phosphates by colloids are together known as phosphate sorption in soil. A large proportion of fertilizer P is made currently unavailable by P sorption. Retention of P by interaction with hydroxides and oxyhydroxides of Fe and Al in dilute solutions of P involves replacement of OH^- by PO_4^{3-}. Since the soil solution is acidic, the surface of these minerals has a net positive charge, although both (+) and (-) sites exist. The predominance of (+) charges readily attracts $H_2PO_4^-$ and other anions. When the orthophosphate ion is bonded through one Al–O–P bond, the $H_2PO_4^-$ is considered labile and the adsorption is reversible (Hingston et al. 1974). This orthophosphate ion can be readily desorbed from the mineral surface to soil solution. When two Al–O bonds with $H_2PO_4^-$ occur, a stable six-member ring is formed and the adsorption becomes irreversible. Very little desorption of P takes place from this complex (Tisdale et al. 1997). In acid soils, P retention also readily occurs on the broken edges of kaolinite clay minerals. Exposed OH^- groups on these edges can exchange for $H_2PO_4^-$ similarly to exchange with OH^- on the surface of Fe/Al oxides. Phosphorus interactions with calcite involve two reactions: The first reaction at low P concentrations consists of adsorption of P by calcite surface, while the second process is a nucleation process of calcium phosphate crystals (Griffin and Jurinak 1974). Phosphate is less strongly bound to $CaCO_3$ than to the hydrous oxides of Fe and Al and hence is more available to crop plants. The general consensus is that in soils where Al oxide and Fe oxide are available, carbonates may play only a secondary role in P sorption. There are a number of P sorption models, including Freundlich (1926), Langmuir (1918), and Temkin models (Tempkin and Pyzhev 1940). Freudlich and Langmuir isotherms for general adsorption processes have been discussed in Sect. 8.4.4.

Desorption is the release of P from a solid phase into the soil solution by processes other than mineral dissolution. Desorption occurs when plant uptake, runoff, or leaching depletes soil solution P concentrations to very low levels or in aquatic systems when sediment-bound P interacts with low P waters. Only a small fraction of sorbed P in most soils is readily desorbable. Most P added to soils in fertilizers or manures is rapidly sorbed but not easily desorbed.

10.6.8 Soil Phosphorus May Affect Water Quality

Phosphorus is an essential element for all organisms, and it is not toxic itself. Still, phosphorus may exert detrimental effects, even in low concentration, on water quality of lakes and other surface water reservoirs. Excess soluble phosphorus in soil contributes to eutrophication of water, a process characterized by increased growth of undesirable algae and aquatic weeds, followed by oxygen shortages as the biomass decomposes. Eutrophication restricts water use for fisheries, recreation, industry, and human consumption.

Phosphorus in soils is almost entirely associated with soil particles. During erosion and runoff, soil particles with bound P are carried to water bodies. Although most soils have a large capacity to retain P, a large addition of fertilizer P may stimulate leaching and transporting P to lakes and streams. In this way, water bodies such as lakes, estuaries, or slow-moving streams become enriched with nutrients, particularly phosphate (and nitrate). Consumption of dissolved water due to enhanced algal growth and decomposition leads to a hypoxic condition. At this point, the color of water becomes bluish in the initial and brownish in the latter stages, and odor becomes obnoxious.

10.7 Sulfur in Soils Come from Mineral and Organic Matter

The SO_4^{2-} ion is the available form of sulfur to plants. It remains in soil solution and as an adsorbed anion on colloidal surface. The concentration of SO_4^{2-} in soil solution is very low—10 to 100 mg L^{-1}. The main sources of sulfate in the soil are sulfur-containing minerals and organic matter. Gypsum ($CaSO_4 \cdot 2H_2O$) is a very common secondary mineral source of sulfur in soils developed from sedimentary parent materials. It is also abundant in soils of the arid and semiarid regions. Pyrite (FeS_2) is another important sulfur-containing mineral associated with shale and coal and in hydromorphic soils of the coastal regions receiving ample sulfate with sea salts and plenty of partially decomposed organic matter for reduction of sulfate. These are acid sulfate or potentially acid sulfate soils. In waterlogged soils, sulfate is reduced to elemental sulfur (S), H_2S, or FeS_2. Upon drainage of the soil, these compounds are again oxidized to sulfuric acids and sulfate salts. Sulfur compounds, thus, participate in redox reactions in soil.

$$2S + 3O_2 + 2H_2O \rightarrow 2H_2SO_4$$

Most sulfur, 60–90%, in soils comes from organic matter, which upon mineralization releases available sulfate slowly. Sulfur is, therefore, concentrated in the topsoil or plow layer. Sulfate mineralization is rapid in warm and well-drained soils. Sulfate can be leached, but the rate of leaching is slower than nitrate. Some sulfate is held by clay colloids.

Atmospheric deposition supplies a considerable amount of plant-available sulfur in soils of industrial countries. The amount of deposition may vary from 8 to 15 kg ha^{-1} $year^{-1}$. Some sulfate is added to soils with other fertilizers such as superphosphates as contaminants. If needed, gypsum and some other salts may be added as sulfur fertilizers.

10.8 Potassium in Soil Remains in Soluble, Exchangeable, Fixed, and Mineral-Bound Forms

Plants absorb K^+ ions from soil solution and exchange complexes. Soils contain soluble K^+, exchangeable K^+, fixed K^+, and K minerals. Soil solution contains only 1–40 mg L^{-1} K^+. There is about ten times higher exchangeable K^+ than solution

K^+; so there is a readily available reservoir of potassium to replace solution K^+. The amount of exchangeable potassium in a soil depends on the CEC of the soil and on the levels of other cations. Soluble K^+ and exchangeable K^+ tend to remain in an equilibrium, and when solution K^+ is removed by plant uptake or by leaching, some K^+ is released into soil solution from exchange complexes. K^+ ions in soil are vulnerable to loss by leaching. This is a problem in soils that do not have a large K^+ fixing capacity and in soils that have a low CEC. Leaching of potassium is common in sandy and silty soils.

Hydrous micas and illite clays contain potassium as part of their structure. These clays have layered structures, and K^+ ions are sandwiched between the layers. This is known as fixed potassium. As the clays weather, the layers peel apart, and the potassium ions are released.

Minerals are the principal sources of potassium in soil. Feldspars (orthoclase, microcline) and micas (muscovite, biotite) are the main K-containing mineral groups found in soil. These minerals release potassium as they weather.

$$\underset{\text{(Orthoclase)}}{KAlSi_3O_8} + H_2O \xrightarrow[\text{Hydrolysis}]{} \underset{\text{(Hydrogen aluminum silic}}{HAlSi_3O_8} + KOH$$

Weathering of feldspar is a very slow process, requiring a long time to release useable K^+. So, mineral potassium is unavailable to plants in the short term, but it is an important reservoir of K^+ in the long term.

10.9 Calcium in Soil Is Both a Nutrient and a Soil Conditioner

Calcium is both a plant nutrient and a soil conditioner. As a soil conditioner, calcium plays an important role in determining physical and chemical characteristics, that is, structure and pH of soil. Divalent Ca^{2+} ions produce effective soil aggregation, forming crumbs or peds. Soils dominated by calcium are friable and well structured, have good internal drainage, and have good tilth. Calcium carbonate or lime may reduce soil acidity and increase soil pH.

Calcium is absorbed by plants as the divalent cation (Ca^{2+}) present in soil solution and as exchangeable Ca^{2+} on soil colloids. Soil solution contains low concentration of Ca^{2+} (10–200 mg L^{-1}). Calcium is held on soil clay and humus particles more tightly than Mg^{2+}, K^+ and other exchangeable cations. Parent materials from which soils are formed also usually contain more calcium than magnesium or potassium. Soil Ca^{2+} does not leach to any appreciable extent because it is strongly attracted to the surface of clay. Soils generally have large amounts of exchangeable calcium (300–5,000 mg kg^{-1}). Normally, Ca^{2+} occupies 70–90% of cation exchange sites above pH 6.0.

Calcium is a part of the structure of several minerals, including apatite [$Ca_3(PO_4)_2$], calcite ($CaCO_3$), dolomite ($CaCO_3{\cdot}MgCO_3$), feldspars [anorthite $Ca_2Al_2Si_2O_8$], pyroxene [$(Na,Ca,Mg,Fe,Al)_2{\cdot}Si_2O_6$], hornblende [$Ca_2,Na(Mg,Fe^{2+})_4(Al,Fe,Ti),(Al,Si)_8O_{22}(O,-OH)_2$], and gypsum ($CaSO_4{\cdot}2H_2O$). Total Ca content of soils can range from 0.1% in highly weathered tropical soils to 30% in calcareous soils.

10.10 Magnesium Behaves Similarly to Calcium in Soil

Magnesium is the eighth most abundant element of the earth's crust. It has also some soil conditioning effect. Actually, Mg^{2+} behaves like Ca^{2+} in soil. Both are divalent cations and they stay in soil solution and exchange complexes. Magnesium also improves soil structure and increases soil pH. Availability of both Ca^{2+} and Mg^{2+} is low in low pH and high in high pH of soil. The primary minerals pyroxene [$(Ca,Na)(Mg,Fe^{2+},Fe^{3+},Al)(Si,Al)_2O_6$], amphibole [$Ca_2,Na(Mg,Fe^{2+})_4$ (Al,Fe,Ti) $(Al,Si)_8O_{22}(O,\text{-}OH)_2$], biotite [$K(Mg,Fe)_3(AlSi_3O_{10}){\cdot}(OH)_2$], and olivine [$(Mg,Fe)SiO_4$] and the secondary mineral dolomite ($CaCO_3{\cdot}MgCO_3$) supply most magnesium in soil. The clay minerals montmorillonite [$(Na,Ca)_{0.3}(Al,Mg)_2{\cdot}Si_4O_{10}(OH)_2{\cdot}nH_2O)$], vermiculite [$(MgFe,Al)_3$ $(Al,Si)_4O_{10}(OH)_2{\cdot}4H_2O$], and serpentine [$(Mg,Fe)_3Si_2O_5(OH)_4$] also contain magnesium.

In coastal regions, small droplets of water containing salts of sea are carried to the atmosphere by coastal winds. The phenomenon is called salt spray. These salts including magnesium are deposited on land as dust (dryfall) or with rain (wetfall). Near the coast, around 10 kg Mg ha^{-1} $year^{-1}$ can be deposited this way.

The availability of Mg^{2+} depends on total Mg content in soil, soil pH, CEC, and other cations. High available Mn can directly reduce Mg uptake. High levels of K or Ca provide less Mg to the crop. Whenever the soil K level is higher or when the soil K–Mg ratio is above 1.5:1, plant Mg levels are reduced.

10.11 Iron Is a Micronutrient and a Marker of Soil Genesis

Iron in soils comes from the weathering of many iron-containing minerals of the parent material. The most common iron-containing minerals are biotite [$K(Mg,Fe)_3(AlSi_3O_{10})(OH)_2$], chlorite, and hornblende [$Ca_2,Na(Mg,Fe^{2+})_4(Al,Fe,Ti)(Al,Si)_8O_{22}(O,-OH)_2$].

The iron in those minerals is largely in the ferrous state. When they are weathered during soil formation, most of the iron is changed to the ferric state and forms different iron compounds. The kind of parent material and the conditions under which it is weathered influence the distribution and kinds of iron compounds formed in the soil.

In humid temperate regions where rainfall is high and temperature is low and an organic layer has accumulated, a large part of the bases (sodium, potassium, calcium, and magnesium) of the parent material is dissolved and leached away and a more acid residue is left in the top horizons of the soil. Most iron is reduced, dissolved, and transported to the lower, less acid horizon, where it is precipitated forming new iron compounds such as iron hydrates, oxides, and some organic iron complexes. On the other hand, in tropical regions where rain is abundant and temperature is high, active base elements are also removed by weathering and leaching, iron is oxidized, and an acid soil is formed. The iron is converted largely to hydrated oxides. Some of the hydrated iron subsequently becomes dehydrated to Fe_2O_3. Because the soils of the arid and semiarid regions are less weathered and leached than those formed under heavier rainfall, many of them are calcareous and alkaline in reaction. The iron is less altered and is more uniformly distributed in their soil profiles.

Sizable amounts of the iron compounds form coatings on clay, silt, and sand particles in many soils. The coating may give them their color. The colors of most soils are due to the presence of the iron compounds.

Iron can exist in two valence states: Fe^{2+} and Fe^{3+}. The reversible oxidation/reduction of iron plays an important role in its translocation and transformation in soils and at the same time influence the redox condition of the soil. Even at low concentration, iron oxides within soils have a high pigmenting power and the color of pigments depends on the oxidation states and the predominance of the oxides. Thus color, and distribution of iron oxides within a profile, is helpful in explaining soil genesis and is used widely in soil classification. Iron oxides may affect soil structure and fabric. Different classes of iron oxides include magnetite (Fe_3O_4), hematite (α-Fe_2O_3), maghemite (γ-Fe_2O_3), goethite (α-FeOOH), lepidocrocite (γ-FeOOH), akaganeite (β-FeOOH), feroxyhite (δ-FeOOH), and ferrihydrite ($Fe_2O_3 \cdot 2FeOOH \cdot 2.6H_2O$).

10.12 Manganese Is an Active Redox Reactant in Soil

Manganese is more abundant than other micronutrients in soil. Common manganese minerals in soil include pyrolusite (MnO_2), manganite (MnOOH), and hausmannite (Mn_3O_4). Total Mn content in soil ranges from 0.02 to 1%, with an average of 0.25%. Volcanic soils are particularly high in Mn; it may be as high as 10% there. Manganese may occur in several ionic forms such as Mn^+, Mn^{2+}, and Mn^{4+}. Only Mn^{2+} is available to plants. Manganese is a fairly mobile element in soil and can be leached, particularly on acid soils. Manganese availability depends on soil pH, organic matter, and waterlogging. Mn is more available in low soil pH; even it may become toxic at pH below 5.5. On the other hand, it may be deficient in alkaline soils. Mn can be "tied up" by the organic matter. So, soils with high organic matter content can be Mn deficient. Under waterlogging, Mn^{2+} can initially be reduced to Mn^+, which is unavailable to plants. Under long-term reducing condition, however, available Mn can be increased. As soil dries, unavailable Mn^+ is oxidized to available Mn^{2+}; even some available Mn^{2+} can be oxidized to unavailable Mn^{4+}. High levels of Fe, Zn, Cu, and Mo in soil can reduce Mn uptake by plants. High NO_3^- and SO_4^{2-} can enhance Mn uptake.

10.13 Weathered, Leached, and Acid Soils Have Relatively Low Copper Levels

Copper is found in many different compounds in soil—sulfates, sulfides, sulfosalts, and carbonates. The most abundant Cu mineral is the chalcopyrite (the fool's gold—$CuFeS_2$). Total copper in soil ranges from 2 to 250 mg kg^{-1}, with an average of 30 mg kg^{-1}.

Copper in soil is inherited from the parent material. Soils developed from sandstones or acid igneous rocks contain lower Cu than those developed from shales and basic igneous rocks. Highly weathered, leached, and acid soils have relatively low Cu levels than other soils. Copper is strongly fixed by soil organic matter, clay, and oxide minerals. It is one of the least mobile trace elements in soils.

Copper concentrations in the soil solution of mineral soils are controlled primarily by soil pH and the amount of Cu^{2+} adsorbed on clay and soil organic matter. Copper is highly available at low pH (below 5). The lowest concentrations of Cu are associated with the organic or peat soils.

10.14 Fine-Textured Soils Generally Contain More Zinc Than Sandy Soils

The various forms of Zn in soil include minerals, organic matter, adsorbed Zn on the surfaces of organic matter and clay, and dissolved Zn in the soil solution. Total Zn in soil ranges from 10 to 300 mg kg^{-1}. Available Zn (DTPA extractable) has been positively correlated with total Zn, organic matter, clay content, and cation exchange capacity and inversely correlated with free $CaCO_3$, soil pH, and base saturation. Soil pH affects Zn availability; as soil pH increases, Zn availability decreases. Zinc availability to plants is lower in organic soils and in mineral soils with significant amounts of organic matter.

Clay fraction of the soil exerts a strong attractive force for zinc ions possibly due to similarity in size and charge between magnesium and zinc ions. Zinc ion reacts with the clay mineral lattice, where it may substitute or exchange for magnesium, thus making it relatively unavailable. High soil phosphorus levels have been known to induce zinc deficiency.

10.15 Molybdenum Availability Increases with Increasing Soil pH

Molybdenum is found in soil in minerals and as exchangeable Mo on the surfaces of Fe/Al oxides and clays and soil organic matter. In soil solution, molybdenum may exist in several forms depending on the pH. At pH >5.0, molybdenum exists primarily as MoO_4^{2-}, which is the available form for plants, but at lower pH the $HMoO_4^-$ and H_2MoO_4 forms dominate.

Molybdenum content in soils varies with parent material, degree of weathering, and organic matter content. Total Mo ranges from <3 to 15 mg kg^{-1}, but most soils contain 0.25 to 5.0 mg kg^{-1}. Unlike most other micronutrients, Mo availability increases with increasing soil pH. Molybdenum may be deficient under very acidic soil conditions. Molybdenum deficiency may be corrected by liming, but over-liming may form insoluble $CaMoO_4$. Unnecessary Mo application may create Mo toxicity to the crops and elevated Mo concentration in cereals and vegetables. Soils higher in organic matter generally have higher molybdenum levels. Soil P and S levels may influence Mo availability. Phosphorus increases Mo availability, whereas S decreases it.

10.16 Tourmaline Is the Main Mineral Source of Boron in Soil

There are four main forms of B in soils: water soluble, adsorbed, organically bound, and fixed in clay and mineral lattices. Much of the total boron in soil is present as a component of the mineral group tourmaline [(Ca,K,Na,I)(Al,Fe,Li,Mg,Mn)$_3$(Al,Cr,Fe,V)$_6$(BO$_3$)$_3$(Si,Al,B)$_6$O$_{18}$(OH,F)$_4$]. Most of the remainder is in secondary, moderately insoluble minerals. As soluble salts, B occurs as calcium and sodium borates. Parent rocks differ in boron contents. For example, sedimentary rocks generally have higher levels of B than igneous rocks. Parent materials and their degree of weathering determine the amount of boron in soil. The average concentration of B in the earth's crust is about 10 mg kg^{-1}; in soil it ranges from 2 to 100 mg kg^{-1}. There are low B contents in soils derived from acid igneous rocks, fresh water sedimentary deposits, and in coarse-textured soils low in organic matter. Boron is highly mobile and subject to leaching losses in coarse-textured soils low in clay and organic matter. Boron availability in soil may also be influenced by soil pH. In general, B becomes more available to plants with decreasing pH.

10.17 Chloride Is the Most Mobile Anion in Soils

Chlorine is found in soil as chloride (Cl^-) ion, which is soluble and is contained in negligible amounts in the mineral, adsorbed and organic soil fractions. Chloride has a high mobility in soils, and it is subject to leaching when rainfall or irrigation exceeds evapotranspiration. Chloride inputs to soil come from rainfall, fertilizers, manures, irrigation water, and with seawater flooding. Atmospheric deposition from rainfall averages around 20 kg Cl ha^{-1} year^{-1} with higher levels in coastal regions. Large quantities of chloride can be added where irrigation water contains high levels of chloride. The anions nitrate, sulfate, borate, and molybdate may be antagonistic to chloride.

10.18 Soils Around Industries Contain the Highest Nickel

Nickel is the 22nd most abundant element of the earth's crust (Sunderman and Oskarsson 1991). It is found primarily as oxides or sulfides in soil.

Nickel can exist in soils in several forms such as inorganic crystalline minerals or precipitates, and complexes and adsorbed on organic or inorganic exchange surfaces, water soluble, and free ion or chelated metal complexes in soil solution. Soil generally contains about 4–80 mg Ni kg^{-1}. Nickel is generally uniformly distributed throughout the soil profile. Soils around industrial complexes and mining areas may contain nickel concentrations that may be toxic to plants. Nickel is also released into the environment from anthropogenic activities such as metal mining, smelting, fossil fuel burning, vehicle emissions, disposal of household, municipal and industrial wastes, fertilizer application, and organic manures.

10.19 Nutrients Interact in Soils

Nutrients undergo complex reactions that lead to changes in their availability in soil. At the same time, available nutrients have also complex interactions. These interactions may be positive; increase in availability of one element may increase the uptake of another. It is stimulation. An example of stimulation is the increased demand of magnesium by increased level of nitrogen. Likewise, if more potassium is used, more manganese is required. There may be negative interactions as well; it is called antagonism. High levels of a particular nutrient in the soil can interfere with the availability and uptake of other nutrients by the plant. Those nutrients which interfere with one another are said to be

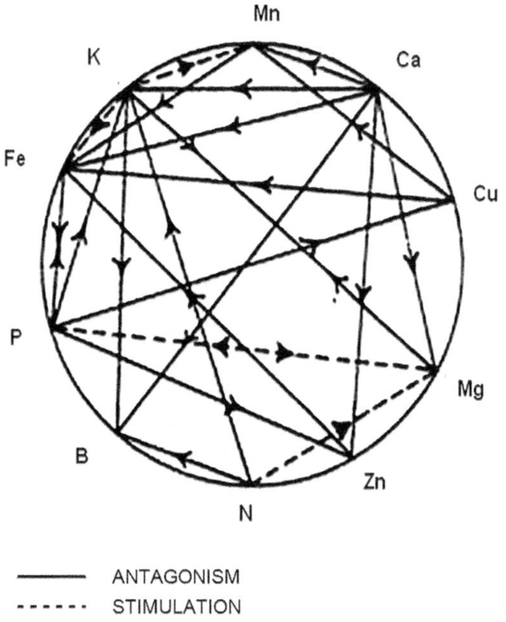

Fig. 10.1 Mulder's chart of nutrient interactions in soil (http://www.montmorillonite.org/Mulders%20chart.htm)

antagonistic. For example, high nitrogen levels can reduce the availability of boron, potassium, and copper; high phosphate levels can influence the uptake of iron, calcium, potassium, copper, and zinc; high potassium levels can reduce the availability of magnesium. Mulder (1953) published a nutrient interaction chart, which is shown in Fig. 10.1. These interactions need to be considered in planning a fertilizer program for crops.

10.20 Any Material That Provides Crops with a Nutrient Is a Fertilizer

When the soil has adequate available nutrient reserves, plants do not need any nutrient supplement. But soils often become deficient in major nutrients, particularly N, P, and K, and sometimes S, Zn, and other micronutrients. Then they need fertilizers. Fertilizers are inorganic or organic, natural or synthesized materials that are added to soil to supply plant nutrients. Fertilizers are added to increase soil fertility and plant growth and yield. Many other materials, including water, lime, pesticide, growth hormones, and soil conditioners, are used to increase plant growth and yield. They do not supply or they are not intended to supply plant nutrients; they are not fertilizers.

Industrially manufactured chemical compounds (mostly inorganic and some organic) with high nutrient contents are popularly used as fertilizers. They are called "commercial fertilizers," "inorganic fertilizers," "chemical fertilizers," and "industrial fertilizers." On the other hand, many natural materials such as composts, manures, oil cakes, fish meal, and guano are used as fertilizers from ancient times. They are called organic fertilizers.

Industrial fertilizers have specific nutrient concentrations; for example, urea contains 46% nitrogen. Industrial fertilizers are required in relatively small quantity. They are easy to store, easy to distribute in market, easy to apply in field, easy to mix with irrigation water, and easy to handle compared to organic fertilizers. To supply 100–150 kg N ha^{-1} to an annual crop, one would need about 217–326 kg urea per hectare. To supply the same amount of nitrogen from organic fertilizers, one would need 15–20 t ha^{-1} compost or farmyard manure because they contain very little nitrogen (the composition of composts and manures is also variable). It is difficult to collect such a huge quantity of organic residues. Processing, storing, marketing, and applying in field are also difficult. For these reasons, farmers usually depend more on commercial fertilizers. Otherwise, organic fertilizers could prove superior because organic fertilizers act as soil conditioners as well. They also contain some micronutrients. However, if applied on equal nutrient quantity basis, both organic and inorganic fertilizers may prove to be equally effective. Modern farmers usually apply organic fertilizers as part of the total requirement. The other part is supplemented with inorganic fertilizers.

10.20.1 Industrial Fertilizers Are Synthetic Products

10.20.1.1 Nitrogen Fertilizers

Ammonia is the basic substance for the production of industrial nitrogen compounds, including fertilizers. Nitrogen from atmosphere and hydrogen from natural gas are used for the production of ammonia in industries. This is called chemical fixation of atmospheric nitrogen which is based on Haber–Bosch process (Tempkin and Pyzhev 1940). The general process may be outlined as

$$N_2 + 3H_2 \xrightarrow[1{,}200^{\circ}C,\,500\,atm]{Catalyst} 2NH_3$$

The overall reaction is

$$7CH_4 + 10H_2O + 8N_2 + 2O_2 = 16NH_3 + 7CO_2$$

The ammonia thus produced can be utilized directly as a fertilizer (anhydrous NH_3). Other nitrogenous compounds are prepared from ammonia. Such as:

1. $NH_3 + O_2 + NH_3 \rightarrow$ Ammonium nitrate

2. $NH_3 + CO_2 \rightarrow$ Urea

3. $NH_3 + H_2SO_4 \rightarrow$ Ammonium sulfate

Table 10.3 Nitrogen fertilizers

Fertilizer	Formula	% N
Anhydrous ammonia	NH_3	82
Ammonia solution	NH_4OH	20
Urea	NH_2–CO–NH_2	46
Ammonium nitrate	NH_4NO_3	33
Ammonium sulfate	$(NH_4)_2SO_4$	21
Monoammonium phosphate (MAP)	$(NH_4)H_2PO_4$	10
Diammonium phosphate (DAP)	$(NH_4)_2HPO_4$	18
Ammonium chloride	NH_4Cl	25
Potassium nitrate	KNO_3	13
Sodium nitrate	$NaNO_3$	16

4. $NH_3 + H_3PO_4 \rightarrow \text{Ammonium phosphate}$

5. $NH_3 + H_2O + NH_4NO_3 \rightarrow \text{Nitrogen solution}$

Several chemical compounds are used as nitrogen fertilizers. Some of them with their percentage nitrogen contents are given in Table 10.3. Fertilizer materials in the table look like pure chemical compounds. Actually commercial fertilizers contain several impurities and contaminants.

10.20.1.2 Choice of Nitrogen Fertilizers

Some of the factors that determine the choice of a fertilizer are price, amount of fertilizer needed, availability in the market, ease of handling and application, ease of storage, loss of fertilizer from field, fertilizer efficiency, etc. Among chemical nitrogen fertilizers, ammonia is the cheapest and the most concentrated material. It is generally a gas but supplied under low pressure as a liquid. Small amount of ammonia may supply the full requirement of N of a crop. But application of ammonia is difficult. It is also corrosive in concentrated form. It may be injected in soil or applied with irrigation water in sprinkler or drip systems. When applied to soil, ammonia is converted to NH_4^+. Ultimately, there remains little difference between effects of ammonia and urea when applied on equal amount basis. Urea is also converted to ammonium. Farmers usually use urea because it contains higher nitrogen, it is cheap, and it is easy to handle.

$$NH_2 - CO - NH_2 + 2H_2O \rightarrow (NH_4)_2CO_3$$
$$(NH_4)_2CO_3 + 2H^+ \rightarrow 2NH_4^+ + CO_2 + H_2O$$

In a well-drained soil, ammonium is rapidly converted to nitrate by the process of nitrification. So, there is also little difference between nitrate and ammonium fertilizers in terms of their effects on crops. MAP and DAP have limited use. They are applied when both nitrogen and phosphorus are needed. Ammonium sulfate (21% N) and ammonium nitrate (33% N) were widely used in the past. These materials are now used in preparing mixed fertilizers.

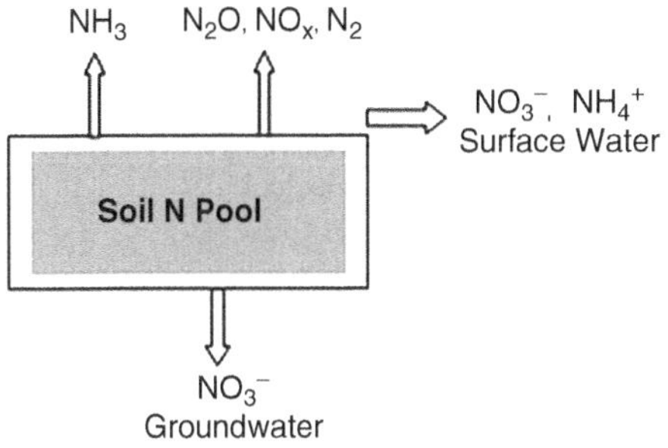

Fig. 10.2 Transfer of soil N to different pools

10.20.1.3 Fate of Applied Nitrogen

Whichever substance is used as nitrogen fertilizers, some of nitrogen is inevitably lost from the applied fertilizers. The chief processes of nitrogen loss are erosion, volatilization, leaching, and denitrification. The pathways of nitrogen loss from soil are shown above (Fig. 10.2).

Leaching loss following nitrification was mentioned in Sect. 10.6.1. Materials containing ammonium such as ammonium sulfate, ammonium nitrate, ammonium phosphate, or substances that produce ammonium after application to soil such as urea behave similarly in this regard. If the pH of soil is high (>7.0), if the soil is calcareous or sodic, and if the soil is saturated with water, considerable nitrogen may be lost by volatilization. Application of ammonia at shallow depths in dry soil can result in considerable loss of ammonia by direct escape from the soil as a gas. Ammonium ions are relatively less mobile than nitrates. Some ammonium may be held by colloids. But NH_4^+ is readily converted to NO_3^- by the nitrification process. Nitrate is highly mobile and easily lost by leaching particularly in coarse-textured soils. When soil does not have sufficient air, some microorganisms use the oxygen from NO_3^- in place of O_2 in soil air and rapidly convert NO_3^- to nitrogen oxide and nitrogen gas (N_2). These gases escape to the atmosphere and are not available to plants. This transformation can occur within 2 or 3 days of water saturation and can result in substantial losses of nitrogen fertilizers. The processes of nitrogen volatilization, nitrification, and denitrification have been described in Sect. 9.4.2.

10.20.1.4 Preventing Nitrogen Losses

The following techniques have been successfully employed to prevent losses of nitrogen from added fertilizers:

1. Granule size of N fertilizer is increased to reduce contact with the soil. Large urea balls are placed at the center of four rice hills to reduce denitrification loss in rice-growing areas of Bangladesh.
2. In many rice-growing areas, urea–mud–neem cake balls are prepared and applied between rows of rice. Neem (*Azadirachta indica*) cake prevents nitrification and denitrification.

Table 10.4 Slow-release and controlled-release nitrogen fertilizers

Fertilizer	Composition	% N content
SRF		
Ureaform	Urea formaldehyde	40
IBDU	Isobutylidene diurea	32
CRF		
SCU	Sulfur-coated urea	31–38
PCSCU	Polymer-coated SCU	30

3. Nitrogen is applied in split dozes. Generally, three splits are done; one-third fertilizer is applied during soil preparation, one-third is applied during most vigorous vegetative growth stage, and the remaining one-third during flower initiation.
4. Fertilizers are applied in narrow bands except being broadcasted. It reduces leaching, volatilization, and denitrification.
5. Nitrification inhibitors such as "nitrapyrin" have been used with variable success.
6. Slow-release (SRF) and controlled-release (CRF) nitrogen fertilizers are used.

10.20.1.5 Slow-Release and Controlled-Release N Fertilizers

There are some slow-release fertilizers (SRF) and controlled-release fertilizers (CRF) (Table 10.4). These materials have low solubility. Some compounds such as sulfur-coated urea prevent nitrification. Nitrogen is released at a slow pace and corresponds to the rate of absorption by roots. These materials are relatively costly and have not yet gained popular use.

10.20.1.6 Residual Effects of N Fertilizers

When part of fertilizers added to the current crop becomes available to the next crop, the fertilizer has a residual effect, we say. Residual effects depend on the rate of N applied, the crop, the amount of leaching, and the soil type. Residual effects of N fertilizers have been observed when deep-rooting cereals such as winter wheat and follow crops such as potatoes to which heavy N dressings have been applied. A significant proportion of the N fertilizer is converted into labile organic N which is readily mineralized during the succeeding crop. However, the proportion of actual fertilizer N remaining in mineral form in the soil by harvest time is very small (Powlson 1997). There will be little residual N from a fertilizer application of 150 kg N/ha to a preceding crop because much of the excess nitrate has been leached out when rainfall is enough.

10.20.1.7 Phosphorus Fertilizers

The natural source of phosphorus is the rock phosphate or phosphate rock that may be primary or secondary but contains the mineral fluorapatite [$Ca_{10}(PO_4)_6(F,OH)_2$]. Phosphate rock is mined and ground to a powder, which can be used directly as a phosphate fertilizer, or is processed for the manufacture of other phosphate fertilizers. Ground rock phosphate is insoluble in soil and is solubilized over time to phosphate ions and absorbed by plants. The acidulation of rock phosphate with sulfuric acid produces ordinary superphosphate (OSP), which is a mixture of monocalcium phosphate and calcium sulfate.

Table 10.5 Phosphorus fertilizers

Fertilizer	Formula	% P
Rock phosphate	$Ca_{10}(PO_4)_6(F,OH)_2$	11–15
Phosphoric acid	H_3PO_4	23
Ordinary superphosphate (OSP)	$Ca(H_2PO_4)_2{\cdot}CaSO_4$	9
Triple superphosphate (TSP)	$Ca(H_2PO_4)_2$	20
Monoammonium phosphate (MAP)	$(NH_4)H_2PO_4$	21
Diammonium phosphate (DAP)	$(NH_4)_2HPO_4$	20
Ammonium polyphosphate	$(NH_4)_4P_2O_7$	15
Potassium dihydrogen phosphate	KH_2PO_4	23

$$Ca_{10}(PO_4)_6F_2 + 7H_2SO_4 = 3Ca(H_2PO_4)_2 + 7CaSO_4 + 2HF$$

The OSP so produced consists of about one-half monocalcium phosphate and one-half calcium sulfate. It contains 9% phosphorus. The phosphorus in monocalcium phosphate is water soluble. OSP is rarely used today because other phosphorus fertilizers are less expensive on an equivalent phosphorus basis. Treatment of phosphate rock with phosphoric acid produces triple superphosphate (TSP), which is monocalcium phosphate with a higher content of phosphorus. The reaction is

$$Ca_{10}(PO_4)_6F_2 + 14H_3PO_4 = 10Ca(H_2PO_4)_2 + HF$$

TSP contains 20% P and is widely used as a phosphate fertilizer. Ammonium phosphates and polyphosphates may also be used, particularly where both N and P are needed. Table 10.5 shows different phosphorus fertilizers with their percentage concentration.

Industrial phosphate fertilizers are water soluble. When they are applied to soils they can undergo reactions with other soil constituents which may render them insoluble. This is known as phosphate sorption. The nature and degree of phosphate sorption depend on physicochemical properties of soil, particularly pH, clay content, type of clay, organic matter, oxides, and hydroxides of Fe, Al, Mn, etc. The result is the reduction in currently available phosphorus. At both low and high soil pH, phosphorus solubility is reduced. Adjustment of soil pH near neutrality may correct phosphate deficiency in many soils. Typically, only 10–20% of fertilizer P is absorbed by plants during the first year; the remainder is nearly all retained as fertilizer reaction products, which become less soluble with time.

10.20.1.8 Potassium Fertilizers

Huge natural potassium deposits are found in places where ancient seas have evaporated leaving the salts as evaporite

Table 10.6 Potassium fertilizers

Fertilizer	Formula	% K
Potassium chloride (muriate of potash, MP)	KCl	50
Potassium sulfate	K_2SO_4	41
Potassium nitrate	KNO_3	38

deposits. These salt deposits were buried under various kinds of sediments and rocks. Several minerals representative of the salts in seawater, mainly salts of sodium and potassium, formed. For example, sylvite is KCl and sylvinite is a mixture of KCl and NaCl. Langbeinite is a mixture of potassium and magnesium sulfates. KCl is separated from the other compounds in the ore; more than 95% of the potassium in fertilizers is KCl. Some of the deposits are relatively concentrated KCl and require little, if any, processing.

The method of processing is known as flotation. The ore is ground, suspended in water, and treated with a flotation agent that adheres to the KCl crystals. As air is passed through the suspension, KCl crystals float to the top and are skimmed off. The material is dried and screened to obtain the proper particle size. Some of the potassium fertilizers with their nutrient contents are given in Table 10.6.

10.20.1.9 Micronutrient Fertilizers

The micronutrients are removed from soil by crops in relatively small amounts. They are added to soils from minerals, dry- and wetfalls, with organic residues and as contaminants of major element fertilizers. Therefore, deficiency of micronutrients Fe, Mn, Zn, Cu, B, and Mo is not as common as those of NPK. Deficiency of Fe, Mn, Zn, and Cu often occurs in soils above pH 7, and deficiency of Mo may occur in very acid soils. However, in deficiencies of micronutrients, suitable salts, as listed below (Table 10.7), may be used as micronutrient fertilizers. Only 10 kg ha^{-1} of each of Mn, Zn, and Cu; 1 kg ha^{-1} B; and 0.2 kg ha^{-1} Mo may be added to soils to correct their deficiencies. Frit is an increasingly popular controlled-release micronutrient fertilizer. It is made by fusing the element in glass that can be crushed and mixed with NPK fertilizers (White 2006).

Table 10.7 Micronutrient fertilizers

Element	Fertilizer	% Nutrient content
Fe	Ferrous sulfate ($FeSO_4{\cdot}7H_2O$)	20% Fe
Mn	Manganous oxide (MnO)	48% Mn
	Manganese sulfate ($MnSO_4{\cdot}nH_2O$)	23–25% Mn
Zn	Zinc sulfate ($ZnSO_4{\cdot}7H_2O$)	35% Zn
Cu	Copper sulfate ($CuSO_4{\cdot}5H_2O$)	25% Cu
Mo	Sodium molybdate ($Na_2MoO_4{\cdot}2H_2O$)	40% Mo
B	Borax ($Na_2B_4O_7{\cdot}10H_2O$)	11% B
Co	Cobalt sulfate ($CoSO_4{\cdot}7H_2O$)	21%
Ni	Sewage sludge	variable

10.20.1.10 Mixed Fertilizers

A single chemical compound may supply one or two nutrients. Urea (NH_2–CO–NH_2) supplies only nitrogen, while triple superphosphate ($Ca(H_2PO_4)_2$) supplies phosphorus and calcium. Urea and TSP are single fertilizers. On the other hand, two or more chemical compounds may be mixed together in a suitable proportion to supply two or more nutrients at a time from the same material. These are called complex fertilizers or mixed fertilizers. A 10-10-10 mixed fertilizer is a fertilizer mixture that contains 10% N, 10% P, and 10% K. [Percentages of P_2O_5 and K_2O instead of P and K are also widely used, although the Soil Science Society of America favors a fertilizer grade based on the percentages of nitrogen, phosphorus, and potassium (Foth 1990). However, the basis is to be clearly specified by the manufacturer on the fertilizer bag. To obtain P_2O_5 from P values, multiply by 2.29 and to obtain K_2O from K, multiply by 1.20; to obtain P from P_2O_5 and K from K_2O, multiply by 0.44 and 0.83, respectively.]

An expression indicating the percentages of plant nutrients in a fertilizer mixture is termed as fertilizer grade, and the relative proportion of major plant nutrients in the mixed fertilizer taking "N" as one is called fertilizer ratio. For example, in a 5-10-5 mixed fertilizer, the fertilizer grade is 5-10-5 and the fertilizer ratio is 1:2:1. Mixed fertilizers have some advantages and disadvantages. A balanced and complete fertilizer mixture suited to the crop may be prepared and applied at a time to the field, and thus, time and labor are saved. Storage and handling costs are reduced. Micronutrients can also be incorporated in the mixture. Residual acidity can be neutralized by using neutralizers in mixture. Soil conditioners may be added as filler materials. However, when mixed fertilizers are used, the cost of plant nutrients becomes higher than straight fertilizers. If only one nutrient is required by the crop, the fertilizer mixtures are not useful. Sometimes farmers may add nutrients in excess when they use fertilizer mixtures. Let us prepare 2 t 5-10-5 (N–P–K) mixed fertilizers by using ammonium sulfate (21% N), triple superphosphate (20% P), and potassium sulfate (41% K) as sources of N, P, and K, respectively.

For 2 t or 2,000 kg 5-10-5 (N–P–K) mixed fertilizer,
the amount of N needed = 5/100 × 2,000 = 100 kg,
the amount of P needed = 10/100 × 2,000 = 200 kg, and
the amount of K needed = 5/100 × 2,000 = 100 kg.
For 100 kg N, amount of ammonium sulfate = 100 × 100/21 = 476 kg.
For 200 kg P, amount of triple superphosphate = 200 × 100/20 = 1,000 kg.
For 100 kg K, amount of potassium sulfate = 100 × 100/41 = 244 kg.

	Total = 1,720 kg
To make 2,000 kg, we need extra material	280 kg
Total 5-10-5 fertilizers	2,000 kg

This extra material is known as "filler material." Wood ash or lime can be used as filler materials for mixed fertilizers.

10.20.1.11 Liquid Fertilizers

The term liquid fertilizer refers to anhydrous ammonia, aqueous ammonia, N solutions, and liquid-mixed fertilizers. They are true solutions. There are also suspensions or slurry-type mixtures of N, P, and K. They are called fluid fertilizers. Urea, urea–ammonium nitrate solution, ammonium polyphosphate solution, and finely ground KCl are used for making fertilizer solutions or suspensions. Micronutrients and pesticides are readily incorporated into liquid or fluid fertilizers. Fluid fertilizers can be applied to irrigation water in drip and sprinkler or underground injection systems and can be used for direct application to plant foliage. Application of nutrients with irrigation water is called fertigation. It requires sophisticated control systems to regulate water and nutrients. The system needs careful monitoring so that the chemicals used do not get precipitated (Sect. 10.22).

Suspensions contain more material added to the water than can dissolve, so it becomes a thin slurry. One to two percent clay (attapulgite, an alumino–Mg–silicate clay) is added to help hold particles in suspension. This allows for preparation of higher grades, as compared with liquid fertilizers. Similar effects may be obtained from both liquid and dry fertilizers if similar amount is applied in the same manner. The solution fertilizers are quite popular and can be applied with herbicides mixed into the solution.

10.20.2 Organic Fertilizers Are Natural Materials

Many natural organic materials have been used as fertilizers from ancient times. Farmyard manures and composts are very familiar organic fertilizers. Other organic fertilizers include poultry manure, blood meal, bone meal, fish meal, oil cakes, worm castings, peat, seaweed, humic acid, and guano.

Organic fertilizers have low nutrient concentrations. Their composition is also variable, and the calculation of the requisite amount of nutrient is difficult. Huge quantities of materials are needed to supply adequate nutrients for crop growth. Nevertheless, there are many advantages of using organic fertilizers. Organic fertilizers contain nitrogen in the organic form that is slowly mineralized in soil, and available nitrogen is released at a pace favorable to plants. Hence, they act as slow-release fertilizers, and leaching and denitrification losses are minimized. Organic fertilizers act like compound fertilizers containing several nutrients at a time, including micronutrients. For example, farmyard manures contain about 0.5% N, 0.25% P, 0.4% K, 0.08% Ca, 0.02% S, 0.4% Fe, 0.003% Cu, 0.004% Zn, and 0.007% Mn. Organic fertilizers improve physical conditions of the soil; they enhance root growth that utilizes nutrients better. Residual effects of organic fertilizers are positive and long-lasting. However, collection, storage, marketing, distribution, and application of huge quantity of organic fertilizers are difficult and cost and labor consuming. Otherwise, organic fertilizers are very effective in increasing growth and yield of crops. They keep the soil biologically active and healthy. Organic fertilizers add multiple nutrients, improve soil structure, increase water-holding capacity, and improve aeration.

Table 10.8 Nutrient concentration in different animal manures

Farmyard manure	% N	% P	% K
Cow manure	2.0–3.0	0.5–1.0	1.0–2.0
Horse manure	1.0–2.0	0.5–1.0	1.0–2.0
Pig manure	2.0–3.0	0.5–1.0	1.0–2.0
Sheep manure	3.0–4.0	0.5–1.0	2.0–3.0
Poultry manure	3.0–4.0	1.0–2.0	1.0–2.0

Source: http://extension.usu.edu/files/publications/factsheet/HG-510.pdf

10.20.2.1 Farmyard Manure

The term farmyard manure was originally applied to well-trodden farm materials including vegetable feed wastes, feces, and urine of yarded animals in a farm along with straw to bind up feces. At present, farmyard manures are obtained by decomposing dung, urine, and feed wastes of farm animals. Generally, excreta and waste materials are gradually gathered and dumped in a pit in a shed at a cool dry place near the cattle shed. Care is taken so that the materials are not completely dried in the sun or wetted by rainwater. It needs 5–6 months for the completion of the decomposition process. By this time, the foul odor is gone and the materials have a crumbly appearance. The quality of material depends on the kind of farm animals, their feed, and the conditions of decomposition. Farmyard manures contain several nutrients. Concentrations of N, P, and K in different farmyard manures are given in Table 10.8.

10.20.2.2 Composts

Farms have many kinds of organic residues, including straw, food wastes, tree leaves, weeds, grass clippings, or other plant and garbage wastes, the disposal of which is a problem. These materials are alternatively composted and used as a fertilizer. For composting, organic residues, mostly plant materials with some animal dung and bedding, are gathered in a pile in a dry shaded place providing favorable water and aeration. As the organic matter decomposes, the bulk of the materials is reduced with the production of carbon dioxide and water. Nutrients are continuously reused and recycled by the microbes and conserved within the composting pile.

Table 10.9 Effect of NPK, poultry manure, and kitchen manure on growth of corn

Treatment	Height of 60-day old-plants, cm	Number of leaves plant^{-1}	Yield of grains, g plot^{-1*}
Control	65	5	503^{d}
NPK	105	10	1,785^{c}
Kitchen compost	100	10	2,117^{b}
Poultry manure	135	14	3,000^{a}

*Figures followed by different letters in the column vary significantly by Duncan's multiple range test

Table 10.10 Miscellaneous organic fertilizers

Materials	% N	% P	% K
Alfalfa meal	2.0–3.0	0.5–1.0	1.0–2.0
Blood meal	12.0	1.0–2.0	0–1.0
Bone meal	1.0–6.0	11.0–30.0	0
Cottonseed meal	6.0	3.0	1.0
Fish meal	6.0–12.0	3.0–7.0	2.0–5.0
Hoof meal	12.0–14.0	1.5–2.0	0
Kelp	1.0–1.5	0.5–1.0	5.0–10.0
Legumes	2.0–4.0	0–0.5	2.0–3.0
Sawdust	0–1.0	0–0.5	0–1.0
Sewage sludge	2.0–6.0	1.0–4.0	0–1.0

Source: http://extension.usu.edu/files/publications/factsheet/HG-510.pdf

Thus, with a loss of carbon, oxygen, and hydrogen, other nutrient elements are concentrated during the process of decomposition. The finished product is rich, dark, crumbly, and sweet smelling.

The compost pile is occasionally, preferably at 15-day intervals, turned up and down to aerate the debris of the compost pile. It hastens decomposition. The low nitrogen content of many composting materials may retard their decomposition. For this reason, some nitrogen fertilizer is sometimes added to the compost pile for rapid decomposition. Since composts are low in N, P, and K contents, sometimes they are enriched by adding NPK fertilizers.

In the crop field of the Soil Science Department of Chittagong University, we prepared composts from kitchen wastes of eight student hostels. The wastes included vegetable scraps, fish scraps and bones, and chicken scraps. Collected materials were piled up in a concrete bin layer by layer. Each 15-cm layer was separated by a 2.5-cm-thick layer of top soil. Wastes were collected for 2 months and decomposed with occasional turning up for 8 months. An experiment was conducted by applying 20 t ha^{-1} each of this compost and poultry manure, and NPK (120, 60, 100 kg ha^{-1}, respectively) separately on corn (*Zea mays*). The height and number of leaves per plant data show that the prepared compost was as good as inorganic fertilizers (Table 10.9).

10.20.2.3 Green Manuring

Green manuring is a technique of plowing under a crop, usually a legume, when it is at a suitable stage of growth. The green manuring plants should have the ability to grow fast, to grow satisfactorily in poorly fertile soils, and to produce enough succulent biomass that adds organic matter after incorporation into the soil and decomposes at a moderate rate. The crop is grown in a period of fallow between two crops. Legume seeds are sown densely in the field after harvest of a crop, and the seedlings are allowed to grow until flowering. The crop utilizes little native soil nitrogen. About 80–90% of the nitrogen need of the crop is met by atmospheric nitrogen fixation. At the flowering stage, when the plant tissue is still tender and the biomass accretion is at maximum, the crop is plowed under or incorporated into the soil, keeping the soil moisture level near field capacity to enhance decomposition. The next crop is planted at least after 15–20 days of incorporation. Green manuring adds 2–7 t ha^{-1} organic matter and 100–300 kg ha^{-1} nitrogen depending on the legume species, the type of soil, and environment of decomposition. Thirty to fifty percent of this nitrogen is utilized by the next crop. That is, green manuring can reduce almost 30–50% of the nitrogen fertilizer need of the next crop. The remainder nitrogen is made available in succeeding years. Besides adding nitrogen, green manuring has other functions. Green manure increases soil organic matter, enhances activity of soil microorganisms, and improves soil structure. Deep-rooted green manure crops bring nutrients to the surface soil. Green manuring helps combat weeds. Important green manure crops are alfalfa (*Medicago sativa*), cowpea (*Vigna catjang*), lablab (*Dolichos lablab*), lentil (*Lens esculenta*), pigeon pea (*Cajanus cajan*), sesbania (*Sesbania aculeata*), soybean (*Glycine max*), sun hemp (*Crotalaria juncea*), sweet clover (*Melilotus officinalis*), etc. These crops produce 3–7 Mg ha^{-1} organic matter and contribute 120–300 kg ha^{-1} nitrogen to soil.

10.20.2.4 Miscellaneous Organic Fertilizers

Many plant and animal products and by-products are used as organic fertilizers. These materials are very popular in organic farms, where farmers use all natural materials, organic and inorganic, as fertilizers, soil conditioners and pesticides. Oil cakes, bone meals, fish meals are used as pot mixes in greenhouses and nurseries. Table 10.10 shows nutrient concentrations of some miscellaneous organic fertilizers obtained from different sources.

10.21 Determining the Kind and Dose of Fertilizer Is Called Fertility Evaluation

There are several methods of determining the kind and amount of fertilizers needed for a crop in a soil. The major techniques include visual diagnosis, analysis of plant tissues

Table 10.11 Plant parts and time of sampling for plant analysis

Crop	Sampling time	Plant part to be sampled
Corn (*Zea mays*)	Initial silk	Ear leaf
Wheat (*Triticum aestivum*)	Prior to heading	Top two leaves
Soybean (*Glycine max*)	Prior to pod set	Most recent leaf
Cucumber (*Cucumis sativus*)	Flower to fruit set	Fifth leaf from growing tip
Tomato (*Lycopersicon esculentum*)	Mid-bloom	Leaf adjacent to top flower cluster
Alfalfa (*Medicago sativa*)	Prior to flowering	Top 15 cm
Pea (*Pisum sativum*)	Prior to flowering	Newest fully developed leaf
Cabbage (*Brassica oleracea*)	Mid-season	Upper mature leaves
Potato (*Solanum tuberosum*)	Prior to flowering	Newest mature leaf
Sorghum (*Sorghum bicolor*)	Prior to heading	Second fully developed leaf

and soils, pot culture experiments, and field trials. However, any of the methods individually may not give satisfactory results. For example, plant test values often need to be correlated with soil test values. Soil tests are most useful before planting to predict fertilizer needs; plant tissue tests are best used during the growing season to monitor plant nutrient uptake. Soil samples supplement tissue samples by providing information on soil fertility, pH, and soluble salts. Comparison of the two samples can help reveal whether nutrition or other factors—such as pests, disease, or environmental stress—are involved.

10.21.1 Visual Symptoms May Indicate Nutrient Deficiency and Toxicity

Ideally deficiency of any one nutrient element in a plant should produce characteristic deficiency symptoms that may be corrected by the application of that element only. However, identification of deficiency and need of a particular nutrient element on the basis of visual symptoms is often difficult and confusing. Deficiencies of multiple elements together make deficiency symptoms more complicated. Often deficiencies of two elements produce closely similar symptoms such as nitrogen and magnesium. Besides this, by the time plant nutrient deficiency symptoms become obvious, it may be too late to correct the deficiency by fertilizers. Significant damage has already been done, and it would be difficult for the crop to recover. However, symptoms are useful in determining which nutrients are deficient in perennial crops, where fertilizers are applied in different seasons. Lawns commonly turn light green or yellow when nitrogen is deficient, and this indication of nitrogen deficiency can be very useful for lawn owners. Many trees, shrubs, and flowers develop iron deficiency symptoms when growing on alkaline soils and these symptoms are very important to many gardeners. Nutrient deficiency symptoms may be confirmed by short-duration pot culture, sand culture, and solution culture experiments.

10.21.2 Tissue Test Indicates Nutrient Status in Plants

There are rapid tissue tests and total plant analysis. Rapid tissue tests are generally carried out on sap from leaf and petiole samples with plant tissue test kits, and to get information about nitrate-N, sulfate-S, or phosphate-P. The use of diphenylamine as an indicator of the concentration of nitrate in the sap of corn leaf has satisfactorily been used to judge sufficiency or deficiency of nitrogen in corn. Simplicity of use and immediate results obtained from tissue tests are beneficial to plant nutritionists who diagnose plant growth problems in the field. On the other hand, plant analysis is the total nutrient analysis of plant samples, preferably leaves. Plant tissue analysis should not be confused with tissue tests. Plant tissue analysis is performed on dried plant tissue that has been processed in a laboratory and analyzed for any one or all of total N, P, K, S, Ca, Mg, Zn, Cl, Cu, Fe, B, Mn, Mo, and Co.

Plant tissue analysis measures concentrations of most essential plant nutrients in leaf tissue. Healthy plants contain predictable concentrations of nutrients that vary with plant type and growth stage. Tissue analysis results are an indicator of plant nutritional health and could be an excellent basis for diagnosing whether existing problems are nutritional in nature. However, plant analytical results differ with plant species, plant parts of the same species, growth stage, age of the leaf, position of leaf in plant, etc. Table 10.11 shows suitable parts and time of sampling for some crop plants.

Plant tissue tests are frequently used for fertilizer recommendations for long duration crops like tea, coffee, grapes, sugarcane, pineapple, and fruit trees. Plant analysis has also been employed for corn, wheat, barley, tomato, tobacco, and soybean. In perennial plants, the nutritional status of the plant can be assessed, and the appropriate fertilizer can be applied long before harvest, which allows plenty of time for the plants to benefit from the fertilization.

Chemical tests of plant tissue are used to determine the degree of deficiency or sufficiency of nutrients in plants. Research data have been used to relate nutrient composition

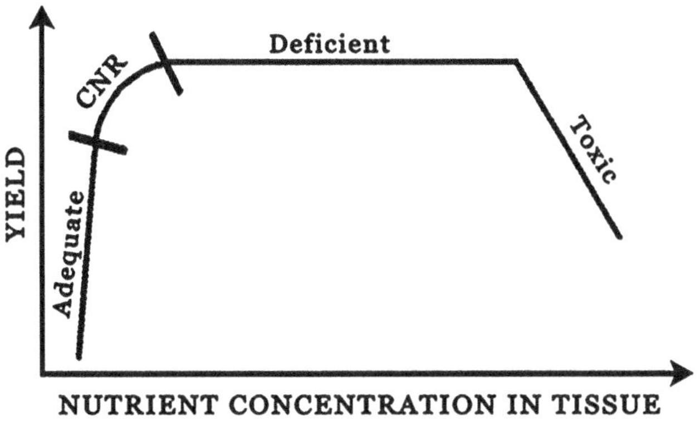

Fig. 10.3 Relationship between nutrient concentration in plant and crop yield

Table 10.12 Critical nutrient ranges and sufficiency ranges for cereals

Nutrients	Critical nutrient ranges	Sufficiency ranges
(%)		
N	1.25–1.75	1.75–3.0
P	0.15–0.25	0.26–0.5
S	0.1–0.15	0.15–0.40
K	1.0–1.5	1.5–3.0
Ca	0.10–0.2	0.2–1.0
Mg	0.1–0.15	0.15–0.50
(mg kg^{-1})		
Fe	15–20	20–250
Zn	10–15	15–70
Cu	2.3–3.7	3.7–25
Mn	10–15	15–100
B	3–5	5–25
Mo	0.01–0.02	0.03–5

Source: Manitoba Provincial Soil Testing Laboratory (1987)

of particular plants to growth or yield. For example, corn plants have sufficient nitrogen for maximum yield when the nitrogen content of the leaf opposite and below the uppermost ear is 3%. Red pine needles with less than 0.35% potassium are deficient in K; maximum growth was obtained when needle K was about 0.5%.

10.21.2.1 Critical Nutrient Ranges in Plants

There are three important threshold markers of nutrient concentrations in plants: (a) The deficiency level, which reflects the nutrient concentration below which deficiency symptoms become visible; (b) the critical nutrient range (CNR), which stands for the nutrient concentration above which the plant is sufficiently supplied with nutrients for achieving the maximum potential yield, and below which growth or yield is considerably reduced and a response to added nutrient occurs; and (c) the toxicological value, which indicates the nutrient concentration above which toxicity symptoms can be observed. Critical nutrient ranges vary among plants and nutrients, but occur somewhere in the transition between nutrient deficiency and sufficiency. The sufficiency range is the nutrient level at which the plant has enough nutrients to function and develop properly, but not so much that it becomes toxic (Fig. 10.3).

For a general understanding, critical nutrient ranges (CNRs) and sufficiency ranges (SRs) of the nutrients for cereals are given in Table 10.12. However, the readers should be aware that a single value or a range may not be applicable to all plants. Critical nutrient ranges for many different crops have been proposed; but establishing well-defined standard critical nutrient ranges or sufficient ranges of all the nutrients for all the crops under widely varied growing conditions is difficult.

10.21.2.2 Diagnosis and Recommendation Integrated System

The critical nutrient concept has many limitations. Critical ranges differ with growth stage, leaf age, leaf position, height of plants from which leaf is sampled, soil conditions, crop management, etc. To avoid these difficulties, a different approach called "Diagnosis and Recommendation Integrated System" (DRIS) was adopted (Beaufils 1973; Walworth and Sumner 1987). DRIS uses nutrient ratios instead of individual nutrient concentrations as a basis of judging nutrient deficiency or sufficiency. The most common ratio is the N–S ratio. For most crops, it should range between 10 and 15. When it exceeds 18, the S becomes limiting in relation to N. The ratios are fairly constant over leaf age, leaf position, growth stage, variety, etc. In addition, ratios take nutrient interactions into consideration. So, ratios are more acceptable than nutrient concentrations or critical nutrient ranges. However, the usage of nutrient ratios has also been criticized on the ground that the same N–S ratio of 10, which indicates a balance between N and S, can be created when both N and S are sufficient, when both are deficient, and when S is deficient. DRIS takes all the nutrient ratios at a time into consideration to calculate DRIS indices.

The DRIS is based on the comparison of crop nutrient ratios with optimum values from a high yielding group (DRIS norms). The DRIS provides a means of simultaneous identifying imbalances, deficiencies, and excesses in crop nutrients and ranking them in order of importance. The major advantage of this approach lies in its ability to minimize the effect of tissue age on diagnosis, thus enabling one to sample over a wider range of tissue age than permissible under the conventional critical value approach.

For basic data based on which DRIS norms are determined: (a) The area for which DRIS norms are to be developed (e.g., region, district, state) is decided; (b) a large number of sites where a particular crop is growing are selected at random in order to represent the whole production area; (c) at each site, plant and soil samples are taken for all essential element analyses; (d) other parameters likely to be related directly or indirectly to yields are also recorded; (e) entire population of observation is divided into two subpopulation (high and low yielders) on the basis of vigor, quality, and yields; (f) each element in the plant is expressed in as many ways as possible (concentrations and ratios); (g) mean of each type of expression for each subpopulation is calculated; and (h) each form of expression which significantly discriminates between the high and low yielding subpopulation is retained as a useful diagnostic parameter. DRIS norms are originated after the reference population definition; in other words, the relation between all the nutrients pairs and their respective standard deviations or coefficients of variation are obtained. The DRIS indices are calculated after obtaining the DRIS norms. For that the nutrient ratios N/P, N/K, Ca/N, Mg/N, K/P, Ca/P, P/Mg, Ca/K, Mg/K, Mg/Ca, and so on, their mean, and coefficient of variations are calculated first, and then the indices are obtained. The readers are referred to Bangroo et al. (2010) for calculation of DRIS indices and their interpretation.

Considering that DRIS uses the nutritional balancing concept (relationship among nutrients), it is postulated that this method might be more precise than the others in the detection of nutritional deficiency or/and excess. Advantages of DRIS over other diagnosis methods include the following: (1) It presents continuous scale and easy interpretation; (2) it allows nutrient classification (from the most deficient up to the most excessive); (3) it can detect yield limitation due to nutrient unbalance, even when none of the nutrients is below the critical level; and (4) it allows diagnosis of the total plant nutritional balance, through an unbalance index (Baldock and Schulte 1996). An additional advantage of DRIS according to some authors is its less sensitivity to tissue aging (Walworth and Sumner 1987). An important limitation of the DRIS method is that the established standard sampling period, in some annual crops, may occur too late in the growing season, so that fertilizer application will not be effective to correct a nutritional problem, or may not match the sudden symptoms of a nutritional disorder (Walworth and Sumner 1987). For this problem, it would be necessary to get nutritional reference values for several maturation stages, but standards have not yet been established for many crops. The main limitation of the DRIS system is the lack of adequate database for the purpose of comparison. This method is not very useful to small farmers. This procedure is of difficult application (Filho 2004). There is a need for precise definition, at the sampling time, of plant growth stages in the field and communicating this information to the person taking care of the diagnosis so that appropriate norms can be selected and used.

10.21.3 Soil Test Is an Efficient Tool of Fertility Evaluation

A soil test can be an important management tool in developing an efficient soil fertility program as well as monitoring a field for potential soil and water management problems. A soil test provides basic information on the nutrient supplying capacity of the soil.

Soil tests are used to determine the levels of available nutrients and probability of response to added fertilizers, particularly P, K, and in some instances Mn, Cu, Zn, and Fe. There is growing interest in extending the use of soil testing to assess the potential impact of soils and soil amendments on other sectors of the environment. Understanding many factors, including the soil test extractants used, the methods of soil sampling, handling, and storage, is required to interpret the quantitative results of a soil test. For fertility evaluation, generally the available nutrient levels together with soil texture, pH, and organic matter contents are determined.

Only about a half kilogram of soil is sampled to determine fertility level of a hectare of land having about 2 million kg soil in the furrow slice. Therefore, extreme care must be taken to make the soil sample representative. All sorts of variations in the field must be taken into consideration. The field may need to be separated into different blocks. Usually, it is better to prepare a single soil sample for each block from several subsamples. For large areas, 20–25 subsamples may be taken and mixed together to form a composite sample. A soil probe or auger is ideal for taking soil samples, but a sharp spade, long knife, or trowel can be used. For cereals and other shallow-rooted crops, soil samples from 0 to 15 cm depth may be suitable. For deep-rooted crops, soil sampling depths may be 0–30 or 0–45 cm.

Several soil extractants are used to determine available nutrients of soil. The choice of the extractant will depend on soil properties (pH, organic matter content, cation exchange capacity, etc.) and the overall purpose of soil testing (agronomic vs. environmental, measurement of total vs. soluble vs. plant available, element concentration, and others). Generally, soil and water in 1:2.5 ratio for pH; 1 N KCl solution for NO_3–N, Bray and Kurtz 1, Mehlich 3, and Olsen extractants for available P; 1 N NH_4OAc for exchangeable K, Ca, and Mg; DTPA (diethylenetriaminepentaacetic acid) for Zn, Fe, and B; and calcium phosphate solution for S are used. These methods are described in detail in Page et al. (1982).

Soil test values are classified into low, medium, and high categories (Table 10.13), and fertilizer recommendations are made accordingly, taking yield targets into consideration. At low levels of soil test extractable nutrients, crop yield is limited by nutrient deficiency, and there will be a positive response to added fertilizers. There may be some crop response at medium levels of nutrients, but at high levels, there will usually be no response.

The amount of nutrients extracted by a soil test must be well correlated with plant nutrient uptake by plants for the test to be useful. Pot and field experiments are conducted for calibration of soil test results with crop response.

Table 10.13 Low, medium, and high categories of soil test values

Element	Low	Medium	High
NO_3–N[a]	<10	10–20	>20
Olsen P[b]	<10	10–20	>20
Extractable K[a]	<150	150–250	>250
Extractable Ca[a]	<1,000	1,000–2,000	
Extractable Mg[b]	<60	60–300	>300
DTPA extractable Fe[c]	<2.5	2.5–4.5	>4.5
DTPA extractable Zn[c]	<0.5	0.5–1.0	>1.0
DTPA extractable Cu[c]	<0.3	0.3–1.0	>1.0
DTPA extractable Mn[c]	<1.0	1.0–2.5	>2.5

Source: [a]Marx et al. (1999), [b]Horneck et al. (2011), [c]Herrera (2000)

10.21.4 Pot Experiments Are Closely Observed Small-Scale Fertilizer Trials

Pot experiments are often conveniently used to find out the effects of added fertilizer on crop plants and to relate crop response to soil test and plant test values. The soil, for which the need of fertilizer or the effect of added fertilizers will be evaluated, is filled in earthen, plastic, or porcelain pots of appropriate size, and the crop is grown after giving the necessary treatments. The pots can be arranged according to an experimental design in a small area within the lawn, nursery shed or experimental block, in shade or open sunlight as needed under close observation, or at completely controlled environment in the greenhouse. The soil of the pots may be analyzed before planting and after the harvest; plant parts may be analyzed at different growth stage, and yield of the plant can be estimated. Results may be used to infer the effect of the fertilizers and the uptake of nutrients by the crop and to construct the crop response curve.

However, pot experiments have several limitations. Pot soils behave differently from field soils. Pot soils tend to remain loose at the onset of the experiment and get compacted at the end. Root growth in pots is often restricted by the size of the pot. Aeration, water movement, and water saturation in pot soils are very different from field soils. Yield of potted crops often becomes magnified and, without correlating to results of a field trial in the soil where the crop will be ultimately grown, cannot give useful results. Therefore, as the conditions of pot experiments are different from those in the field, the results are not always directly applicable to large-scale farming.

10.21.5 Field Trials Are Large-Scale Fertilizer Experiments

Once the soil test or plant test values are obtained, they are correlated with crop response from field trials. By field trials, we get how much fertilizers will be needed by a particular crop in a particular soil. The expected response of the crop at any given soil test level is what determines the recommended level of each nutrient. The amount of fertilizer recommended is determined from many field trials on different soils over many years. The approach is based on a research data base that adequately predicts a crop response under good or normal conditions. The research base must be sufficient for each crop on the existing soil types under most of the prevailing weather conditions. Let us think of a field trial with six fertilizer combinations preconceived from reference database. We have to take at least three replications. A suitable area of the field or the entire field will then be divided into 18 equal plots, may be in 3 blocks. Six fertilizer dozes will be randomly and separately assigned to each of the plots. This is a randomized complete block field design. A crop will be grown in the plots with equal management inputs. After harvest the dose of fertilizer for the maximum economic yield will be selected for that crop in that and similar soils. Maximum economic yield may differ from maximum yield.

Field experiments are the more reliable methods, but being time-consuming and expensive, they are conducted mainly by the research farms and research organizations. Farmers wishing to use field experiments as a valid means of determining the fertility status of their soils should seek the advice of agronomists.

Crops, Soils, and Human Nutrition

Humans get carbohydrates, proteins, vitamins, alkaloids, and mineral nutrients mainly from plant products. Only 17 plant species provide approximately 90% of human food requirement. Eight cereal grains including wheat, corn (maize), rice, barley, sorghum, oats, rye, and millet provide 56% of the food energy and 50% of the protein consumed on earth. Three cereals: wheat, maize, and rice together comprise at least 75% of the world's grain production. Cereals, the main sources of carbohydrates, lack a number of nutrients which are essential for human health and well-being. They contain numerous vitamins and minerals with low biological availability (Cordain 1999). Fruits and vegetables supplement necessary vitamins and minerals to humans. Besides carbon, nitrogen, hydrogen, and oxygen, humans need 29 mineral nutrients, including aluminum, arsenic, boron, bromine, calcium, chlorine, chromium, cobalt, copper, fluorine, germanium, iodine, iron, lead, lithium, magnesium, manganese, molybdenum, nickel, phosphorus, potassium, rubidium, selenium, silicon, sodium, sulfur, tin, vanadium, and zinc, for their life processes (McDowell 2003). Food materials, including cereals, vegetables, fruits, nuts, meat, and fish, contain minerals in widely

(continued)

(continued)
varying amounts. Since plant materials constitute more than 90% of the food, they are the main sources of mineral nutrients to human. Plants absorb these nutrients from the soil. Therefore, soil is the principal source of human nutrients as well. But during analysis of plant and soil samples in the laboratory, we hardly remember this fact; we only consider essential nutrients for plants. Some examples of how deficiency of minerals in soil affects human health are noted here. Iodine deficiency is the leading cause of brain damage and mental retardation in the world. Iodine deficiency also causes endemic goiter, cretinism, dwarfism, and muscular disorder. More than 300 million people are suffering from goiter in the South Asian region including Bangladesh. Although coastal saline areas of this country are sufficient, the alluvial and piedmont soils away from the Bay of Bengal are deficient in iodine. Iodine deficiency is met by iodine supplementation in common salt in Bangladesh (Yusuf et al. 2008). Again, selenium deficiency to human occurs in the regions of the world containing low-selenium soils. These regions include Keshan Province in China (from which the name of the Keshan disease was taken), New Zealand, and Finland. Although "Keshan" disease was eliminated from China in the 1970s by an aggressive selenium supplementation program, Se deficiency still occurs in many parts of China (Sunde 2006). Serious Se deficiency in soils, crops, and serum samples of human has been reported from Yugoslavia (Maksimovic et al. 1992).

10.22 Fertilizer Application Methods Depend on Crops and Fertilizers

The time and methods of fertilizer application vary with the kind of fertilizers and the type of crops. For example, broadcasting may be suitable for basal dressings and closely growing crops but unsuitable for orchards. Liquid fertilizers may be applied by spraying over foliage or injecting into the soil or with irrigation water. Bulky organic manures should be applied well ahead of sowing so that the preliminary decomposition takes place before the seeds germinate. Otherwise, decomposing microorganisms may compete with the crop plants for nutrients. Considerable immobilization may take place after addition of partially decomposed manures and composts. Failing pre-sowing application, they may be applied any time after the seedlings have established themselves and in growing perennial crops. On the basis of time of application, fertilizers may be applied as basal dressing and top dressing:

1. *Basal dressing*: It is the application of fertilizers during field preparation or together with the seed. During field preparation, all the phosphate and potash fertilizers and a part of nitrogen fertilizer are applied through broadcasting and incorporated with the soil by harrowing. Fertilizers may be applied along with the seed by seed and fertilizer drillers.
2. *Top dressing*: It is the application of fertilizers to a standing crop. A part of nitrogen fertilizers is applied during maximum vegetative growth stage, and another part is given prior to flowering. Application of nitrogen fertilizers in such splits is known as split application. The top dressing of P and K is ordinarily done only in orchards and pasture lands which occupy the land for several years. In some countries, airplanes are employed for fertilizer application in hilly terrains, where large amount are to be applied and where deficiency develops over large tracts of land. In such application, fertilizer and pesticides may be combined together. It saves labor and time but may be expensive. On the basis of application in field, methods of fertilizer application may be of several types:
   (a) *Broadcasting*: Spreading of fertilizers by hand or mechanical spreaders over the entire surface of field during cultivation, prior to sowing of seeds, or in a standing crop is termed as broadcasting. Broadcasting may be as basal dressing or top dressing. Broadcasting distributes large amounts of fertilizers, usually N, P, and K, evenly throughout the field. However, broadcasting micronutrient fertilizers has the risk of uneven distribution due to their small amounts.
   (b) *Placement*: When fertilizers are not distributed throughout the whole field but are restricted to a place with reference to the seed or root is known as fertilizer placement. There are several placement methods:
      (i) *Plow-sole placement*: The fertilizer is placed in a continuous band on the bottom of the furrow during the process of plowing. Each band is covered as the next furrow is turned. By this method, fertilizer is placed in moist soil where it can become more available to growing plants during dry seasons. It results in less fixation of P and K than that which occurs normally when fertilizers are broadcast over the entire soil surface.
      (ii) *Deep placement or subsurface placement*: In this method, fertilizers are placed in the reduction zone as in paddy fields, where it remains available to the crop during the active vegetative period. It ensures better distribution in the root zone and prevents any loss by surface runoff. Fertilizers may be applied under the plow furrow in the dry soil before flooding the land and making it ready for transplanting.

(iii) *Localized placement*: It refers to the application of fertilizers into the soil close to the seed or root. It is usually employed when relatively small quantities of fertilizers are to be applied. It has some advantages. The roots of the young plants are assured of an adequate supply of nutrients; localized placement promotes a rapid early growth, and it reduces loss of applied fertilizers. It may have some limitations too. Seeds or roots may be damaged and the application may need special equipment.

(iv) *Combined drilling placement*: It refers to the drilling of seed and fertilizer together while sowing. It places the seed and small quantities of fertilizers in the same row. This is found useful in cereal crops, cotton, and grasses but not for pulses and legumes. This may affect the germination of the seed, particularly in legumes, due to excessive concentration of soluble salts.

3. *Band placement*: Fertilizer is placed in narrow bands which may be continuous or discontinuous to the side of seedling, some distances away from it and either at level with the seed, above the seed level, or below the seed level. There are two types of band placement: It includes hill and row placement. Hill placement is practiced in widely spaced (row to row about 1 m) crops. Row placement is done when plants are sown in close rows.
4. *Pellet placement*: In this method, mainly urea is applied in the form of pellets 2.5–5.0 cm deep between the rows of paddy crop. Fertilizer is mixed with soil in the ratio of 1:10 and made into dough. Small pellets of a convenient size are then made and deposited in the soft mud of paddy fields.
5. *Foliar application*: Roots are the main absorbing organs of the plants, but however, they can absorb small amount of nutrients from dilute solutions sprayed on their leaves. Sometimes fertilizers, particularly micronutrient fertilizers which are needed in small quantities, are applied as foliar sprays. Foliar sprays are a convenient way to apply additional nutrients to plants, supplementing those taken up from the soil. Foliar sprays also provide a quick means to help overcome an existing nutrient deficiency. When insecticides and fungicides are sprayed, fertilizers may be added in the same spray. The application of fertilizer with insecticides and fungicides reduces labor costs and machinery costs leading to less cost of crop production.
6. *Fertigation*: Fertigation is the application of fertilizer to standing crops together with irrigation water preferably in drip and sprinkler irrigation systems. Fertilizers are dissolved in irrigation water in appropriate concentrations, and the water is released at a controlled rate to keep pace with the nutrient uptake rate of the crop. Principal advantages of fertigation are increase in yield by 25–30%, saving fertilizer by 25–30%; precise application and uniform distribution of fertilizers, reducing nutrient losses; and major- and micronutrients may be applied at the same time, saving of cost, labor, energy, and risk of environmental contamination. Sandy soils can be brought under cultivation in this system. The system needs careful monitoring of the concentration of nutrients in solution and requires sophisticated control and distribution systems which may not be affordable by small farm holders.

Study Questions

1. What do you mean by soil fertility and productivity? How do they differ? What are the criteria of essentiality of elements to plants? What are the forms of nutrients that are absorbed by plants?
2. Discuss the physiological roles of N, P, and Mg. All of them affect carbohydrate metabolism—how? Mention the sources of N and P in soils. How do the pools of P in soil interact?
3. Discuss the environmental impacts of N and P fertilizers. How can we minimize fertilizer loss from soils? Sandy soils can be effectively fertilized by fertigation—explain.
4. Sandy, acid, and calcareous soils are prone to micronutrient deficiency—why?
5. What do you mean by availability of nutrients? How can you evaluate fertility of soil by soil test methods? Discuss fertilizer placement.

References

Amesz J (1993) The role of manganese in photosynthetic oxygen evolution. Biochem Biophys Acta 726:1–12

Anderson G (1980) Assessing organic phosphorus in soils. In: Khasawneh FE et al (eds) The role of phosphorus in agriculture. ASA/CSSA/SSSA, Madison

Arnon DI, Stout PR (1939) The essentiality of certain elements in minute quantity for plants with special reference to copper. Plant Physiol 14:371–375

Auld DS (2001) Zinc coordination sphere in biochemical zinc sites. Biometals 14:271–313

Bai C, Reilly CC, Wood BW (2006) Nickel deficiency disrupts metabolism of ureides, amino acids, and organic acids of young pecan foliage. Plant Physiol 140:433–443

Baldock JO, Schulte EE (1996) Plant analysis with standardized scores combines DRIS and sufficiency range approaches for corn. Agron J 88:448–456

Bangroo SA, Bhat MI, Ali T, Aziz MA, Bhat MA, Wani MA (2010) Diagnosis and Recommendation Integrated System (DRIS) – a review. Int J Cur Res 10:84–97

Beaufils ER (1973) Diagnosis and recommendation integrated system (DRIS). Soil Science Bulletin no. 1. University of Natal, Natal

Blevins DG, Lukaszewski KM (1998) Boron in plant structure and function. Ann Rev Plant Physiol Plant Mol Biol 49:481–500

Bonilla I, Blevins D, Bolaos L (2009) Boron functions in plants: looking beyond the cell wall. In: Taiz L, Zeiger EA (eds) Companion to plant physiology, 4th edn. Sinauer Associates Inc., Sunderland

Brady NC, Weil RR (2002) The nature and properties of soils, 13th edn. Pearson Education Inc., New Delhi

Bray RH, Kurtz LT (1945) Determination of total, organic, and available forms of phosphorus in soils. Soil Sci 59:39–45

Chen C, Huang D, Liu J (2009) Functions and toxicity of nickel in plants: recent advances and future prospects. Clean 37(4–5):304–313

Chen W, He ZL, Yang XE, Mishra S, Stoffella PJ (2010) Chlorine nutrition of higher plants: progress and perspectives. J Plant Nutr 33(7):943–952

Collard JM, Corbisier P, Diels L, Dong Q, Jeanthon C, Mergeay M, Taghavi S et al (1994) Plasmids for heavy-metal resistance in Alcaligenes-eutrophus Ch34 – mechanisms and applications. FEMS Microbiol Rev 14(4):405–414

Conrad R (1996) Soil microorganisms as controllers of atmospheric trace gases (H_2, CH_4, OCS, N_2O and NO). Microbiol Rev 60:609–640

Cordain L (1999) Cereal grains: humanity's double-edged sword. In: Simopoulos AP (ed) Evolutionary aspects of nutrition and health. Diet, exercise, genetics and chronic disease. World Rev Nutr Diet Basel Karger 84:L19–73

Cowan CE, Zachara JM, Resch CT (1991) Cadmium adsorption on iron-oxides in the presence of alkaline-earth elements. Environ Sci Technol 25:437–446

Cui Y, Zhao N (2011) Oxidative stress and change in plant metabolism of maize (*Zea mays* L.) growing in contaminated soil with elemental sulfur and toxic effect of zinc. Plant Soil Environ 57(1):34–39

Deak S (1985) Health hazards from nitrates in drinking water. Report on a WHO meeting, Copenhagen, 5–9 March 1984. WHO, Geneva

Ewing MC, White RMM (1951) Cyanosis in infancy from nitrates in drinking water. Lancet 260:931

Fageria VD (2001) Nutrient interactions in crop plants. J Plant Nutr 24(8):1269–1290

Filho FAAM (2004) DRIS: concepts and applications on nutritional diagnosis in fruit crops. Sci Agricn (Piracicaba Braz) 61(5):550–560

Fixen PE, Grove JH (1990) Testing soils for phosphorus. In: Westerman RL (ed) Soil testing and plant analysis. SSSA, Madison

Foth HD (1990) Fundamentals of soil science, 8th edn. Wiley, New York

Frolking SE, Mosier AR, Ojima DS, Li C, Parton WJ, Potter CS, Priesack E, Stenger R, Haberbosch C, Dorsch P, Flessa H, Smith KA (1998) Comparison of N_2O emissions from soils at three temperate agricultural sites: simulations of year-round measurements by four models. Nutr Cycl Agroecosyst 52:77–105

Freundlich H (1926) Colloid and capillary chemistry. Methuen, London

Goldbatch HE, Huang L, Wimmer MA (2007) Boron functions in plants and animals: recent advances in boron research and open questions. Adv Plant Anim Boron Nutr Part 1:3–25. Springer

Griffin RA, Jurinak JJ (1974) Kinetics of the phosphate interaction with calcite. Soil Sci Soc Am Proc 38:75–79

Hall SJ, Huber D, Grimm NB (2008) Soil N_2O and NO emissions from an arid, urban ecosystem. J Geophys Res 113. doi:10.1029/2007JG000523

Hepler PK (2005) Calcium: a central regulator of plant growth and development. Plant Cell 17:2142–2155

Herrera E (2000) Soil test interpretations. Guide A-122. http://aces.nmsu.edu/pubs/_a/a-122.pdf. Accessed 22 Dec 2011

Hingston FJ, Posner AM, Quirk JP (1974) Anion adsorption by goethite and gibbsite. 2. Desorption of anion from oxide surfaces. J Soil Sci 25:16–26

Hopkins B, Ellsworth J (2005) Phosphorus availability with alkaline/calcareous soil. Western Nutr Manage Conf 6:88–94

Horneck DA, Sullivan DM, Owen JS, Hart JM (2011) Soil test interpretation guide. http://www.sanjuanislandscd.org/Soil_Survey/files/page18_3.pdf. Accessed 3 Jan 2012

http://www.montmorillonite.org/Mulders%20chart.htm. Accessed 27 Aug 2011

IPCC (1992) Climate change 1992: the supplementary report to the IPCC scientific assessment. Cambridge University Press, Cambridge

IPCC (1995) Climate change 1994: radiative forcing of climate change and an evaluation of the IPCC IS92 emission scenarios. Cambridge University Press, Cambridge

Jiang BF, Gu YC (1989) A suggested fractionation scheme of inorganic phosphorus in calcareous soils. Sci Agric Sin 22(3):58–66 (in Chinese)

Kuo S (1996) Phosphorus. In: Sparks DL (ed) Methods of soil analysis: Part 3 Chemical methods. SSSA, Madison

Langmuir I (1918) The adsorption of gases on plane surfaces of glass, mica and platinum. J Am Chem Soc 40:1361–1402

Maksimovic ZJ, Djujic I, Jovic V, Rsumovic M (1992) Selenium deficiency in Yugoslavia. Biol Trace Elem Res 33(1–3):187–196

Manitoba Provincial Soil Testing Laboratory (1987) http://www1.agric.gov.ab.ca/$department/deptdocs.nsf/all/agdex3791/$file/540-.pdf?OpenElement. Accessed 23 Dec 2011

Marschner H (1993) Mineral nutrition of higher plants, 2nd edn. Academic, London

Marx ES, Hart J, Stevens RG (1999) Soil test interpretation guide. http://ir.library.oregonstate.edu/xmlui/bitstream/handle/1957/14361/ec1478.pdf;jsessionid=3B8749E2C71F904D4699 7A5669BF6029?sequence=1. Accessed 12 Jan 2012

McBride MB (2000) Chemisorption and precipitation reactions. In: Sumner ME (ed) Handbook of soil science. CRC Press, Boca Raton

McDowell LR (2003) Minerals in animal and human nutrition, 2nd edn. Elsevier, Amsterdam

McKenzie R (1992) Micronutrient requirement of crops. Alberta Agriculture, Food and Rural Development http://www1.agric.gov.ab.ca/$department/deptdocsf/all/agdex713/$file/531-1.pdf?OpenElement. Accessed 25 Jan 2012

Mehlich A (1953) Determination of P, Ca, Mg, K, Na, and NH4. North Carolina Soil Test Division (Mimeo), Raleigh, NC

Mehlich A (1984) Mehlich 3 soil test extractant: a modification of Mehlich 2 extractant. Commun Soil Sci Plant Anal 15:1409–1416

Mendel RR, Hansch RJ (2002) Molybdoenzymes and molybdenum cofactor in plants. J Exp Bot 53(375):1689–1698

Mishra SN, Jaiwal PK, Singh RP, Srivastiva HS (1999) Rhizobium-legume association. In: Srivastava HS, Singh RP (eds) Nitrogen nutrition and plant growth. NH Science Publishers, Inc., Enfield

Mulder (1953) Les elements mineurs en culture fruitière, presented at the 1° Congegno Nazionale de Frutticotura, Montan de Saint Vincent. http://www.apal.com.au/site/DefaultSite/filesystem/documents/APAL%20PLANT%20NUTRIENT%20INTERACTIONS%20July%2008.pdf. Accessed 24 Oct 2011

Nakagawa H, Jiang C-J, Sakakibara H, Kojima M, Honda I, Ajisaka H, Nishijima T, Koshioka M, Homma T, Mander LN, Takatsuji H (2005) Overexpression of a petunia zinc-finger gene alters cytokinin metabolism and plant forms. Plant J 41:512–523

Olsen SR, Cole CV, Watanabe FS, Dean LA (1954) Estimation of available phosphorus in soils by extraction with sodium bicarbonate. USDA Circ 939

Page AL, Miller RH, Keeney DR (1982) Methods of soil analysis. Part 2. Chemical and microbiological properties, 2nd edn. American Society of Agronomy Inc./Soil Science Society of America Inc., Madison

Pierzynski GM, Sims JT, Vance GF (2005) Soils and environmental quality. CRC Press, New York

Powlson DS (1997) Integrating agricultural nutrient management with environmental objectives – current state and future prospects. The Fertiliser Society, York

Power JF, Prasad R (1997) Soil fertility management for sustainable agriculture. CRC Press, London

Reid R (2007) Update on boron toxicity and tolerance in plants. In: Xu F, Goldbach HE, Brown PH, Bell RW, Fujiwara T, Hunt CD, Goldberg S, Shi L (eds) Advances in plant and animal boron nutrition. Springer, Dordrecht

Shikanai T, Muller-Moule P, Munekage Y, Niyogi KK, Pilon M (2003) PPA1, a P-type ATPase of *Arabidopsis*, functions in copper transport in chloroplasts. Plant Cell 15:1333–1346

Sims JT, Hodges SC, Davis J (1998) Soil testing for phosphorus: current status and uses in nutrient management programs. In: Sims JT (ed) Soil testing for phosphorus: environmental issues and implications. Southern Cooperative Series Bulletin no. 389. University of Delaware, Delaware

Solis P, Torrent J (1989) Phosphate fractions in calcareous vertisols and inceptisols of Spain. Soil Sci Soc Am J 53:462–466

Srivastava PC (1997) Biochemical significance of molybdenum in crop plants. In: Gupta UC (ed) Molybdenum in agriculture. Cambridge University Press, New York

Sunde RA (2006) Selenium. In: Bowman BA, Russell RM (eds) Present knowledge in nutrition, 9th edn. ILSI Press, Washington, DC

Sunderman FW, Oskarsson A (1991) Nickel. In: Merian E (ed) Metals and their compounds in the environment. VCH, Weinheim

Takano K (2003) Findings of field-diagnosis based on the Brix of peaches. Kikan Hiryo 94:55–61 (in Japanese)

Tamimi YN, Silva JA, Yost RS, Hue NV (1997) Adequate nutrient levels in soils and plants in Hawaii, Honolulu (HI). University of Hawaii (Agronomy and Soils, AS-3)

Tempkin MI, Pyzhev V (1940) Kinetic of ammonia synthesis on promoted iron catalysts. Acta Physiochim 12:327–356

Tisdale SL, Nelson WL, Beaton JD, Havlin JL (1997) Soil fertility and fertilizers, vol 5. Macmillan Publishing Company, New York

Tran TS, Simard RR (1993) Melich III – extractable elements. In: Carter MR (ed) Soil sampling and methods of analysis. Lewis Publishers, Boca Raton

Turner BL, Paphazy M, Haygarth PM, McKelvie ID (2002) Inositol phosphates in the environment. Philos Trans R Soc Lond Ser B 357:449–469

USEPA (2007) Ground water & drinking water. Consumer fact sheet on nitrates/nitrites. http://www.epa.gov/safewater/contaminants/dw_contamfs/nitrates.html. Accessed 9 Jan 2012

Walworth JL, Sumner ME (1987) The diagnosis and recommendation integrated system (DRIS). In: Stewart BA (ed) Advances in soil science, vol 6. Springer, New York

Wang C, Zhang SH, Wang PF, Hou J, Zhang WJ, Li W, Lin ZP (2009) The effect of excess Zn on mineral nutrition and antioxidative response in rapeseed seedlings. Chemosphere 75(11):1468–1476

Wang J, Liu WZ, Mu HF, Dang TH (2010) Inorganic phosphorus fractions and phosphorus availability in a calcareous soil receiving 21-year superphosphate application. Pedosphere 20(3):304–310

White RE (2006) Principles and practice of soil science: the soil as a natural resource, 4th edn. Blackwell Publishing, Malden

Wild A (1996) Soils and the environment. Low Price edn. Cambridge University Press, Cambridge

Yienger JJ, Levy H II (1995) Empirical model of global soil biogenic NO_x emissions. J Geophys Res 100(11):11447–11464

Yusuf HKM, Rahman AKMM, Chowdhury FP, Mohiduzzaman M, Banu CP, Sattar MA, Islam MN (2008) Iodine deficiency disorders in Bangladesh, 2004–05: ten years of iodized salt intervention brings remarkable achievement in lowering goitre and iodine deficiency among children and women. Asia Pac J Clin Nutr 17(4):620–628

Yruela I (2005) Copper in plants. Braz J Plant Physiol 17(1):145–146

11 Problem Soils and Their Management

Some soils have serious physical and chemical limitations to cultivation. They are problem soils. These soils need special management for satisfactory crop production. Physical limitations can be managed by irrigation, drainage, mulching, manuring, tillage, and soil conservation measures such as terracing, contouring, and cover crops whichever is appropriate. Use and disturbance of soils on steep slopes and in wetlands should be avoided because of the risk of their degradation. They should be left under natural conditions. Saline soils are managed through adequate irrigation, leaching, draining, and growing salt-tolerant crops. Sodic soil management needs chemical amendments such as application of sulfur and gypsum along with leaching and draining. Management of acid soils involves liming, leaching, and safe disposal of acid-wash water along with cultivating acid-tolerant crops.

11.1 Problem Soils Have Limitations to Cultivation

Soils that have serious constraints to cultivation and that need special management techniques are called problem soils. These limitations may be physical such as dryness, wetness, steepness, and extreme textures and chemical such as acidity, salinity, sodicity, and lack of fertility. The following categories of problem soils have been identified by the expert consultation of the Asian Network on Problem Soils at Bangkok, Thailand, 29 August to 1 September 1989 (FAO/AGL 2000):

1. *Cold soils*: Land areas with a 24-h mean temperature of less than 5 °C during the growing period
2. *Dry soils*: Desert and semidesert soils with growing periods which are rainless-dry
3. *Steep soils*: Soils which have steep slopes in excess of 30%
4. *Shallow soils*: Soils which have depth limitations within 50 cm of the surface caused by the presence of coherent and hard rock or hard pans
5. *Poorly drained soils*: Soils which are waterlogged and/or flooded for a significant part of the year
6. *Coarse-textured soils* (*sandy soils*): Soils having coarse texture with less than 18% clay and more than 65% sand, or have gravel, stones, boulders, or rock outcrops in surface
7. *Heavy cracking clays* (*Vertisols*): Soils which crack at least 1 cm wide at 50-cm depth at some period in most years
8. *Poorly fertile soils*: Soils which exhibit deficiencies in plant nutrients
9. *Saline/sodic soils*: Soils with high salt content or exchangeable sodium saturation within 100 cm of the surface
10. *Acid sulfate soils*: Soils in which sulfidic materials have accumulated under permanently saturated, generally brackish water conditions
11. *Peat soils* (*Histosols*): Soils in which more than half of the upper 80 cm is composed of organic materials

A problem soil is always related to a specific land use situation. A soil giving problems to a certain land use may not be problematic for another land use. For example, some saline soils bear excellent halophytic forest vegetation, and salinity is not a problem there; but when these areas are cleared for agricultural crop production, they do not give satisfactory yields. Soil problems may be natural or acquired.

11.2 Dryland Soils Need Sustainable Management for Food Security

11.2.1 Aridity and Salinity Are the Problems of Dryland Soils

According to UNCCD (2000), drylands are characterized by a precipitation to potential evapotranspiration ratio between 0.05 and 0.65 (very low rainfall, very high evapotranspiration). According to FAO (2000), drylands include regions classified climatically as arid, semiarid, and dry subhumid. Rainfall is low, erratic, uncertain, and unevenly distributed, and there are frequently high temperatures. Annual rainfall is less than 500 mm, and evaporation is higher than precipitation for a large part of the year. According to Molden and

K.T. Osman, *Soils: Principles, Properties and Management*,
DOI 10.1007/978-94-007-5663-2_11, © Springer Science+Business Media Dordrecht 2013

Oweis (2007), as much as 90% of the rainfall in arid environments evaporates back into the atmosphere.

Drylands are distributed in Asia, Africa, Australia, Canada, Central America, and the USA. In India, dryland agriculture accounts for nearly two-thirds of the total cropped area (Rosegrant et al. 2002), and in China, arid and semiarid lands cover 52% of the total land (Li Shengxiu and Xiao Ling 1992). Sub-Sahelian West Africa is largely semiarid, with 300–800-mm annual rainfall. Here, the soils are mostly Aridisols, Inceptisols, and Alfisols (Lal 1993). Dregne (1976) mentioned that Entisols form the most common soil order of the dry regions, followed by Aridisols, Mollisols, Alfisols, and Vertisols (Chap. 4). The major characteristics of dryland soils are the following: a generally low level of organic matter, alkaline to slightly acidic soil reaction in surface, accumulation of salts/lime/gypsum, coarse to medium texture, and low biological activity (Arnon 1992). In addition to soil moisture deficit, dryland soils are highly susceptible to wind and water erosion. Estimates of annual soil loss range from 10 to 50 t ha^{-1} for wind erosion alone. Many soils of drylands have become saline due to high evaporation and limited or no leaching. Land use for cropping in drylands is problematic; sources of water for irrigation are scarce, and rainfall often does not come when needed most.

Dryland farming occupies a key position in the context of food security. It contributes significantly to the production of cereals, pulses, and livestock. Although dryland crop yields are lower than those of irrigated agriculture, the lower cost of land and equipment make it feasible in some places. The prairies in Canada and the Great Plains, Pacific Northwest, and parts of the southwest and intermountain areas in the United States of America constitute the major dryland-farming areas of North America. Large-scale cereal and livestock production is common in parts of Argentina, Australia, Canada, the former Soviet Union, and the United States of America. In India, around 300 million people depend for their sustenance on dryland agriculture, of whom 30–40% are poor. Improving dryland crop yields is important both to maintain food security and to improve the livelihoods of the poor (Ryan and Spencer 2001).

11.2.2 Some Crops Are Suitable for Drylands

Many plants may adapt to soil dryness by several mechanisms. Some plants have short life cycle; they can germinate and grow during a very short period of available moisture. There are plants with deep or extensive root systems which enable them to draw water from a large volume of soil. Some plants store water in their tissues and release it very slowly. Plants with very narrow leaves transpire at a low pace. Dryland crops may have any or a combination of such mechanisms. Creswell and Martin (1998) ranked some crops from 0 to 3 according to their tolerance to water stress (0 no tolerance, 1 slight tolerance, 2 moderate tolerance, and 3 high tolerance). According to this scheme, the following crop plants have moderate to high tolerance to soil water stress (tolerance rank 2 and more).

Crop types	Crops
Cereals:	Pearl Millet (*Pennisetum americanum*)
Grain legumes:	Pigeon Pea (*Cajanus cajan*), Lablab Bean (*Dolichos lablab*), Mung Bean (*Vigna radiata*), Tepary Bean (*Phaseolus acutifolius*), Mat Bean (*Vigna aconitifolia*), Marama Bean (*Tylosema esculentum*)
Leafy vegetables:	Chaya (*Cnidoscolus chayamansa*), Horseradish Tree (*Moringa oleifera*), Leucaena (*Leucaena leucocephala*)
Root crops:	Cassava (*Manihot esculenta*), African Yam Bean (*Sphenostylis stenocarpa*)
Fruit trees:	Karanda (*Carissa carandas*), Dove Plum (*Dovyalis abyssinica*), Pomegranate (*Punica granatum*), Cashew (*Anacardium occidentale*), Prickly Pear (*Opuntia* sp.), Date Palm (*Phoenix dactylifera*)
Oil plants:	Shea Butter (*Butyrospermum paradoxum*)
Feed legumes:	Mesquite (*Prosopis sp.*) Leucaena (*Leucaena leucocephala*), Apple-Ring Acacia (*Acacia albida*), Umbrella Thorn (*Acacia tortilis*), Jerusalem Thorn (*Parkinsonia aculeata*)
Fiber plants:	Henequen (*Agave fourcroydes*), Sisal (*Agave sisalana*)
Timber plant:	Umbrella Thorn (*Acacia tortilis*)

Crops are also classified as cool-weather crops and warm-weather crops. Cool-weather crops for semiarid regions include wheat, barley, chickpeas, horsebeans, lentils, linseed, oats, peas, sugar beets, and vetches. Warm-weather crops are beans, corn (maize), cotton, cowpeas, groundnuts, millets, pigeon pea, sorghum, sunflowers, sesame, etc. (Arnon 1992).

11.2.3 Supplemental Irrigation Reduces the Risk of Crop Failure

Supplemental irrigation (SI) is defined as the addition of small amounts of water to essentially rainfed crops during times when the rainfall fails to provide sufficient moisture for normal plant growth, in order to improve and stabilize yields (Oweis 1997). To ascertain the time and amount of supplemental irrigation well in advance is very difficult due to the uncertainty of rains, but SI can be effective during critical periods of water shortages. It is an effective way of alleviating the adverse effects of soil water stress on the yield of dryland crops. Often soil moisture stress develops in the dry rainfed areas during the most sensitive crop growth stages such as flowering and grain filling. Then, it substantially

Fig. 11.1 Straw mulching in a corn (maize) field (With permission from Livinghistory Farm)

reduces the yield of crops. Supplemental irrigation, with a limited amount of water during the critical crop growth stages, results in a substantial improvement in yield and water productivity (Oweis et al. 2000). A cubic meter of water applied in supplemental irrigation at the optimum time with appropriate management could produce 2.0–3.5 kg of additional grain over that of rainfed production (Ilbeyi et al. 2006). In a study, barley genotypes were irrigated at different levels to replenish 33, 66, and 100% of the soil moisture deficit in the crop root zone in an area under a Mediterranean climate with total rainfall of only 186 mm. The mean grain yield of barley was 0.26 t ha^{-1} (rainfed), 1.89 t ha^{-1} (33% SI), 4.25 t ha^{-1} (66% SI), and 5.17 t ha^{-1} (100% SI) (Oweis and Hachum 2012).

If the soil is dry, selecting suitable crops, timely irrigation, and soil moisture conservation become the primary concerns. If the soil is saline at the same time, more irrigation water is needed for leaching the salts beyond the root zone. Sprinkler irrigation system works well in most crop types including orchards and field crops, but its installation cost is high (Chap. 6). Drip irrigation along with mulching is good for row and sparsely populated crops. Furrow irrigation may be applied, but inappropriate furrow irrigation may lead to the enhancement of soil salinity.

11.2.4 Mulching Reduces Evaporation Loss of Soil Water

Mulch is any type of material that is spread or laid over the surface of the soil as a covering (Fig. 11.1). There are several types of organic and inorganic cover mulches. Types of organic mulches are compost, composted manure, grass clippings, newspaper, straw, shredded leaves, straw, etc. Types of inorganic mulches are gravels, pebbles and crushed stones, plastic sheets, woven ground cloth, aluminum-coated plastic foil, ground rubber tires, "dust mulch," etc. For dryland farming, organic mulches are preferred. They are natural and cheap, and along with protection, they improve soil fertility. Mulches protect soils, increase infiltration, improve water-holding capacity, reduce weed growth, and prevent soil compaction. Organic mulches are gradually decomposed and release plant nutrients. Organic mulches release polysaccharides, polyuronides, and other cementing agents which improve soil structure. Mulched plants have more roots than plants that are not mulched. However, organic mulches pose some problems as well. Organic mulches may be invaded by slime molds and other fungi and may harbor insect pests. Nitrogen and phosphorus may be immobilized by decomposing microorganisms.

Campbell (1907) was one of the early promoters of dryland farming. He recognized that a "dust mulch" on the surface of soil conserved substantial amounts of soil water. Dust mulching is a common practice in India and China (Xi Chengfan 1961). Dust mulches are created by intensively hoeing the soil surface, making a fine textured layer of soil. Dust mulching breaks the soil capillarity and removes the continuity of pores, reducing the evaporative loss of soil moisture.

11.3 Steep Soils Should Be Left Under Natural Condition

Steeplands are characterized by slope inclinations of more than 12%. Soils of steeplands are often very shallow because of active erosion processes. Steep soils are usually extremely variable due to site-specific differences in geology, geomorphology, vegetation, and rainfall. The soils on slopes are often young because of natural and accelerated erosion processes. The topmost layers are lost continually, or they continually remain overlain by the deposition of materials moving downslope. Mass movements such as landslides and mud flows are specific to steeplands.

Steeplands are often under natural vegetation, restricted to forestry use only or used for extensive grazing. For arable cropping, steeplands have severe physical limitations with a high risk of erosion occurrence. There exists a vast range of varying land use combinations in steeplands due to differing agroecological conditions and socioeconomic circumstances over relatively short distances. If steeplands are utilized in a slash and burn system, they degrade after a few years with severe soil erosion symptoms. Steeplands can be stabilized by terracing and conservation-effective agricultural practices. Soil conservation methods for erosion-prone steep slopes have been discussed in relation to soil erosion (Chap. 12). It is, however, wise to keep the steeplands undisturbed in their natural state. Under intensive management, some steeplands can be used for improved forestry or orchard husbandry. In already degraded steeplands, the following physical and vegetational measures may be undertaken.

In steepland farming, physical structures such as rock barriers, contour bunds, waterways (diversion ditches, terrace channels, and grass waterways), stabilization structures (dams), windbreaks, and terraces (diversion, retention, and bench) are often necessary. Since building these structures is usually labor intensive, both construction and maintenance require long-term collaborative effort by farmers, the local community, and the government.

Vegetative barriers using grasses or trees planted along contours can be as effective as physical structures in controlling soil erosion. They can be established by farmers for little or no cost. Trees and grasses can also be harvested periodically for fuel and fodder. However, it is often useful to combine vegetation conservation measures with physical structures such as planting useful trees along the downslope side of rock barriers. Once established, the trees can strengthen the stability of the rock wall. Some farming systems suitable for sloping lands are described in Chap. 12.

11.4 Low Water Retention Capacity and Low Fertility Are the Problems of Sandy Soils

Sandy soils are characterized by less than 18% clay and more than 65% sand in the first 100 cm of the solum. In the World Reference Base (WRB) soil classification system, sandy soils may occur in the following reference soil groups: Arenosols, Regosols, Leptosols, and Fluvisols, Arenosols being the most common. In the Soil Taxonomy, most sandy soils are in the suborder Psamments.

Sandy soils can be divided in three categories: residual sands, shifting sands, and recently deposited sands. Sandy soils may occur from arid to humid and from extremely cold to extremely hot climates. The main problems with sandy soils are poor water-holding capacity, low organic matter content, and poor soil fertility. Sandy soils do not hold much water, and, at the same time, applied water is rapidly drained away. They do not hold much nutrients, and, for low nutrient retaining capacity, native and added nutrients are lost quickly by leaching. So, proper irrigation, enhancing water retention, and sufficient fertilizer application should be the most important aspects of sandy soil management. However, timing and amount of irrigation water applied are crucial decisions. Applying too much water means increased pumping costs, reduced water efficiency, and increased potential for fertilizers leaching below the root zone and into the groundwater. Enhancing the organic matter content is the key to alleviating the soil moisture and nutrient retention problems in sandy soils. Crop rotations, cover cropping, reduced or no-tillage, and shelter belts may effectively reduce wind erosion. In sandy soils, farmers are often tempted to use high dose of fertilizers and irrigation water. It causes leaching of fertilizers and contamination of groundwater.

Many sandy soils are nonused wastelands. Sandy soils may be used for extensive grazing in the dry zone; and if irrigation can be given, arable cropping can be profitable. In temperate areas, mixed arable cropping with supplemental irrigation is done during dry spells. Sandy soils in temperate region are useful for intensive horticultural crops because they allow mechanical harvesting when other clay-rich soils may be too wet and plastic during times of heavy rainfall. In the perhumid tropics, sandy soils are used for field crops with irrigation and fertilizer application. Satisfactory crop yields may be obtained from sandy soils if some general management rules are adopted. These are use of organic residues, mulching and frequent low-intensity irrigation, as well as fertilizer application. Fertigation with sprinkler and drip systems may be employed. Clay deposits may be applied if feasible.

11.5 Deep and Wide Cracks and Undesirable Consistence Are the Problems of Vertisols

Soils with deep cracking clays are classified as Vertisols (Fig. 11.2). Eswaran and Cook (1988) defined Vertisols as soils that are clayey in texture, have cracking structure, and are dark colored. FAO (1998) referred to Vertisols as soils having a vertical horizon within the 100 cm from the soil surface. Vertisols cover an area of 257 million ha worldwide (Dudal and Eswaran 1988). The major areas of Vertisols are found in Australia (70.5 Mha), India (70 Mha), Sudan (40 Mha), Chad (16.5 Mha), and Ethiopia (10 Mha). In different regions of the world, they have different names, including Dark Clays, Black Earths, Black Cotton Soils, Dark Cracking

Fig. 11.2 Wide cracks in a Vertisol (Photo courtesy of Professor Paul McDaniel, University of Idaho)

Soils, Grumusols, and Regurs (Dudal 1965). Vertisols have high clay contents (30–95%). The predominant clay is the 2:1 type smectites (montmorillonite) that expand and contract with changes in soil moisture content (Elias et al. 2001).

Vertisols become very hard when dry and sticky when wet. Tillage is hampered in Vertisols by stickiness when wet and hardness when dry (Dudal 1980). It is extremely difficult to plow the hard dry soil with a wooden country plow and some simple harrows drawn by a pair of oxen. Vertisols are regarded as rather marginal for arable cropping. However, much evidence has been produced that Vertisols can be transformed into productive cropland if their management addresses their inherent physical problems. Agricultural use of Vertisols in Africa ranges from grazing, firewood production, and charcoal burning through smallholder post-rainy season crop production (millet, sorghum, cotton) to small-scale (rice) and large-scale irrigated crop production (cotton, wheat, barley, sorghum, and sugarcane). Duchaufour (1998) considered Vertisols as the most fertile soils of the seasonally dry tropics.

Freebairn et al. (1986) conducted a research to determine the best tillage practices to be adopted for large-scale production of food crops in a Vertisol in Queensland, Australia. The tillage treatments consisted of conventional tillage (four or five cultivations to 10-cm depth with disc or tined implements), early blade (one cultivation to 10-cm depth after harvest with an implement fitted with a 180-cm-wide sweeps), late blade (weeds controlled initially by herbicides, one cultivation to 10-cm depth with 180-cm-wide sweeps), and no-till (weed control was achieved entirely by boom spray applications of herbicides). Best seedling establishment and vigor were obtained us ing a power-driven saw-tooth coulter. However, it did not seem economically viable. The most practical design in terms of cost, versatility, and establishment was a planter that included a smooth coulter, a narrow (spear point) tine opener, and a single-ribbed, variable weight press wheel. Willcocks and Twomlow (1992) reported that Saluka method of planting was the commonest in Sudan. The method involves the use of a spear-like stick to stab a hole in the surface at an interval equivalent to the stride distance of the operator and another person following behind place seed in the hole. Application of hydrated lime to swelling soils is a common practice, and it is effective in prevention of clay expansion. Sodium in the clay surface is replaced with calcium by this method, thereby reducing the ability to swell. If the expansive soil occurs as a thin layer on the surface, it may be replaced with non-expansive fill. This method is labor consuming and expensive and is not always feasible.

11.6 Acid Soils Need Liming and Acid-Tolerant Crops

Acid soils are characterized by a pH which is strongly (4.5–5.5) to extremely acid (<4.5), a low cation exchange capacity and a low base saturation. Acid soils occur in the tropics and subtropics as well as in moderate climates. Their formation depends on specific conditions of climate, topography, vegetation, parent material, and time for soil formation. Acid sulfate soils occur in the tropics in low-lying coastal land formerly occupied by mangrove swamps. If these soils are drained, sulfuric acid is produced by oxidation of pyrite (FeS_2) resulting in extremely acid soil conditions.

Despite major difficulties for agricultural use, acid soils can be very productive if lime and nutrients are applied at proper time and quantity. Management of acid soils involves the adjustment of soil pH to a desired level by liming, adequate irrigation and drainage, and selecting suitable crops.

A list of plants including crop plants and their suitable pH ranges is given below.

List of Some Acid-Tolerant Plants

Suitable range of pH	Plants
Vegetable crops	
pH 4.5–5.5:	Radish, sweet potato
pH 5.5–6.5:	Endive, parsley, pepper, rhubarb, soybean
Fruit crops	
pH 4–5.5:	Blueberry, cranberry, raspberry
pH 5–6.5:	Apple, grape, strawberry
Woody plants	
pH 4.0–5.0:	Spruce, black, azalea
pH 4.5–6:	White birch, heather, rhododendron, fir, balsam, hemlock, pine, jack
pH 5.0–6.5:	Beech, oak, pine, tamarack
Houseplants	
pH 4.5–5.5:	Achmines, Adiantum, African violet, Aloe, Amarylis, Aphelandra, Araucaria Norfolk Pine, Azalea, Begonia, Caladium, Calathea, Crossandra, Cyclamen, Dieffenbachia, Epiphyllum, Gardenia, Hydrangea, Impatiens, Maranta, Peperomia, Pilea, Polypodium, Primula, Rechsteineria, Saxifraga, Scindapsus, Streptocarpus, Syngonium, Zygocactus
pH 5.5–6.5:	Anthurium, Bromeliad, Cattleya, Columnea, Cymbidium, Cypripedium, Daffodil, Gladiolus, Hyacinth, Iris, Narcissus, Phalaenopsis, Platycerium, Thipsalidopsis, Tulip, Vanda

Source:http://www.coopext.colostate.edu/TRA/PLANTS/acidlove.shtml

11.6.1 Liming Increases Soil pH

Liming has been an efficient way to control soil acidity since long. It increases soil pH and reduces the toxicity of iron, aluminum, manganese, and molybdenum. Liming also stimulates biological activity in soils and increases cycling of nitrogen, phosphorus, and sulfur. It enhances P availability to plants. Lime improves soil structure, porosity, aeration, and water movement in soils. Overliming may, however, reduce the availability of phosphorus and cause deficiencies of micronutrients such as manganese, zinc, copper, and molybdenum.

11.6.1.1 Liming Materials

Lime or liming materials are any materials that contain Ca or Mg and are able to neutralize soil acidity. Liming materials are generally carbonates and bicarbonates, sometimes oxides and hydroxides of calcium and magnesium.

Carbonates

Carbonates are the most widely used liming materials. They are generally less expensive and easier to handle than other lime materials. Ground high-grade limestone or calcite is concentrated calcium carbonate ($CaCO_3$). Dolomite ($MgCO_3+CaCO_3$) is also a common source of agricultural lime. Dolomite usually costs a little more than calcite and changes the soil pH more slowly, but it has the advantage of containing Mg in addition to Ca. Together, calcite and dolomite account for more than 90% of the lime used in the United States. Both materials are naturally occurring rocks that are mined and ground for agricultural use.

Marl and oyster shells are also carbonate materials. Marl is a naturally occurring mixture of clays, carbonates of Ca and Mg, and shell remnants. Oyster shells also contain high proportion of calcium carbonate but are important only in some coastal regions.

Oxides

Oxide liming materials include burned lime, unslaked lime, and quicklime. Oxides are made by baking crushed calcite limestone or dolomite limestone in a furnace, thereby driving off carbon dioxide (CO_2) to form a concentrated oxide (CaO or MgO). This material is of low molecular weight and reacts rapidly in the soil to raise the pH. Oxides are the most efficient of all liming materials on a pound-for-pound basis. But oxides are caustic; that is, they react with moisture and are difficult to handle. In addition, their cost is high relative to carbonate materials. One ton of calcium oxide has the neutralizing power of 1.8 t of calcite.

Hydroxides

Hydroxides are simply oxide materials with water added. They are also known as hydrated lime, slacked lime, or builders lime. These materials are similar to oxides because they are powdery, quick acting, and unpleasant to handle. Hydroxides are also more expensive than carbonate materials. Many by-products of mining, refining, processing, and manufacturing processes are used as liming materials. Slags from blast furnaces and electric furnaces as well as fly ash and bottom ash from coal-burning plants are often applied as lime. Lime sludges from sugar beet processing plants (sugar lime), paper mills, ore processing, and water-softening plants are sometimes used to raise soil pH. Wood ashes from wood stoves or fireplaces also may be used. Active gradients in these materials are quite variable, and their requirement cannot be precisely assessed. Their contaminants also restrict their use.

Fluid Lime

Lime can be applied in liquid form as a suspension. Lime is not dissolved but is suspended in water. Liquid lime may be prepared by pouring 50% water to 48% lime solids and 2% clay to maintain a suspension. The material used in this suspension should pass a 100-mesh sieve. The advantages of using liquid lime include (1) good application uniformity, (2) high-quality lime material (mesh size), and (3) quick soil pH change.

Table 11.1 Calcium carbonate equivalence of common liming materials

Liming material	Composition	CCE
Calcitic limestone	$CaCO_3$	100
Dolomitic limestone	$CaMg\,(CO_3)_2$	109
Calcium oxide	CaO	179
Calcium hydroxide or magnesium hydroxide	$Ca(OH)_2$	136
Slag	$CaSiO_3$	80

Source: http://learningstore.uwex.edu/assets/pdfs/a3671.pdf

11.6.1.2 Quality of Lime

Two factors primarily affect the quality of liming materials: (1) chemical composition or purity and (2) particle size or fineness.

Chemical Composition

Acid neutralizing capacity of a liming material is determined by its calcium carbonate equivalence (CCE). It is the acid neutralizing capacity of the material compared to pure calcium carbonate. In CCE comparisons, pure calcium carbonate has been assigned a value of 100. CCE values of some common liming materials are given in Table 11.1.

Dolomite has a slightly greater CCE than calcium carbonate due to the lower atomic weight of Mg compared to Ca. Oxide materials have the highest CCE values since CO2 is removed in the burning process. Marl and by-product materials have low CCE values because of the presence of impurities.

Fineness

The neutralizing capacity of a liming material is determined by its particle size. 100% of lime particles that pass a 100-mesh sieve will react within the 1st year, while only 60% of the liming materials that pass a 20-mesh sieve (but held on 100-mesh sieve) will react within a year of application. Materials that do not pass the 20-mesh sieve will not react within a 1 year after application. So, for practical purpose, neutralizing power of a liming material needs adjustment for the fineness of the material.

To determine the fineness factor of a liming material of which 70% passes a 100-mesh sieve and 97% passes a 20-mesh sieve, the following calculations need to be done:

(a) To subtract the % passing, a 100-mesh sieve from the % passing a 20-mesh sieve and multiply this difference with 0.60.

(b) To add the % passing, the 100-mesh sieve and divide the sum by 100.

Thus, the fineness of a material of which 70% passes a 100-mesh sieve and 97% passes a 20-mesh sieve is $\{(97-70)\times 0.60+70\}/100=0.86$.

Effective Neutralizing Value (ENV)

The ENV is the fraction of the liming material's CCE that will react with soil acidity in the first year of application. The ENV is calculated by multiplying the CCE with the fineness of the material. For example, a liming material with CCE of 90% and a fineness of 0.86 has an ENV of $90\times 0.86=77.4$.

11.6.1.3 Mechanism of Lime Action

Lime dissolves slowly in soil solution to release Ca and bicarbonate ions:

$$CaCO_3 + H_2O + CO_2 = Ca^{2+} + 2HCO_3^-$$

The bicarbonate ions neutralize the hydronium ions in soil solution, and the Ca^{2+} ions replace aluminum and hydronium ions from the exchange sites of the colloids. Once displaced into solution, these aluminum and hydronium ions are also neutralized by bicarbonate ions.

$$HCO_3^- + H_3O^+ = CO_2 + 2H_2O$$

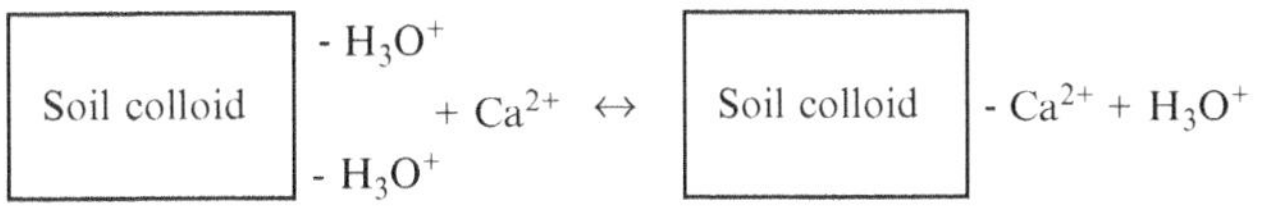

11.6.1.4 Lime Requirement

Even if two different soils have the same level of acidity (say pH 5.5) and the cropping pattern needs to raise its pH to the same level, say 6.5, they will require different amounts of lime because the soils may differ in clay content, cation exchange capacity, and base saturation percentage. These properties give soils a unique capacity known as buffer capacity or the capacity to resist change in soil pH. Some of applied lime will neutralize acids in soil solution and some will be consumed for replacing exchangeable H^+ and Al^{3+}. In other words, doses of lime cannot be estimated from the pH value alone. Lime requirement of each of the acid soils need to be determined separately. Nowadays, computer programs are available to estimate lime requirement from existing pH, desired pH, clay content, CEC, exchangeable Al^{3+} and H^+, and percent base saturation.

There are a number of lime requirement determination methods for different types of soils. Two methods are generally used—the SMP (Shoemaker–McLean–Pratt) and Adams–Evans buffer methods. The SMP method was designed for use with soils that have large lime requirements and significant reserves of exchangeable Al (Shoemaker et al. 1961). The Adams–Evans buffer was designed for soils that are coarse-textured, with low cation exchange capacities and organic matter contents, and thus low lime requirements (Adams and Evans 1962). The Mehlich lime buffer was developed for use on Ultisols, Histosols, Alfisols, and Inceptisols (Mehlich 1976). To avoid hazardous chemicals, all three buffers have been modified. All three modifications correlated well with the original buffer, and no changes in calibration were necessary for any of the modified methods (Hoskins 2005; Huluka 2005; Sikora 2006). The original SMP buffer or its modification (Sikora buffer) is the most commonly used method.

Lime Requirement Procedure (SMP Buffer Method)

SMP Buffer

1. Weigh 32.4 g of para-nitrophenol, 54.0 g of K_2CrO_4, and 955.8 g $CaCl_2{\cdot}2H_2O$ into an 18-L bottle. Add 9 L distilled water, shaking vigorously during addition.
2. Weigh 36.0 g of $Ca(OAc)_2$ into a large container and dissolve in 5 L of distilled water.
3. Combine the two solutions, shaking during mixing and every 15–20 min for 2–3 h.
4. Add 45 ml triethanolamine, shaking during addition and periodically thereafter until completely dissolved.
5. Dilute to 18 L with distilled water, adjust to pH 7.50 using 15% NaOH, and filter.
6. Store in a container with the air inlet protected by Drierite and Ascarite to prevent contamination by water vapor and carbon dioxide. Avoid excessive agitation of the solution after pH adjustment.

Procedure

1. Add 10 ml of SMP buffer to the soil–water slurry used for pH determination.
2. Place in a mechanical shaker, close tightly, shake at 250 excursions/min for 10 min, and let it sit for 20 min.
3. Swirl, insert electrodes, and read the pH. Read to the nearest 0.01 pH unit.
4. Determine lime requirement from soil–buffer pH and calibration data for local soils.

11.7 Acid Sulfate Soils Need Liming and Soil Washing

Acid sulfate soils contain iron sulfides in the reduced and sulfuric acid in the oxidized conditions. These soils commonly occur in coastal lowlands, estuaries, floodplains, wetlands, and mangrove environments. There are two types of acid sulfate soils: actual and potential. Actual acid sulfate soils are drained soils where the sulfide is oxidized to sulfuric acid giving the soil an extremely acid condition, the pH value being as low as 3.0. Potential acid sulfate soils are saturated soils with pyrite (FeS_2) contents. Potential acid sulfate soils become actual acid sulfate soils upon draining and oxidation. Sometimes the sulfidic materials are exposed to the surface, while there are also many buried sulfidic horizons. When exposed to the surface, these soils become a threat to the environment. Exposed acid sulfate soils are often characterized by the presence of spotted yellow coloration of jarosite [$KFe_3(SO_4)_2(OH)_6$]. Rain and floodwater can flush sulfuric acid formed by oxidation of pyrite into nearby waterways, killing fish, other aquatic organisms, and vegetation. Strong acidity and Al^{3+} toxicity are the problems associated with use of acid sulfate soil for crop production. Some shrimp farms have been established in acid sulfate soils in the coastal regions of Bangladesh. Unsatisfactory growth of phytoplankton, shrimp kills, damage of the gills, and obnoxious odor are problems in shrimp farming in acid soils. Potential acid sulfate soils can be profitably used for paddy rice cultivation if fresh irrigation water can be provided. The management options for acid sulfate soils are discussed in the following points.

Acid sulfate soils are harmless when left in waterlogged and undisturbed conditions. These soils may better be kept in their natural state. Deeply seated sulfidic horizons must not be exposed for shrimp pond construction.

1. Minimum or shallow disturbances may be undertaken; existing drains to be redesigned so that they are shallower and wider and do not penetrate sulfidic layers; groundwater fluctuations should be avoided.
2. Oxidation of pyrite and removal of acids by irrigation and drainage have been employed for reclamation of acid sulfate soils with varying success. Aeration, flooding, and disposal of acid water should be done with great caution so that reclamation work does not lead to degradation of environment.
3. Acids in acid sulfate soils may be neutralized by liming well ahead of utilization for cropping. Thoroughly mixing the appropriate amount and type of lime into disturbed acid sulfate soils will neutralize any acid produced.
4. Hydrated lime is often more appropriate for treating acid waters due to its higher solubility. Once the acid water has been treated to pH 6.5–8.5 and metals have been reduced to appropriate levels, it can usually be safely released from the site at a controlled rate to prevent significant changes to the quality of off-site waters. Quicklime, sodium bicarbonate dolomite, and some industry by-products such as basic slag may also be used.
5. Hydraulic separation is suitable for coarse-textured soils containing iron sulfides. Sluicing or hydrocycloning is used to hydraulically separate the sulfides from the coarser textured materials. This could be an effective form of management when the sediments contain less than 10–20% clay and silt and have low organic matter content. The separated sulfidic material extracted via the process requires special management involving either neutralization or strategic reburial.
6. Ex-situ soil oxidation, neutralization, and washing have been tried in some instances. This method is very much costly, and farmers find little interest.

Readers are referred to Bloomfield and Coulter (1973) for a better understanding of the "Genesis and Management of Acid Sulfate Soils."

11.7.1 Applying Lime

Lime can be applied anytime between the harvest of a crop and the planting of the next. Lime is usually broadcast on the soil during tillage and incorporated into the soil. Lime should be mixed to tillage depth in order to effectively neutralize soil acidity in the primary root zone. On moderately acid soils (pH 5.2–5.7), a single application of lime before tillage will usually give good results. For strongly acid soils (pH 5.0 and lower) that have very high lime requirements, it may be desirable to apply one-half of the lime before tillage and the remaining half after tillage. For large areas that have high lime requirements (3–4 t/acre), it may be best to apply half of the required lime in a first-year application and the remainder in the second year.

11.8 Saline and Sodic Soils Are Common in Arid and Coastal Regions

All soils contain some soluble salts. Some of these soluble salts are plant nutrients. But some soils accumulate salts in levels that are harmful for germination of seeds and growth of crops. Then the soils are called saline. The criterion for a soil to be saline is its electrical conductivity of the saturation extract because electrical conductivity is proportional to the concentration of salts in solution. Saturation extract is obtained by adding water to the soil up to its maximum water-holding capacity and then drawing the soil solution under suction. If the electrical conductivity of the saturation extract (ECe) at 25°C is >4 dS m^{-1} (decisiemens per meter), the soil is saline (Richards 1954). Some soils, saline or nonsaline, contain high exchangeable sodium. These soils are called sodic soils. The criteria are exchangeable sodium percentage (ESP) and sodium adsorption ratio (SAR) (see Chap. 7 for definition of these terms). If the ESP is >15 (or SAR >13), the soil is sodic. Thus, soils may be classified on the basis of ECe and ESP (or SAR) into four categories (US Salinity Laboratory Staff 1954) (Table 11.2).

11.8.1 Saline Soils Accumulate Natural Salts

Saline soils, both natural and human induced, are found to develop under three environmental settings—arid–semiarid regions, coastal areas, and humid regions. Their main occurrences are in the arid and semiarid regions where evapotranspiration greatly exceeds precipitation. Salts released from weathering of rocks cannot be leached due to water scarcity. Moreover, capillary rise of water brings salts to the surface; water evaporates, leaving the salts in the surface soil. Inadequate irrigation and inefficient drainage have converted more areas into salt-affected soils. In coastal areas, seawater floods some soils regularly. In other areas, farmers irrigate the soils with saline water. Rise of the groundwater table due to high rainfall, low base flow of water, and improper drainage is the main cause of soil salinization in humid regions. Over 20 million ha of land is severely affected by salinity worldwide (Rhoades and Loveday 1990). Table 11.3 gives the classes of saline soils.

Table 11.2 Categories of saline and sodic soils

Category of soil	ECe, dS m^{-1}	ESP	Common name
Saline nonsodic	>4	<15	Saline soil
Saline–sodic	>4	>15	Saline–sodic soil
Nonsaline sodic	<4	>15	Sodic soil
Nonsaline nonsodic	<4	<15	Normal soil

Natural salts in saline soils include mainly chlorides (Cl^-) and sulfates (SO_4^{2-}) of sodium (Na^+), calcium (Ca^{2+}), magnesium (Mg^{2+}), and potassium (K^+), although NaCl dominates. There may also be some carbonates (CO_3^{2-}), bicarbonates (HCO_3^-), and nitrates (NO_3^-). All these cations and anions may exert toxic effects and create nutritional imbalances if present in excess quantities. These are specific ion effects. Excess salts in soil create a general osmotic problem. Salts decrease the osmotic potential of soil water and reduce water absorption by plants. Plants suffer from both water stress and salt stress in saline soils. Effects of soil salinity are manifested in loss of stand, reduced plant growth, reduced yields, and, in severe cases, crop failure. The pH of saline soils, if they are not sodic at the same time, remains around neutrality. There is a white salt crust on the surface of some saline soils (earlier known as white alkali soils).

11.8.1.1 Management of Saline Soils

For managing saline soils, salts may either be diluted to the tolerable limit or removed by leaching. If there is a salt crust on the surface of the soil, salt farming may be more profitable than cropping. For preparing for cropping, decrusting may be done mechanically and with soil flushing. However, decrusting followed by flushing has not been very successful for crop yields. But some soils may need decrusting before leaching. Some soils are only slightly saline. For shallow-rooted

Table 11.3 Soil salinity classes

Soil salinity class	ECe, dS m^{-1}	Effect on crop plants
Nonsaline	0–2	Salinity effects negligible
Slightly saline	2–4	Yield of sensitive crops may be restricted
Moderately saline	4–8	Yield of many crops are restricted
Strongly saline	8–16	Only tolerant crops yield satisfactorily
Very strongly saline	>16	Only a few tolerant crops yield satisfactorily

Source: http://www2.vernier.com/sample_labs/AWV-09-COMP-soil_salinity.pdf

crops, the salts may be driven below the root zone by temporary leaching. This technique will need less water than normal leaching. Furrow irrigation with suitable seedling placement may also be satisfactory. Drip irrigation at the root area also dilutes salts and keeps the salts apart. If the soil is considerably saline, it needs removal of excess soluble salts by leaching. No chemical amendment is needed here. However, a reliable estimate of the quantity of water required to accomplish salt leaching is required. The salt content of the soil, salinity level to achieve, depth to which reclamation is desired, and soil characteristics are important factors determining the amount of water needed for reclamation. A useful rule of thumb is that a unit depth of water will remove nearly 80% of salts from a unit soil depth. Thus, 30-cm water passing through the soil will remove approximately 80% of the salts present in the upper 30 cm of soil. To leach soluble salts in irrigated soils, more water than required to meet the evapotranspiration needs of the crops must pass through the root zone to leach excessive soluble salts. This additional irrigation water has typically been expressed as the leaching requirement (LR). Leaching requirement was originally defined as the fraction of infiltrated water that must pass through the root zone to keep soil salinity from exceeding a level that would significantly reduce crop yield under steady-state conditions with associated good management and uniformity of leaching (Rhoades 1974).

$$\mathrm{LR} = \frac{EC_{\mathrm{iw}}}{EC_{\mathrm{dw}}},$$

where LR is leaching requirement, EC_{iw} is electrical conductivity of irrigation water, and EC_{dw} is the electrical conductivity of drainage water. Several leaching requirement models have been proposed by Corwin et al. (2007). Leaching is most often accomplished by ponding freshwater on the soil surface and allowing it to infiltrate. Effective leaching occurs when the salty drainage water is discharged through subsurface drains that carry the leached salts out of the area under reclamation. Leaching may reduce salinity levels when there is sufficient natural drainage, that is, the ponded water drains without raising the water table. Leaching should preferably be done when the soil moisture content is low and the groundwater table is deep. Several methods of irrigation and drainage may be employed for salt and water management in saline soils. Drip and sprinkler irrigation methods are suitable in slightly saline soils. Flooding irrigation, where water is available, followed by draining through deep surface drains or subsoil drains is suitable for leaching. Irrigation and drainage methods are discussed in detail in Chap. 6.

11.8.1.2 Salt-Tolerant Crops

Some crops are tolerant to a level of soil salinity (Table 11.4). It means that the yield of the crop is not reduced more than 25% in the ECe level. Some crops are slightly tolerant, others are highly tolerant. Farmers should choose crops or crop sequences on the salinity values of their soils. Salinity tolerance of some field crops is given in Table 11.4.

Table 11.4 Salinity tolerance of some important crops

		Yield loss		
	Threshold value	10%	25%	50%
Crop	ECe (dSm^{-1})			
Barley	8.0	9.6	13.0	17.0
Beans	1.0	1.5	2.3	3.6
Canola	2.5	3.9	6.0	9.5
Corn	2.7	3.7	6.0	7.0
Oat	5.2	6.7	9.0	12.8
Rye	5.9	7.7	12.1	16.5
Safflower	5.3	8.0	11.0	14.0
Sorghum	4.0	5.1	7.1	10.0
Sugar beet	6.7	8.7	11.0	15.0
Sunflower	2.3	3.2	4.7	6.3
Triticale	6.1	8.1	12.0	14.2
Wheat	4.7	6.0	8.0	10.0

Source: https://extension.usu.edu/files/publications/publication/AG-SO-03.pdf

11.8.2 Sodic Soils Have High Exchangeable Sodium

Sodic soils are low in total soluble salts but high in exchangeable sodium. The combination of high levels of sodium and low total salts tends to disperse soil particles and humus, making sodic soils of poor tilth. These soils are generally structureless, sticky when wet, and nearly impermeable to water. As they dry, they become hard, cloddy, and crusty. The pH of sodic soils ranges from 8 to 10. Hydrolysis of exchangeable Na^+ and $CaCO_3$, and $MgCO_3$ and resulting OH^- gives this high soil pH. Sodic soils are detrimental to growth of most plants. High pH, low micronutrient availability, and Na^+ toxicity are the constraints of their agricultural use. The main approach of sodic soil management is the replacement of exchangeable sodium by Ca or H, lowering of pH, and improving soil physical conditions. Reclamation of sodic soil is difficult, time consuming, labor intensive, and expensive. In some sodic soils, humus is dispersed, and the surface soil becomes black. They are called black alkali soils. Plant growth is adversely affected in sodic soils due to one or more of the following factors:

1. Poor physical conditions give poor tilth and adverse soil moisture–aeration relationships.
2. High pH reduces availability of P, Fe, Mn, and Zn; these nutrients may become deficient.
3. Excess sodium, molybdenum, and boron may become toxic.

11.8.2.1 Sodic Soil Management

Sodicity cannot be removed by leaching alone. Chemical amendment is needed here to replace the exchangeable sodium from colloidal complexes, lower the pH, and improve poor physical conditions. Whether a sodic soil would be reclaimed or not is best dictated by local conditions, available resources, and the kind of crops to be grown on the reclaimed soils. Initially reclamation cost is high and cropping for first few years may not be profitable. Ultimately, reclamation processes convert unproductive sodic soils into productive croplands.

Chemical Amendments

Chemical amendments for sodic soil reclamation can be broadly grouped into the following categories: (1) soluble calcium salts, for example, gypsum and calcium chloride, and (2) acids or acid-forming substances, for example, sulfuric acid, iron sulfate, aluminum sulfate, lime sulfur, sulfur, and pyrite. Gypsum ($CaSO_4{\cdot}2H_2O$) is a white mineral that occurs extensively in natural deposits. It is relatively cheap, available, and effective. It is ground before application to the soil. Gypsum is slowly soluble in water and is a direct source of soluble calcium. Gypsum reacts with both Na_2CO_3 and the adsorbed sodium as follows:

$$Na_2CO_3 + CaSO_4 = CaCO_3 + Na_2SO_4\,(\text{leachable})$$
$$Na-Clay-Na + CaSO_4 \Leftrightarrow Ca-Clay + Na_2SO_4\,(\text{leachable})$$

Calcium Chloride

Calcium chloride ($CaCl_2{\cdot}2H_2O$) is a highly soluble salt which supplies soluble calcium directly. Its reactions in sodic soil are similar to those of gypsum.

$$Na_2CO_3 + CaCl_2 = CaCO_3 + 2NaCl\,(\text{leachable})$$
$$Na-Clay-Na + CaCl_2 \Leftrightarrow Ca-Clay + 2NaCl\,(\text{leachable})$$

Sulfuric Acid

Upon application of sulfuric acid (H_2SO_4) to soils containing calcium carbonate, it immediately reacts to form calcium sulfate and thus provides soluble calcium indirectly.

$$Na_2CO_3 + H_2SO_4 = CO_2 + H_2O + Na_2SO_4\,(\text{leachable})$$
$$CaCO_3 + H_2SO_4 = CaSO_4 + H_2O + CO_2$$
$$Na-Clay-Na + CaSO_4 \Leftrightarrow Ca-Clay + Na_2SO_4\,(\text{leachable})$$

Iron Sulfate and Aluminum Sulfate

Both of iron sulfate ($FeSO_4{\cdot}7H_2O$) and aluminum sulfate [$Al_2(SO_4)_3{\cdot}18H_2O$] are solid granular materials soluble in water. When applied to soils, these compounds hydrolyze to form sulfuric acid.

$$FeSO_4 + 2H_2O = H_2SO_4 + Fe(OH)_2$$
$$H_2SO_4 + CaCO_3 = CaSO_4 + H_2O + CO_2$$
$$Na-Clay-Na + CaSO_4 \Leftrightarrow Ca-Clay + Na_2SO_4\,(\text{leachable})$$

Aluminum sulfate acts similarly.

Sulfur

Sulfur (S) is not soluble in water and does not supply calcium directly for replacement of adsorbed sodium. When applied to soils, sulfur undergoes oxidation to form sulfuric acid and gives reactions as shown below:

$$2S + 3O_2 = 2SO_3,$$
$$SO_3 + H_2O = H_2SO_4\,(\text{Sulfuric acid performs as above})$$

Pyrite

Pyrite (FeS_2) is another possible amendment for sodic soil reclamation. The following series of reactions occur after applying pyrite to soil:

Step I $2FeS_2 + 2H_2O + 7O_2 = 2FeSO_4 + 2H_2SO_4$

Step II $4FeSO_4 + O_2 + 2H_2SO_4 = 2Fe_2(SO_4)_3 + 2H_2O$

Step III $Fe_2(SO_4)_3 + FeS_2 = 3FeSO_4 + 2S$

Step IV $2S + 3O_2 + 2H_2O = 2H_2SO_4$ (Sulfuric acid performs as above)

Choice of Amendments

The choice of amendments depends on:

(a) Relative effectiveness as judged from improvement of soil properties and crop growth
(b) Relative costs involved
(c) Risk of handling materials
(d) Time required for an amendment to react in the soil and effectively replace adsorbed sodium

Iron and aluminum sulfates are usually too costly and have not been used for any large-scale improvement of sodic soils in the past. Because amendments like sulfur and pyrite must first be oxidized in soil to sulfuric acid, the amendments are relatively slow acting. Being the cheapest and most abundantly available, gypsum is the most widely used amendment.

Quantity of Amendment

Amount of exchangeable sodium (and amount of Na_2CO_3 present in some cases) that must be replaced determines the quantity of an amendment necessary to reclaim sodic soil. It depends on soil texture, the type of clay, and the crops intended to be grown. The relative tolerance of a crop to exchangeable sodium and its normal rooting depth will largely determine the soil depth up to which excess adsorbed sodium must be replaced.

The quantity of pure gypsum required to supply half a cmole of calcium per kg soil for the upper 15-cm soil depth will be

$$\frac{\text{molecular weight of gypsum}}{200} = \frac{172}{200} = 0.86\text{g kg}^{-1}$$
$$= 86 \times 10^{-5}\,\text{kg kg}^{-1}\,\text{soil}$$
$$= 86 \times 10^{-5} \times 2.24 \times 10^{6}\,\text{kg ha}^{-1}$$
$$= 1{,}926\ \text{kg or } 1.96\ \text{Mg ha}^{-1}$$

The amount of other amendments required may be estimated from their gypsum equivalence. Taking gypsum equivalence of gypsum ($CaSO_4{\cdot}2H_2O$) as 1.00, gypsum equivalence of sulfuric acid (H_2SO_4), iron sulfate ($FeSO_4{\cdot}7H_2O$), and pyrite becomes 0.57, 1.62, and 0.63, respectively.

11.8.3 Saline–Sodic Soils Are Both Saline and Sodic

Saline–sodic soils are saline and sodic at the same time. They contain excess soluble salts and excess exchangeable sodium. Their pH remains between 7.5 and 8.5. However, as long as there are excess salts, the physical conditions of the saline–sodic soils (soil structure, aeration, infiltration, drainage) are satisfactory. Caution must be taken in reclamation of saline–sodic soils so that these soils are not converted into sodic soils during leaching of salts. Sodic soils are more problematic than saline–sodic soils. Most sodic soils are left idle or they are used for extensive grazing only. Some salt-resistant crops as mustard and sorghum may be cultivated on sodic soils with a humus-rich surface horizon without amelioration. After reclamation, sodic soils can be productive for arable crop production especially in cooler climates, for instance, for wheat production. For reclamation of saline–sodic soils, both chemical amendment and leaching are required.

11.9 Peat May Be Productive, but Reclaiming Peat Soil Is Risky

Peat soils are classified as Histosols in both Soil Taxonomy and WRB soil classification systems (Chap. 4). Histosols are characterized by high organic matter contents. The amount of organic matter content of a soil to become an organic soil depends on its clay content. A soil having 60% clay must contain 18% by weight of organic matter strictly to be an organic soil (Lindbo and Kozlowski 2005). Histosols have specific characteristics: low bulk density, low bearing capacity, and low soil fertility. The floristic composition and evolving decomposition products determine the nature of organic matter which forms the Histosols. Histosols accumulate only in slightly decomposed plant residues due to varying combinations of waterlogging, low temperatures, and severe acidity. In the zone of Histosols, natural drainage is very poor. Histosols form in wet depressions where waterlogging prevents the rapid decomposition of the plant litter.

If carefully reclaimed and managed, peat soils can be productive under capital-intensive forms of arable cropping and horticulture. Largest areas of farmed organic soils in Europe are found in Russia (70,400 km^2), Germany (12,000 km^2), Belarus (9,631 km^2), Poland (7,620 km^2), and Ukraine (5,000 km^2). In comparison, cultivated organic soils of the United States and Canada cover 3,080 km^2 altogether. In Europe, cultivation of organic soils takes 14% of the total peatland area (Ilnicki 2002).

Artificial drainage along with liming and fertilization is necessary to bring peat into agriculture. The greatest problem associated with agricultural activities in peat soils is subsidence or the lowering of the surface of reclaimed peat. Subsidence is mainly due to drainage. The waterlogged and anaerobic peat in the natural swamp becomes aerobic when drained. It leads to biological oxidation of the organic deposits. The main processes of subsidence of peat according to Lucas (1982) are the following: (1) processes causing the removal of organic materials: oxidation, burning, wind erosion, and water erosion; (2) processes causing consolidation of materials: compaction, shrinkage, and dehydration; and (3) geological subsidence. Subsidence of peat also occurs due to a loss in volume in addition to oxidation. Schothorst (1977) recognizes the following components in subsidence: (1) shrinkage due to physical processes, the withdrawal of moisture from the surface layers by evapotranspiration may cause high moisture tensions in the root zone resulting in a decrease in volume of those layers; (2) oxidation through biochemical processes; and (3) consolidation or compression due to a mechanical process. Lucas (1982) reported from Canada that wind erosion has been one of the predominant causes of wastage and losses of peat.

Millions of hectares of peatlands have been "reclaimed" in the temperate zone. In many instances, this "reclamation" initiated gradual degradation and ultimately the loss of these precious ecosystems. Therefore, Histosols should better be kept under natural vegetation. In many wetlands, beautiful freshwater swamp forests flourished in Histosols. They are habitats and breeding grounds of many fish and bird species. There is a growing tendency worldwide not to further reclaim peats and bogs. These unique ecosystems should better be preserved.

Study Questions

1. What do you mean by soil problems? What are the different categories of problem soils? Dryland soils and sandy soils have some common problems; how can you overcome them?
2. Write the characteristics of deep cracking soils. Discuss the land use problems of Vertisols. Explain that steepland soils should better be left under natural conditions.
3. What do you mean by acid soils? What is an acid sulfate soil? Discuss the problems associated with acidity of soils. Liming alone is not enough for managing acid soils—explain.
4. Mention different liming and acidifying materials. What are calcium carbonate equivalence and gypsum equivalence? What is the significance of particle size of a liming material? Suppose a soil has a pH 4.5. You like to increase its pH to 6.5. How will you determine the lime requirement?
5. Explain soil salinity and sodicity. How would you reclaim a sodic soil? What are the suitable crops for saline soils?

References

Adams F, Evans CE (1962) A rapid method for measuring lime requirement of red-yellow podzolic soils. Soil Sci Soc Am Proc 26:355–357
Arnon I (1992) Agriculture in dry lands. Principles and practice. Elsevier, Amsterdam
Bloomfield C, Coulter JK (1973) Genesis and management of acid sulfate soils. Adv Agron 25:265–326
Campbell HW (1907) Soil culture manual: a complete guide to scientific agriculture as adapted to the semi-arid regions. The Campbell Soil Culture Co. (Inc.), Lincoln
Corwin DL, Rhoades JD, Simunek J (2007) Leaching requirement for soil salinity control: steady-state versus transient models. Agri Water Manage 9:65–180
Creswell R, Martin FW (1998) Dryland farming: crops and techniques for arid regions. ECHO, 17391 Durrance Rd., North Ft. Myers FL 33917, USA
Dregne HE (1976) Soils of arid regions. Elsevier, Amsterdam
Duchaufour P (1998) Handbook of pedology. Soils-vegetation-environment. Masson, Paris
Dudal R (1965) Dark clay soils of tropical and subtropical regions. FAO Agricultural Development paper no. 83. FAO, Rome
Dudal R (1980) Soil-related constraints to agricultural development in the tropics. In: Soil related constraints to food production in the tropics. IRRI/Cornell University, Los Baños
Dudal R, Eswaran H (1988) Distribution, properties and classification of vertisols. In: Wilding LP, Puentes R (eds) Vertisols: their distribution, properties, classification and management. Texas A&M University Printing Center, College Station
Elias EA, Salih FM, Salih AA, Alaily F (2001) Selected morphological characteristics of soils from Gezira Vertisols, with particular reference to cracking. Int Agrophys 15:79–86
Eswaran H, Cook T (1988) Classification and management related properties of vertisols. In: Jutzi SC, Haque I, Mcintire J, Stares JES (eds) Management of vertisols in Sub-Saharan Africa. Proceedings of a conference held at ILCA, Addis Ababa, 31 Aug–4 Sept 1987
FAO (1998) Ethiopia soil fertility initiative concept paper. Report no. 98/028 CP-ETH. FAO, Rome
FAO (2000) Land resource potential and constraints at regional and country levels. World Soil Resources report no. 90, Rome
FAO/AGL (2000) Agricultural problem soils (FAO/AGL problem soil database: www.fao.org/ag/agl/agll/prosoil/calc.htm). Accessed 5 Dec 2011
Freebairn DM, Ward LD, Clark AL, Smith CD (1986) Research and development of reduced tillage systems for vertisols in Queensland, Australia. Soil Tillage Res 8:211–229
Hoskins B (2005) Modification of the Mehlich lime buffer test for maine. Soil Science Society of American Abstracts, ASA-CSSA-SSSA international meetings, 6–10 Nov 2005. Salt Lake City, Utah
http://learningstore.uwex.edu/assets/pdfs/a3671.pdf. Accessed 18 Dec 2011
http://www.coopext.colostate.edu/TRA/PLANTS/acidlove.shtml. Accessed 22 Nov 2011
http://www.extension.usu.edu/files/publications/publication/AG-SO-03.pdf. Accessed 13 Jan 2012
http://www.livinghistoryfarm.org/farminginthe50s/machines_11.html. Accessed 5 Nov 2011
http://www2.vernier.com/sample_labs/AWV-09-COMP-soil_salinity.pdf. Accessed 20 Jan 2012
Huluka G (2005) A modification of the Adams-Evans soil buffer determination solution. Com Soil Sci Plant Anal 36:2005–2014
Ilbeyi A, Ustun H, Oweis T, Pala M, Benli B (2006) Wheat water productivity and yield in a cool highland environment: effect of early sowing with supplemental irrigation. Agric Water Manage 82:399–410
Ilnicki P (2002) Torfowiska i Torf. Warszawa
Lal R (1993) Soil erosion and conservation in West Africa. In: Pimental D (ed) Soil erosion and conservation. Cambridge University Press, Cambridge
Li Shengxiu, Xiao Ling (1992) Distribution and management of drylands in the People's Republic of China. Adv Soil Sci 18:148–302
Lindbo DL, Kozlowski DA (2005) Histosols. In: Lal R (ed) Encyclopedia of soil science, 2nd edn. CRC Press, Boca Raton
Lucas RE (1982) Organic soils (histosols): formation, distribution, physical and chemical properties and management for crop production. Research report 435. Farm Science, Michigan State University
Mehlich A (1976) A new buffer pH method for rapid estimation of exchangeable acidity and lime requirement of soils. Com Soil Sci Plant Anal 7:637–652
Molden D, Oweis TY (2007) Pathways for increasing agricultural water productivity. In: Molden D (ed) Water for food, water for life. Earthscan/International Water Management Institute, London/Colombo
Oweis T (1997) Supplemental irrigation – a highly efficient water-use practice. ICARDA-037/1000
Oweis T, Hachum A (2012) Supplemental irrigation a highly efficient water-use practice, Revised and extended 2nd edn. ICARDA (International Center for Agricultural Research in the Dry Areas, Aleppo, Syria
Oweis TH, Zhang H, Pala M (2000) Water use efficiency of rainfed and irrigated bread wheat in a Mediterranean environment. Agron J 92:231–238
Rhoades JD (1974) Drainage for salinity control. In: van Schilfgaarde J (ed) Drainage for agriculture, Agronomics monograph 17. SSSA, Madison
Rhoades JD, Loveday J (1990) Salinity in irrigated agriculture. In: Stewart BA, Nielsen DR (eds) Irrigation of agricultural crops, Agronomics monograph 30. SSSA, Madison

Richards LA (1954) Diagnosis and improvement of saline and alkali soils, USDA agricultural handbook 60. USDA, Washington, DC
Rosegrant M, Cai X, Cline S, Nakagawa N (2002) The role of rainfed agriculture in the future of global food production, EPTD discussion paper 90. IFPRI, Washington, DC
Ryan J, Spencer D (2001) Future challenges and opportunities for agricultural land in the semi-arid tropics. ICRISAT, Patancheru
Schothorst CJ (1977) Subsidence of low moor peat soils in the western Netherlands. Geoderma 17:265–291
Shoemaker HE, McLean EO, Pratt PF (1961) Buffer methods of determining lime requirements of soils with appreciable amounts of extractable aluminum. Soil Sci Soc Am Proc 25:274–277
Sikora FJ (2006) A buffer that mimics the SMP buffer for determining lime requirement of soil. Soil Sci Soc Am J 70:474–486
UNCCD (2000) An introduction to the United Nations Convention to combat desertification. Available at http://www.unccd.int. Accessed 3 Oct 2011
US Salinity Laboratory Staff (1954) Diagnosis and improvement of saline and alkali soils. Agriculture handbook no. 60, USDA
Willcocks TJ, Twomlow SJ (1992) An evaluation of sustainable cultural practices for rainfed sorghum production in east Sudan. Soil Tillage Res 24:183–198
Xi Chengfan (1961) Soil moisture and conservation of water for resisting drought. Soils 4:27–29

Soil Resources and Soil Degradation

12

The total land area of the world is estimated to be 130,575,894 km^2, including rocky surfaces, deserts, ice-covered areas, and lands with soil. Only about 12% land is suitable for agricultural crop production without much limitation, 24% is used for grazing, and 31% is occupied by forests. The remaining 33% has too many constraints for most uses. Not all agricultural soils are fertile and productive. Some soils are naturally unproductive; some are arid and saline; some are very sandy and dry; some are wet and waterlogged for most of the growing season. Some soils have been degraded by human activity. Recent estimates suggest that land degradation affects 3,500 M ha or 23.5% of the Earth's land area and impacts 1.5 billion people. GLASOD recognizes five types of soil degradation processes—water erosion, wind erosion, chemical deterioration, physical deterioration, and degradation of biological activity. Soil erosion along with compaction and surface sealing is considered to be a physical degradation process. There are on-site and off-site effects of water and wind erosion. These processes may also lead to desertification in arid and semiarid regions. Many soils have been salinized by soil mismanagement, including modification of hydrology through irrigation and drainage. Some soils have been polluted by organic and inorganic pollutants. Degraded soils need sustainable management.

12.1 Soil Is a Natural Resource

The European Charter on Soil was declared in 1972 by the European Council emphasizing the need for management and protection of soil resources (Tolgyessy 1993). It cites the following:

(1) Soil is one of the most valuable of man's properties. It enables the life of plants, animals, and humans on Earth. (2) Soil is a limited resource which is easily destroyed. (3) The industrial society employs soil for agriculture as well as for industrial and other purposes. The policy of regional planning must combine considerations of the soil properties and the present and future needs of the society. (4) Agriculturalists and foresters must apply methods which protect the soil quality. (5) Soil must be protected against erosion. (6) Soil must be protected against pollution. (7) Development of urbanization must be planned so that the neighboring areas are exposed to the minimum possible damage. (8) During the planning of engineering projects, the effects on the soil must be evaluated so that the price includes provision for adequate protective measures. (9) A list of soil resources is a prerequisite for any planning. (10) To ensure the extensive use and protection of soil, further research and interdisciplinary cooperation are required. (11) Attention at all professional levels as well as ever-increasing attention of the general public should be paid to soil protection, and (12) governments and state authorities must carefully plan and nurture, and interdisciplinary cooperation are required.

Of the total global land area of 130,575,894 km^2 (Eswaran et al. 1999), some lands such as deserts and ice-covered areas do not contain soil. Where there is soil, there is some sort of vegetation—forests, grasslands, or croplands. The natural vegetation in many areas has been removed to use the soil for cropping or some other purposes, including urban and industrial use. Some vegetation has been retained in its original form or disturbed, and some have been exploited, legally or illegally, judiciously or indiscriminately, for our need. Now, soil resources provide about 96% of world food, and the remaining food comes from rivers, lakes, and seas (Pimental and Hall 1989).

About 38.5 million km^2 or 29.45% of the Earth's ice-free land surface is too dry (deserts and dryland soils—the Aridisols) for sustainable human habitation (Beinroth et al. 1994). About 20.2 million km^2 or 15.46% of the land occur in the cold tundra zone (Arctic regions, permafrosts, Gelisols). Saline (Salids/Solonchaks) and alkaline (Calcids/Calcisols) soils occupy 3.11 million km^2 or 2.4% of the land surface, and soil acidity affects 18.42 million km^2 or 14.1% of the total land (mainly Oxisols, Ultisols, and Spodosols) (Eswaran et al. 1997). Not all agricultural soils are fertile and productive. There is only 11–12% land area suitable for agricultural

K.T. Osman, *Soils: Principles, Properties and Management*,
DOI 10.1007/978-94-007-5663-2_12, © Springer Science+Business Media Dordrecht 2013

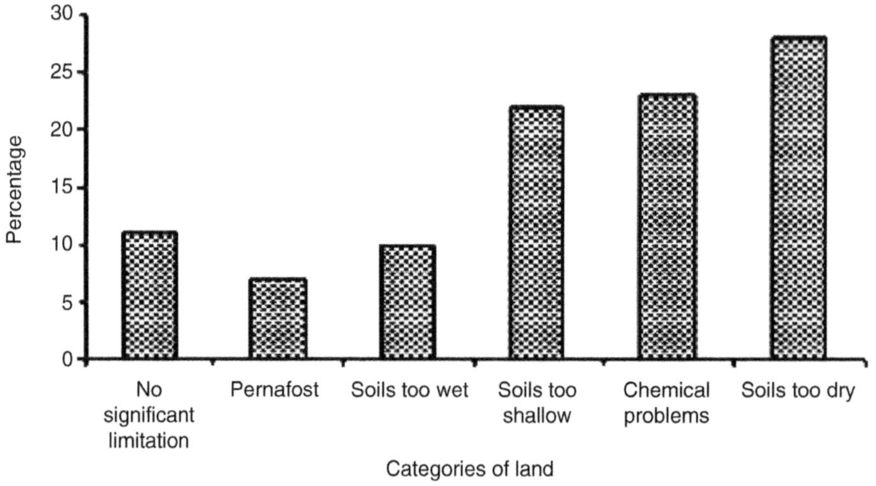

Fig. 12.1 Global soil conditions (Based on data from FAO 1998)

crop production without much limitation. Soils of only about 3% of the total land area of the Earth have a high level of productivity (some Alfisols and Mollisols) (Aswathanarayana 1999). Figure 12.1 shows different categories of land.

12.1.1 There Are Nine Land Quality Classes

Eswaran et al. (1999) identified three soil performance classes (low, medium, and high) and three soil resilience classes (low, medium, and high). Combining them, they divided world soils into nine land quality classes. They defined land quality, soil resilience, and soil performance as follows:

Land Quality: The ability of the land to perform its function of sustainable agricultural production and enable it to respond to sustainable land management.

Soil Resilience: The ability of the land to revert to a near original production level after it is degraded, as by mismanagement. Land with low resilience is permanently damaged by degradation.

Soil Performance: The ability of the land to produce (as measured by yield of grain, or biomass) under moderate levels of inputs in the form of conservation technology, fertilizers, and pest and disease control.

The properties of the different land quality classes are shown in Table 12.1.

Class I lands or prime lands occur in the United States, Argentina, Uruguay, southern Brazil, Europe, Northern China, and South Africa. Such lands are not found in the tropics. Tropical soils (Oxisols, Ultisols, some Alfisols, Entisols, Inceptisols, and Vertisols—Chap. 4) are generally of low productivity. Class II and III lands are extensive in the tropics and in the temperate areas. Most of these lands are under some forms of agriculture and irrigated, when possible, in the semiarid parts. Only about 3% of the global land surface can be considered as prime or Class I land. Another 8% of land is in Classes II and III. A major part of Class IV, V, and VI lands, particularly the Class V lands, are in the tropics. In the Amazon basin, Central Africa, and Southeast Asia, these form large tracts of forests. Class IV, V, and VI lands occupy a significant part of the Earth's surface (36.8 million km^2) and also support more than 50% of the world's population. Class IV or poorer-quality lands may be managed with high inputs and can be made productive (Buol and Eswaran 1994). From a sustainability point of view, Classes VII, VIII, and IX are not suitable for agriculture.

Eswaran et al. (1999) estimated that the total global land area suitable for cultivation is 60.2 million km^2. WRI (1997) estimated that about 49.77 million km^2 were under cultivation in 1995. This suggests that the amount of globally available land is only about 10 million km^2 for future cropping, forestry, and for other uses. The percentage of cropland in Asia, Africa, South America, North America, and Europe are 15.2, 6.3, 6.0, 13.0, and 6.0, respectively. The proportions of pasture in the corresponding regions are 20, 22, 14, 08, and 23% (WRI 1997). Area of land in different quality classes are shown in Fig. 12.2.

12.2 Soil Degradation Is the Exhaustion of Soil's Potential to Serve Desired Function

Land degradation sets in when the potential productivity associated with a land-use system becomes non-sustainable or when the land within an ecosystem is no longer able to perform its environmental regulatory function of accepting, storing, and recycling water, energy, and nutrients. On the other hand, soil degradation has been described as the measurable loss or reduction of the current or potential capability

Table 12.1 Land quality classes and their properties

Land quality class	Properties
I	This is prime land. Soils are highly productive with few limitations. Soil temperature and moisture conditions are ideal for annual crops. Soil management consists largely of sensible conservation practices to minimize erosion, appropriate fertilization, and use of best available plant materials. Risk for sustainable grain crop production is generally <20%
II and III	The soils are good and have few problems for sustainable production. Care must be taken to reduce degradation, particularly for Class II soils. The lower resilience characteristics of Class II soils make them more risky, particularly for low-input grain crop production. However, their productivity is generally very high, and consequently, response to management is high. Conservation tillage is essential, buffer strips are generally required, and fertilizer use must be carefully managed. Due to the relatively good terrain conditions, the land is suitable for national parks and biodiversity zones. Risk for sustainable grain crop production is generally 20–40%, but risks can be reduced with good conservation practices
IV, V, VI	If there is a choice, these soils must not be used for grain crop production, particularly soils belonging to Class IV. All three classes require important inputs of conservation management. In fact, no grain crop production must be contemplated in the absence of a good conservation plan. Lack of plant nutrients is a major constraint, and so a good fertilizer use plan must be adopted. Soil degradation must be continuously monitored. Productivity is not high, and so low-input farmers must receive considerable support to manage these soils or be discouraged from using them. Land can be set aside for national parks or as biodiversity zones. In the semiarid areas, they can be managed for range. Risk for sustainable grain crop production is 40–60%
VII	These soils may only be used for grain crop production if there is a real pressure on land. They are definitely not suitable for low-input grain crop production; their low resilience makes them easily prone to degradation. They should be retained under natural forests or range, and some localized areas can be used for recreational purposes. As in Class V and VI, biodiversity management is crucial in these areas. Risk for sustainable grain crop production is 60–80%
VIII, IX	These are soils belonging to very fragile ecosystems or are very uneconomical to use for grain crop production. They should be retained under their natural state. Some areas may be used for recreational purposes but under very controlled conditions. In Class IX, which is largely confined to the Boreal area, timber harvesting must be done very carefully with considerable attention to ecosystem damage. Class VIII is mainly the deserts. Risk for sustainable grain crop production is >80%

Adapted from Eswaran et al. (1999)

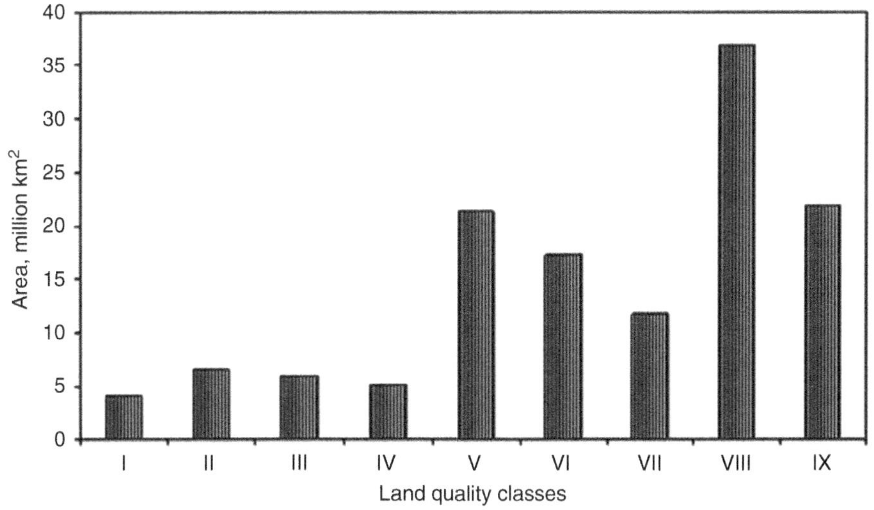

Fig. 12.2 Area of land under different quality classes

of soils to produce plant materials of desired quantity and quality. Comparatively, land degradation is a broader term and encompasses lowering of the current or potential capability of a land area to serve a desired function including agriculture, transport, construction, and recreation (Blaikie and Brookfield 1987; Blum 1997; Chisholm and Dumsday 1987). In the following sections, land degradation and soil degradation will be used interchangeably. Bai et al. (2008) estimated that land degradation affects 3,500 M ha or 23.5% of the Earth's land area and impacts 1.5 billion people. According to GLASOD (Oldeman 1991), half of the degraded soil area is moderately degraded. Strong and extreme degradation covers more than 300 million ha. Losses have continued to mount since 1990, and additional 5–6 million ha were lost to severe soil degradation annually (UNEP 1997; WRI 2001). Mechanisms of land degradation include physical, chemical, and biological processes (Lal 1994). Important among physical processes are decline in soil structure, leading to crusting, compaction, erosion, and desertification. Significant chemical processes include acidification, leaching, salinization, reduction in cation retention capacity, pollution, and fertility depletion. Biological processes include reduction in total and biomass carbon and decline in land biodiversity. According to Beinroth et al. (1994), land degradation results from a mismatch between land quality and land use.

12.2.1 There Are Five Main Types of Soil Degradation

Soil degradation may be of five main types according to GLASOD. These types are water erosion, wind erosion, chemical deterioration, physical deterioration, and degradation of biological activity (Table 12.2) (Oldeman 1991, 2000).

Land area so far degraded by each type of soil degradation is shown in Fig. 12.3.

12.2.2 Soil Degradation May Be Due to Natural and Anthropogenic Causes

Soil degradation may result from natural causes and human-induced causes. Topographic and climatic factors such as steep slopes, frequent floods and tornadoes, storms and high-velocity wind, high intensity rains, leaching in humid regions, and drought in dry regions are among natural causes. Deforestation and overexploitation of vegetation, shifting cultivation, soil desurfacing, overgrazing, unbalanced fertilizer use and lack of soil conservation practices, overextraction of groundwater are some anthropogenic causes of soil degradation.

Table 12.2 Types, degree, and causative factors of soil degradation according to GLASOD

Types of degradation		
W	*Water erosion*	
Wt	Loss of topsoil	
Wd	Terrain deformation/mass movement	
Wo	Off-site effects	
	Wor	reservoir sedimentation
	Wof	flooding
	Woc	coral reef and seaweed destruction
E	*Wind erosion*	
Et	Loss of topsoil	
Ed	Terrain deformation	
Eo	Overblowing	
C	Chemical deterioration	
Cn	Loss of nutrients or organic matter	
Cs	Salinization	
Ca	Acidification	
Cp	Pollution	
Ct	Acid sulfate soils	
Ce	Eutrophication	
P	Physical deterioration	
Pc	Compaction, sealing and crusting	
Pw	Waterlogging	
Pa	Lowering of water table	
Ps	Subsidence of organic soils	
Po	Other physical activities such as mining and urbanization	
B	Degradation of biological activity	
Degree of degradation		
Light	Somewhat reduced agricultural activity	
Moderate	Greatly reduced agricultural activity	
Strong	Unreclaimable at the farm level	
Extreme	Unreclaimable and impossible to restore	
Causative factors in soil degradation		
f	Deforestation and removal of natural vegetation	
g	Overgrazing	
a	Agricultural activities	
e	Overexploitation of vegetation for domestic use	
i	(Bio)industrial activities	

12.3 Physical Degradation of Soil Includes Compaction, Surface Sealing, and Erosion

Soil compaction, desertification, and erosion are the major types of physical degradation of soil.

12.3.1 Soil Compaction Is the Consolidation Under Pressure

Soil compaction is the physical consolidation of the soil by an applied force, usually by tillage implements, that destroys soil structure, compresses soil volume, increases bulk density, reduces porosity, and limits water and air movement. Seedling emergence, root extension, and water and nutrient uptake by plants are reduced due to soil compaction. As a result, growth and yield of crops are also reduced.

Stunted plant growth, badly formed plant roots, standing water, formation of large clods after tillage, and physically dense soil are some signs of soil compaction. It is a global problem, especially associated with mechanized agriculture. It has caused yield reductions of 25–50% in some regions of Europe (Eriksson et al. 1974) and North America, and between 40 and 90% in West African countries (Charreau 1972; Kayombo and Lal 1994). On-farm losses through soil compaction in the USA have been estimated at US$1.2 billion per year (Gill 1971). Oldeman et al. (1991) suggested

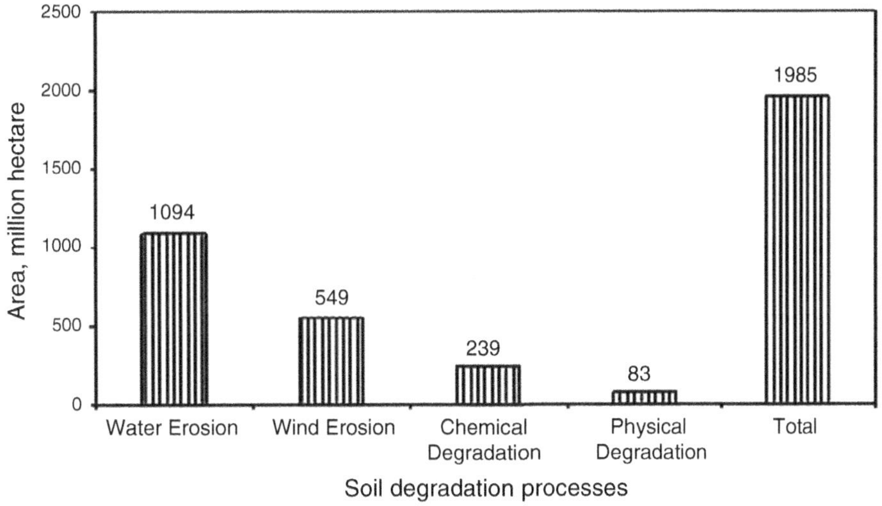

Fig. 12.3 Land area so far degraded by different processes (Data from Oldeman 1994)

Human Development Activity Is a Major Cause of Soil Degradation

People have been the main cause of worldwide soil degradation from prehistorical times. Acceleration of natural processes like erosion, leaching, evaporation and decomposition, and desperate actions like deforestation, biomass burning, overgrazing, disposal of hazardous wastes, and soil desurfacing are some of the human involvements in soil degradation. Human development activities in urban areas of developing countries have become a major threat to the loss of agricultural land as a whole, and reduction in soil productivity in particular. For example, brick is an important element of urban development. But most brick fields are located in fertile alluvial lands because there the soils are silty clay loam or of similar textures suitable for brickmaking (Fig. 12.4). You will see tens to hundreds of brickfields in agricultural lands along both sides of the river Buriganga near Dhaka, Bangladesh. They occupy large tracts of croplands, and for brickmaking, more or less a uniform layer of soil, 15–20 cm thick, is truncated from the surface of huge areas of paddy fields. A five-story building of 225 m^2 floor space needs 4,50,000 bricks equivalent to about 1 ha furrow slice soil (100 m × 100m × 15 cm). Such a mechanical removal of topsoil is termed as soil desurfacing. The surface is the most fertile part of the soil; it contains the highest organic matter and nutrients and possesses the most suitable physical and chemical conditions for plant growth. In a study, soil desurfacing was found to decrease organic matter, nutrients, and available water, and the exposed subsoil was compact. Crop yields were reduced significantly (Grewal and Kuhad 2002). It is almost impossible to recover the ecological losses of soil desurfacing. You may remember that the formation of only a centimeter of topsoil may need more than a century (Chap. 3). Should we spare such a precious resource as topsoil for bricks?

Fig. 12.4 Brick fields in agricultural land (Photo courtesy of Dr. Animesh Biswas)

that compaction is by far the most important type of physical deterioration of agricultural soils.

12.3.1.1 Surface Sealing and Surface Crust

Soil structure destruction and dispersion of soil particles lead to surface sealing. Soil aggregates collapse by the load of heavy farm machineries and/or hoof strokes of farm animals. Dispersion also occurs due to rainfall impact, plow pressure, or by soluble Na in irrigation water. Dispersed particles clog the pores and make the soil impervious, particularly in the surface. This is surface sealing which, on further consolidation, gives rise to surface crusts.

Further deterioration leads to soil compaction. In the advanced stage, soil particles are compressed by pressure of tillage implements such as moldboard plows, disks, and sweep-type tools so that the air in pores is expelled and finer soil particles are forced into the larger pores. As a result, the soil volume is reduced and the bulk density is increased. Deterioration of surface soil structure generally results in surface crusting and hardsetting. The term hardsetting was first used by Northcote (1960) in his soil classification system of Australia. Hardsetting is a characteristic of soil horizons, usually cultivated seedbeds, which contain unstable soil aggregates. Soil aggregates collapse, and the seedbed slumps when the soil is wet, and a hard, structureless mass of soil results upon drying. Although soil properties under surface crusting and hardsetting are similar and the processes are also similar, the phenomena apply to different depths of the seedbeds. Crusting occurs when surface aggregates break down on wetting to form a hard, dense, structureless mass at the surface, often only a few millimeters thick. Hardsetting, however, involves a much greater thickness of material, which commonly includes not only the A1 or Ap horizon but also the E horizon (Greene 2005).

Some soils are naturally and genetically hardsetting. They are abundant in tropical areas (Fabiola et al. 2003). Naturally hardsetting soils are unable to develop water-stable aggregates. Hardsetting condition can occur in soils with high exchangeable sodium percentage. Some hardsetting soils are impervious, compacted as well as cemented. Amorphous silica and imogolite-like aluminosilicates may act as cementing agents (Chartres et al. 1989).

There are two types of soil compaction—shallow and deep. Shallow compaction occurs near the soil surface and can be broken up by normal tillage. Deep compaction is caused mainly by axle load. It is extremely difficult to break the deeply compacted layers. Special techniques are needed to overcome such compaction problems. Compacted layers below the plow depth are known as tillage pans or plow pans. Extensive areas have undergone subsoil compaction (van den Akker and Schjonning 2004).

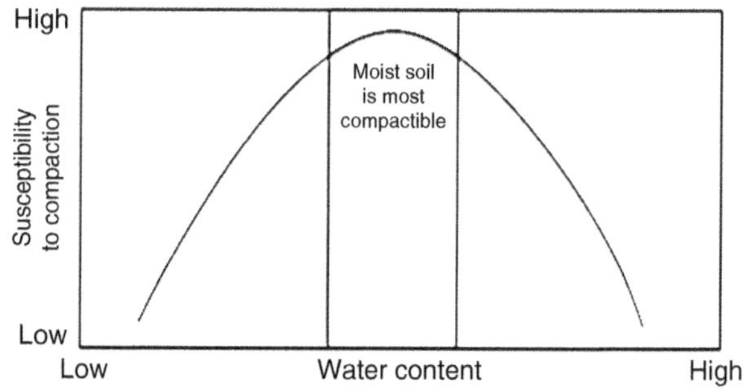

Fig. 12.5 Soil moisture level and susceptibility to compaction

12.3.1.2 Factors Affecting Soil Compaction

Soil compaction is a problem mainly associated with tillage. The types of tillage implements and mode of their operation are largely related with soil compaction. Weight and design of the implement determine the load on soil. Moreover, the number of axles and tires and number of trips all affect soil compaction. A pressure of 4 psi or more on most soils can produce considerable compaction. A pickup truck will impose about 50 psi pressures on the soil, and a liquid manure spreader will produce 70–90 psi. For a tractor of a given weight, four-wheel drive is better than two-wheel drive because the weight can be evenly distributed to all tires. Soil texture, structure, organic matter, and water are important soil factors that determine susceptibility of soil to compaction. Soils made up of particles of about the same size compact less than soil with a variety of particle sizes. Smaller particles can fill the pores between larger particles, resulting in a denser soil. A sandy loam soil is more easily compacted than a sandy or clayey soil.

A soil with a higher level of organic matter generally has a better structure. Hard, dense, low organic matter soils suffer more from compaction than loose, friable, high organic matter soils. A dry soil is not easily compacted. Soil compaction increases with the increase in soil water content during tilling. A soil saturated with water resists compaction because water occupies the pores and water is not compressible. Soils are easily compacted at soil water contents slightly higher than field capacity. Figure 12.5 shows the relationships of soil water with soil compaction.

12.3.1.3 Remediation of Soil Compaction

Prevention is better than cure; it is even truer for remedies of soil compaction. To prevent compaction, tilling the land at moist condition (so moist that a ball can be made of soil by rolling) should be avoided. Selection of tillage implement is important in relation to the formation of tillage pans. Once compacted, it is very difficult to decompact the soil. To alleviate compaction, subsoiling with heavy duty shanks to depths of 30–50 cm or even deeper and usually

Heavy Farm Machineries May Cause Severe Soil Compaction

Soils are usually not much compacted by manual or oxen-driven tillage operations. But farming in developed countries is altogether mechanized. Agricultural tractors and other implements such as combines are used for tilling, planting, and harvesting. These 20–200-HP gasoline and diesel 2-wheel drive and 4×4 machines usually have power steering, dual-range transmissions, and two-speed live power takeoff (Fig. 12.6). There is also a wide range of farm implements such as disks, plows, tillers, scrapers, and diggers that are normally not powered. Public right-of-way regulations require axle loads to be no greater than 20,000 lb for most vehicles. The front axle load of a large combine with attached head and full grain tank or the axle load of a full grain cart or liquid manure tank may exceed this value by 50% or more. Tractors alone have weights of about 20 t. These heavy agricultural machineries result in more permanent damage to the soil than it was previously believed. Soil compaction, induced by wheel traffic from large agricultural equipment, has been shown to extend well below the depth of plowing, reducing crop growth and yield in many situations. Estimates suggest that the area of soil degradation due to compaction in Europe may have exceeded 33 million ha.

Controlled traffic farming (CTF) is adopted in Australia to reduce the risk of soil compaction since the early 1990s involving production of grain crops under a production system where the high-powered tractors, large harvesters, and heavy trucks all have the same track settings of 3 m. In the field, the width of the combine harvesters and planters are matched at perhaps 9 m, allowing the 3-m tracks to be centrally spaced at the same distance. The advantages are that the tractor can be less powerful because the wheels run on packed soil, while only minimum or no tilling is required on the non-compacted beds in between nor is there any need to subsoil every few years. In CTF, water penetration is better, there are no ruts or ridges left over from previous operations, the fuel saved is substantial, and the crops yield much better because of zero compaction.

Fig. 12.6 Heavy farm machineries

spaced about 75 cm apart is often done. It is an expensive operation, and in the long run, it may not be completely satisfactory. Subsoiled fields may recompact to a greater degree. Recompaction of decompacted soils may be prevented by generous addition of organic residues which will make the soil more load bearable and improve soil structure (van Es and Hill 1995). Compacted soils are poorly drained. Changing tillage tools, reducing tillage and traffic over the field, converting to a no-till program, and rotating the field to a deep-rooted forage crop for several years may be some of the approaches for management of compacted soils. Selection of tolerant crops (crops adapted to poor drainage) may be another approach.

12.3.2 Desertification Occurs Mainly in Arid and Semiarid Regions

Desertification occurs mainly in the arid and semiarid regions due to human actions on ecosystems combined with adverse climatic conditions. Dregne (1977) defined, "Desertification is the impoverishment of terrestrial ecosystems under the impact of man. It is a process of deterioration in these ecosystems that can be measured by reduced productivity of desirable plants, undesirable alterations in the biomass and the diversity of the micro and macro flora and fauna, accelerated soil deterioration, and increased hazards for human occupancy." The formal definition of desertification adopted by the United Nations

Convention on Desertification in 1994 is "Land degradation in arid, semiarid, and dry subhumid areas resulting from various factors, including climatic variations and human activities." Due to desertification, forests, grasslands, shrublands, and croplands are converted to desertlike lands.

About 33% of the Earth's land surface is vulnerable to desertification, and these areas are distributed over the semiarid to weakly aridic areas of Africa, India to Turkey and Ukraine, countries of Central Asia, and much of the winter-rain areas of the Mediterranean (Eswaran and Reich 1998). Desertification affects over two billion people and around 100 countries across all 5 continents (Bied-Charreton 2008). Ten-million ha productive land with 24 billion t of topsoil is lost every year due to desertification. The principal worldwide problems caused by desertification are loss of biological productivity, loss of economic productivity, and loss of complexity in the landscape (UNCCD 1994).

12.3.2.1 Natural- Versus Human-Induced Desertification

Causes of desertification are complex, frequently local, and vary from one part of the world to another. The severity of impact also varies with less-developed countries experiencing greater human misery.

There has been much debate on the causes and processes of desertification in the past due to differences in approaches of experts of many different disciplines such as geography, ecology, environmental science, economics, and sociology. Some scientists believe that desertification is a natural phenomenon occurring due to irregular fluctuation of short- and long-term drought, as experienced in African Sahel (Hill and Peter 1996). To others, desertification is human induced, resulting from pressures on land resources, unwise development policies, and misuse of land (Graetz 1996). However, the vulnerable regions are undeveloped, poverty-stricken, and poor in biomass resources. In these arid and semiarid regions, soil and vegetation are fragile. Human pressure on vegetation and land is enormous, but the support of technology, soil, and climate is inadequate. Culet (2002) suggested that overgrazing, deforestation, and other agricultural activities may, respectively, contribute 35, 30, and 28% to desertification. Deforestation, overgrazing, overexploitation, illegal felling, and excessive fuelwood collection gradually turned forestlands and grasslands into wastelands. Unsustainable agricultural practices, including excessive use of chemicals (fertilizers, pesticides, and herbicides), inappropriate technologies, wrong choice of crops, inadequate and inefficient irrigation practices associated with salinity, overabstraction of groundwater, etc., have led to problems of desertification in vulnerable regions.

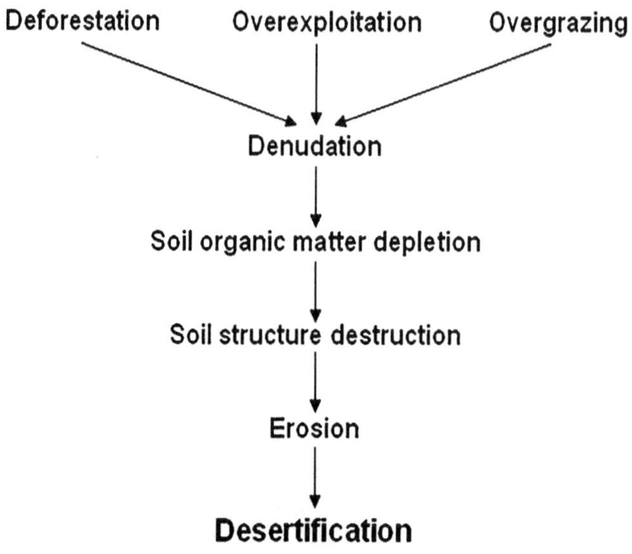

Fig. 12.7 Desertification scheme

12.3.2.2 Mechanism of Human-Induced Desertification

Desertification is a result of complex interactions of diverse factors prevailing in arid and semiarid regions.

Denudation of land due to deforestation, overexploitation of biomass, and overgrazing appear to mark the initiation of desertification. In the latter stages, soil organic matter is depleted and soil structure is destroyed. Soil particles are detached from aggregates, dust clouds are formed by suspending fine to very fine soil particles in air, and soil materials are eroded primarily by wind. Thus, once forested lands, scrublands, grasslands, and croplands are converted into desertlike lands, reversing, which is extremely difficult, is not impossible. A hypothetical desertification scheme is presented in Fig. 12.7.

12.3.3 Soil Erosion Is the Detachment and Transport of Soil Particles

Erosion occurs in various land surfaces ranging from floodplains to mountains and by various agents, including gravity, water, wind, glacier, and frost. No external force is needed on steep slopes to set the loosened detritus in motion. In many cases, frost and high temperature separate pieces of weathered rock, and the loose material moves downhill to form piles of hillside waste, debris cones, outwash fans, and other formations (Zachar 1982). Erosion occurs even in undisturbed landscapes by natural forces; it is called normal erosion, natural erosion, or geological erosion. The degree and frequency of erosion are accelerated by human activity—activity related to economical, social, cultural, agricultural, and industrial development. Generally, accelerated erosion is considered to be soil erosion proper. Figure 12.8 shows the rate of soil erosion in different continents.

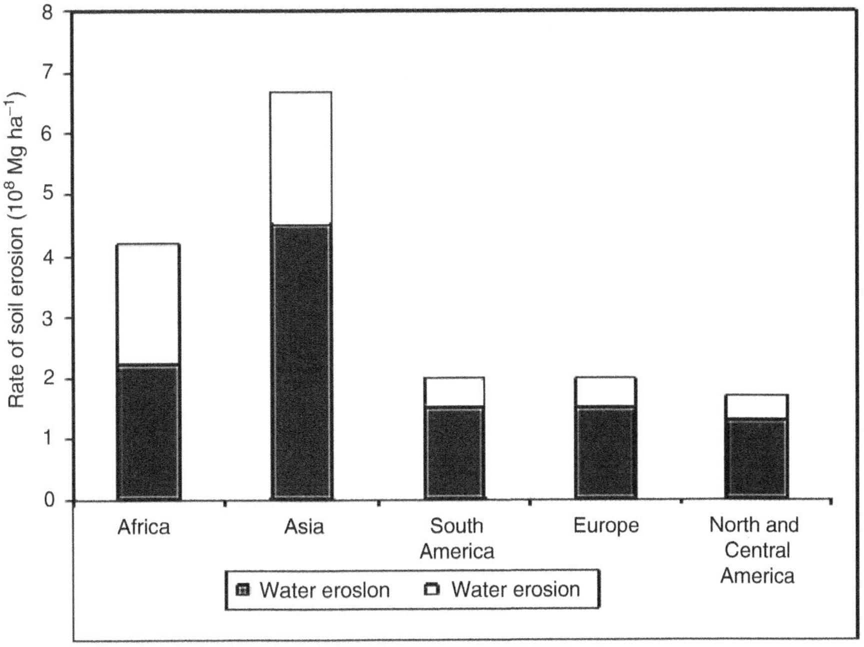

Fig. 12.8 Rates of soil erosion in different continents (Based on data of WRI 1997)

Soil erosion involves the detachment and transport of soil particles. Detachment of soil particles occurs by rainfall impact, flowing water, or by any mechanical force such as tillage. Detached particles are transported by water and wind. Natural soil erosion is of little concern because the amount of soil erosion is low and can be compensated by the amount of soil formation. Depending on agents of erosion, there are two types of soil erosion—water erosion and wind erosion, and they have two types of impacts—on-site and off-site. Causes of soil erosion can be summarized as follows: overgrazing, burning of native grassland/tussock and scrub vegetation, forest fires, shifting cultivation, forest harvesting with skidder hauling machinery, burning of cereal crop stubble and removal of beneficial organic matter, cultivation up- and downslope, cultivating slopes steeper than 20°, growing similar crops/pasture continuously, compaction of soil by stock and machinery, formation of compaction pans which reduces normal water infiltration and increases overland flow, and tilling the soil to a very fine tilth.

12.3.3.1 Water Erosion

Water erosion is caused by water—water that comes in rain and runs off the land as overland flow or stream flow. Detachment of soil particles from aggregates by the impact of falling raindrops or flowing water is the initial stage of water erosion; this is followed by transport of the detached particles by runoff water. Runoff water laden with suspended particles also detaches more soil particles as it moves across the surface. Soil erosion is a process of soil loss, particularly from the surface, but sometimes a large mass of soil may be lost, as in landslides and riverbank erosion. Kohnke and Bertrand (1959) classified water erosion into sheet erosion, internal erosion, and channel erosion. Channel erosion was further divided into rill erosion, gully erosion, and stream erosion. However, the following four types of water erosion are generally recognized: splash erosion, sheet erosion, rill erosion, and gully erosion. Splash and sheet erosion are sometimes called interrill erosion.

12.3.3.2 Splash Erosion

At the start of a rain event, falling raindrops detach soil particles which fill up the large soil pores. Consequently, the infiltration capacity of the soil is reduced, water does not enter the soil, and soon a thin film of water covers the ground. Further, raindrops beat the water and splash the suspended soil particles away. Soil particles are transported to some distance by the splashing. Processes of splash erosion involve raindrop impact, splash of soil particles, and formation of craters (Ghadiri 2004). Actually, splash erosion is the beginning of other types of soil erosion, particularly sheet erosion.

12.3.3.3 Sheet Erosion

When an almost uniform layer of soil is removed by water from the whole slope, it is called sheet erosion. It is the most dangerous type of soil erosion because it occurs almost silently leaving little or no signs of soil removal. When erosion has greatly advanced, exposure of tree roots or abrupt change of soil color due to exposure of light-colored subsoil indicates loss of soil. Sheet erosion generally occurs on level and smooth soil surfaces. Most soils are not so smooth, and

Fig. 12.9 Formation of a gully in a degraded forest soil (Photo courtesy of Dr. Animesh Biswas)

sheet erosion is the preliminary state of rill erosion. According to Zachar (1982), the first phase of sheet erosion is soil removal by raindrop action–raindrop erosion. The second phase is laminar erosion or sheet erosion.

12.3.3.4 Rill Erosion

When rainfall exceeds the rate of infiltration, water accumulates on the surface, and if the land is sloping, it moves along the slope. On gently sloping lands, with standing crops or in fields that have been recently tilled, moving water concentrates along tiny channels called rills. The cutting action of flowing water detaches soil particles, and runoff water carries them away. The amount of soil loss may be high, but the small channels do not usually interfere with tillage implements. The rills may be leveled by normal tillage operations.

12.3.3.5 Gully Erosion

Gullies develop when large quantities of water accumulate and run through a single channel in relatively steep slopes (Fig. 12.9).

Gullies may also develop by the gradual deepening of rills. There are two types of gullies: ephemeral and permanent. Ephemeral gullies form shallow channels that can be readily corrected by routine tillage operations. On the other hand, permanent gullies are very large and cannot be smoothed by regular tillage (Blanco and Lal 2008). There are several forms of gullies: flat, narrow, broad, and round. Gullies may be U-shaped formed in loose to non-cohesive, uncompacted soils, and V-shaped where hillslope watercourses cut through surficial regolith. Uncontrolled gullies form streams.

12.3.3.6 Landslide/Landslip/Mudflow

The downward and outward movement of a large block of soil and regolith caused by gravity are called landslides (Fig. 12.10) and landslips.

Landslides are deep-seated mass movement, and soil slip is a shallow and rapid sliding or flowing movement of the soil. There are different forms of landslides, including mudflows, mudslides, debris flows, rock falls, and rockslides. Slides move in contact with the underlying surface. Flows are plastic or liquid movements in which landmass breaks up in water and flows during movement. Landslides are caused by unstable geological conditions, steep slopes, intense rainfall, weak soils, earthquakes, and human-induced changes of land forms. Human-induced causes are excavation, loading, deforestation, irrigation, mining, vibrations, and water impoundment. The volume of soil detached by landslides depends on the geology and the mechanism of landslide. For example, landslides initiated by rainfall are smaller, while those initiated by earthquakes are very large. There are two categories of landslide: (1) rapidly moving and (2) slow-moving. Rapidly moving landslides (debris flows and earth flows) present the greatest risk to human life.

12.3.3.7 River Bank/Stream Bank Erosion

Stream/riverbank erosion occurs due to bank scour and mass failure. The direct removal of bank materials by the physical action of flowing water is called bank scour. It is often dominant in smaller streams and the upper reaches of larger streams and rivers. Mass failure occurs when large chunks of bank material become unstable and topple into the stream or river (Fig. 12.11). Riverbank erosion can be accelerated by lowering streambed, inundation of bank soils followed by rapid drops in water flow, saturation of banks from off-stream sources, removal of protective vegetation from stream banks, poor drainage, readily erodible material within the bank profile, wave action generated by boats, excessive sand and gravel extraction, and intense rainfall.

12.3.3.8 Effects of Water Erosion

As far as damage is concerned, erosion may be either harmless (benignant) or harmful (malignant). In harmless erosion, the rate of soil removal is less than the rate of soil formation. The estimated average tolerance (T) level of soil erosion used in soil and water conservation planning in the USA is 11 Mg ha^{-1} $year^{-1}$. The T value is the amount of soil erosion that does not significantly decrease soil productivity. The specific rates of maximum tolerable limits of erosion vary with soil type (Blanco and Lal 2008). On a global scale, it is

Fig. 12.10 Landslide on a hillslope (Photo courtesy of Dr. Animesh Biswas)

Fig. 12.11 Riverbank erosion (Photo courtesy of Mr. Muhibbullah)

estimated that about 1,960 M ha of land are prone to erosion, which represents about 15% of the Earth's total land area, of which 50% is severely eroded, and much of that is being abandoned (Lal et al. 2004). On-site effects of water erosion include loss of soil particles, loss of organic matter and nutrients (10-mm topsoil loss equals to 350 kg ha^{-1} N, 90 kg ha^{-1} P, and 1,000 kg ha^{-1} K; Hicks and Anthony 2001), surface sealing and soil compaction, exposure of roots, terrain deformation, difficulty in tillage operations, exposure of subsoil, reduction in growth and yield of crops, loss of growing crops, decline in soil quality, and reduced capability of ecosystem functions. Off-site effects of water erosion include burrowing of crops and households, siltation on lands, sedimentation of water reservoirs, floods, eutrophication of water bodies, pollution of land and water, damage of roads and railways and other installations, and lowering of water quality.

12.3.3.9 Universal Soil Loss Equation (USLE)

Wischmeier and Smith (1965) proposed an equation for estimating the amount of soil loss due to water erosion, particularly sheet and rill erosion on the basis of farm-level experimental data on various soils at many different locations in the United States using the same standard conditions. Erosion plots were 22.6 m long on 9% slopes and were subjected to the same soil management practices. This equation has been applied satisfactorily in many other areas than the USA and, through revisions of the factors included in the equation (Wischmeier and Smith 1978), has become universally accepted. It is known as the Universal Soil Loss Equation (USLE) and is written below:

$$A = R.K.L.S.C.P$$

Here, A = soil loss expressed in tons/acre (×2.24 Mg ha^{-1} $year^{-1}$), R = rainfall and runoff erosivity factor, K = soil erodibility factor, L = slope length factor, S = slope gradient factor, C = vegetation cover factor, which represents the ratio of soil loss under a given crop to that from bare soil, and P = soil conservation practice factor, which represents the ratio of soil loss where contouring and strip cropping are practiced to that where they are not. The rainfall erosivity factor (R) is calculated from the kinetic energy of a 30-min period of rain. Soil erodibility factor (K) expresses the quantity of eroded soil resulting from a constant amount of rain energy on a 9% slope gradient of 22.6 m slope length. The K factor varies for different soils ranging from negligible values (0.03) for the most resistant soil types up to 0.69 for soil types most susceptible to erosion (Zachar 1982).

12.3.3.10 Revised Universal Soil Loss Equation (RUSLE)

Several process-based erosion models have been incorporated in the revised soil loss equation. Now, R factor includes rainfall and runoff erosivity factor (runoff erosivity also includes snowmelt where runoff is significant). The C factor is based on computation of sub-factor called soil loss ratios (SLR). The SLR depends on sub-factors, namely, prior land use, canopy cover, surface cover, surface roughness, and soil moisture (Renard et al. 1997).

12.3.3.11 Factors Affecting Water Erosion

Rainfall

Erosivity is the capacity of rain to cause erosion. It is the capacity of raindrops to detach and of runoff to detach and transport soil particles during erosion. Characteristics of rainfall affecting erosivity are the following: amount, intensity, terminal velocity, drop size, and drop size distribution. The same amount of rain may have different effects on the amount of erosion depending on the intensity and soil surface conditions. A period of 30 min of maximal intensity rain is taken as the basis for the assessment of the erosivity of rain (Wischmeier and Smith 1978). According to Wischmeier and Smith 1978, the product of the energy and the intensity of a downpour expressed in terms of the index EI_{30} varies from 100 to 10,000 Jm^{-2} for different downpour conditions. By dividing these values by 100, a range of values for the erosivity of downpours ranging from 1 to 100 is obtained. By summing the EI values for all downpours occurring during a given period, the erosivity of the rain in that period is obtained (Zachar 1982). Wind velocity increases the erosivity of rain. Measurement of kinetic energy, E of raindrops, is difficult under natural rain. Electronic sensors based on optical and laser devices have been used for direct measurements (Lovell et al. 2002). Intensity is the most important rainfall property that determines the amount of erosion. A short-duration high intensity rainfall is more efficient to cause erosion than a long-duration low intensity rainfall. However, when high amount is combined with high intensity, rain produces high erosion.

Applied water in rain, snowmelt, and irrigation is absorbed by the soil; fills up the soil pores and surface soil depressions; is stored in surface detention ponds if in place; and accumulates on the soil surface at a given depth, and the excess runs off. Runoff is also known as overland flow or surface flow.

Runoff = INPUT − OUTPUT = (rain, snowmelt, irrigation) − (infiltration, evaporation, rain interception by canopy, water absorption, transpiration, surface retention) (Blanco and Lal 2008). Similar to the rainfall erosivity, runoff erosivity is the ability of runoff to cause soil erosion.

Soil

Erodibility is the susceptibility of soil to detachment by rain and transport by runoff water. Soil factors affecting erodibility by water include (1) those factors that affect infiltration rate, permeability, and total water retention capacity and (2) those factors that allow soil to resist dispersion, splashing, abrasion, and the transporting forces of rainfall and runoff. Erodibility depends on texture, structure, organic matter content, and infiltration capacity. The most erodible soils are characterized by low clay and organic matter contents and poor structural stability. Soils with an intermediate texture (fine sand to coarse loam) are more erodible than both coarse-textured sandy soils, where particle size and mass are large, and fine-textured clayey soils, which are more cohesive. Sandy soils absorb water more rapidly than clayey soils. Infiltration capacity is positively correlated with amount of coarse soil particles and negatively with that of fine particles (Wuest et al. 2006). Under low intensity rains, sandy soils produce less runoff than clayey soils. Clay soils absorb less water, and most of the rain water in such soils runs off. So, if clay soils are not well structured, they are sensitive to erosion. Poorly structured soils are more detachable, unstable, and susceptible to erosion.

There are several soil erodibility indices. Readers are referred to Song et al. (2005) for methods of calculating the soil erodibility factor.

Slope Length and Gradient

The distance from the point of origin of overland water flow to a point where the slope decreases to the extent that deposition occurs is known as the length of slope. Runoff from the upper part of a slope adds to the total amount of runoff that occurs on the lower part of the slope. Therefore, more erosion occurs at the lower part of the slope. Water erosion increases as the 0.5 power of the slope length. This is used to calculate the slope length factor, L. About 1.3 times greater average soil loss occurs per acre for every doubling of slope length. On the other hand, velocity of runoff increases as the degree of slope (slope gradient S) increases. A doubling of runoff water velocity increases erosive power 4 times and causes a 32-time increase in the amount of material of a given particle size that can be carried in the runoff water (Blanco and Lal 2008).

Vegetation

Vegetation intercepts rain and impedes raindrop impact, reduces concentration of water in soil surface, increases infiltration capacity, retains some water, and absorbs water from the soil, making soil more water absorptive. A vegetative cover, therefore, reduces soil erosion. However, vegetation types widely differ in their efficiency to counteract soil erosion. Characteristics of vegetation affecting soil erosion include canopy characteristics such as leaf size, leaf area, leaf orientation, and canopy cover; root characteristics such as soil binding capacity; and plant density. Some cropping systems effectively reduce erosion, but some others may enhance it. For example, erosion on plots 22.6 m long with an 8% slope had about 1,000 times more erosion when in continuous corn as compared with continuous bluegrass (Foth 1990). Bluegrass offers a continuous plant cover, while continuous corn cultivation leaves the land bare for a long time of the year. Crop rotations including different cover types, rooting depths, and tillage requirements protect soil better than unplanned cropping patterns.

12.3.4 There Are Vegetative and Engineering Methods of Erosion Control

The principles of water erosion control are the following: to reduce the raindrop impact, to improve the stability of soil aggregates, to increase infiltration and decrease amount of runoff, to reduce the velocity of runoff and concentration of water in channels, to carry surface runoff water safely out of the field, to maintain a healthy, dense vegetative cover over soil for as long as possible, and to improve soil organic matter.

Approaches of control of soil erosion by water include the following: control of grazing and burning, minimum disturbance of the soil during rains, minimum tillage required for sowing and transplanting, selection of right kinds of crops and adoption of crop rotation, avoiding monocultures, application of manures/composts, use of mulch as and when possible, and use of earthwork only under emergency.

12.3.4.1 Conservation Tillage

Tillage aims at loosening soil, incorporating residues, preparing a seedbed, controlling weeds, and mixing lime, fertilizer, and other chemicals. Cropping without tillage or limited tillage has not become as popular as with tillage. Often, tillage results in negative impacts, including exposure of the soil surface to wind and water erosion and loss of soil organic matter through oxidation. To avoid these negative effects of tillage, conservation tillage practices, including no-tillage, minimum tillage, and strip tillage have been developed. These management strategies have proved effective for controlling soil erosion and improving soil quality. Several studies have demonstrated that conservation tillage can reduce soil erosion and associated P loss (McDowell and Sharpley 2001).

Growing crops with minimal disturbance to the soil surface is called conservation tillage. Conservation tillage techniques include minimum tillage, mulch tillage, and no-tillage. In no-tillage, crops are planted directly in the residues of the previous crop with no prior tillage. For row crops, a slit is made in the soil in which the seed is sown. Minimum tillage involves the minimum manipulation of soil. It is actually a localized tillage. Minimum and no-tillage leave more residues on the soil surface than conventional tillage, resulting in reduced runoff and soil erosion for which they are called conservation tillage. In strip-tillage system, a strip 30–45 cm wide is tilled in the row between undisturbed spaces. Strip-till is less effective than no-till and subsoil systems, because bare soil exposed in the tilled strip is susceptible to erosion. It can be made effective by covering the exposed part with organic residues. Conservation tillage systems have some disadvantages too. Yield of crop tends to be lower, and weed infestation is a major problem.

12.3.4.2 Mulching

The use of mulch in soil moisture conservation has been discussed in Chap. 11. Mulching has multiple advantages. It reduces the impact of solar radiation and raindrops. It protects soil aggregates from detachment by raindrops. It reduces evaporation and loss of soil moisture; it increases infiltration and reduces amount and velocity of runoff. It decreases surface sealing, crusting, and compaction. Stubble mulching is frequently recommended for reducing runoff and erosion. Residues of wheat or stalk of corn of the previous crop are retained during tilling the land for next crop without turning them under. Stubble mulch is a very efficient protector of wind erosion.

12.3.4.3 Contour Cropping

Contour farming is the plowing and planting crop in the contour (Fig. 12.12), that is, across the slope. Contours are arbitrary lines drawn roughly perpendicular to the direction of slope. So, contour farming is a cross-slope farming system. Contours reduce velocity of runoff, give accumulated water more time to infiltrate, and deposit detached soil particles. Contour cropping contrasts with up- and downslope farming. Furrows in an up- and downslope direction become channels of concentrated runoff, forming rills or even gullies. In contour farming, ridges and furrows are formed by tillage, planting, and other farming operations to change the direction of runoff from directly downslope to around the hillslope. Contour farming is most effective on slopes between 2 and 10%. Contour farming is not well suited to rolling topography having a high degree of slope irregularity.

Several factors influence the effectiveness of contour farming to reduce soil erosion. They are rainfall intensity, slope steepness, soil properties, ridge height, cover and roughness, and the critical slope length. Cover, roughness, and ridge height can be influenced by management. This practice will be less effective in achieving the stated purpose(s) on slopes exceeding 10% and in areas with 10-year, 24-h rainfall of

Fig. 12.12 Contour cropping (Photo courtesy of USDA NRCS)

Fig. 12.13 Strip cropping (Photo courtesy of USDA NRCS)

about 6.5 in. Spacing of contour lines is chosen on the basis of slope, soil, rainfall, and crop type. Annual and perennial crops are planted in the ridges or furrows of the contours. Contour farming can be combined with strip cropping.

12.3.4.4 Strip Cropping

The process of growing crops in alternate strips is known as strip cropping (Fig. 12.13). Crops of the strips vary in their root/shoot characteristics and cultural requirements. Crop strips break sloping landscapes in wide segments with diverse vegetative cover which intercepts runoff and promotes water infiltration, thereby reducing runoff and soil erosion. In strip cropping, sod forming crops may be alternated with cereals, legumes with nonlegumes, and root crops with vegetables. Crop yields between strip cropping and monocultures are not significantly different. Carefully planned strip cropping systems may retard soil erosion effectively. Cropping in strips perpendicular to the slope reduces soil erosion as compared to bare soil or up- and downslope cropping. Thus, strip cropping may be successfully combined with contour farming.

The width of the strips depends on soil slope, erosion potential, crop type, and equipment size. Narrow strips reduce flow lengths more effectively than wide strips. The width of strips must match the equipment turn or width for cultivation. On gentle slopes of up to 5%, a strip width of about 30 m is recommended, while on steeper slopes, the width must be less than 20 m (Bravo and Silenzi 2002).

12.3.4.5 Contour Strip Cropping

Contour strip cropping is planting row crops in strips on the contour. It is more efficient in erosion control than contour farming and strip cropping alone because of the plant and crop diversity. The grass, legumes, or small grains used in strips slow runoff and trap sediment, leaving row crops.

Fig. 12.14 Outline of a SALT farm (Source: http://blog.agriculture.ph/tag/upland-farming-philippines)

Permanent grass/legume strips must be maintained between strips in soils with severe erosion. These strips can be used as traffic lanes for cultural operations (Blanco and Lal 2008).

12.3.4.6 Sloping Agricultural Land Technology (SALT)

SALT is a technology package of soil conservation and food production that integrates several soil conservation measures (Tacio 1993). It involves planting field crops and perennial crops in bands 3–5 m wide between double rows of nitrogen-fixing shrubs and trees planted along the contour. Thus, it is a modification of the contour strip cropping, but it may be practiced in land of slope more than 10%. Field crops include legumes, cereals, and vegetables, while the main perennial crops are cacao, coffee, banana, citrus, and fruit trees (MBRLC 1988). The nitrogen-fixing trees are thickly planted in double rows to make hedgerows (Fig. 12.14). When a hedge is 1.5–2 m tall, it is cut down to about 75 cm, and the cuttings are placed in alleyways to serve as organic fertilizers. SALT establishes a diversified stable ecosystem. SALT was developed on a marginal site in the Philippines by the Mindanao Baptist Rural Life Center (MBRLC) in 1971. There are ten steps in establishing a SALT farm.

Step 1 Making the A-frame: A-frame is a wooden frame of two legs in the shape of English alphabet "A" having a water or sprit level on the middle arm.

Step 2 Drawing contour lines: Contour lines are drawn by joining the points of equal elevations selected by the A-frame in a slope.

Step 3 Cultivating the contour lines: 1-m strips along the contour lines should be plowed and harrowed to prepare for planting.

Step 4 Planting seeds of different nitrogen-fixing trees and shrubs: Along each prepared contour line, two furrows should be laid out. Leguminous tree seeds (or shrubs) are sown in double rows 12 cm apart. Tree or shrub seedlings will form dense hedgerows which will intercept runoff and trap soil materials. Suitable hedgerow species are *Flemingia macrophylla* (syn. *congesta*), *Desmodium rensonii*, *Calliandra calothyrsus*, *Gliricidia sepium*, *Leucaena diversifolia*, *L. leucocephala*, etc.

Step 5 Growing crops between hedgerows: Crops are grown in the space between the hedgerows. It is called a strip or alley. Cultivation is done on alternate strips.

Step 6 Planting permanent crops: Permanent crops such as coffee, cacao, banana, and citrus of the same height may be planted in cleared spots of hedgerows. Weeding is employed until the hedgerows are large enough to hold the soil in place. Permanent crops are planted in every third strip. Tall crops should be planted at the bottom of the farm, while the short ones are planted at the top.

Step 7 Planting short-term crops: Short-and medium-term cash crops (pineapple, ginger, sweet potato, peanut, mung bean, melon, sorghum, corn, upland rice, etc.) should be planted between the strips of permanent crops.

Step 8 Trimming of nitrogen-fixing trees: Every 30–45 days, the growing hedgerows are cut to a height of 1.0–1.5 m from the ground. The prunings are piled on the soil around the crops as organic fertilizer.

Step 9 Practicing crop rotation: A good way of rotating is to plant cereals such as corn or upland rice, tubers, and other crops on strips where legumes were planted previously.

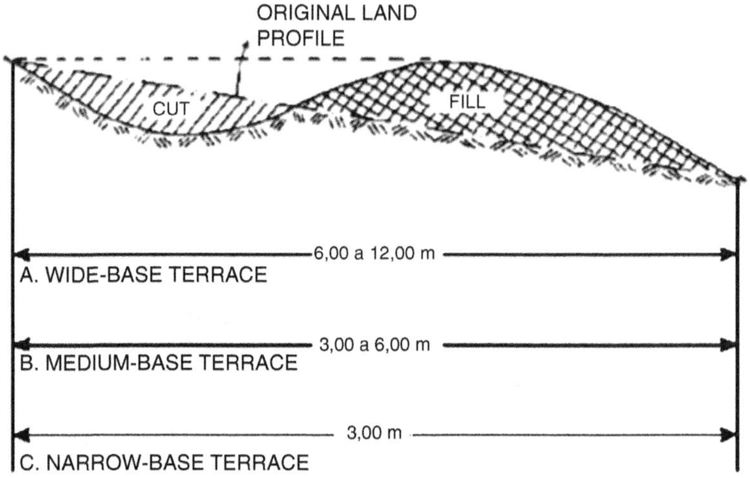

Fig. 12.15 Common terrace forms

Step 10 Building green terraces: To enrich the soil and effectively control erosion, organic materials such as straw, stalks, twigs, branches, and leaves, and also rocks and stones, are piled at the base of the rows of nitrogen-fixing trees. By the passage of time, permanent effective terraces will be formed.

SALT has been applied in many other countries with variable success. It may be an alternative to shifting cultivation in degraded hilly lands.

12.3.4.7 Terracing

Terracing is defined as building a mechanical structure such as an earthen ridge or a stone wall in direction usually perpendicular to the slope of a land. It reduces both length and degree of slope, that is, decreases slope steepness and divides the slope into short, gently sloping sections (Morgan 1986). Terraces are constructed to intercept surface runoff, encourage it to infiltrate, evaporate, or be diverted toward a predetermined and protected safe outlet at a controlled velocity to avoid soil erosion (FAO 2000).

There are various types of terraces depending on their main function, construction process, size of base, and shape. On function, terraces may be retention terrace or level terrace and diversion terrace. Terraces may be channel or Nichols terrace constructed by excavating soil from the upper side only to form a channel and depositing it downhill to form a bank, or ridge or Mangum terrace, a long low ridge of earth with gently sloping sides and a shallow channel along the upper side. Mangum terraces are constructed by excavating the soil from both sides of the embankment. Nichols terraces may be constructed on slopes >10%, while Mangum terraces are recommended for slopes less than 10%. On size of terrace base, they may be narrow base, medium base, and wide base (Fig. 12.15). On shape, terrace types are normal terraces (generally wide/broad base) and bench terraces (Fig. 12.16). Baryła and Pierzgalski (2008) discussed utility of different types of ridged terraces.

According to ICIMOD (1998), there are some limitations of terracing. The disturbance of the soil strata and considerable decline in soil fertility in the first several years, considerable soil loss during construction, need of labor, and investment for construction and maintenance are among some of the limitations. Terraces are not always stable and not suitable for coarse-textured soils and on very steep slopes. However, very steep slopes are best retained in their original state without any sort of interference.

12.3.5 Wind Erosion Is Caused by Blowing Wind

Wind erosion is a regular event in arid and semiarid regions where rainfall is low and irregular and temperature is high. Soil particles are detached by wind velocity and by abrasion of suspended particles in air. In the United States, wind erosion damages from 0.4 to 6 million ha of land annually, and about 2 million ha are moderately to severely damaged each year (Gregory and Borrelli 1986). Wind erosion not only removes soil but also damages crops, buildings, fences, and highways.

There are three types of soil particle movement during wind erosion. These are saltation, suspension, and surface creep. Between 50 and 75% of the soil is carried by saltation, 3–40% by suspension, and 5–25% by surface creep. These three processes of wind erosion occur simultaneously. Saltation causes other particles to move in suspension and surface creep. Neither creep nor suspension can occur without saltation.

12.3.5.1 Saltation

In saltation, fine soil particles (0.1–0.5 mm in diameter) are rolled over the soil surface by direct wind pressure to some distance and then abruptly jump up vertically to a height of 20–30 cm. Lifted particles gain in velocity and then descend in an almost straight line at an angle 5–12° from the horizontal.

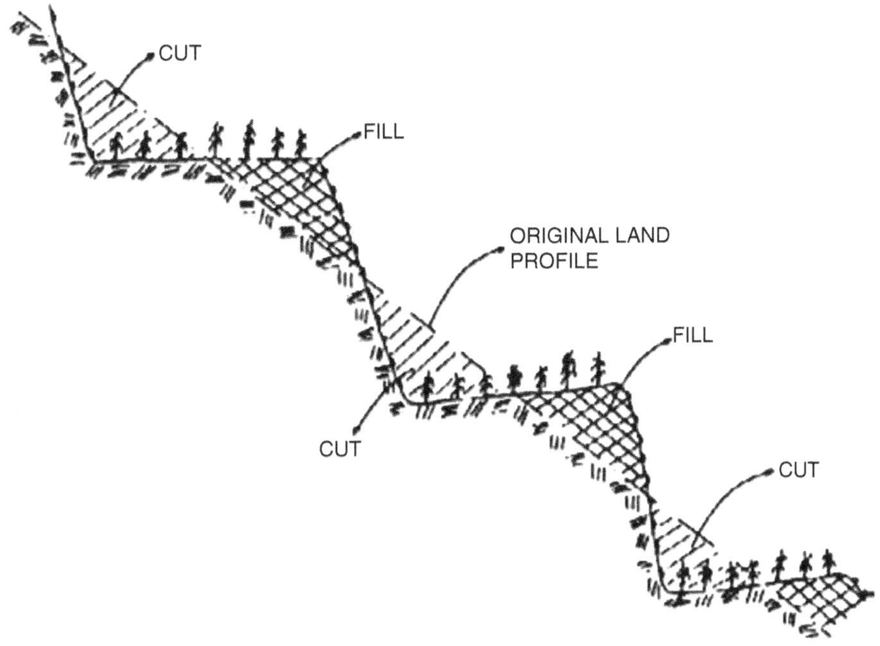

Fig. 12.16 A sketch of a typical bench terrace

The horizontal distance traveled by a particle is four to five times the height of its jump. On striking the surface, the particles may rebound into the air or knock other particles into the air before coming to rest.

12.3.5.2 Suspension

Very small soil particles (<0.1 mm) are lifted into the air by the impact of rolling soil particles of saltation. These particles are carried away by wind to distances depending on its velocity and in its direction. They may be carried very high and over long distances via suspension.

12.3.5.3 Surface Creep

Relatively large sand particles (between 0.5 and 1.0 mm in diameter) are too heavy to be lifted by wind action, but they can be rolled or pushed along the surface by the impact of particles in saltation. The movement is called surface creep.

12.3.5.4 Factors of Wind Erosion

The factors of wind erosion can be shown in the following equation:

$$E = f(I, C, K, L, V)$$

where, E = amount of soil loss Mg ha^{-1} $year^{-1}$, I = soil erodibility index, K = surface roughness factor, C = climate factor (wind velocity, wind angle, etc.), L = length of field, and V = vegetative cover. So, soil, climate, and vegetation are the main factors of soil erosion by wind. Erodibility of soil by wind is related to soil texture, structure, organic matter, and surface roughness. Aggregation into large stable peds occurs at higher contents of clay and organic matter. Soil particles smaller than 0.05 and larger than 1 mm diameter are less susceptible to wind erosion. Therefore, fine sandy soils suffer more from wind erosion. Rough soil surfaces reduce wind erosion by reducing the surface wind velocity and by trapping soil particles. Tillage operations may increase surface roughness. Wind frequency and velocity and rains during high-wind periods are important climatic elements of wind erosion. Climate affects the amount of vegetative cover. Wind erosion is inversely related to the degree of vegetative cover.

12.3.5.5 Wind Erosion Control

The principles of wind erosion control are as follows: to improve soil structure, to keep the soil moist, to reduce the velocity of wind at the soil surface, and to trap soil particles. The following steps can be taken for control of wind erosion. Soil particles can be built up into stable peds too large in size to be moved by saltation. This can be achieved by application of large amounts of organic residues. The wind velocity near the soil surface may be reduced, and moving soil particles can be trapped by ridging the land, by vegetative cover such as current crop, stubbles, and mulches, or by tilling the land into clods. Windbreak vegetation reduces the velocity of wind and traps the soil particles.

Windbreaks are typically linear plantings of single or multiple rows of trees or shrubs in boundaries of crop fields and

Fig. 12.17 Windbreak (Photo courtesy of USDA NRCS)

farm holdings (Fig. 12.17). They are arranged usually perpendicular to the direction of the prevailing wind. They reduce the velocity of the wind at the ground level to a distance approximately ten times of the height of plants. Windbreaks modify the climate in the areas they shelter through (1) moderated soil and air temperatures, (2) increased relative humidity, (3) reduced evaporation and increased soil moisture, and (4) modified distribution of snow. The ultimate benefits are (1) reduction of soil erosion by wind (and water), (2) reduction in crop loss due to storms, (3) increase in yield, and (4) protection of property and livestock. Winter wheat, barley, rye, alfalfa, vegetable, and fruit yields were found to increase when fields are sheltered. Sudmeyer et al. (2007) suggested that windbreaks of trees, or tree and shrub combinations, can offer many benefits on wheat-belt farms, in addition to protection of soil, stock, crops, and pastures. Additional benefits include reduced evaporation from farm dams, reduced groundwater recharge, their use as nature conservation corridors and habitats, increased biodiversity including predators of crop and pasture pests, fire control, tree products, and improved lifestyle and aesthetics. However, these effects are determined by height, length, density, location, and species of trees or shrubs in the windbreak. A list of shrubs and trees suitable for windbreaks is given below. These species can also be used in agroforestry (Chap. 14) schemes.

Shrubs: American cranberry bush (*Viburnum trilobum*), dogwoods (*Cornus sericea*) (*Cornus racemosa*) (*Cornus alternifolia*) (*Cornus amomum*), silky elderberry (*Sambucus canadensis*), American hazelnut (*Corylus americana*), willow sandbar (*Salix interior*), silver buffaloberry (*Shepherdia argentea*), etc.

Small Trees: American plum (*Prunus americana*), black chokeberry (*Aronia melanocarpa*), chokecherry (*Prunus virginiana*), crab apples (*Malus* sp.), Hawthorn, Arnold (*Crataegus arnoldiana*), nanking cherry (*Prunus tomentosa*), pin cherry (*Prunus pensylvanica*), etc.

Deciduous Trees: River birch (*Betula papyrifera*), autumn splendor (*Catalpa speciosa*), cottonwood (*Populus deltoides*), American elm (*Ulmus americana*), silver maple (*Acer saccharinum*), red oak (*Quercus rubra*), white oak (*Quercus alba*), black walnut (*Juglans nigra*), etc.

Conifers: Eastern red cedar (*Juniperus virginiana*), Eastern white pine (*Pinus strobus*), ponderosa pine (*Pinus ponderosa*), red pine (*Pinus resinosa*), white cedar (*Thuja occidentalis*), white spruce (*Picea glauca*), etc.

12.4 Chemical Soil Degradation Results in Undesirable Changes

Chemical degradation of soil refers to the unwanted changes in soil chemical properties such as acidity, alkalinity, nutrient availability, salinity, and sodicity, leading to loss of soil fertility and productivity. The most widespread types of chemical degradation in soils are acidification, salinization, and nutrient depletion (Fig. 12.18).

Acidification can occur naturally in soils developed from acid parent materials, in high-rainfall areas, at higher elevations, and under intensive weathering over a long period of time. But agricultural practices may accelerate or create soil acidity within a short time. Acidification in agricultural soils may be due to application of nitrogenous fertilizers, removal of produce, and buildup of soil organic matter (Upjohn et al. 2005). Removal of bases by erosion and leaching may also contribute to soil acidification.

12.4.1 Anthropogenic Activities May Lead to Soil Acidification

Surface soils are generally acidified first; continued acidification penetrates the subsoil later. Nitrate leaching from topsoil is a

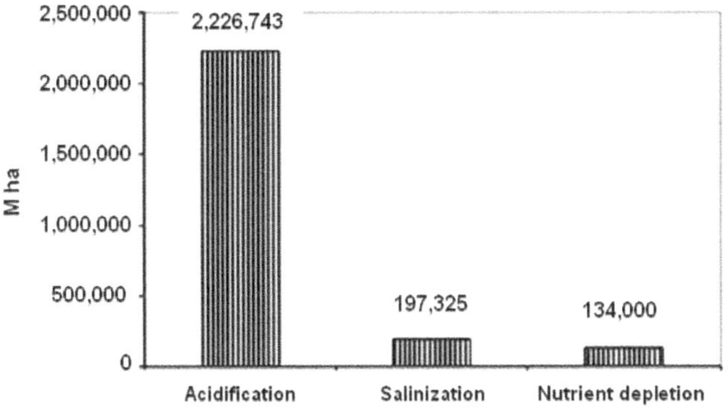

Fig. 12.18 Total land area affected by different types of chemical degradation (Based on data from Derici 2006)

Table 12.3 Acidification potential of different fertilizers

Fertilizer	Acidification potential
Calcium nitrate, potassium nitrate	Low
Ammonium nitrate, urea, animal manure	Medium
Ammonium sulfate, MAP, DAP	High

major cause of topsoil acidification, whereas subsoil acidification appears to result mainly from excess cation over anion uptake by roots (Tang and Rengel 2003). The following problems are associated with soil acidification: (1) Activity of soil microorganisms are reduced; (2) nutrient release from organic matter is reduced, (3) phosphorus in the soil may become less available to plants; (4) Calcium, magnesium, and molybdenum may be deficient, (5) Fe, Al, and Mn may become toxic to plants, (6) uptake of heavy metals such as cadmium by crops and pastures may increase, and (7) growth and yield of crops may be reduced.

Soils may be acidified by fertilizers, particularly nitrogen, when used for long time in excess of crop needs. Acidity is also caused by differential cation–anion uptake by crops. When base-demanding crops are repeatedly grown in a field, base export may result in soil acidification. Legumes usually have a greater capacity to excrete protons and less ability to take up nitrate during growth compared with cereals. They also have higher excess cations or ash alkalinity in their tissues, and thus, higher amounts of alkalinity can be removed in their products (Tang and Rengel 2003). Acidification potential of some fertilizers is shown in Table 12.3.

Watmough and Dillon (2003) reported that depletion of nutrient-base cations (e.g., Ca, Mg, K, and others) in soils occurs due to acid deposition. Acidification and depletion of nutrient cations due to acid deposition have been reported for several forested regions in North America.

Continuing regional inputs of nitrogen and sulfur are of concern because of leaching of base cations, increased availability of soil Al, and the accumulation and ultimate transmission of acidity from forest soils to streams (Fenn et al. 2006). Acid deposition is known as acid rain. It occurs when emissions of acidic materials such as sulfur dioxide (SO_2), ammonia (NH_3), nitrogen oxides (NOx), and acidic particles from the combustion of fossil fuels and other industrial processes undergo complex chemical reactions in the atmosphere and fall to the Earth as wet deposition (rain, snow, cloud, fog) or dry deposition (dry particles, gas). Dry deposition can be broadly defined as the transport of particulate and gaseous contaminants from the atmosphere onto surfaces in the absence of precipitation (Davidson and Wu 1990). The effect of acid rain on the soil depends greatly on the ability of soils to neutralize the acid. Calcareous soils are most effective in neutralization of acidity in acid rains and prevent lowering of soil pH. Anthropogenic soil acidification due to acid rain is connected to high soil solution concentrations of SO_4^{2-} and NO_3^- (Matzner and Davis 1996).

Principles of acidified soil management include use of less acidifying fertilizers, improving nutrient use efficiency, growing acid-tolerant crops and reducing nutrient-leaching losses (Gregan et al. 1989; Helyar 1991). Management of acid soils has been discussed in considerable details in Sect. 11.6. However, the following steps may be taken for the management of acidified soils: Selection of crops suitable for the current soil pH, use of lime to raise pH to the desired level, regular use of maintenance dose of lime, use of optimum nitrogen fertilizer, and maintenance of irrigation and drainage.

12.4.2 Soil Salinization May Result from Soil Mismanagement

A soil is said to be saline if the electrical conductivity of its saturation extract (EC_e) (saturating the soil with water and taking the extract by suction) is greater than 4 dS m^{-1} (decisiemens per meter). The electrical conductivity is proportional to the concentration of salts in solution. Thus, saline soils contain excess soluble salts, generally chlorides and sulfates, with some carbonates and bicarbonates, of sodium, potassium, calcium, and magnesium. Soil salinity is harmful for plants except the halophytes; it causes water stress through osmotic disturbances in plant tissue and by toxicity of some salt constituents. Seed germination, plant growth, and yield of crops are considerably reduced by soil salinity. Sometimes it causes crop failure. Some soils are naturally saline. They are formed by process called primary salinization or natural salinization. Some soils are made saline by mismanagement of soil and crop, particularly improper irrigation and drainage, that is, changing the hydrologic balance. This is known as secondary salinization or human-induced salinization. Salinization, both natural and human-induced, may occur in two climatic settings—arid, semiarid, and humid regions. In arid and semiarid regions, inadequate availability of water or application of less water than the required amount is responsible for salinization. In these areas, potential evapotranspiration

Table 12.4 Global extent of human-induced salinization

Continent	Light 10^6 ha	Moderate 10^6 ha	Strong 10^6 ha	Extreme 10^6 ha	Total 10^6 ha
Africa	4.7	7.7	2.4	–	14.8
Asia	26.8	8.5	17.0	0.4	52.7
South America	1.8	0.3	–	–	2.1
North and Central America	0.3	1.5	0.5	–	2.3
Europe	1.0	2.3	0.5	–	3.8
Australia	–	0.5	–	0.4	0.9
Total	34.6	20.8	20.4	0.8	76.6

Oldeman et al. (1991) with permission

exceeds the amount of irrigation water so that the salts released by weathering cannot leach down. Moreover, there is a net capillary rise of water which brings salts to the surface soil. In humid areas, on the other hand, excess irrigation or poor drainage causes the groundwater table to rise to the root zone of plants and makes the soil saline.

Oldeman et al. (1991) estimated that worldwide 76.6 M ha of lands are affected by human-induced salinization, but they did not differentiate salinity in the irrigated and nonirrigated rainfed areas (Table 12.4). A survey made by Ghassemi et al. (1995) revealed that vast areas of India, China, Pakistan, and Central Asian countries are affected by salinity due to over-irrigation and lack of drainage facilities. Some soils have become saline due to use of saline water for irrigation (Paul et al. 2008).

Two important approaches to soil salinity management are (1) selection of salt-tolerant crops and (2) removal of excess salts from the root zone. Several salt-tolerant crops are available. Removal of salts by irrigation and drainage and drawing the salty water safely may be financially impracticable in some situation. Whether the reclamation will be cost-effective or not depends on salinity level, cost of water and labor, crop return, etc. Reclamation is relatively easy if the soil is saline alone, but difficult if it is saline–sodic or sodic. Reclamation processes for saline soils are discussed in connection with management of problem soils in Sect. 11.8.

12.4.3 Soils Are Often Polluted by Human Activities

Pollution is defined as the introduction of any substance to the environment that adversely affects the usefulness of a resource. The substance that is introduced and that adversely affects the environment is the pollutant. According to van der Perk (2006), pollution and contamination are used synonymously to mean the introduction into the environment by humans of substances that are harmful or poisonous to people and ecosystems. According to many other authors, however, meanings of soil pollution and soil contamination are different. To them, soil contamination is the introduction to the soil of a chemical substance which was not originally present (Haan and van Riemsdijk 1986), and soil pollution is the presence of a chemical substance in soil in concentrations higher than natural as a result of human activity and that has a detrimental effect on the environment and its components (Kabata-Pendias 2001). Soil pollution is a global problem that threatens the life of microorganisms, plants, animals, and humans. Soil pollution arises mainly from disposal of solid and liquid wastes in the ground; disposal of chemical wastes from the industries into the soil; indiscriminate use of agricultural chemicals, including fertilizers and pesticides; application of sewage sludge and compost without judging metal contamination; leakage of underground septic tanks into the soil; use of non-decomposable materials; mining activity; and irrigation with polluted water.

12.4.3.1 Sources of Soil Pollutants

Wastes (industrial, municipal, agricultural, domestic, and nuclear) are the chief sources of soil pollution (Alloway 1995). Materials that have reached the end of their useful life are called wastes (Hill 2010). A waste is not necessarily a pollutant but many wastes are. Excreta of animals are excellent manures. Many wastes can be recycled into useful materials again, but most wastes are considered useless and often disposed of carelessly. Municipal, hospital, and industrial wastes and mining as sources of soil pollutants are discussed below.

Municipal Wastes

Municipal wastes include domestic and kitchen wastes, market wastes, hospital wastes, livestock and poultry wastes, slaughterhouse wastes, metals, and glass and ceramic wastes. Municipal wastes can be grouped into five different categories: biodegradable (food and kitchen waste such as meat trimmings or vegetable peelings, yard or green waste, and paper), recyclable materials (glass, plastic bottles, other plastics, metals, and aluminum cans), inert waste (construction and demolition wastes), composite waste (clothing and plastics), and hazardous wastes (medicines, paint, batteries, light bulbs, fertilizer and pesticide containers, and e-waste like old computers, printers, and cellular phones). Some of these wastes are nonbiodegradable materials such as polyethylene and plastic sheets, bags, and bottles. These materials clog soil pores and hamper natural drainage, tillage, and planting operations. Municipal wastes may contain nonbiodegradable organics and heavy metals. Disposal of municipal wastes in rear urban areas in open dumps or as landfills contaminates agricultural land nearby. Aerial and seepage transport of the volatiles and soluble materials pollute large urban land areas in underdeveloped countries where the waste disposal systems are not environment friendly.

Sewage Sludge

Sewage sludge is usually a liquid mixture, composed both of solids and of dissolved organic and inorganic materials. The water is separated from the solid part by a number of treatments before it is environmentally safe for discharge into streams or lakes. The solid residue left is often used as composts in crop fields because it contains considerable amounts of plant nutrients. The content of major nutrients and micronutrients in sewage sludge depends on the source. Biosolids, the product generated from tertiary treatment of waste-activated sludge, have been in use in the UK, European, and Chinese agriculture for about a century. Substantial increase in crop yield due to sewage sludge application has been observed. But for heavy metal contamination of soil and public outrage, Austria, Switzerland, Sweden, and some other European countries have imposed a ban on the use of sewage sludge as a fertilizer.

Hospital Wastes

A huge quantity of waste is generated in health-care facilities. These wastes include general wastes, sharps, non-sharps, blood, body parts, chemicals, pharmaceuticals, medical devices, and radioactive materials. Potentially infectious waste includes all waste items that are contaminated with or suspected of being contaminated with body fluids. Examples include blood and blood products, used catheters and gloves, cultures and stocks of infectious agents, waste from dialysis and dentistry units and from isolation units, wound dressings, nappies, discarded diagnostic samples, and contaminated materials (swabs, bandages, and gauze) and equipment (disposable medical devices). Wastes containing chemical substances are laboratory chemicals, empty bottles of lab or pharmacy chemicals, disinfectants that have expired or are no longer needed, solvents, diagnostic kits, poisonous and corrosive materials, and cleaning agents and others. Genotoxic waste consists of highly hazardous, mutagenic, teratogenic, or carcinogenic waste containing substances with genotoxic properties. Radioactive hospital wastes include unused liquids from radiotherapy or laboratory research; contaminated glassware, packages, or absorbent paper; and urine and excreta from patients treated or tested with radioactive substances. These wastes are often disposed with municipal wastes in open dumps and on lands.

Industrial Wastes

Industrial solid wastes are the major sources of soil pollution by toxic organic and inorganic chemical compounds and heavy metals. All industrial emissions, for example, smoke from brickfields and dust from power plants, ultimately reach the soil and water bodies as fallout. The major generators of industrial solid wastes are the thermal power plants producing coal ash, the integrated iron and steel mills producing blast furnace slag and steel melting slag, nonferrous industries like aluminum, zinc, and copper producing red mud and tailings, sugar industries generating press mud, pulp and paper industries producing lime and fertilizer, and allied industries producing gypsum.

Agrochemicals

Pollutants from agrochemical sources include fertilizers, manures, and pesticides. Fertilizers and manures introduce some heavy metals such as arsenic, cadmium, manganese, uranium, vanadium, and zinc.

Phosphate fertilizers contain considerable amount of Cd. Poultry and pig manures add some zinc, arsenic, and copper. Pesticides, another group of agrochemicals, include insecticides, herbicides, and fungicides. Some organophosphorus compounds, organochlorines, the carbamates, and the pyrethroids are used as pesticides (Figs. 12.19 and 12.20). They are highly toxic chemicals, and some are persistent. Pesticides leave their residues or their derivatives in soil for a long time. Organochlorides such as DDT, aldrin, chlordane, endrin, dieldrin, heptachlor, mirex, and toxaphene represent an early generation of pesticides that are persistent and highly toxic. They are banned in many countries, and modern pesticides are mostly organophosphates and carbamates. Organophosphates such as malathion, methyl parathion, and diazinon have low persistence.

Atmospheric Deposition

Atmospheric deposition refers to substances that are deposited on land from the air. It includes gas, smoke, ashes, and particulates and deposited to soil in precipitation, also called wet deposition and as dry deposition or dry fall. The oxides of sulfur and nitrogen (SO_x, NO_x), chlorides, fluorides, ammonium, etc., are the principal gas component of aerosols. Aerosols contain dust, smoke, ashes, sea salt, water droplets, and some organic materials. Volcanic eruption and burning of biomass and fossil fuel produce aerosol that contains sulfate compounds which fall as acid rain. The occurrence of aerosols is the highest in the Northern Hemisphere where industrial activity is concentered. Suspended dust particles in air are carried to hundreds of kilometers by the wind and settle on soil. Atmospheric deposition may also contain persistent organic pollutants as PCBs, and PAHs and heavy metals as As, Cr, Hg, Ni, and Pb (Luo et al. 2009).

Mining

Mining is an important source of soil pollution, especially of heavy metals. The disposal of Cd from Zn mines in the water of Jinzu River in Japan and contamination of soils and rice grains which were responsible for the itai-itai disease were mentioned earlier. Gold mining is carried out by individuals rather than companies in Gunung Pongkor, West Java. They use traditional methods for separating the gold from the raw material. The main waste product from this process is mud

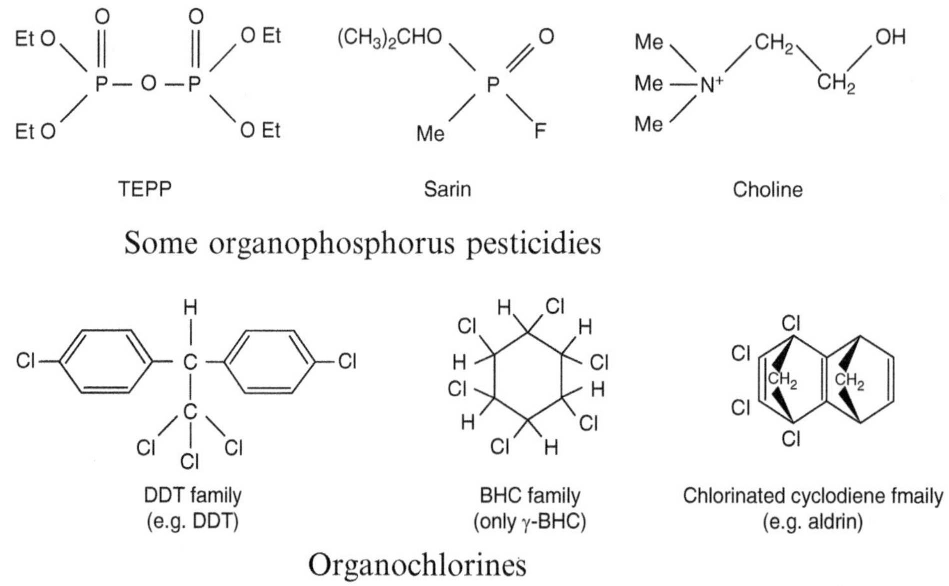

Fig. 12.19 Some organophosphorus and organochlorine pesticides

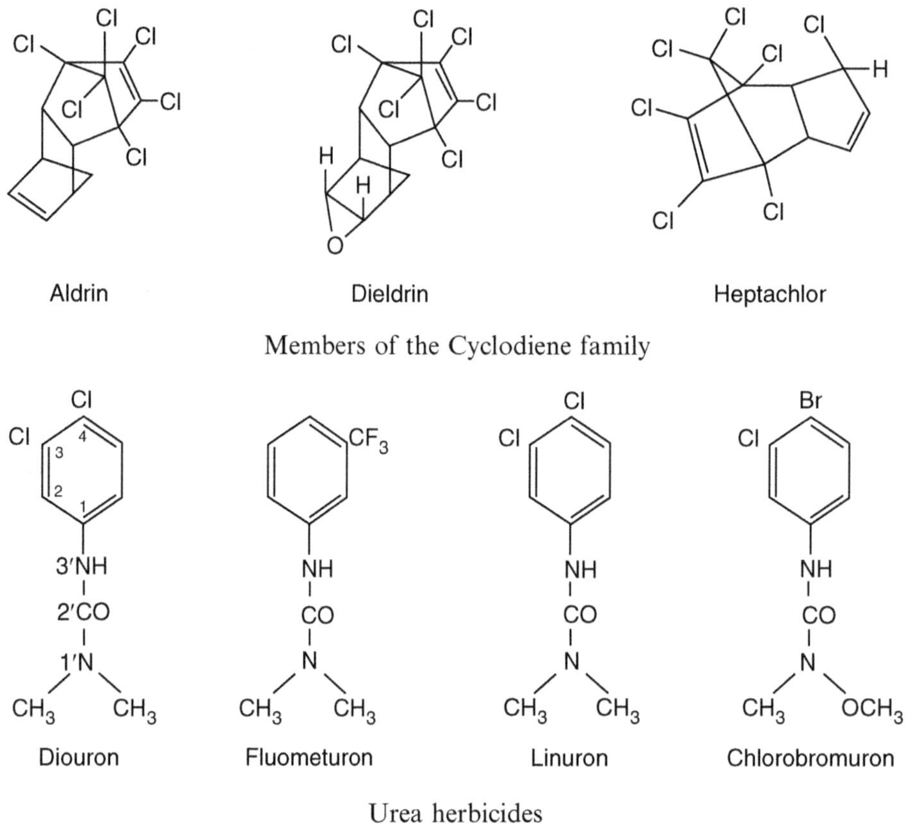

Fig. 12.20 Members of the cyclodiene family and urea herbicides

and rubble which contain a high concentration of mercury. These wastes are disposed of directly in the Cikaniki River, which is also used as a source of irrigation water in the lowland rice fields around the mining areas. A soil survey conducted by Kurnia et al. (2000) in this area found that the soil surrounding the traditional mining was polluted by mercury. The pollution covered the land around six villages. A high concentration of mercury was found in rice straw and rice grain. All of the values were higher than the maximum permitted level of Hg in soils (0.5 mg kg^{-1}).

Table 12.5 Sources of pollutants in soil

Source	Materials	Pollutants
Agricultural	Phosphate fertilizers	As, Cd, Mn, U, V, and Zn
	Pig and poultry manures	As, Cu
	Pesticides	As, Cu, Mn, Pb, Zn, organics
	Oil spillages	Hydrocarbons
Electricity generation	Fallout, ash	Si, SO_x, NO_x, heavy metals, coal dust
Gasworks	Tars	As, Cd, Cu, Pb, $S^{=}$, SO_4, hydrocarbons, phenols, benzenes, xylene, naphthalene, PAHs, etc.
Metalliferous mining and smelting	Dispersed tailings, ore separates	Metals, including heavy metals, cyanides
Metallurgical industries	Metal wastes, solvents, residues, aerosols	Metals, including heavy metals (As, Cd, Cr, Cu, Ni, Pb, Mn, etc.)
Chemical industries	Particulates, effluents, scraps, damaged parts	Heavy metals (As, Cd, Cr, Cu, Ni, Pb, Mn, etc.) and organics, including PAHs
Waste disposal	Sewage sludge, scrap heaps, coal ash, waste incinerators, landfills	NH_4^+, PAGHs, PCBs, NO_3^-, metals (As, Cd, Cr, Cu, Ni, Pb, Mn, etc.)
Transport	Particulates, acid deposits, solvents, oils	Pb, Br, Cd, Zn, P, As

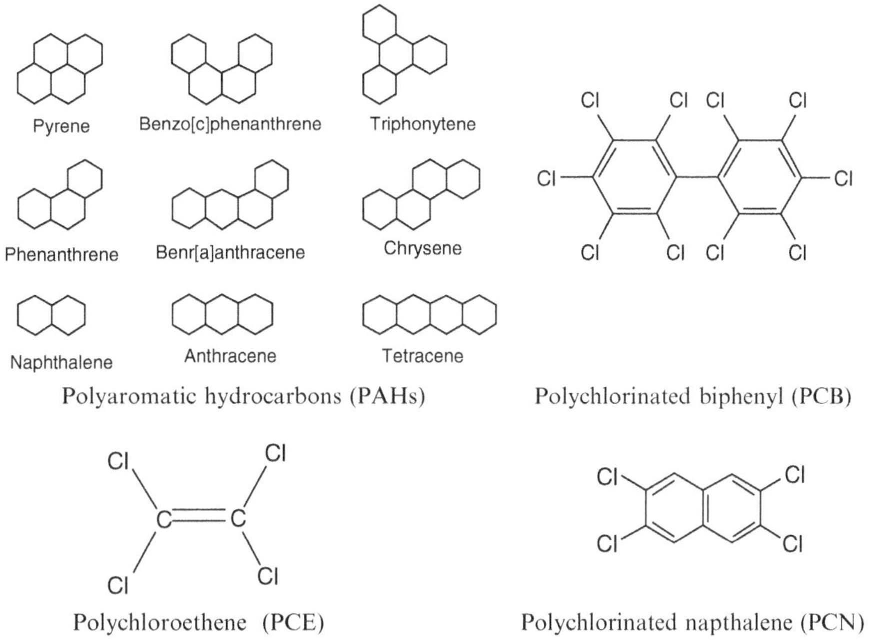

Fig. 12.21 Structure of PAH, PCB, PCE, and PCN

Miscellaneous Sources of Soil Pollutants

Lead contamination is common in the vicinity of roads and highways because of the emission from vehicle fuels. Traffic is a source of atmospheric deposition of benzene and PAHs; they are constituents of petrol. Nuclear tests in the open atmosphere caused global atmospheric radioactive fallout of various radionuclides (e.g., radiocesium ^{134}Cs and ^{137}Cs and radiostrontium ^{90}Sr). In addition, there are many thousand synthetic chemicals used in industries that can reach the soil by direct or indirect means. Most environmental pollutants are xenobiotic (man-made) chemicals. Some sources of soil pollutants are summarized in Table 12.5.

12.4.3.2 Organic Pollutants in Soil

Organic pollutants in soil include short-chain alkanes; chlorinated solvents such as polychloroethylene (PCE, CCl_2CCl_2), trichloroethylene (TCE, C_2HCl_3), dichloroethylene (DCE, CHClCHCl), and vinyl chloride or chloroethylene (VC, CH_2-CHCl); monoaromatic hydrocarbons; chlorinated aromatic compounds; polynuclear aromatic hydrocarbons (PAHs); and polychlorinated biphenyls (PCBs). Monoaromatic hydrocarbons include benzene, toluene, ethylbenzene, and xylenes (also collectively known as BTEX). Chlorinated aromatic compounds include hexachlorobenzene and pentachlorophenol.

Persistent Organic Pollutants

Persistent organic pollutants (POPs) are organic compounds that, to a varying degree, resist photolytic, biological, and chemical degradation (Pedro et al. 2006). POPs include phthalate esters (PAEs), nonylphenol ethoxylates (NPEs), PAHs, PCBs, and bisphenol A. (Figs. 12.21 and 12.22) added to soils with wastewater and sewage sludge.

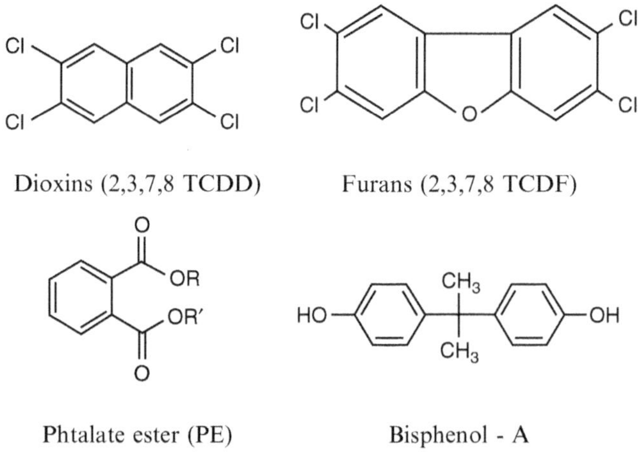

Fig. 12.22 Structures of dioxins, furans, phthalate ester, and bisphenol A

Dioxins and furans are some toxic organic pollutants that are added to soil with herbicides and wastes from paper and wood industries. Of all the dioxins and furans, 2,3,7,8-tetrachloro-p-dibenzo-dioxin (2,3,7,8 TCDD) is considered the most toxic. Crookes and Howe (1993) reported that the abundance of PCB in a highly sludge-amended soil was about eight times higher than in sludge itself, indicating a buildup of PCBs during many years of heavy sludge amendment at the site. PCNs (polychlorinated naphthalenes) are newly recognized persistent pollutants. Halogenated hydrocarbons, including dioxins and furans, are by far the most important group of POPs.

Hazards of Organic Pollutants

Human health risks of organic pollutants include allergies, damage to the nervous system, disruption of the immune system, reproductive disorders, and cancer (van der Perk 2006). Some volatile organic compounds such as tetrachloroethylene, trichloroethylene, dichloroethylene, and vinyl chloride or chloroethylene, which may contaminate the soil, are potential carcinogens. Chlorinated aromatic compounds, such as hexachlorobenzene and pentachlorophenol, polychlorinated biphenyls, and PAHs are also potential carcinogens. They are lipophilic and bioaccumulate in fatty human tissues. The PAHs tend to concentrate in food chains (McElroy et al. 1989) and create acute toxicity (Heitkamp and Cerniglia 1988); some PAHs (e.g., benzo[a]pyrene) are recognized mutagens. Although organophosphates such as malathion, methyl parathion, and diazinon are not much persistent in the environment, they are more toxic to humans and can be absorbed through the skin, lungs, and intestines during handling and application. The NPEs and PAEs exert estrogen-like, and PCBs and dioxins have antiandrogenic activity. The most severe effects ascribed to estrogen-like substances seem to be on the human reproductive health (Toppari et al. 1995, Meyer et al. 1995). Some pesticides, including malathion and parathion, and PCBs are neurotoxic.

Remediation of Organic Pollutants

Significant progress has been made in regulating soil pollution during the past decade. There are physical, chemical, and biological methods of remediation chosen on the basis of site and soil characteristics, type of pollutants to be removed, concentration, and the end use of the contaminated soil. However, the approach includes isolation, immobilization, toxicity reduction, physical separation, extraction, and decomposition (Sheoran et al. 2008). There are *in situ* (on-site) and *ex situ* (off-site) remediation methods. *In situ* methods are used directly at the contamination site so that soil does not need to be excavated, and therefore the chance of causing further environmental harm is minimized. Most thermal or chemical processes are economically not sound and are time-consuming (Ward et al. 2003), but hydrogen peroxide oxidation has been successfully employed for the remediation of organic pollutants in soil. It is relatively fast, taking only days or weeks; the contaminants are treated *in situ* and converted to harmless substances (e.g., H_2O, CO_2, O_2, halide ions). Hydrogen peroxide can be electrochemically generated on-site, which may further increase the economic feasibility and effectiveness of this process for treatment of contaminated sites. Natural iron oxide minerals (hematite Fe_2O_3, goethite FeOOH, magnetite Fe_3O_4, and ferrihydrite) present in soil can catalyze hydrogen peroxide oxidation of organic compounds. Disadvantages include difficulties controlling *in situ* heat and gas production. Low soil permeability, incomplete site delineation, and soil alkalinity may limit the applicability of the hydrogen peroxide oxidation technique (Goi et al. 2009). Attempts of soil washing and solvent extraction have also been taken. In these techniques, water and other solvent mixtures, including dichloromethane, ethanol, methanol, and toluene, have been utilized (Rababah and Matsuzawa 2002). Additionally, surfactants have also been tried. Surfactants such as Tween 40, Tween 80, Brij 30, Dowfax 8390, and Steol 330 have shown to be effective for PAH removal from soil (Ahn et al. 2008). Recent studies have advocated the use of vegetable oil as a nontoxic, biodegradable, and cost-effective alternative to these conventional solvents and surfactants (Gong et al. 2006). Successful removal of PAHs from soil with efficiencies above 80% has been reported. Pizzul et al. (2007) reported the use of rapeseed oil on the degradation of polycyclic aromatic hydrocarbons in soils by *Rhodococcus wratislaviensis*.

Bioremediation

Bioremediation is defined as the elimination, attenuation, or transformation of polluting substances by the use of biological processes (Shukla et al. 2010). It uses relatively low-cost, low-technology techniques, which generally has a high public acceptance and can often be carried out on site (Vidali 2001). It employs biological agents, mainly microorganisms, for example, yeast, fungi, or bacteria, to clean up contaminated soil. This technology includes biostimulation (stimulating viable native microbial population), bioaugmentation (artificial introduction of viable population), bioaccumulation (live cells), biosorption (dead microbial biomass), phytoremediation (plants), and rhizoremediation (plant and microbe interaction). The first patent for a biological remediation agent was registered in 1974, which was a strain of *Pseudomonas putida* (Prescott et al. 2002) able to degrade petroleum. Now, more than 100 genera of microbes are used for organic pollution control of soil. These organisms belong to at least 11 different prokaryotic divisions (Glazer and Nikaido 2007). Rhizoremediation is a process of bioremediation which involves the removal of specific contaminants by mutual interaction of plant roots and suitable microbial flora. The fungi *Phanerochaete chrysosporium* and *Trametes versicolor*, most commonly used in lignin biodegradation studies, have been implicated in the degradation of a variety of contrasting aromatic xenobiotics, including chlorophenols, pesticides, and dyes (Rubilar et al. 2007). The white-rot fungus *Anthracophyllum discolor* presents a high ligninolytic activity, mainly through manganese peroxidase activity; it has a great potential for xenobiotic degradation (Tortella et al. 2009), especially polycyclic aromatic hydrocarbons (Acevedo et al. 2010).

12.4.3.3 Heavy Metal Pollution of Soils

Heavy metals are metals (or metalloid, e.g., As) that have a density greater than 5 g cm^{-3} and an atomic mass exceeding that of calcium. The most common environmentally important heavy metals are zinc (Zn), copper (Cu), lead (Pb), cadmium (Cd), mercury (Hg), chromium (Cr), nickel (Ni), tin (Sn), and silver (Ag) and the metalloid arsenic (As). Some heavy metals play an essential role in plant and animal physiology and are thus required in small amounts for normal healthy growth (Zn, Cu, and Ni for plants; Zn, Cu, Se, and Cr for animals). They are essential micronutrients. Other heavy metals are not essential and have no nutritional value (e.g., Cd, Pb, Hg, Sn, and As). All these metals are highly toxic at high concentrations or at concentrations higher than required. Some soils are naturally enriched with heavy metals. Some other soils are contaminated by anthropogenic routes. Soils enriched naturally or anthropogenically with metals are called metalliferous soils. Some plants are adapted to metalliferous conditions. They are either metal tolerant or have the capacity to accumulate very high content of metals. Metalliferous soils contain abnormally high concentrations of some of the elements that are normally present as minor constituents (200–2,000 mg kg^{-1}, e.g., Mn) or trace constituents (0.01–200 mg kg^{-1}, e.g., Zn, Cu, Ni, Cr, Pb, As, Co, Se, Cd).

The most important types of natural metalliferous soils are (1) serpentine soils, developed from Fe- and Mg-rich ultramafic rocks, which are significantly enriched in Ni, Cr, and Co; (2) the calamine soils, enriched in Pb and Zn, that contain elevated Cd also and sometimes carry high concentrations of As and/or Cu; (3) Cu and Co containing soils such as those of the Shaban Copper Arc in the Democratic Republic of Congo, derived from argillites and dolomites containing sulfides of these metals, together with important secondary minerals (oxides, basic carbonates, silicates); and (4) soils derived from various Se-rich rocks. Serpentine soils occur over extensive areas in many countries of the world such as Japan, the Philippines, Indonesia, the USA, Turkey, Greece and the Balkan lands, parts of the European Alps, central Brazil, Cuba, Puerto Rico, and parts of Central America. Notable Se-rich soils have arisen particularly from the Cretaceous shales of the Midwestern United States (Mudgal et al. 2010).

Heavy metals find their way to the soil from such anthropogenic activities as mining and smelting, road metalling, application of sewage sludge and phosphate fertilizers and pesticides, manuring, burning of coal and gasoline, leaching from building materials, and direct domestic or industrial discharges and disposals. Commonly, sewage sludge-based organic fertilizers contain variable quantities of heavy metals that contaminate soils. The US Environmental Protection Agency has set limits of heavy metal application in sewage sludge to soil. Maximum allowable concentrations in sewage sludge for application to soils of As, Cd, Cr, Cu, Pb, Hg, Mo, Ni, Se, and Zn are 75, 85, 3,000, 4,300, 420, 840, 57, 75, 100, and 7,500 mg kg^{-1}, respectively. Soil properties that affect the behavior of heavy metals in soil include depth, slope, texture, porosity, pore size distribution, infiltration, permeability, microbial population and diversity, organic matter content, cation exchange capacity, soil pH, redox potential, and temperature. Heavy metals enter into a variety of reactions, as soil conditions permit, with the physical and chemical components of soil. These reactions affect the mobility, bioavailability, distribution, and retention of heavy metals in soil. Adsorption and precipitation are the principal reactions that metals undergo in soil. Many soil constituents, particularly the organic and inorganic colloids, control the sorption of heavy metals. When dissolved, Hg, Cd, Pb, Ni, Cu, Zn, Cr, and Mn form cations and As, Mo, Se, and B form anions. Clay particles, humus, iron, and aluminum oxides and hydroxides adsorb these ions on their surfaces and reduce their mobility. Sulfide, carbonate, and phosphate compounds cause precipitation of heavy metals. These reactions in soil result in the formation of soluble and insoluble metal ions and compounds as well as metal–mineral and metal–organic complexes. Metals may form complexes with dissolved organic matter. This process of ligand formation generally increases with decreasing pH. At high pH, heavy metals may precipitate as carbonates and hydroxides. Heavy metals may also be coprecipitated with iron, aluminum, and manganese oxyhydroxides. Soil pH regulates adsorption, complexation, and precipitation and thus determines the mobility of heavy metals in oxidized conditions. Heavy metals form sulfide compounds with sulfide ions in reduced condition and are precipitated.

Chemical Behavior of Heavy Metals in Soils

There are several species of heavy metals in soils. Speciation of a heavy metal can broadly be defined as the identification and quantification of the different, defined species, forms, or phases in which an element can occur. The species and forms of metals and their reactions (sorption/desorption, precipitation/dissolution) are governed by a set of edaphic and metal characteristics. The kinetics and equilibria of the reversible transfer of metals between the solid and the liquid phases also depend on these characteristics.

Heavy metals in soil may be found in one or more of the following forms (Aydinalp and Marinova 2003):

(a) Dissolved (in soil solution)
(b) Exchangeable (on organic and inorganic colloidal surfaces)
(c) As structural components of the lattices of soil minerals
(d) As insoluble precipitates with other soil components as in other metal oxides and carbonates.

The first two forms are mobile or labile, while the other two are currently unavailable and potentially available in the longer term. The easily soluble fractions are of the greatest interest, as their highest mobility, bioavailability, or toxicity can influence the quality of environment. Extractants including distilled water and several salt solutions (KCl, $CaCl_2$, NH_4OAc, NH_4NO_3, EDTA, and DTPA) have been used to extract the mobile or bioavailable forms of heavy metals with single extraction procedures (Karczewska et al. 1998). The mobility of metals is strongly affected by soil properties and may vary significantly in relation to changes in soil pH and organic carbon content. Other soil properties such as cation exchange capacity (CEC); organic matter content; quantity and type of clay minerals; the content of the oxides of iron (Fe), aluminum (Al), and manganese (Mn); and the redox potential determine the soil's ability to retain and immobilize heavy metals.

Mobility of heavy metals in soil depends greatly on soil pH. Fuller (1977) considered that in acid soils (pH 4.2–6.6) the elements Cd, Ni, and Zn are highly mobile, Cr is moderately mobile, and Cu and Pb are practically immobile, and in neutral to alkaline soils (pH 6.7–7.8), Cr is highly mobile, Cd and Zn are moderately mobile, and Ni is immobile. Humic substances form metal–organic complexes which influence metal mobilization in soils. Using gel permeation chromatography, the important role of organic matter in solubilizing heavy metals in sewage sludges and in agricultural soils amended with liquid manure can be clearly shown. Aluminum mobilization in Spodosols was demonstrated to be largely determined by the formation, solubilization, migration, and subsequent precipitation of organic complexes. Organic matter plays an important role not in forming complexes but also in retaining heavy metals in an exchangeable form. Copper is bound and rendered unavailable chiefly through the formation of complexes, while Cd is retained in an exchangeable form and is more readily available. When mobility increases, the quantities of heavy metals available to plants increase, resulting in the appearance of toxicity phenomena.

War, Ammunitions, and Soil Pollution

War needs weapons and ammunitions. War ends but its effect on soil, water, and air continues. Firing ranges often leave soils contaminated with metals from spent bullets. Lead is the primary soil contaminant of concern in these ranges. The normal operation of a range can produce lead concentrations of several percent (1% equals to 10,000 mg kg^{-1}; safe level is only 35 mg kg^{-1}) in soils located behind and adjacent to targets. In abandoned delaboration sites around the former World War I front in Belgium and France, soils are contaminated by arsenic, heavy metals, explosives, and chemical warfare-related compounds. The most important contaminants found during the investigation were arsenic and lead. The battlefields of Northern France still pose a problem even 90 years after the end of hostilities.

Depleted uranium is an important source of soil pollution. It is highly toxic and has been used in active combat by the US and UK forces. It is used in combination with other metals to penetrate armor. It often ignites, spreading fine dust around the affected land area. This material has as much as 75% of the strength of enriched uranium, so it brings lasting health problems with it wherever it is used. Two hundred eighty-six metric tons of depleted uranium were used during the First Gulf War in 1991.

(continued)

(continued)

Napalm and Agent Orange were used with other "agents" in the Vietnam War. Agent Orange was one of a cluster of harmful chemicals the "rainbow herbicides"—Agent Pink, Agent Green, Agent Purple, Agent Blue, and Agent White used in Operation Ranch Hand. The purpose of using these agents was to strip trees of their leaves so that the enemy could be seen more easily. These defoliants continue to cause death and environmental problems today. Agent Orange contained dioxin which is one of the most potent chemicals. In Vietnam, there are areas where the levels of dioxins in the soil exceed internationally accepted levels by as much as 100 times. A large number of diseases have been linked with the use of Agent Orange. Operation Ranch Hand was a deliberate program of chemical herbicide use designed to kill the crops of the Viet Cong. The chemicals were sprayed from air.

Arsenic

Arsenic (As) has an atomic number 33 and an atomic mass 74.92, boiling point 602.8°C, and melting point 816.8°C. Arsenic produces arsenic acid with concentrated nitric acid, arsenious acid with dilute nitric acid, and arsenic trioxide with concentrated sulfuric acid. Arsenic occurs usually in conjunction with sulfur and metals and also as a pure elemental crystal. Arsenic forms the oxides As_2O_3 and As_2O_5, which are hygroscopic and readily soluble in water. Arsenic (V) acid is a weak acid. Its salts are called arsenates, which are dissolved in arsenic-contaminated groundwater. All rocks contain some arsenic, typically 1–5 mg kg^{-1}. Some igneous and sedimentary rocks may contain higher As concentrations. Principal arsenic-bearing minerals are arsenopyrite (AsFeS), realgar (AsS), and orpiment (As_2S_3). The sources of As in soils are both geochemical and anthropogenic. Metallic arsenic is mainly used for strengthening alloys of copper and lead. Arsenic is used in semiconductor electronic devices. Arsenic compounds are used for production of pesticides, herbicides, and insecticides. Several organoarsenic compounds were developed as chemical warfare agents during World War I, including vesicants such as lewisite and vomiting agents such as adamsite. Residues of these materials have found their way to the soil (Smith et al. 1998). Arsenic is a notorious pollutant of groundwater which is used for irrigation of extensive rice fields in Southeast Asia. Groundwater pollution of As and its consequences are mentioned in detail in Chap. 6 in connection with quality of irrigation water.

Soils may contain 1–40 mg As kg^{-1}, with an average of 5–6 mg kg^{-1}. Indiscriminate use of arsenical pesticides during the early to mid-1900s has led to extensive contamination of soils worldwide. Contamination in excess of 1,000 mg As kg^{-1} has been recorded at many sites throughout Australia. Similar contaminated sites also exist in the United States, Africa, and other parts of the world (Smith et al. 1998). The main sources of As in soils is the parent materials from which the soil is derived (Yan-Chu 1994). The distribution of As in soils may vary with soil type, depending on the nature of the parent material. Background concentrations do not generally exceed 15 mg As kg^{-1} (NRCC 1978), although concentrations ranging from 0.2 to 40 mg As kg^{-1} soil have been reported (Walsh et al. 1977). Dudas and Pawluk (1980) reported background As concentrations that averaged 5 mg As kg^{-1} in 78 soil samples in Alberta. Much higher As concentrations have been reported in acid sulfate soils developed on pyritic parent material. For instance, Dudas (1987) attributed elevated As concentrations that ranged from 8 to 40 mg As kg^{-1} in Canadian acid sulfate soils to the weathering of pyrites in the parent material.

Arsenic forms a variety of inorganic and organic compounds in soils (Vaughan 1993) and is present mainly as inorganic species, either As (V) or As (III) (Masscheleyn et al. 1991). Under oxic soil conditions (Eh > 200 mV; pH 5–8), As is commonly present in the +5 oxidation state. However, As (III) is the predominant form under reducing conditions (Masscheleyn et al. 1991). Both As (V) and As (III) species have been reported to be subject to chemical and microbial oxidation–reduction and methylation reactions in soils and sediments (Brannon and Patrick 1987).

Cadmium

Cadmium (Cd) is a lustrous, silver-white, ductile, very malleable, and soft metal. It is soluble in acids but not in alkalis. It has an atomic number 48, atomic mass 112.4 g mol^{-1}, density 8.7 g cm^{-3} at 20 °C, melting point 321 °C, and boiling point 767 °C. Cadmium has a relatively low crustal abundance, although it occurs ubiquitously in rocks and soils (Alloway 1995). It is rarely found in its elemental form. The main minerals containing cadmium are greenockite (CdS), octavite (CdSe), and monteponite (CdO) (Kabata-Pendias and Mukherjee 2007). Cadmium is often found in association with zinc ores and zinc-bearing lead ores. Its compounds almost exclusively involve the +2 oxidation state. Cadmium forms simple salts with oxygen, sulfur, and many common anions such as chloride, nitrate, and carbonate. Cadmium often forms simple hydrated hxdroxyl ions such as $[Cd(OH)(H_2O)x]^+$ in aqueous solution. It also has appreciable coordination chemistry with ligands including halides, hydroxides, cyanides, and nitrate (Alloway 1995). Organocadmium compounds are rather reactive and unstable.

Cadmium is regarded as one of the most toxic trace elements in the environment. The increased emissions with its persistence and its relatively rapid uptake and accumulation by food chain crops contribute to its potential environmental hazards. According to Heinrichs et al. (1980), the average concentration of cadmium in the lithosphere is 0.098 mg kg^{-1}. Most mafic and granitic rocks contain a very low concentration of cadmium (<0.1 mg kg^{-1}). It has relatively high concentrations (about 10 mg kg^{-1}) in sedimentary rocks. Some shales are unusually high in Cd (>100 mg kg^{-1}).

Amounts of indigenous cadmium in cultivated and noncultivated soils are determined by the quantities of Cd in the parent materials together with amounts added through atmospheric deposition, fertilizers, pesticides, and irrigation water. Generally, soils of similar age derived from igneous rocks contain the lowest Cd, soils derived from metamorphic rocks have intermediate, and soils derived from sedimentary rocks contain the largest quantities of Cd. Noncontaminated soils contain very low concentration of Cd. In a survey, a total of 3,305 soil samples were collected from crop-producing areas in 36 states of the USA. Concentrations of Cd ranged from 0.005 to 2.40 mg kg^{-1}, with mean and median values of 0.27 and 0.20 mg kg^{-1}, respectively (Holmgren et al. 1993). Surface soils around mining areas contain high concentrations of Cd. For example, Munshower (1977) reported 29 mg Cd kg^{-1} in surface soil at 2 km northeast of a lead–zinc smelter in Montana. The Cd concentration of surface soil decreased with distance and reached near background levels 24 km from the source. Buchauer (1973) reported concentrations of Cd in the surface organic horizon as great as 750 mg kg^{-1}. The soils contaminated by mining operations will remain high in Cd more or less permanently because Cd is quite immobile in soil. Cadmium concentration in soil solution is relatively low ranging from 0.2 to 6.0 μg L^{-1} (Kabata-Pendias and Pendias 2000).

As with all cationic metals, the chemistry of cadmium in the soil environment is to a great extent controlled by pH. Under acidic conditions, Cd solubility increases and very little adsorption of Cd by soil colloids, hydrous oxides, and organic matter takes place. At pH values greater than 6 units, Cd is adsorbed by the soil solid phase or is precipitated and the concentrations of dissolved Cd are greatly reduced. Cadmium forms soluble complexes with inorganic and organic ligands, in particular with chloride ions. The formation of these complexes will increase Cd mobility in soils. Chloride can be expected to form a soluble complex with Cd^{2+} as $CdCl^{+}$, thereby decreasing the adsorption of Cd^{2+} to soil particles. In contrast to inorganic ligand ions, Cd^{2+} adsorption by kaolinite could be enhanced by the presence of organic matter via the formation of an adsorbed organic layer on the clay surface (Adriano et al. 2005). Cadmium mobility and bioavailability are higher in noncalcareous than in calcareous soils. Liming of soil raises the pH, increasing cadmium adsorption to the soil and reducing bioavailability. A general trend emerges that toxicity increases in soil when mobility of cadmium is higher, that is, soil toxicity increases as soil pH, or soil organic matter decrease.

Cadmium may be adsorbed by clay minerals, carbonates, or hydrous oxides of iron and manganese or may be precipitated as cadmium carbonate, hydroxide, and phosphate. Adsorption mechanisms may be the primary source of cadmium removal from soils. In soils and sediments polluted with metal wastes, the greatest percentage of total cadmium was associated with the exchangeable fraction. Cadmium concentrations have been shown to be limited by cadmium carbonate in neutral and alkaline soils.

Cadmium Contamination of Soils and Crops, and the Itai-Itai Disease in Japan

Concentrations of Cd in crops grown on soils elevated in Cd, either naturally or from anthropogenic sources, may accumulate substantially greater than 1.0 mg kg^{-1} Cd. Consumption of cadmium-contaminated food grains may cause severe Cd poisoning (Page et al. 1987). The well-known episode of Cd poisoning of humans, the itai-itai disease, was reported in Japan in the mid-1950s (Tsuchiya 1978). The name of the disease came from the screams due to the severe pains in the bones caused by cadmium poisoning. It first impairs kidney function and progressively causes osteomalacia. Victims suffer from calcium deficiency as occurs with old age, malnutrition, hormone imbalance, and when pregnant or breast feeding. The outbreak of itai-itai started around 1912 and continued until 1945. The disease was thought to be nutritional and pathological at first. After many years of investigation, the Ministry of Health and Welfare officially announced in 1968 that the disease was due to chronic Cd poisoning and the source of cadmium was the mines in the mountainous areas of Toyama Prefecture. Mining operations were being done for centuries in the Toyama Prefecture of Japan. In the early 1900s, the Kamioka Mining Co. Ltd. was extracting zinc ores in these mines for a long time. Wastewater of these operations was released in the Jinzu River, the water of which became seriously contaminated with cadmium. This water was used by local community people for their domestic need and irrigation in paddy fields. Soils and rice grains, grown in these paddies, were also contaminated with cadmium. Since the per capita dietary consumption of rice among the Japanese population is high, the impact of Cd enrichment in the rice on the indigenous population was proportionally magnified (Page et al. 1987).

Lead

Lead (Pb) is a bright and silvery metal which is soft, ductile, malleable, and resistant to corrosion. It has a density of 11.32 g cm^{-3}, melting point of 327.5 °C, and boiling point of 1,749 °C. There are two stable oxidation states of Pb, Pb(II), and Pb(IV), but it occurs mainly as Pb^{2+}. Three oxides of lead are known: PbO, Pb_3O_4, and PbO_2. Lead is amphoteric in nature, that is, it is soluble in both acids and bases. In solution at high pH, Pb may exist as an anion, whereas the cationic form is present at lower pH values (Roy et al. 1993).

Lead is a widespread nonbiodegradable chemical contaminant found in the soil receiving disposal of city wastes, sewage, and industrial effluents. It is found in low concentrations in all rocks, soils, and dusts, and its content in soils usually ranges from 2 to 200 mg kg^{-1}. Three main sources of elevated lead levels in soils include lead-based paint on exterior surfaces such as the walls of buildings; point source emitters such as smelters, batteries, or mine tailings; and leaded gasoline emissions from automobiles. Soil-lead levels within 25 m of roadways are typically 30–2,000 mg kg^{-1} higher than natural levels and can sometimes be as high as 10,000 mg kg^{-1} (USEPA 1996). Total Pb contents of 195 soil samples from agricultural lands of the United States varied from 7 to 13,240 mg kg^{-1}. Lead-based paints and lead-loaded gasoline were the sources of elevated Pb in soil (USEPA 1998). A study in England and Wales has shown that in surface soils lead ranged between 50 and 106 mg kg^{-1} with a geometric mean of 42 mg kg^{-1} (Davis et al. 1993). A median lead value of only 11 mg kg^{-1} for over 3,000 surface soils from croplands in the USA was reported by Holmgren et al. (1993).

The proportion of total soil Pb that exists in the soil solution is very small. Alloway et al. (1985) found Pb concentrations ranging from 61 to 12,537 μg L^{-1} in soil solutions enriched from a weathering Pb ore and soil contaminated by Pb/Zn mining and sewage sludge. Lead may exist in the soil solution as the free metal ion, Pb^{2+}, and soluble organic and inorganic complexes. The speciation in the soil solution depends to a large extent on pH, dissolved organic carbon, and the presence of ligands. Lead activity decreases with increasing soil pH. It is widely acknowledged that Pb^{2+} is the most significant Pb species in solution at pH values less than either 7 (Wang and Benoit 1996) or 8 (Lindsay 1979). The dominant form of Pb in solution at pH values greater than 7 is $PbCO_3$ (Wang and Benoit 1996). Other minor Pb species include $PbOH^+$, $Pb(OH)_2$, $Pb(OH)_3$, $PbCl_3$, $PbCl^+$, $PbNO_3^+$, and $Pb(CO_3)_2$ (Wang and Benoit 1996). Lead exists as Pb^{2+} in solution at pH 4 (Markus and McBratney 2000).

Mercury

Mercury (Hg) has an atomic number 80. It is the only metal that is liquid at standard conditions for temperature and pressure. It is a silvery white liquid that has a freezing point of −38.83 °C and boiling point of 356.73 °C. Almost all commercial production of mercury comes from the sulfide ore, cinnabar, which results from hydrothermal mineralization associated with volcanic activity. Mercury abundance in the Earth's crust is very low being in the range of 0.02–0.06 mg kg^{-1}, although it is likely to be more concentrated in argillaceous sediments and in coal (Kabata-Pendias and Mukherjee 2007). The total mercury concentrations in soils of UK were reported to range from 0.07 to 1.22 mg kg^{-1} with a mean value of 0.13 mg kg^{-1}. Urban soils were found to contain higher total mercury concentrations ranging from 0.07 to 1.53 mg kg^{-1}, with a mean of 0.35 mg kg^{-1} (Environment Agency 2007).

Mercury readily forms amalgams with sodium and zinc. Mercury forms inorganic compounds in both the Hg (I) and Hg (II) valent states. A large number of organomercury compounds are also known and can be synthesized by the action of sodium amalgam or Grignard reagent and $HgCl_2$ with halogenated hydrocarbons (Greenwood and Earnshaw 1997). Mercury is most commonly encountered in the environment in elemental form, as inorganic mercuric compounds or as monomethylmercury compounds with the general formula CH_3HgX. The most important source of mercury is the naturally occurring mineral cinnabar (HgS). Monomethylated mercury compounds are most likely to be found in soil as a result of natural microbial transformation of inorganic mercury.

Elemental Hg is stable in the soil environment. It has a strong tendency to form complexes with other anions (such as Cl^-, OH^-, and S^{2-}) and humic matter. The mercuric cation (Hg^{2+}) is rarely found in soil solution under natural conditions, and the major fraction is bound in soil minerals or adsorbed to either inorganic mineral surfaces or to organic matter. Yin et al. (1996) observed in experimental studies that soil adsorption decreased significantly above pH 5 as a result of increasing amounts of dissolved organic matter and the tendency for mercury to complex strongly to organic carbon.

Inorganic mercury can be methylated by abiotic and microbial processes in soil systems and is the primary source of methylmercury compounds in soil (ATSDR 1999). In surface soils, about 1–3% of total mercury is in the methylated form with the rest predominantly as mercuric compounds (Kabata-Pendias and Mukherjee 2007). Dimethylmercury is a highly toxic and volatile compound and is readily lost from soil to air. Monomethylated mercury compounds (CH_3HgX) are also volatile, and due to their relatively high mobility compared with inorganic forms, they are the most important mercury species for environmental pollution (ATSDR 1999).

Safe Limits of Heavy Metals in Soil

Having a universally accepted safe limit of heavy metals in soil is difficult. The "maximum allowable concentrations" of Cd, Cu, and Hg in Chinese agricultural soils are 0.30, 50.00,

Table 12.6 Threshold levels of heavy metals in soil

Heavy metals	Threshold levels in soils of concern for human health, mg kg^{-1}	Heavy metals	Threshold levels in soils of concern for human health, mg kg^{-1}
Arsenic	0.07	Molybdenum	380
Cadmium	1.70	Nickel	1,600
Chromium	100,000	Selenium	380
Cobalt	660	Silver	380
Copper	3,000	Vanadium	530
Lead	80	Zinc	23,000
Mercury	18		

Source: California Office of Human Health Assessment, http//:oehha.ca.gov/risk/chhsltable.html

and 0.30 mg kg^{-1}, respectively (Zhou et al. 2008). According to Coskun et al. (2006), "world limits" of Cd, Hg, Pb, and As in soil are 0.35, 0.30, 35.00, and 10.00 mg kg^{-1}, respectively. The toxic limits in soils of As, Cd, Cr, Cu, Pb, Hg, Mo, Ni, Se, and Zn are 1–4, 5–20, 5–30, 20–100, 5–30, 1–3, 10–100, 30–300, 5–10, and 100–400 mg kg^{-1}, respectively (Alloway 1995). Threshold levels of heavy metals in soils of concern for human health are given in Table 12.6.

Absorption of Heavy Metals by Plants

Plants absorb heavy metals through roots from soil and by leaves from air. Once absorbed, there are a variety of mechanisms to distribute metals within the plant body. A portion of absorbed metals may be retained in roots, barred by membranes, precipitated and accumulated in vacuoles, or translocated to shoots where they perform physiological functions if they are nutrients or beneficial elements or interfere with normal physiological functions leading to toxicities. Some plants prevent metal toxicity by metal compartmentalization and binding to intracellular ligands. In addition to the organic acids, plants have phytochelatins and metallothioneins. Phytochelatins are a family of peptides, while metallothioneins are similar to phytochelatins in being Cys-rich, metal complexing ligands. According to Cho et al.(2003), the sensitivity of plants to heavy metals depends on (1) uptake and accumulation of metals through binding to extracellular exudates and cell wall constituents; (2) efflux of heavy metals from cytoplasm to extranuclear compartments, including vacuoles; (3) complexation of heavy metal ions inside the cell by various substances, for example, organic acids, amino acids, phytochelatins, and metallothioneins; (4) accumulation of osmolytes and osmoprotectants and induction of antioxidative enzymes; and (5) activation or modification of plant metabolism to allow adequate functioning of metabolic pathways and rapid repair of damaged cell structures. Yet, metals are accumulated in plants and become toxic, and for these reasons, the yield of crop is reduced and the crop is rendered unsafe if grown in a contaminated soil.

Heavy Metals in Plant Food

The levels of heavy metals in edible plant parts collected from crop fields and markets of vegetables, cereals, fruits, and spices have considerably been investigated. Metal contamination of plant materials has been reported from metal-contaminated places such as mining and industrial areas and in crop fields amended with sludge-based fertilizers or irrigated with contaminated water.

Lead concentration in polished rice from paddy grown in a contaminated site of China reached 0.69 mg kg^{-1}, which is higher than the maximum allowable concentration by the safety criteria for milled rice for China (0.20 mg kg^{-1}). Cadmium contents in 31% of the rice samples exceeded the national maximum allowable concentration (Fu et al. 2008). Levels of Cd, Zn, Pb, and Cr in roots, stems, and leaves of the vegetable plants *Amaranthus viridis* and *Talinum triangulare* planted on poultry dumpsite of Nigeria were 0.62–2.74, 50.67–102.98, 2.27–7.21, and 0.64–4.45 mg kg^{-1}, respectively. Some of these values were above the safe levels (Adefila et al. 2010). Metal levels in vegetables, including parsley, onion, lettuce, garlic, peppermint, spinach, broad bean, chard, purslane, grapevine leaves; and fruits, including tomato, cherry, grape, and strawberry, grown in Manisa region of Turkey ranged from 0.56 to 329.7, 0.01 to 5.67, 0.26 to 30.68, 0.001 to 0.97, and 0 to 0.06 mg kg^{-1} for Fe, Cu, Zn, Pb, and Cd, respectively (Bagdatlioglu et al. 2010).

Contamination with heavy metals of fruits, including orange, mango, almond, lemon, sweet orange, grapefruits, chico, papaya, muskmelon, apple, quince, grape, strawberry, banana, pineapple, papaya, carambola, longan, wampee, date palm, and apricot, has been reported (Davarynejad et al. 2010). Concentrations of Cu, Ni, Zn, Fe, Pb, and Hg of 15 common spices available at local markets in the Kumasi Metropolis of Ghana ranged from 9 to 21, 31 to 50, 59 to 74, 110 to 494, 96 to 115, and 0.001 to 0.025 mg kg^{-1}, respectively (Nkansah and Amoako 2010). Permissible limits of heavy metals in edible parts of plants are shown in Table 12.7.

Table 12.7 The permissible limits of some heavy metals in plant food

Metal	Plant food material	Permissible limit, (mg kg^{-1})	References
Copper	All foods	10.00	CAC (1993)
Cadmium	All foods	0.05	Walker (1988)
	Vegetables and fruits, excluding leafy vegetables, fresh herbs, mushrooms	0.05	Graffham (2006)
	Leafy vegetables, fresh herbs, celeriac and all cultivated fungi	0.20	Graffham (2006)
	Stem vegetables, root vegetables, and potatoes (peeled)	0.10	Graffham (2006)
Lead	Cereals and legumes	0.20	CAC (2003)
	Vegetables, excluding brassica, leafy vegetables, fresh herbs, and all fungi	0.10	Graffham (2006)
	For potatoes, the maximum level applies to peeled potatoes		
	Brassica, leafy vegetables, and all cultivated fungi	0.30	Graffham (2006)
Zinc	Grains	50.00	USDA (2003)
	Beans	100.00	

WHO permissible limits in spices for Cu, Ni, Zn, Fe, Pb, and Hg are 50, 50, 100, 300, 100, and 10 mg kg^{-1}, respectively (Nkansah and Amoako 2010)

Table 12.8 Toxicities of heavy metals in plants

Heavy metals	Common toxicities of heavy metals on plants
Arsenic	Detrimental to mung bean seed germination
Cadmium	Inhibition of the biochemical processes—photophosphorylation, ATP synthesis, mitochondrial NADH oxidation, and electron-transport system; it affects seed germination and seedling growth. Rice plants are particularly vulnerable
Chromium	Toxic at a tissue concentration of ~0.1 mmol kg^{-1}; toxic effects mainly on roots and a direct or indirect effect on leaves, resulting in intense growth inhibition; symptoms of Cr toxicity appear as severe wilting and chlorosis of plants
Copper	Concentrations of <4 ppm are deficient, and >20 ppm are toxic. Cu is a constituent of a number of plant enzymes, which trigger a variety of physiological processes in plants such as photosynthesis, respiration, cell wall metabolisms, and seed formation
Lead	Inhibition of seed germination and seedling growth
Mercury	Toxic effects of Hg on plants include retardation of growth and premature senescence. Hg has been shown to inhibit synthesis of proteins in plant leaves and to reduce photosynthetic activity as it has a strong affinity for sulfhydryl or thiol groups, which are involved in enzymatic reactions. Hg also impairs cell division
Molybdenum	Mo is an important component of several enzymes that catalyze unrelated reactions. In plants, the most important functions of Mo are associated with N metabolism, for example, with nitrogenase and nitrate reductase enzymes. Mo toxicity in plants has not been observed under field conditions
Nickel	Ni toxicity causes pale yellow stripes longitudinally in leaf. In extreme cases, the entire plant may turn white with marginal necrosis (burn). In dicots, Ni toxicity causes an interveinal chlorosis that looks very similar to manganese deficiency

Toxicity of Heavy Metals

Organisms are exposed to soil contaminants by direct contact, handling, absorption, or ingestion and food intake. Heavy metals are bioaccumulated by incorporating into tissues, and as they are transferred through the food chain, they are biomagnified at higher trophic levels.

Heavy Metal Toxicity in Plants

Plants absorb heavy metals, essential or nonessential, from soils. Copper, molybdenum, zinc, and nickel are essential trace elements required by plants in extremely small amounts. If absorbed in relatively large amounts, all essential or nonessential heavy metals become toxic to plants. According to Dan et al. (2008), heavy metals are potentially toxic for plants; phytotoxicity results in chlorosis, weak plant growth, yield depression, reduced nutrient uptake, disorders in plant metabolism, and, in leguminous plants, a reduced ability to fix molecular nitrogen. Metals inhibit seed germination, seedling growth, photosynthesis, and enzyme activity, but the effects vary with the metals, their concentrations as well as speciation, and the plant species. In a study, Fargasova (1994) observed that Cr, Cd, Hg, Pb, and As all reduced seed germination of mustard (*Sinapis alba*), but As was the most inhibitory. Cadmium was found to be less toxic for germination of *Sinapis alba* seeds, but it is highly toxic to mung bean (*Vigna radiata*) seeds. Spinach, soybean, and curly cress, for instance, were sensitive to Cd, whereas cabbage and tomato were resistant. Toxicity symptoms of some heavy metals to plants are summarized in Table 12.8.

Effect of Heavy Metals on Human

Toxicity of heavy metals involves either compounds with intrinsic toxicity or activated metabolites. It may be acute or chronic depending on the type of exposure. Changes in physiological

Table 12.9 Toxicities of heavy metals in human

Heavy metals	Common toxicities of heavy metals in human
Arsenic[a]	Arsenicosis is the most common disease caused by drinking As-contaminated water for a long time. Inorganic arsenic compounds are "known to be human carcinogens." Arsenic may cause cancers of the bladder, kidney, skin, liver, lung, colon, and lymph. Darkening of the skin and the appearance of small "corns" or "warts" on the palms, soles, and torso
Cadmium[a]	The well-known itai-itai disease was caused by the consumption of rice grown by irrigating with Cd-contaminated river water in Japan after 1945. Cd toxicity causes renal tubular dysfunction, high blood pressure, lung damage, and lung cancer. Acute Cd inhalation may lead to pneumonitis and pulmonary edema. Chronic exposure via inhalation may cause emphysema. One of the most widely known toxic effects manifested by Cd poisoning is nephrotoxicity
Chromium[b]	Impaired growth, altered immune function, disturbances in aortic plaque and size, corneal lesion formation, and decrease in reproductive functions; both Cr(III) and Cr(VI) are potent human carcinogens. The major target organ for Cr(III) and Cr(VI) is the respiratory tract. Classical symptoms are perforations and ulcerations of the septum, bronchitis, decreased pulmonary function, and pneumonia
Copper[b]	Copper is an essential element. Cu toxicity in humans is very rare and is usually associated with long-term intake of cow's milk or with severe malnutrition in infants and young children
Lead[a]	Lead damages the central nervous system, kidneys, and lungs. It causes anemia; nausea, anorexia, and abdominal cramps; muscle aches and joint pain; difficulty in breathing, asthma, bronchitis, and pneumonia; and damage to the fetus and miscarriage. Children are more vulnerable to Pb poisoning than adults. Pb may cause male infertility
Mercury[a]	Severe mercury poisoning occurred in Minamata of Japan from contaminated fish and in Iraq from contaminated wheat. Wheat was sprayed with MeHg fungicide. In the outbreak, more than 6,000 children and adults had been poisoned, with nearly 500 deaths. Symptoms include paresthesias, ataxia, dysarthria, and deafness. Chronic Hg poisoning may result from exposure to small amounts of Hg over long periods; symptoms include salivation, loss of appetite, anemia, gingivitis, excessive irritation of tissues, nutritional disturbances, and renal damage accompanied by proteinuria. Exposure to Hg may cause nausea, severe gastrointestinal irritation, and abdominal pain
Molybdenum[b]	Toxic effects of Mo include increased blood xanthine oxidase, increased concentrations of uric acid in blood and urine, and a high incidence of gout
Nickel[a]	Exposure to nickel sulfate and nickel chloride can cause vomiting and headaches, and the metal can cross human placental barrier, affecting the fetus. Ni-sensitive individuals often develop contact dermatitis. Inhalation of Ni compounds has been considered responsible for lung, sinonasal, and laryngeal carcinomas. Other illnesses include pneumonoconiosis and emphysema. Ni may cause kidney cancer. Among different types of Ni compounds, crystalline nickel subsulfide is carcinogenic, whereas amorphous NiS is not
Selenium[b]	It can cause nausea, vomiting, and diarrhea. Chronic oral exposure to high concentrations can produce selenosis. Major signs of selenosis are hair loss, nail brittleness, and neurological abnormalities. Brief exposures to high levels in air can result in respiratory tract irritation, bronchitis, difficulty breathing, and stomach pains. Longer-term exposure can cause respiratory irritation, bronchial spasms, and coughing
Zinc[c]	Zinc is considered to be relatively nontoxic, especially if taken orally. However, excess amount can cause system dysfunctions that result in impairment of growth and reproduction. The clinical signs of zinc toxicosis have been reported as vomiting, diarrhea, bloody urine, icterus, liver failure, kidney failure, and anemia (Fosmire 1990)

[a]Yu (2005), [b]Bradl (2005), and [c]Duruibe et al. (2007)

and biochemical processes may be manifested in impaired central nervous system (CNS) function and oxidative metabolism, injury to the reproductive system, or altered DNA leading to carcinogenesis. Heavy metals can induce adverse effects in human by disruption or destruction of cellular structure, chemical combination with a cell constituent, inhibition of enzymes, initiation of a secondary action, free-radical-mediated reactions, and disruption of reproductive function (Yu 2005). Toxicities due to most common heavy metal contaminants are listed in Table 12.9.

Effect of Heavy Metals on Soil Microorganisms

Some heavy metals are essential micronutrients (cobalt, chromium, nickel, iron, manganese, and zinc) for soil microorganisms (Bruins et al. 2000, Collins and Stotzky 2001). They are involved in redox processes in order to stabilize molecules through electrostatic interactions, as catalysts in enzymatic reactions, and regulating the osmotic balance (Hussein et al. 2005). On the other hand, some other heavy metals (cadmium, mercury, and lead) have no biological role and are detrimental to the organisms even at very low concentrations. High levels of both of the essential and nonessential metals are toxic to the soil microorganisms (Pawloska and Charvat 2004). Heavy metals adversely affect growth, morphology, and biochemical activities of microorganisms and ultimately reduce their biomass and diversity (Roane and Pepper 2000). Heavy metals can damage the cell membranes, alter enzymes specificity, disrupt cellular functions, and damage the structure of the DNA. Toxicity of these heavy metals occurs through the displacement of essential metals from their native binding sites or through ligand interactions (Bruins et al. 2000). Also, toxicity can occur as a result of alterations in the conformational structure of the nucleic acids and proteins and interference with oxidative phosphorylation and osmotic balance (Bruins et al. 2000). Heavy metals bring about change of the diversity, population size, and overall activity of the soil microbial communities (Kelly et al. 2003). Leita et al. (1995) studied influence of Pb, Cd, and Ti on microbial biomass, survival, and activity during a laboratory incubation of soil. In comparison to uncontaminated soil, the microbial biomass C decreased sharply in soil contaminated with Cd and Ti, whereas the addition of Pb did not have any significant inhibitory effect on the level of microbial biomass C. Long-term heavy metal contamination of soils reduces microbial respiration (Doelman and Haanstra 1984). Many reports have shown large reductions in microbial activity due to short-term exposure to toxic metals too (Hemida et al. 1997). Bacterial activity, measured by thymidine incorporation technique, had been shown to be very sensitive to metal pollution both under laboratory and field conditions (Diaz-Ravina and Baath 1996). The size of mycorrhizal roots has been found to decrease in soils containing high concentrations of heavy metals such as Cu, Pb, and Zn (Bell et al. 1988). Wastewaters containing high concentrations of heavy metals have also been shown to inhibit mycorrhizal infection of soybean, especially in soils with low pH (Angle and Heckman 1986). Lead and copper are less mobile than Zn and Cd. Ni, Cd, and Zn are potentially more serious contaminants of soil solutions than Cu and Pb (Biddapa et al. 1982). Absence of nitrogen fixation in clover crops grown on soils contaminated with heavy metals over a long period of time has been found to result from a survival only of ineffective rhizobial strains (Giller et al. 1989).

12.4.3.4 Heavy Metal Remediation of Soil

A variety of methods are employed for heavy metal remediation of soils. For the sake of brevity, a brief description of the methods is given in the following sections. There are some good reviews of the heavy metal remediation techniques (USEPA 1997; Stegmann et al. 2001; Bradl and Xenidis 2005; Hamberg 2009; Wang et al. 2010). However, the choice of the method depends on the degree and kind of contamination, contaminating metal(s), cost involved, farm facilities, technology available, etc. As mentioned in relation to organic pollutant remediation, the methods may be on-site or off-site, *in situ* or *ex situ*, and physical, chemical, and biological. Physical methods include soil washing, encapsulation, vitrification, and electrokinesis; chemical methods include solidification, precipitation, and ion exchange; and biological methods use microorganisms or plants to remove heavy metals.

Physical Remediation

Soil Washing

In soil washing, the coarse materials such as residues, woods, and stones are removed by sieving first, and then magnetic materials are removed using magnets. Soil materials are mechanically mixed well with water usually containing additives such as surfactants, oxidizing agents, and acids, and settled for fractionation of coarse sand and gravel from the fine fraction. Less contaminated coarse fraction can be reused, while the highly contaminated fine fraction undergoes additional treatment. The fine fraction may be cleaned by flotation, leaching, or high-gradient magnetic separation.

Soil Flushing

In soil flushing, an aqueous solution is injected into the contaminated zone of soil followed by extraction of groundwater and elutriate (flushing solution mixed with the contaminants) and aboveground treatment and discharge. The goal of this *in situ* flushing is to enhance the solubility or mobility of contaminants and accelerating the remediation process. Flushing solutions may include water, surfactants, cosolvents, acids, bases, oxidants, chelants, and solvents which percolate through the soil, and soluble compounds present in the soil are dissolved. The elutriate is pumped out of the contaminated zone into a water treatment system to remove pollutants.

Encapsulation

The basic principle is the underground construction of an impermeable vertical barrier to allow the containment of gases and liquids. A variety of construction methods such as cutoff slurry walls using mainly cement–bentonite–water slurries, thin walls, sheet pile walls, bored-pile cutoff walls, jet grouting curtains, injection walls, and frozen barriers have been developed.

Vitrification

Vitrification involves the conversion of contaminated soil materials into glass-like substances. Vitrification can be used both as an *in situ* and as an *ex situ* technique. However, the processing is easier as an *in situ* process, but it is disadvantageous due to greater exposure if radioactive or dispersive contaminants are treated. When heated at a temperature of 1,000–2,000 °C, contaminated soil is melted; the hot liquid, upon cooling, solidifies into glass.

Electrokinesis

Electrokinetic decontamination or electroremediation of polluted soil is an *in situ* treatment technology for soils of variable textures. In this method, electrodes are placed into the soil and a direct electrical current is applied. This induces movements of ions to their respective electrodes through electroosmosis, electromigration, and electrophoresis. Transport of contaminants in the electric field are electromigration and electroosmosis in fine-grained soils and electromigration and electrophoresis in coarse-grained soils.

Chemical Remediation

Heavy metals occur as dissolved ions or bound to colloidal particles in soils. They are either precipitated or flocculated and separated. Precipitation is a process where ionic components transform into a non-soluble phase. In flocculation, small undissolved solids of colloidal size are aggregated into larger solid flocks. Precipitates and flocculates are then separated mechanically from the fluid by sedimentation, centrifugation, or flotation.

Solidification: Solidification and stabilization reduce the mobility of contaminants by immobilizing them within the soil through physical and chemical means. Solidification is described as the process by which sufficient quantities of solidifying material are added to wastes or contaminated soils to convert them into solids of a high structural integrity. Solidification is often followed by encapsulation.

Precipitation: The most important chemical parameters in the precipitation process are pH and concentration of metal ions. Generally, heavy metals tend to be present in ionic form at low pH levels and precipitate when pH is raised. Heavy metals can be precipitated as insoluble hydroxides, sulfides, carbonates, and others by adding precipitants (digested sludge, Fe salts, calcium hydroxide, Al salts, etc.). They are then separated by different solid/liquid separation techniques. The precipitation of metal sulfides has been proven to be very effective. Adsorptive materials such as iron oxides, clay minerals, zeolites, and organic matter are sometimes used to immobilize the metals in soil.

Ion Exchange: Ion exchanger is a matrix or resin laden with dissociable counter ions. Common ion exchangers are made of interlaced polystyrene and polyacrylate or condensation resins made from phenol and formaldehyde. When a solution containing dissolved metal ions are passed through the resin column, the metal ions are exchanged with adsorbed ions on resin surfaces. For regeneration, acids such as HCl and H_2SO_4 and alkaline solution (NaOH) are used. Wastewater produced in the process may be treated for precipitation or flocculation of metal ions.

Bioremediation

In bioremediation, organisms are employed in extraction and removal of metals from the contaminated soil. Organisms include microorganisms and higher plants.

Microbial Remediation

Microorganisms interact with toxic metals in very diverse processes. There are three general categories of biotechnological process for treating soil containing toxic metals: biosorption, extracellular precipitation, and uptake by purified biopolymers and other specific molecules derived from microbial cells. Biosorption is a physicochemical process of metal binding to microorganisms, and bioaccumulation is an active process using natural or recombinant microbial biomass to absorb metal ions. This is a new approach of metal remediation, and active research is going on in this respect.

Phytoremediation

Phytoremediation uses the ability of plants to adsorb, degrade, volatilize, or accumulate contaminants in soil, sediments, surface, or groundwater. According to Ghosh and Singh (2005), the chief processes of phytoremediation are phytostabilization, phytodegradation, phytoaccumulation, phytovolatilization, rhizodegradation, and evapotranspiration. Phytoaccumulation or phytoextraction is the most commonly and popularly used process of phytoremediation of heavy metals from contaminated soils. Phytoextraction utilizes the metal-hyperaccumulating ability of some plants in their aboveground parts. Selected hyperaccumulating plants are grown in contaminated soils, and harvested aboveground parts are treated in different ways (reduction in volume and weight, composting, compaction, burial, thermal treatment).

12.4.3.5 Metal Hyperaccumulator Plants

Some metal-tolerant plants are characterized by the ability to accumulate very high concentrations of metals, far in excess of normal physiological requirements (if any) and far in excess of the levels found in the majority of other species tolerant of metals. These plants are known as metal "hyperaccumulator" plants. The term hyperaccumulation was first applied by Jaffre et al. (1976) for nickel concentrations in the New Caledonian tree species *Sebertia acuminata*. According to Reeves (1992), a hyperaccumulator of Ni is a plant in which a Ni concentration of at least 1,000 mg kg^{-1} has been recorded in the dry matter of any aboveground tissue in at least one specimen growing in its natural habitat. Later, 1,000 mg kg^{-1} criterion was also taken for hyperaccumulation of Cu, Co, and Pb. A concentration >10,000 mg kg^{-1} is applied to Mn and Zn hyperaccumulation. Many plants have been identified as metal hyperaccumulators. However, a list of some important metal hyperaccumulator plants obtained from Reeves and Baker (2000) is given below.

Metal	Hyperaccumulator plants
Chromium	*Alyxia rubricaulis*, *Maytenus bureaviana*, *M. pancheriana*, *M. sebertiana*, *Garcinia amplexicaulis*, *Austromyrtus bidwillii*, *Eugenia clusioides*, *Eugenia* sp., *Beaupreopsis paniculata*, *Macadamia angustifolia*, *M. neurophylla*, *Astragalus stanleya*, *Haplopappus*, *Machaeranthera*
Cobalt and Copper	*Pandiaka metallorum*, *Anisopappus davyi*, *Cyanotis longifolia*, *Ascolepis metallorum*, *Bulbostylis pseudoperennis*, *Phyllanthus williamioides*, *Crotalaria cobalticola*, *Vigna dolomitica*, *Aeollanthus subacaulis* var. *linearis*, *Haumaniastrum robertii*, *Eragrostis racemosa*, *Actiniopteris* sp., *Buchnera henriquesii*, *Sopubia neptunii*, *Triumfetta dekindtiana*, *T. welwitschii* var. *descampii*, *Xerophyta retinervis* var. *equisetoides*

(continued)

(continued)

Metal	Hyperaccumulator plants
Cadmium, Lead, and Zinc	*Arabidopsis halleri, Thlaspi caerulescens, T. brachypetalum, T. ochroleucum, T. cepaeifolium, T. praecox, T. stenopterum, Polycarpaea synandra, Dichapetalum gelonioides, Armeria maritima, Agrostis tenuis, Arrhenatherum elatius, Festuca ovina, Rumex acetosa, Viola calaminaria*
Manganese	*Vaccinium myrtillus, Austromyrtus bidwillii*
Nickel	*Berkheya coddii, Pentacalia* (10 species), *Senecio* (9 species), *Alyssum* (52 taxa), *Bornmuellera* (6 taxa), *Cochlearia aucheri, Peltaria emarginata, Streptanthus polygaloides, Thlaspi* (23 taxa)

A concentration higher than 100 mg kg^{-1} is taken for Cd hyperaccumulation (Reeves and Baker 2000). Some hyperaccumulator plants accumulate very high amounts of heavy metals. For example, *Thlaspi calaminare* and *Phyllanthus serpentinus* were reported to accumulate 39,600 mg kg^{-1} Zn and 38,100 mg kg^{-1} Ni, respectively, in their leaves (Siegel 2002).

12.4.3.6 Radionuclides in Soil

Nuclides of any element that have atomic number greater than bismuth-83 is unstable and therefore radioactive (Igwe et al. 2005). They are called heavy nucleons or radionuclides. Radionuclides disintegrate or change spontaneously with a loss of energy in the form of ionizing radiation (van der Perk 2006). There are more than 60 radionuclides in nature. They can be placed in three general categories: primordial, cosmogenic, and anthropogenic. Examples of some naturally occurring radionuclides are americium-241, cesium-137, cobalt-60, iodine-129 and iodine-131, plutonium, radium, radon, strontium-90, technetium-99, tritium, thorium, and uranium. Cesium-137 is produced in nuclear fission, and it has a radioactive half-life of 30.17 years (Mster 1996). Primordial radionuclides are left over from the creation of the Earth. They typically have half-lives of hundreds of millions of years. Examples include uranium-235, uranium-238, thorium-232, and potassium-40. Primordial radionuclides end up in soil as part of the rock cycle.

Nuclear fission for atomic weapon testing and nuclear power generation provides some of the sources of soil contamination with anthropogenic radionuclides. To the naturally occurring radionuclides in soil such as ^{40}K, ^{87}Rb, ^{14}C, ^{235}U, ^{238}U, and ^{232}Th, a number of fission products have been added. However, only two of these are sufficiently long-lived to be of significance in soils: strontium-90 and cesium-137 with half-lives of 29.1 and 30 years, respectively. The average levels of these nuclides in soil in the United States are about 388 mc km^{-2} for 90Sr and 620 mc km^{-2} for 137Cs (Holmgren et al. 1993). These levels of the fission radionuclides in soil are not high enough to be hazardous (Igwe et al. 2005). However, artificial radioactivity may be released into the environment during the normal operations of nuclear facilities and installations such as nuclear ore processing, uranium enrichment, fuel fabrication, reactor operations, and application of radioisotopes in the fields of nuclear medicine, research, industry, and agriculture. Soils may receive some radionuclides from radioactive waste materials that have been buried for disposal (Knox et al. 2000). Plutonium, uranium, americium, neptunium, curium, and cesium are among the elements whose nuclides occur in radioactive wastes. Uranium mining activities produce large volume of residues for ore processing. These wastes contaminate adjacent land areas. They release other toxic pollutants as well. The radionuclides released from these wastes can give rise to human exposure by transport through the atmosphere, aquatic systems, or through soil subcompartments.

A catastrophic accident occurred at the Chernobyl Nuclear Power Plant in Ukraine on 26 April 1986. The explosion caused radiation to be released into the atmosphere over a 9-day period, with prevailing winds sending the plume generally in a north to northeasterly direction. The plume eventually spread over Europe, resulting in significant fallout of radiation associated with precipitation events in Austria, Switzerland, Germany, and Sweden. High levels of radioactive deposition were also reported in Italy and Britain. The releases during the accident contaminated about 125,000 km^2 of land in Belarus, Ukraine, and Russia with radiocesium levels greater than 37 kBq/m^2 and about 30,000 km^2 with radiostrontium greater than 10 kBq m^{-2}. Soil and water pollution from the accident is extensive, having been recorded in 22 oblasts of Russia and considerable areas of some other countries. In these areas, radionuclides have been measured in the soil to depths up to 25 cm. This quarter meter represents the vertical zone in which crop cultivation takes place. As a result, in Northern Ukraine, over 100,000 ha of agricultural land containing some of the world's richest soils have had to be abandoned (Savchenko 1995).

The movement of radionuclides in soil varies with soil type, pH, rainfall, and agricultural tilling. Radiocesium is generally confined to particles with a matrix of uranium dioxide, graphite, iron–ceramic alloys, rare-earth silicate, and silicate combinations of these materials. Cesium-137 behaves like potassium, and strontium-90 behaves like calcium in soil. Cesium-137 is not adsorbed strongly in soil and is readily taken up by plants. The movement of these radionuclides in the soil depends also on the chemical breakdown of these complexes by oxidation to release more mobile forms. The bulk of the fission products are distributed between organomineral and mineral parts of the soil largely in humic complexes. Radionuclides deposition can be a significant pathway to human exposure by ingestion of contaminated pasture by animals and then by the ingestion of contaminated animal products (meat and milk). Plant foods also accumulate radionuclides.

Study Questions

1. Discuss soil as a natural resource. Explain the need for management and protection of soil. What are the main types of human-induced soil degradation? Distinguish between physical and chemical degradation of soil with examples.
2. Soil compaction is mainly the product of mechanized farming. Explain. Discuss the factors affecting soil compaction. Write on the remedy of soil compaction.
3. Soil erosion is a natural process; it is accelerated by human. Explain. Describe the main types of water erosion. Discuss the suitable farming systems in sloping lands.
4. What is desertification? Discuss the principal causes of desertification. Describe the effects of desertification. How is desertification related with land use and erosion?
5. What are the environmentally important heavy metals? What are the effects of heavy metals on plants, humans, and microorganisms? Mention the methods of remediation of metal-polluted soils.

References

Acevedo F, Pizzul L, Castillo MP, González ME, Cea M, Gianfreda L, Diez MC (2010) Degradation of polycyclic aromatic hydrocarbons by free and nanoclay-immobilized manganese peroxidase from *Anthracophyllum discolor*. Chemosphere 80(3):271–278

Adefila EO, Onwordi CT, Ogunwande IA (2010) Level of heavy metals uptake on vegetables planted on poultry droppings dumpsite. Arch Appl Sci Res 2(1):347–353

Adriano DC, Bolan NS, Vangronsveld J, Wenzel WW (2005) Heavy metals. Encyclopedia of soils in the environment. Elsevier, Amsterdam

Ahn CK, Kim YM, Woo SH, Park JM (2008) Soil washing using various nonionic surfactants and their recovery by selective adsorption with activated carbon. J Hazard Mater 154:153–160

Alloway BJ (1995) Heavy metals in soils. Blackie Academic and Professional, an Imprint of Chapman & Hall, London

Alloway BJ, Tills AR, Morgan H (1985) The speciation and availability of cadmium and lead in polluted soils. In: Hemphill DD (ed) Trace substances in environmental health, vol 18. University of Missouri, Columbia

Angle JS, Heckman JR (1986) Effect of soil pH and sewage sludge on VA mycorrhizal infection of soybeans. Plant Soil 93(3):437

Aswathanarayana U (1999) Soil resources and the environment. Oxford/IBH Publishing Co. Pvt. Ltd, New Delhi

ATSDR (1999) Toxicological profile for mercury. Agency for Toxic Substances and Disease Registry. Available from: http://www.atsdr.cdc.gov/toxprofiles/tp46-p.pdf

Aydinalp C, Marinova S (2003) Distribution and forms of heavy metals in some agricultural soils. Pol J Environ Stud 12(5):629–633

Bagdatlioglu N, Nergiz C, Ergonul PG (2010) Heavy metal levels in leafy vegetables and some selected fruits. J Consum Prot Food Saf 5:421–428

Bai ZG, Dent DL, Olsson L, Schaepman ME (2008) Proxy global assessment of land degradation. Soil Use Manage 24:223–234

Baryła A, Pierzgalski E (2008) Ridged terraces – functions, construction and use. J Environ Eng Landsc Manage 16(2):Ia–If

Beinroth FH, Eswaran H, Reich PF, Van Den Berg E (1994) Land related stresses in agroecosystems. In: Virmani SM, Katyal JC, Eswaran H, Abrol IP (eds) Stressed ecosystems and sustainable agriculture. Oxford/IBH, New Delhi

Bell R, Evans CS, Roberts ER (1988) Decreased incidence of mycorrhizal root tips associated with soil heavy-metal enrichment. Plant Soil 106(1):143–145

Biddapa CC, Chino M, Kumazava K (1982) Migration of heavy metals in two Japanese soils. Plant Soil 66(3):299–316

Bied-Charreton M (2008) Integrating the combat against desertification and land degradation into negotiations on climate change: a winning strategy. www.csf-desertification.org. Accessed 21 Apr 2011

Blaikie P, Brookfield H (1987) Land degradation and society. Methuen, London

Blanco H, Lal R (2008) Principles of soil conservation and management. Springer, Dordrecht/London

Blum WEH (1997) Soil degradation caused by industrialization and urbanization. In: Proceedings of the international conference on problems of anthropogenic soil formation, Moscow, 16–21 June 1997

Bradl HB (2005) Heavy metals in the environment. Elsevier, Amsterdam

Bradl HB, Xenidis A (2005) Remediation techniques. In: Bradl HB (ed) Heavy metals in the environment, vol 6. Elsevier/Academic, Amsterdam

Brannon JM, Patrick WH Jr (1987) Fixation, transformation, and mobilization of arsenic in sediments. Environ Sci Technol 21:450–459

Bravo O, Silenzi JC (2002) Strip cropping in the semi-arid region of Argentina: control of wind erosion and soil water accumulation. Soil Sci 167:346–352

Bruins MR, Kapil S, Oehme FW (2000) Microbial resistance to metals in the environment. Ecotoxicol Environ Saf 45:198–207

Buchauer MJ (1973) Contamination of soil and vegetation near a zinc smelter by zinc, cadmium, copper and lead. Environ Sci Technol 7:131–135

Buol SW, Eswaran H (1994) Assessment and conquest of poor soils. In: Maranville JW et al (eds) Adaptation of plants to soil stresses, INTSORMIL publication 94–2. University of Nebraska, Lincoln

CAC (Codex Alimentarius Commission) (1993) Joint FAO/WHO Food Standards Program

CAC (Codex Alimentarius Commission) (2003) Evaluation of certain food additives and contaminants. FAO/WHO, Codex stan 230–2001, Rev 1–2003, Rome

Charreau C (1972) Problemes Poses Par L'utilization Agricole Des Sols Tropicaux Par Des Cultures Annuelles. Agronomie Tropicale 27:905–929

Chartres CJ, Kirby JM, Raupach M (1989) Poorly ordered silica and aluminosilicates as temporary cementing agents in hard-setting soils. Soil Sci Soc Am J 54(4):1060–1067

Chisholm A, Dumsday R (1987) Land degradation: problems and policies. Cambridge University Press, Cambridge

Cho M, Chardonnens AN, Dietz KJ (2003) Differential heavy metal tolerance of *Arabidopsis halleri* and *Arabidopsis thaliana*, a leaf slice test. New Phytol 158:287–293

Collins YE, Stotzky G (2001) Influence of heavy metals on the electrokinetic properties of bacteria. Am Soc Microbiol 39:967–980

Coskun M, Steinnes E, Viladimirovna F (2006) Heavy metal pollution of surface soil in the Thrace region, Turkey. Environ Moni Assess 119:545–556

Crookes MJ, Howe PD (1993) Environmental hazard assessment. Halogenated naphthalenes. Department of Environment, London

Culet P (2002) Desertification. Encyclopedia of life support systems. EOLSS Publishers Co Ltd., Oxford

Dan T, Hale B, Johnson D, Conard B, Stiebel B, Veska E (2008) Toxicity thresholds for oat (*Avena sativa* L.) grown in Ni-impacted agricultural soils near Port Colborne, Ontario, Canada. Can J Soil Sci 88:389–398

Davarynejad GH, Vatandoost S, Soltesz M, Nyeki J, Szabo Z, Nagy PT (2010) Hazardous element content and consumption risk of 9 apricot cultivars. Int J Hortic Sci 16(4):61–65

Davidson CI, Wu Y-L (1990) Dry deposition of particles and vapors. In: Lindberg SE, Page AL, Norton SA (eds) Advances in environmental science: acidic precipitation, vol 3, Sources, deposition, and canopy interactions. Springer, New York

Davis A, Drexler JW, Ruby MV, Nicholson A (1993) Micromineralogy of mine wastes in relation to lead bioavailability, Butte, Montana. Environ Sci Technol 27:1415–1425

Derici MR (2006) Degradation: chemical. In: Lal R (ed) Encyclopedia of soil science. Taylor & Francis, New York

Diaz-Ravina M, Baath E (1996) Development of metal tolerance in soil bacterial communities exposed to experimentally increase metal levels. Appl Environ Microbiol 62:2970–2977

Doelman P, Haanstra L (1984) Short-term and long-term effects of Cd, Cr, Cu, Ni, Pb, and Zn on microbial respiration in relation to abiotic soil factors. Plant Soil 79:317–321

Dregne HE (1977) Desertification of arid lands. Econ Geogr 53:329

Dudas MJ (1987) Accumulation of native arsenic in acid sulphate soils in Alberta. Can J Soil Sci 67:317–331

Dudas MJ, Pawluk S (1980) Natural abundances and mineralogical partitioning of trace elements in selected Alberta soils. Can J Soil Sci 60:763–771

Duruibe JO, Ogwuegbu MOC, Egwurugwu JN (2007) Heavy metal pollution and human biotoxic effects. Int J Phys Sci 2(5):112–118

Environment Agency (2007) UK soil and herbage pollutant survey. Report no. 7: Environmental concentrations of heavy metals in UK soil and herbage. Environment Agency, Bristol

Eriksson J, Hakansson I, Danfors B (1974) The effect of soil compaction on soil structure and crop yields. Bulletin 354. Swedish Institute of Agricultural Engineering, Uppsala

Eswaran H, Reich P (1998) Desertification: a global assessment and risks to sustainability. In: Proceedings of 16th International Congress Soil Science, Moutpellier, France

Eswaran H, Reich P, Beinroth F (1997) Global distribution of soils with acidity. In: Moniz AZAMC, Furlani RE, Schaffert NK, Fageria CA, Rosolem, Cantarella H (eds) Plant-soil interactions at low pH: sustainable agriculture and forestry production. Proceedings 4th international symposium on plant-soil interactions at low pH, Belo Horizonte, Minas Gerais, Brazil

Eswaran H, Beinroth F, Reich P (1999) Global land resources and population supporting capacity. Am J Altern Agric 14:129–136

Fabiola N, Giarola B, da Silva AP, Imhoff S, Dexter AR (2003) Contribution of natural soil compaction on hardsetting behavior. Geoderma 113(1–2):95–108

FAO (1998) FAO Production yearbook, vol 52. FAO statistics series no. 148, Rome

FAO (2000) Manual on integrated soil management and conservation practices. FAO Land and Water Bulletin 8, Rome

Fargasova A (1994) Effect of Pb, Cd, Hg, As, and Cr on germination and root growth of *Sinapis alba* seeds. Bull Environ Contam Toxicol 52:52

Fenn ME, Huntington TG, Mclaughlin SB, Eagar C, Gomez A, Cook RB (2006) Status of soil acidification in North America. J For Sci 52:3–13

Fosmire GJ (1990) Zinc toxicity. Am J Clin Nutr 51:225–227

Foth HD (1990) Fundamentals of soil science, 8th edn. Wiley, New York

Fu J, Zhou Q, Liu J, Liu W, Wang T, Zhang Q, Jiang G (2008) High levels of heavy metals in rice (*Oryza sativa* L.) from a typical E-waste recycling area in southeast China and its potential risk to human health. Chemosphere 71:1269–1275

Fuller WH (1977) Movement of selected metals, asbestos and cyanide in soil: application to waste disposal problem. Solid and Hazardous Waste Research Division, U.S. Environmental Protection Agency, Cincinnati

Ghadiri H (2004) Crater formation in soils by raindrop impact. Earth Surf Process Landf 29:77–89

Ghassemi F, Jakeman AJ, Nix HA (1995) Salinization of land and water resources: human causes, extent, management and case studies. Centre for Resource and Environmental Studies, The Australian National University, Canberra

Ghosh M, Singh SP (2005) A review on phytoremediation of heavy metals and utilization of its byproducts. Appl Ecol Environ Res 3(1):1–18

Gill WR (1971) Economic assessment of soil compaction. ASAE Monograph, St. Joseph

Giller KE, Mcgrath SP, Hirsch PR (1989) Absence of nitrogen fixation in clover grown on soil subject to long-term contamination with heavy metals is due to survival of only ineffective Rhizobium. Soil Biol Biochem 21(6):841–848

Glazer AN, Nikaido H (2007) Microbial biotechnology: fundamentals of applied microbiology, 2nd edn. Cambridge University Press, Cambridge, New York

Goi A, Trapido M, Kulik N (2009) Contaminated soil remediation with hydrogen peroxide oxidation. World Acad Sci Eng Technol 52:185–189

Gong Z, Wilke B-M, Alef K, Li P, Zhou Q (2006) Removal of polycyclic aromatic hydrocarbons from manufactured gas plant-contaminated soils using sunflower oil: laboratory column experiments. Chemosphere 62:780–787

Graetz RD (1996) Empirical and practical approaches to land surface characterisation and change detection. In: Hill J, Peter D (eds) The use of remote sensing for land degradation and desertification monitoring in the Mediterranean Basin. European Commission, Brussels

Graffham A (2006) EU legal requirements for imports of fruits and vegetables (a suppliers guide). Fresh Insights No. 1, DFID/IIED/NRI

Greene SBR (2005) Hardsetting soils. Encyclopedia of soil science, 2nd edn. CRC Press, Boca Raton

Greenwood NN, Earnshaw A (1997) Zinc, cadmium and mercury, 2nd edn, Chemistry of the elements. Butterworth-Heinemann, Oxford

Gregan PD, Hirth JR, Conyers MK (1989) Amelioration of soil acidity by liming and other amendments. In: Robson AD (ed) Soil acidity and plant growth. Academic, Sydney

Gregory JM, Borrelli J (1986) The Texas tech wind erosion equation. Am Soc Agric Eng 86:2528

Grewal MS, Kuhad MS (2002) Soil desurfacing–impact on productivity and its management. In: 12th ISCO conference, Beijing

Haan FAM, van Riemsdijk WM (1986) Behaviour of inorganic contaminants in soil. In: Assink JW, Van Den Brink WJ (eds) Contaminated soil. Martinus Nijhoff Publishers, Dordrecht

Hamberg R (2009) In situ and on-site soil remediation techniques – a review. Bachelor thesis, Department of Civil and Environmental Engineering, Division of Waste Science and Technology Lulea University of Technology

Heinrichs H, Schulz-Dobrick B, Wedepohl KH (1980) Terrestrial geochemistry of Cd, Bi, Ti, Pb, Zn and Rb. Geochim Cosmochim Acta 44:1519–1532

Heitkamp MA, Cerniglia CE (1988) Mineralization of polycyclic aromatic hydrocarbons by a bacterium isolated from sediments below an oil field. Appl Environ Microbiol 54:1612–1614

Helyar KR (1991) The management of acid soils. In: Wright RJ, Baligar VC, Murrmann RP (eds) Plant–soil interactions at low pH. Kluwer Academic, Dordrecht

Hemida SK, Omar SA, Abdel-Mallek AY (1997) Microbial populations and enzyme activity in soil treated with heavy metals. Water Air Soil Pollut 95:13–22

Hicks DH, Anthony T (2001) Soil conservation technical handbook. The Ministry for the Environment, Wellington

Hill MK (2010) Understanding pollution, 3rd edn. Cambridge University Press, Cambridge
Hill J, Peter D (1996) The use of remote sensing for land degradation and desertification monitoring in the Mediterranean Basin. European Commission, Brussels
Holmgren GGS, Meyer MW, Chaney RL, Daniels RB (1993) Cadmium, lead, zinc, copper and nickel in agricultural soils of United States of America. J Environ Qual 22:335–348
http//:oehha.ca.gov/risk/chhsltable.html. Accessed 16 May 2011
Hussein H, Farag S, Kandil K, Moawad H (2005) Tolerance and uptake of heavy metals by Pseudomonads. Process Biochem 40: 955–961
ICIMOD (1998) Bioterracing & soil conservation. Issues in Mountain Development ICIMOD. Kathmandu, Nepal. [online] http://www.icimod.org.np/publications/imd/imd98-7.htm. Accessed 17 Jan 2012
Igwe JC, Nnorom IC, Gbaruko BC (2005) Kinetics of radionuclides and heavy metals behaviour in soils: implications for plant growth. Afr J Biotechnol 4(13):1541–1547
Jaffre T, Brooks RR, Lee J, Reeves RD (1976) Sebertia acuminata: a hyperaccumulator of nickel from New Caledonia. Science 193:579–580
Kabata-Pendias A (2001) Trace elements in soils and plants. CRC Press, New York
Kabata-Pendias A, Mukherjee AB (2007) Trace elements from soil to human. Springer, Berlin
Kabata-Pendias A, Pendias H (2000) Trace elements in soil and plants. CRC Press, Boca Raton
Karczewska A, Szersze L, Kabała C (1998) Forms of selected heavy metals and their transformation in soils polluted by the emissions from copper smelters. Adv Geo-Ecol 31:705
Kayombo B, Lal R (1994) Response of tropical crops to soil compaction. In: Sloane BD, van Ouwerkkerk C (eds) Soil compaction in crop production. Elsevier, Amsterdam
Kelly JJ, Haggblom MM, Tate RL (2003) Effects of heavy metal contamination and remediation on soil microbial communities in the vicinity of a zinc smelter as indicated by analysis of microbial community phospholipids fatty acid profiles. Biol Fertil Soils 38:65–71
Knox AS, Seamans JC, Mench MJ, Vangronseveld J (2000) Remediation of metals and radionuclides. Contaminated soil using in situ stabilization techniques. Macmillan Publishers, New York
Kohnke H, Bertrand AR (1959) Soil conservation. McGraw-Hill Company, New York
Kurnia U, Sutono S, Anda M, Sulaeman AM, Kurniawansyah, dan SH, Talaohu (2000) Pengkajian baku mutu tanah pada lahan pertaniah. Laporan Akhir Kerjasama Penelitian Bapedal-Puslitbangtanak (in Bahasia Indonesia)
Lal R (1994) Tillage effects on soil degradation, soil resilience, soil quality, and sustainability. Soil Tillage Res 27:1–8
Lal R, Sobecki TM, Iivari T (2004) Soil degradation in the United States: extent, severity, and trends. Lewis Publishers, Boca Raton
Leita L, De-Nobil, Muhlbachova M, Mondini GC, Zerbi G (1995) Bioavailability and effects of heavy metals on soil microbial biomass survival during laboratory incubation. Biol Fertil Soil 19:103–108
Lindsay WL (1979) Lead. In: Chemical equilibria in soils. Wiley, New York
Lovell DJ, Parker SR, Van Peteghem P (2002) Quantification of raindrop kinetic energy for improved prediction of splash-dispersed pathogens. Phytopathology 92:497–503
Luo L, Ma Y, Zhang S, Wei D, Zhu Y (2009) Inventory of trace element inputs to agricultural soils in China. J Environ Manage 90:2524–2530
Markus J, McBratney AB (2000) A review of the contamination of soil with lead. I. Origin, occurrence and chemical form of soil lead. Progress Environ Sci 2(4):291–318
Masscheleyn PH, Delaune RD, Patrick WH Jr (1991) Effect of redox potential and pH on arsenic speciation and solubility in a contaminated soil. Environ Sci Technol 25:1414–1418
Matzner E, Davis M (1996) Chemical soil conditions in pristine Nothofagus forests of New Zealand as compared to German forest. Plant Soil 186:285–291
MBRLC (1988) A manual on how to farm your hilly land without losing your soil. Mindanao Baptist Rural Life Center, Davao del Sur
McDowell RW, Sharpley AN (2001) Approximating phosphorus release from soils to surface runoff and subsurface drainage. J Environ Qual 30:508–520
McElroy AE, Farrington JW, Teal JM (1989) Bioavailability of polycyclic aromatic hydrocarbons in the aquatic environment. In: Varanasi U (ed) Metabolism of polycyclic aromatic hydrocarbons in the aquatic environment. CRC Press, Boca Raton
Meyer O, Muller J, Meyts ER-D, Scheike T, Sharpe R, Sumpter J, Skakkebæk NE (1995) Male reproductive health and environmental chemicals with estrogenic effects. Environ Health Perspect 104(suppl 4):741–803
Morgan RPC (1986) Soil erosion and conservation. Longman, Essex
Mster GMA (1996) Soil introduction to environmental engineering and science. Prentice-Hall Inc., New York
Mudgal V, Madaan N, Mudgal A (2010) Heavy metals in plants: phytoremediation: plants used to remediate heavy metal pollution. Agric Biol J N Am 1(1):40–46
Munshower FF (1977) Cadmium accumulation in plants and animals of polluted and nonpolluted grasslands. J Environ Qual 6:411–413
Nkansah MA, Amoako CO (2010) Heavy metal content of some common spices available in markets in the Kumasi metropolis of Ghana. Am J Sci Ind Res 1(2):158–163
Northcote KH (1960) A factual key for the recognition of Australian soils. Divisional report no. 4/60. CSIRO Division of Soils, Australia
NRCC (1978) Effects of arsenic in the Canadian environment. National Research Council Canada Publication no NRCC 15391
Oldeman LR (1991) Global extent of soil degradation. ISRC annual report, Wageningen
Oldeman LR (1994) Global extent of soil degradation. In: Greenland DJ, Szaboles I (eds) Soil resilience and sustainable land use. CAB International, Wallingford
Oldeman LR (2000) GLASOD classification of soil degradation. ESCAP environment statistics course (draft)
Oldeman LR, Hakkeling RTA, Sombroek WG (1991) World map of the status of human-induced soil degradation: an explanatory note. International Soil Reference and Information Center, Wageningen
Page AL, Chang AC, El-Amamy M (1987) Cadmium levels in soils and crops in the United States. In: Hutchinson TC, Meema KM (eds) Lead, mercury, cadmium and arsenic in the environment. Wiley, New York
Pawloska TE, Charvat I (2004) Heavy metal stress and developmental patterns of Arbuscular Mycorrhizal Fungi. Appl Environ Microbiol 70(11):6643–6649
Pedro J, Alvarez JA, Illman W (2006) Bioremediation and natural attenuation. Wiley, Hoboken
Pimental D, Hall CW (1989) Food and natural resources. Academic, San Diego
Pizzul L, Castillo MP, Stenstrom J (2007) Effect of rapeseed oil on the degradation of polycyclic aromatic hydrocarbons in soils by *Rhodococcus wratislaviensis*. Int Biodeter Biodegr 59:111–118
Prescott LM, Harley JP, Klein DA (2002) Microbiology. Fund Appl Microbiol 2:1012–1014
Rababah A, Matsuzawa S (2002) Treatment system for solid matrix contaminated with fluoranthene. I – Modified extraction technique. Chemosphere 46(1):39–47
Reeves RD (1992) Hyperaccumulation of nickel by serpentine plants. In: Proctor J, Baker AJM, Reeves RD (eds) The vegetation of ultramafic (serpentine) soils. Intercept Ltd., Hampshire

Reeves RD, Baker AJM (2000) Metal-accumulating plants. In: Raskin I, Ensley BD (eds) Phytoremediation of toxic metals. Wiley, New York
Renard KG, Foster GR, Weesies GA, Mc Cool DK, Yoder DC (1997) Predicting soil erosion by water: a guide to conservation planning with the revised USLE, USDA hand book no. 703. USDA, Washington, DC
Roane TM, Pepper IL (2000) Microbial responses to environmentally toxic cadmium. Microbial Ecol 38:358–364
Roy WR, Krapac IG, Steele JD (1993) Sorption of cadmium and lead by clays from municipal incinerator ash-water suspensions. J Environ Qual 22:537–543
Rubilar O, Feijoo G, Diez MC, Lu-Chau TA, Moreira MT, Lema JM (2007) Biodegradation of pentachlorophenol in soil slurry cultures by *Bjerkandera adusta* and *Anthracophyllum discolor*. Ind Eng Chem Res 46:744–6751
Savchenko VK (1995) The ecology of the Chernobyl catastrophe: scientific outlines of an international programme of collaborative research. UNESCO, Paris
Sheoran AS, Sheoran V, Poonia P (2008) Rehabilitation of mine degraded land by metallophytes. Min Eng J 10(3):11–16
Shukla KP, Singh NK, Sharma S (2010) Bioremediation: developments, current practices and perspectives. Gen Eng Biotechnol J 3:1–20
Siegel FR (2002) Environmental geochemistry of potentially toxic metals. Springer, Berlin
Smith E, Naidu R, Alston AM (1998) Arsenic in the soil environment: a review. Adv Agron 64:149–195
Song Y, Liu L, Yan P, Cao T (2005) A review of soil erodibility of water and wind erosion research. J Geo Sci 15(2):167–176
Stegmann R, Brunner G, Calmano W, Matz G (2001) Treatment of contaminated soil. Springer, Berlin
Sudmeyer R, Bicknell D, Coles N (2007) Tree windbreaks in the wheatbelt. Department of Agriculture and Food, Government of Western Australia
Tacio HD (1993) Sloping Agricultural Land Technology (SALT): a sustainable agroforestry scheme for the uplands. Agrofor Sys 22:145–152
Tang C, Rengel Z (2003) Role of plant cation/anion uptake ratio in soil acidification. In: Rengel Z (ed) Handbook of soil acidity. Marcel Dekker, Inc., New York
Tolgyessy J (1993) Chemistry and biology of water, air and soil: environmental aspects. Elsevier, Amsterdam
Toppari J, Larsen JC, Christiansen P, Giwercman A, Grandjean P, Guilette Jr. LJ, Jegou B, Jensen TK, Jouannet P, Keiding N, Leffers H, McLachlan JA (1995) Male reproductive health and environmental chemicals with estrogenic effects, Miljoprojekt nr 290. Report of the Ministry of Environment and Energy, Danish Environmental Protection Agency, Copenhagen, Denmark
Tortella GR, Rubilar O, Cea M, Avendano M, Fernandez-Alberti S, Diez MC (2009) Chlorpyrifos degradation in a biomix of biobed system with allophonic top soil. Bio MicroWorld, Lisbon, Portugal, 2–4 Dec 2009
Tsuchiya K (1978) Cadmium studies in Japan-a review. Elsevier/North Holland Biomedical Press, Amsterdam
UNCCD (United Nations Convention to Combat Desertification) (1994) The convention to combat desertification. http://unccd.int/action-programmes/northmed/northmed.php. Accessed 24 Sept 2011
UNEP (1997) Report of the UNEP/RIVM/PE workshop on global and regional modeling of food production and land use and the long-term impact of degradation of land and water resources, Bilthovan, The Netherlands
Upjohn B, Fenton G, Conyers M (2005) Soil acidity and liming agfact AC.19, 3rd edn. NSW Department of Primary Industries. http://www.agric.nsw.gov.au/reader/soil-acid/2991-soil-acidity-and-liming-.pdf
USDA (2003) Zinc in foods-draft for comments. Foreign Agricultural Service (GAIN report) # CH3043
USEPA (1996) Distribution of soil lead in the nation's housing stock. U.S. Environmental Protection Agency, Washington, DC
USEPA (1997) Engineering bulletin: technology alternatives for the remediation of soils contaminated with As, Cd, Cr, Hg, and Pb, EPA/540/S-97/500. Office of Research and Development, US Environmental Protection Agency, Cincinnati
USEPA (1998) Sources of lead in soil: a literature review. Environmental Protection Agency, Washington, DC
Van den Akker JJH, Schjonning P (2004) Subsoil compaction and ways to prevent it. In: Schjønning P, Elmholt S, Christensen BT (eds) Managing soil quality: challenges in modern agriculture. CAB International, Wallingford
van der Perk M (2006) Soil and water contamination. Taylor & Francis, London
Van Es HM, Hill RL (1995) Soil compaction and soil-structure degradation. In: Crop residue management to reduce erosion and improve soil quality. Conservation research report no. 41. U.S. Department of Agriculture, Agricultural Research Service
Vaughan GT (1993) Investigation report CETLHIR148: the environmental chemistry and fate of arsenical pesticides in cattle tick dip sites and banana plantations. CSIRO, Division of Coal and Energy Technology, Centre for Advanced Analytical Chemistry, Sydney
Vidali M (2001) Bioremediation: an overview. Pure Appl Chem 73:1163–1172
Vlek PLG, Hillel D, Braimoh AK (2008) Soil degradation under irrigation. In: Braimoh AK, Vlek PLG (eds) Land use and soil resources. Springer, Dordrecht
Walker JM (1988) Regulation by other countries in foods and the human environment. In: Proceeding no. 2 "Cadmium Accumulation in Australian Agriculture". National symposium, Canberra, 1–2 Mar 1988. Australian Government Publishing Service, Canberra
Walsh LM, Summer ME, Keeney DR (1977) Occurrence and distribution of arsenic in soils and plants. Environ Health Perspect 19:67–71
Wang EX, Benoit G (1996) Mechanisms controlling the mobility of lead in the Spodosols of a northern hardwood forest ecosystem. Environ Sci Technol 30:2211–2219
Wang LK, Hung YT, Shammas NK (2010) Handbook of advanced industrial and hazardous wastes treatment. CRC Press, Boca Raton
Ward W, Singh A, van Hamme J (2003) Accelerated biodegradation of petroleum hydrocarbon waste. J Ind Microbiol Biotechnol 30:260
Watmough SA, Dillon P (2003) Calcium losses from a forested catchment in south-central Ontario, Canada. Environ Sci Technol 37:3085–3089
Wischmeier H, Smith DD (1965) Predicting rainfall-erosion losses from cropland east of the Rocky Mountains, Agriculture handbook 282. USDA, Washington, DC
Wischmeier WH, Smith DD (1978) Predicting rainfall erosion losses: a guide to conservation planning, USDA agriculture handbook 537. U.S. Government Printing Office, Washington, DC
WRI (World Resources Institute) (1997) World resources 1996–1997. World Resources Institute, Washington, DC
WRI (World Resources Institute) (2001) Disappearing land: soil degradation. Sustainable Development Service: Global Trends, Washington, DC
Wuest SB, Williams JD, Gollany HT (2006) Tillage and perennial grass effects on ponded infiltration for seven semi-arid loess soils. J Soil Water Conserv 61:218–223
Yan-Chu (1994) Arsenic distribution in soils. In: Arsenic in the environment Part I: Cycling and characterization. Wiley, New York
Yin Y, Allen HE, Li Y, Huang CP, Sanders PF (1996) Adsorption of mercury (II) by soil: effects of pH, chloride, and organic matter. J Environ Qual 25:837–844
Yu MH (2005) Environmental toxicology biological and health effects of pollutants. CRC Press, Boca Raton
Zachar D (1982) Soil erosion. Elsevier, Oxford
Zhou JFQ, Liu J, Liu W, Wang T, Zhang Q, Jiang G (2008) High levels of heavy metals in rice (Oryza sativa L.) from a typical E-waste recycling area in southeast China and its potential risk to human health. Chemosphere 71:1269–1275

Wetland Soils

13

Lands that remain waterlogged or inundated for most of the time of the year so that conditions become conducive to the development of hydrophytic vegetation are called wetlands. Wetlands may be natural or man-made. There are many different types of wetlands, including coastal and inland, tidal and swamp, and saline and freshwater. Bogs, swamps, marshes, fens, and peatlands are some typical wetland types. Wetlands perform important ecological functions, including storage and filtration of water, groundwater recharge, settling of sediments, and removal of pollutants. Wetlands are important reservoirs of carbon, and disturbance of wetlands may increase emission of greenhouse gases stemming from the decomposition of organic matter accumulated in an anaerobic environment over a long time. Many wetlands have been degraded, and in 1971 world nations in Ramsar Convention have agreed to preserve the wetlands. Wetlands have interesting biogeochemistry which is governed mainly by saturation with water, aeration, groundwater fluctuation, and E_h. These transformations are inextricably linked to the functions of the wetlands.

13.1 Wetlands Are Defined in Many Different Ways

According to Cowardin et al. (1979), there is no single, correct, indisputable, ecologically sound definition for wetlands, primarily because of the diversity of wetlands and because the demarcation between dry and wet environments lies along a continuum. The United States Environmental Protection Agency (USEPA 1993) defines wetlands as, "Those areas that are inundated or saturated by surface or ground water at a frequency and duration sufficient to support, and that under normal circumstances do support prevalence of vegetation typically adapted for life in saturated soil conditions." The Ramsar Convention provides a very wide definition, "Wetlands are areas of marsh, fen, peatland or water, whether natural or artificial, permanent or temporary, with water that is static or flowing, fresh, brackish or salt, including areas of marine water the depth of which at low tide does not exceed six meters" (Ramsar 1987). According to Reddy and DeLaune (2008), "Wetlands consist of a biologically active soil or sediment in which the content of water in or the overlying floodwater is great enough to inhibit oxygen diffusion into the soil/sediment and stimulate anaerobic (oxygen-free) biogeochemical processes and support hydrophytic vegetation."

13.2 Wetlands Are Kidneys of Nature

Swamps and marshes were at one time thought to be obstructions to industrial development. Many wetlands were drained and replaced by farmlands and railroads. In the United States, 50% of the wetlands were drained by 1970. Now, particularly after the Ramsar Convention in 1971 in Iran, people have come to recognize the values of wetlands. Wetlands, which occupy 6% of the Earth's land and freshwater surface, play a major role in maintaining the stability of the global environment. Wetlands are of huge practical importance: in global element cycles, as centers of biodiversity and, in global food production, for unique biogeochemistry and as habitat of diverse organisms (Kirk 2004).

Wetlands are recognized as kidneys of nature because they perform cleansing function of the environment. They store and filter water before it enters the surface and groundwater reservoirs; they remove nutrients and pollutants from water, and they settle the sediments. Wetland soil and vegetation filter contaminants out of water as it percolates through, returning cleaner water to rivers, lakes, and underground aquifers. Huge quantities of nutrients are carried to wetlands from surrounding upland watersheds. Wetland plants absorb these nutrients and keep the concentration of nutrients in water at a low level which reduces algal blooms and fish kills that could be caused by these nutrients.

Wetlands slow and retain surface water, providing water storage and shoreline stabilization. Water storage refers to their ability to temporarily retain heavy rain, surface water, and

K.T. Osman, *Soils: Principles, Properties and Management*,
DOI 10.1007/978-94-007-5663-2_13, © Springer Science+Business Media Dordrecht 2013

floodwaters. Shoreline stabilization refers to the stabilization of soil and vegetation along bodies of water. Wetlands contribute to flood control, for example, by collecting excess rainfall and releasing it slowly over time rather than in a torrent.

Wetlands provide wildlife habitat for a wide range of terrestrial, semiaquatic, and aquatic animals and numerous plant species. They are spawning ground of numerous fishes and nesting areas of many birds. The mix of aquatic and terrestrial habitats within a wetland provides a variety of food sources and habitat quality. They are often highly productive, and they provide abundant food for migratory waterfowl and wading birds. Resident wildlife is found utilizing wetlands in abundance such as beaver, muskrat, and river otter. Most amphibians live in or at least reproduce in wet habitat, so wetlands are vital to hundreds of species of frogs and salamanders. Many species of reptiles exist primarily in and around water, including alligators, water snakes, and numerous species of turtles. Thus, wetlands perform tremendous ecological functions. Wetlands are rich in biodiversity. Many species of plants and animals are wetland-dependent—they cannot simply survive without wetlands.

Wetlands contribute to climate regulation. Land cover can affect local temperature and precipitation; wetland ecosystems may affect greenhouse gas sequestration and emissions, or affect the timing and magnitude of runoff and flooding, for example. Wetlands also improve water quality through mechanical, physical, physicochemical, biological, and biochemical processes (Vymazal 2010). Therefore, wetlands must be preserved, restored, and conserved. That was the aim of the Ramsar Convention.

13.3 Many Wetlands Have Already Been Lost

Wetlands are lost by such human actions as urbanization, industrialization, farmland development, constructing roads and railroads, etc. Wetlands have been degraded by chemical contamination, increased nutrient inputs and eutrophication, hydrologic modification, and sedimentation. All these impacts affect the structure and functions of wetlands.

A major factor in the decline of wetlands is the concentration of human populations along coastlines. About three-quarters of the world's people live near oceans, rivers, inland seas, and lakes, and many coastal marshes and bogs have been filled in as part of the relentless quest for buildable land. The waste products of human habitation, particularly sewage and highway runoff, contaminate the remaining wetland tracts and endanger wildlife.

Alteration of wetland hydrology can change the soil chemistry and the structure of plant and animal community. Modifying the quantity of water entering a wetland or the period of saturation and inundation can change wetlands to upland systems or, conversely, to riverine or lacustrine systems. The principal ways of human-induced hydrologic alterations are drainage, dredging, stream channeling, ditching, levying, filling, stream diversion, groundwater withdrawal, and impoundment.

Urbanization and industrialization cause both loss and degradation of wetlands. Degradation occurs by changing water quality, quantity, and flow rates and increasing pollutant inputs and vegetation disturbance. Urbanization and industrialization in wetland areas provide sediments, nutrients, oxygen-demanding substances, road salts, heavy metals, hydrocarbons, bacteria, and viruses (USEPA 1994). Moreover, land development activities increase the amount of impervious surface which prevent rainfall from infiltrating into the soil and increase runoff. Runoff water carries sediments, organic matter, wastes, pesticides and fertilizers, heavy metals, hydrocarbons, road salts, and debris into streams and neighboring wetlands. Increased salinity, turbidity, and toxicity, and decreased dissolved oxygen, all affect aquatic life and, therefore, the food web. Excessive inputs of nutrients can lead to eutrophication or result in the release of pollutants from wetlands into adjacent water resources (USEPA 1993).

13.4 There Are Different Types of Wetlands

Cowardin et al. (1979) put forward the following systematic classification scheme of wetlands of the United States. In this scheme, there are five wetland systems: marine, estuarine, riverine, lacustrine, and palustrine. Marine and estuarine systems each have two subsystems, subtidal and intertidal; the riverine system has four subsystems, tidal, lower perennial, upper perennial, and intermittent; the lacustrine has two, littoral and limnetic; and the palustrine has no subsystems.

There are classes within the subsystems based on substrate material and flooding regime or on vegetative life-form. There are six classes on the basis of substrate and flooding regime: (1) rock bottom with a substrate of bedrock, boulders, or stones; (2) unconsolidated bottom with a substrate of cobbles, gravel, sand, mud, or organic material; (3) rocky shore with the same substrates as rock bottom; (4) unconsolidated shore with the same substrates as unconsolidated bottom; (5) streambed with any of the substrates; and (6) reef with a substrate composed of the living and dead remains of invertebrates (corals, mollusks, or worms). The bottom classes, (1) and (2), are flooded all or most of the time, and the shore classes, (3) and (4), are exposed most of the time.

There are five classes on the basis of the life-form of the dominant vegetation: (1) aquatic bed, dominated by plants that grow principally on or below the surface of the water; (2) moss–lichen wetland, dominated by mosses or lichens; (3) emergent wetland, dominated by emergent herbaceous angiosperms; (4) scrub–shrub wetland, dominated by shrubs or small trees; and (5) forested wetland, dominated by large

Fig. 13.1 A man-made wetland in Dhaka with water lily (*Nymphaea* sp.) (Photo courtesy of K. M. Rafee)

trees. There are modifying terms applied to the classes or subclasses for the system. In tidal areas, there are four water regime modifiers: subtidal, irregularly exposed, regularly flooded, and irregularly flooded. In nontidal areas, eight regimes are used: permanently flooded, intermittently exposed, semipermanently flooded, seasonally flooded, saturated, temporarily flooded, intermittently flooded, and artificially flooded.

According to the Ramsar classification (Ramsar 1987), there are three main classes of wetlands:

1. Marine/coastal wetlands
2. Inland wetlands
3. Human-made wetlands

These classes may further be subdivided on the type of water (fresh/saline/brackish/alkaline) into subclasses and may further be classified on the basis of substrate type or other characteristics.

Marine/coastal wetlands include 12 subclasses, namely, permanent shallow marine waters; marine subtidal aquatic beds; coral reefs; rocky marine shores; sand, shingle, or pebble shores; estuarine waters; intertidal mud, sand, or salt flats; intertidal marshes; intertidal forested wetlands; coastal brackish/saline lagoons; coastal freshwater lagoons; and karst and other subterranean hydrological systems.

Inland wetlands have 20 subclasses, namely, permanent inland deltas, permanent rivers/streams/creeks, seasonal/intermittent/irregular rivers/streams/creeks, permanent freshwater lakes (over 8 ha), seasonal/intermittent freshwater lakes (over 8 ha), permanent saline/brackish/alkaline lakes, seasonal/intermittent saline/brackish/alkaline lakes, permanent saline/brackish/alkaline marshes/pools, seasonal/intermittent saline/brackish/alkaline marshes/pools, permanent freshwater marshes/pools, seasonal/intermittent freshwater marshes/pools on inorganic soil, non-forested peatlands, alpine wetlands, tundra wetlands, shrub-dominated wetlands, freshwater tree-dominated wetlands, forested peatlands, freshwater springs, geothermal wetlands, and karst and other subterranean hydrological systems, inland.

Man-made wetlands include ten subclasses, namely, aquaculture ponds, ponds, irrigated lands, seasonally flooded agricultural land, salt exploitation sites, water storage areas, excavations, wastewater treatment areas, canals and drainage channels, and karst and other subterranean hydrological systems. Mitsch and Gosselink (1993) described the following wetland types: bog, fen, mire, marsh, playa, slough, swamp, wet meadow, and open water. Keddy (2000) described four principal wetland types: swamp, marsh, bog, and fen, as well as two others that support truly aquatic plants (i.e., shallow water at least 25 cm in depth and wet meadow). Figure 13.1 is a shallow man-made wetland.

13.5 Different Wetland Types Have Different Characteristics

Tidal salt marshes: Tidal salt marshes occur along coastlines. They are characterized by salt-tolerant plants such as *Spartina*. Soils can vary from mineral to organic but mainly mineral, with distinct features of saline and sodic soils. Sometimes there may be sulfide accumulation. As a result of nutrient loading from adjacent watersheds and from tidal water, these ecosystems exhibit high rates of primary productivity.

Tidal freshwater marshes: Tidal freshwater marshes are typically found upstream of estuaries. Water levels in these ecosystems are influenced by tides. These marshes are characterized by emergent macrophytes that are not tolerant to salinity. Floral diversity is high, with common species, including cattails, pickerelweed, wild rice, and arrowhead.

Fig. 13.2 Tidal mangrove swamp forest, Sundarbans, Khulna, Bangladesh (Pneumatophores in the foreground. Photo courtesy of Dr. Animesh Biswas)

Fig. 13.3 A freshwater marsh (Photo courtesy of M. A. Hannan)

Mangrove wetlands: Mangrove wetlands are tidally flooded saline water ecosystems found along the coasts of subtropical and tropical regions of the world. Vegetation is characterized by salt-tolerant trees, shrubs, and other plants adapted to brackish and saline tidal waters. Trees have simultaneous xerophytic, hydrophytic, and halophytic adaptations. These wetlands are also found at the interface of saltwater and freshwater. Three types of mangroves have been identified: black mangroves (dominant species *Avicennia* sp.) characterized by their ability to grow inland and having breathing roots, the pneumatophores; red mangroves (dominant *Rhizophora* sp.) characterized by their distinctive arching roots; and white mangroves characterized by their ability to grow farther inland with no visible modified root structure. Figure 13.2 shows a view of the mangrove forest of Bangladesh.

Freshwater marshes: Freshwater marshes are characterized by either mineral or organic soils and emergent macrophytic vegetation. They occur generally along streams in poorly drained depressions and in shallow water along the boundaries of lakes and ponds. Depth of water level varies from a few centimeters to 1 m. Some marshes can completely dry out in periods of drought but retain the characteristic hydrophytic vegetation. Some examples of freshwater marshes are wet meadows, prairie potholes, playa lakes, and vernal pools. Wet meadows occur in low-lying areas of the landscape such as farmlands and transitional areas with water-tolerant grasses, sedges, and rushes. Playas are round depressional areas on the landscape fed by freshwater from rainfall. Figure 13.3 is a portion of a freshwater marsh.

Freshwater swamps: Freshwater swamps are depressions in landscape fed primarily by rainwater. Floodwater reaches the depressions from surrounding watersheds (Fig. 13.4). These swamp areas are dominated by woody plants ranging from the forested red maple (*Acer rubrum*) of the northeastern United States to bottomland hardwood forests in the southeastern United States consisting of tupelo gum (*Nyssa* sp.), oak (*Quercus* sp.), and bald cypress (*Taxodium distichum*). Soils can be mineral and organic and are characterized by

Fig. 13.4 Freshwater swamp forest in Tangua Haor, Bangladesh. Dominant tree species *Barringtonia acutangula* (Photo courtesy of Dr. Kamrul Huda)

high organic matter content. Soils are saturated or flooded, creating highly anaerobic conditions.

Riparian wetlands: Riparian wetlands are adjacent to perennial, intermittent, and ephemeral streams, lakes, or rivers. Riparian areas receive water from (1) groundwater discharge, (2) overland and shallow subsurface flow from adjacent uplands, and (3) adjacent surface water body. These are floodplain areas with alluvial mineral soils.

Bogs: Bogs are characterized by spongy peat deposits and sphagnum moss vegetation. Precipitation is the primary source of water. Bogs are formed as a result of (1) gradual development of sphagnum in ponds and lakes and (2) sphagnum growth on wet areas of uplands that creates poorly drained conditions by holding water. Acidic peat deposits can build up to a considerable depth. Soils are usually organic, but may sometimes be mineral soils high organic matter and acidic conditions.

Fens: Fens are peat-forming wetlands that receive water and nutrients from surrounding watershed through drainage and surface runoff. They are less acidic and often more nutrient enriched and eutrophic and support more diverse vegetation than bogs. Fens support sedges, grasses, shrubs, and trees.

13.6 There Are Discharge and Recharge Wetlands

Wetlands develop on many different topographic conditions ranging from flatlands to sloping seashores. Runoff is minimum and water retention is higher in a flat topography where conditions are favorable for the development of broad areas of wet soils in humid regions. In such areas, most wetland depressions are groundwater discharge areas where groundwater emerges to become surface water. On the other hand, wetland develops only within localized depressions and valleys in a rolling topography. In depressions on uplands, the water table is deep, which are sites of focussed infiltration; soils there may exhibit increased leaching, more organic matter production, and stronger soil development. Many of these areas are recharge sites where surface water infiltrates and becomes groundwater (Schaetzl and Anderson 2005).

13.7 Wetland Soils Are Varied as Varied Are Wetland Types

Wetland soils are characterized by water saturation and anoxic situation, organic matter accumulation, gleying, mottling, iron/manganese segregation, oxidizing root channels and soil pore linings, and production of a reduced soil matrix. Ponnamperuma (1972) included wetland soils in the category of submerged soils. Submerged soils embraced waterlogged (gley) soils, marsh soils, paddy soils, and subaquatic soils. Wetland soils represent a diverse group of soils known as hydric soils. "A hydric soil is a soil that in its undrained condition is saturated, flooded, or ponded long enough during the growing season to develop anaerobic conditions that favor the growth and regeneration of hydrophytic vegetation" (National Technical Committee for Hydric Soils 1985). This definition was revised by deleting the phrase "in its undrained condition" and any direct reference to hydrophytic vegetation (USDA/NRCS 2006). Now hydric soil is defined as, "A hydric soil is a soil that is saturated, flooded, or ponded long enough during the growing season to develop anaerobic conditions in the upper part." For a good understanding of different aspects of wetland soils, the reader is referred to Richardson and Vepraskas (2001).

Wetland soils/sediments have several variants. The following categories may be recognized:

1. Wetlands with non-soil bottom
   (a) Wetlands with rocky bottom

   (b) Wetlands with gravelly and bouldery bottom
   (c) Wetlands with bottom of living or dead corals and mollusks
   (d) Wetlands with sandy bottom
2. Wetlands with soil bottom
   (a) Wetlands with mineral soils
   Soils with aquic suborders (Aqualfs, Aquands, Aquents, Aquepts, Aquolls, Aquods, Aquults, and Aquerts), albolls suborder, aquic subgroups (more than 100 subgroups), and aquisalids, pachic, or cumulic subgroups (National Technical Committee for Hydric Soils 1985 and later modifications—Federal Register 1995). Soils are poorly to very poorly drained. The list suggests enormous variation in characteristics of wetland or hydric soils.
   (b) Wetlands with organic soils (saprists, fibrists, and hemists suborders of the order Histosols)
   Saprists: About two-thirds of the material is well decomposed, and <1/3 of the plant material is identifiable.
   Fibrists: About one-third of the material is well decomposed, and >2/3 of the plant material is identifiable.
   Hemists: About half of the material is well decomposed, and the other half contains identifiable plant material.
3. Soils that are ponded during any part of the growing season (National Technical Committee for Hydric Soils 1985)
4. Soils that are frequently flooded for long duration during the growing season (National Technical Committee for Hydric Soils 1985)

13.7.1 Wetland Mineral Soils Have Mineral Parent Materials

The sequence of horizons in most wetland mineral soils may be A, E, Bt/Bg/Bh/Btg, Cg, and C. The t subscript stands for tont meaning clay, g for gleying, and h stands for humus accumulation. In cultivated wetlands such as paddy soils, there may be a top Ap horizon. In some mineral soils, an organic horizon, O, of variable thickness and differentiation may be present. This O horizon rests on the underlying mineral soil. Sub-horizons include Oi (fibric—little decomposed organic matter), Oe (hemic—intermediately decomposed organic matter), and Oa (sapric—highly decomposed organic matter) sub-horizons. The subordinate horizons of wetland soils are listed in Table 13.1.

While the deeper sediments are generally anoxic, a thin layer of oxidized soil usually exists at the soil–water interface. This layer may be aerated by the aerenchyma tissue of the hydrophytic grasses, sedges, and reeds. The oxidized layer is important since it permits the oxidized forms of prevailing ions to exist. This is in contrast to the reduced forms occurring at deeper levels of soil. The presence of oxidized

Table 13.1 Subordinate horizons of interest in wetland soils

Horizon	Significance
Oi	Fibric organic matter (little decomposition)
Oe	Hemic organic matter (intermediate decomposition)
Oa	Sapric organic matter (high decomposition)
Ap	Plowed A horizon
Bw	Weakly developed B horizon
Bt	Increase in illuvial clay in B horizon
Bg	Gleyed B horizon
Btg	Increase in illuvial clay and gleying
Bh	Humus-rich subsoil, spodic horizon

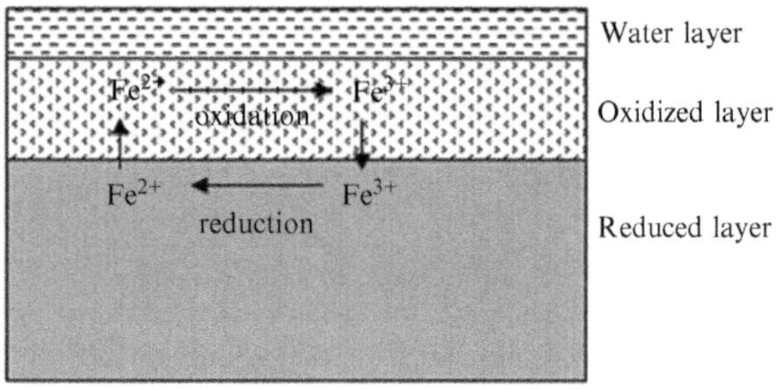

Fig. 13.5 Alternate oxidation–reduction of iron in wetland soil

ferric iron gives the overlying wetland soil a brown coloration, while reduced sediments have undergone "gleying," a process by which ferrous iron gives the underlying sediment a blue-gray tint (Scholz 2006).

Oxidized Fe^{3+} and Mn^{4+} ions of the upper oxidized layer diffuse into the underlying reduced layer, get reduced there into Fe^{2+} and Mn^{2+}, respectively, and again diffuse to the oxidized layer (Fig. 13.5).

The A horizon is usually the darkest layer in the soil (moist value/chroma darker than 3.5/2). Most natural A horizons vary in thickness from approximately 5–30 cm, but some are thicker. Plowing may obscure A horizon features because of mixing with subsoil materials.

The E horizon, which is an eluvial horizon, is lighter in color (grayish to whitish) than the rest of the soil above and below. Most E horizons have a coarser texture than the A horizon. E horizons contain frequent reddish mottles.

E horizons are underlain by a layer having a higher content of clay (Bt horizon) or transported organic material (Bh horizon). The B horizon is also the layer into which material translocates from the overlying E and A horizons. Wetland B horizons are grayer due to reduction and removal of iron pigmenting minerals. Hydric soils have chemically reduced anaerobic subsoil horizon, a condition conducive to gleying which produces a Bg horizon. Matrix colors of Bg horizons are usually gray, with chromas of 2 or less and values of 4 or more, usually with reddish mottles.

C horizon is the unconsolidated material in which soils have formed. C horizons retain the structure and color of the original parent material. Many soils in fluvial settings have only an A and a C horizon, entirely lacking O, E, and B horizons.

13.7.2 Wetland Organic Soils Have Organic Soil Materials

Several terms were used earlier to describe organic soils. Some of these terms are peat, bog, swamp, fen, and mire. Peat is a general term to describe the soils of wetlands with various amounts of undecomposed plant remains. Peat is still used to indicate a high fiber content in the organic soil. Peatland is used as a general landscape term to describe the expansive area of occurrence of peat. Peatlands generally form in depressions, slopes, and raised bogs. Actually most wetlands contain peat. Peat accumulations can be tens of meters deep (Rydin and Jeglum 2006).

Bogs are formed from shrub or moss vegetation. Bogs are areas of peat that are acid. Bogs are peat-accumulating wetlands that have no significant inflows or outflows and support acidophilic mosses such as *Sphagnum*. Fens are sedge and grasslike plant dominated areas of organic materials that contain considerable bases such as calcium. Swamps are formed under woody vegetation with variable amounts of tannin. Mire is a common term used in Europe to indicate the ecosystems in which waterlogged peat has accumulated, usually in raised areas (Collins and Kuehl 2001).

There are two types of organic formations: peat and muck. Peat is an organic soil material in which the original plant remains are recognizable. Peats are fibric materials. They are undecomposed to partially decomposed organic matter. Muck represents organic material in which the original plant remains are not recognizable. It is more decomposed, and contains more mineral matter, and is usually darker in color than peat.

In anaerobic soils, decomposition of organic material occurs slowly; under some conditions (e.g., low temperatures), decomposition may be so slow that the annual addition of organic material is greater than the annual removal of organic material by decomposition. The upper soils in peat wetlands are made up of almost 100% organic material, and individual plant parts can easily be identified even after centuries (Roth 2009).

Organic soils contain organic soil materials in more than half of the upper 80 cm (Soil Survey Staff 2003). An organic soil material, if saturated with water, has an organic carbon content not less than 12% (by weight) if the mineral part of the soil contains no clay; at least 18% organic carbon if the mineral part of the soil contains 60% or more clay; and proportional intermediate organic carbon for intermediate clay. If not saturated with water, soil materials must contain more than 20% (by weight) organic carbon to be organic (Soil Survey Staff 2003). Organic soil materials include peat and muck.

Organic soils develop from organic parent materials. The O horizon is designated for layers dominated by organic soil material. As stated earlier, O horizons have three subordinate horizon designations, a (sapric), e (hemic), and i (fibric), on the degree of organic matter decomposition. The degree of decomposition of the organic matter is determined by estimating the rubbed and unrubbed fiber content of the soil materials. Fibers are defined as pieces of plant tissue (excluding live roots) large enough to be retained on a 100-mesh sieve when the materials are screened after dispersion in sodium hexametaphosphate (Soil Survey Staff 1998).

The most highly decomposed organic material is the sapric material. It contains the lowest amount of fiber. The rubbed fiber content is < one-sixth of the soil volume and < one-third before rubbing. The bulk density of the sapric material is higher and the water content lower than hemic or fibric soil materials. Sapric soil materials are usually black or very dark gray in color. The bulk density of sapric soil materials is usually >0.2 g cm^{-3}, and the maximum water content when saturated is normally <450% on an oven-dry basis (Soil Survey Staff 1998). A general horizon sequence for a soil formed in sapric organic soil materials is Oa1, Oa2, etc.

Hemic material is intermediate in degree of decomposition and has a rubbed fiber content of one-sixth to two-fifths by volume. In an unrubbed sample of hemic material, the fibers can be seen and range from one-third to two-thirds of the volume. Colors of hemic soil materials commonly range from dark grayish brown to dark reddish brown. Bulk density of hemic soil materials is commonly between 0.07 and 0.18 g cm^{-3}. Their maximum water content when saturated commonly ranges from 450 to 850% or more. The horizon of hemic soil materials is designated by Oe.

Fibric soil material is the least decomposed organic soil material, and it contains two-fifths to three-fourths or more fibers by volume after rubbing. Fibric horizon is designated by Oi. Fibric materials are most common in the colder climates. Fibric soil materials are widespread in the raised bogs of the boreal forest zones. They have very low bulk densities (<0.1 g cm^{-3}), and the water content, when the soil is saturated, ranges from about 850 to >3,000% of the weight of the oven-dry material (Soil Survey Staff 1998). The colors of fibric soil materials are commonly brown, light yellowish brown, dark brown, or reddish brown.

13.7.3 Rice-Based Cropping Patterns Are the Most Important Land Use in Wetland Soils

A variety of crops can be grown productively on wetland soils with adequate water management. However, most of wetland areas in the world are used for rice-based production systems. Although rice can be grown in varying soil moisture conditions such as in permanently but shallowly flooded soils, seasonally flooded soils, poorly drained soils, and well-drained soils, rice production from wetland ecologies accounts for more than 95% of the world's rice output. Rice is the staple food of over half the world's population. It is the predominant dietary energy source for 17 countries in Asia and the Pacific, 9 countries in North and South America, and 8 countries in Africa. Rice provides 20% of the world's dietary energy supply, while wheat supplies 19% and corn (maize) 5% (FAO 2004). Wetland rice cultivation extends from 45° North to 40° South. In most rice-growing countries, there are distinct wet and dry seasons. Rice is grown without irrigation where monsoon rain is enough and well distributed. In most low lands, rice is grown with irrigation in the dry season. Rice areas in South and Southeast Asia may, in general, be classified into irrigated, rainfed upland, rainfed shallow water lowland, and rainfed deepwater lowland areas. In many countries, three rice crops are grown with supplemental irrigation. Modern high-yielding dwarf rice varieties have a growing period of about 90–110 days, and some of the varieties are day-neutral.

Most wetland rice fields are on alluvial soils in river valleys, basins, deltas, estuaries, lake fringes, and coastal plains. The surface of most of these areas is either level or gently undulating. Wetland rice is grown on all types of soils from sandy loam to heavy clay, heavy soils on river valleys being better suited. For growing rice, land is inundated, plowed, puddled, leveled, and transplanted. After transplanting seedlings, a depth of water ranging from 5 to 15 cm is maintained on the land usually for the whole growing period. After harvest the land is drained again. Puddling is done to level rice fields and to conserve water. Land preparation is aimed at retaining soil water and controlling weeds. The standing water after transplantation of rice seedlings prevents growth and regeneration of weeds except of some hydrophytic grasses and herbs. Land preparation also facilitates better water management and to a lesser extent fertilizer management. The rice crop can be established either by direct seeding or by transplanting of seedlings which were raised in separate nursery beds. Rice fields are flooded for varying depths and duration during most of the growing period. After harvest, the land is left wet or dry until the next season. This practice is responsible for the development of a surface Ap horizon. Continuous puddling of surface 15–20 cm soil layers may create hard pans below the plow layer, which is usually compact and has low permeability. In Australia, Egypt, and the USA where per hectare yield is higher, rice is generally planted in rotation with other crops. On a given field, rice usually is planted only one in every 2–3 years in Australia and only one in every 1–2 years in Egypt and the USA. The continuous planting of rice on the same field is practiced in many irrigated areas in tropical Asia and Africa. Yields are declining under intensive monoculture rice cultivation in these regions (FAO 2004).

Fertilizer rates for rice in European countries, Japan, and the USA are in general higher than those applied to wetland rice in tropical Asia and Africa. Farmers apply more fertilizers to irrigated rice than to rainfed rice. In the developing countries, farmers usually apply more major nutrient elements, especially nitrogen and to a lesser extent phosphorus and potassium. On the other hand, balanced fertilizer application based on soil tests is practiced in developed countries. Unbalanced fertilizer application in intensive rice production, however, has led to increasing deficiency in other nutrient elements which in turn limits rice yield. Sulfur deficiency has been reported from Bangladesh, Burma, Brazil, Indonesia, India, Nigeria, Philippines, and Thailand (Jones et al. 1982). In Bangladesh, 20 kg ha^{-1} S is generally recommended (Bhuiyan and Islam 1989). Application of 30 kg ha^{-1} S per crop has also been advocated in India (Singh 2004).

13.7.3.1 Wetland Rice Production Systems

The International Rice Research Institute convened in 1982 an International Rice Research Conference to classify the rice-growing environments. Factors affecting rice production were used (IRRI 1984), and wetland rice was classified as follows:

Irrigated rice: Irrigated rice areas have adequate water supply throughout the growing season. In much of the irrigated rice areas, rainfall supplements irrigation water. Irrigated areas are subdivided into three categories: (1) irrigated rice areas with favorable temperature, (2) irrigated rice areas with low temperature in tropical zone, and (3) Irrigated rice areas with low temperature in temperate zone.

Lowland rice: Rainfed lowland rice areas have a great diversity of growing conditions that vary by amount and duration of rainfall, depth of standing water, duration of standing water, flooding frequency, time of flooding, soil type, and topography. Rainfed lowland has five categories: (1) shallow and favorable rainfed lowland areas; (2) shallow and drought-prone rainfed lowland rice areas; (3) shallow, drought, and submergence rainfed lowland rice areas; (4) shallow, submergence-prone rainfed lowland rice areas; and (5) medium-deep, waterlogged rainfed lowland rice areas (stagnant for 2–5 months because of impeded drainage, depth of water 25–50 cm).

Deepwater rice: Deepwater rice areas are flooded with a water depth which is above 50 cm for major part of the growing season.

Tidal wetland rice: Tidal wetland rice areas are near to the seacoasts and inland estuaries that are directly or indirectly influenced by tides. Tidal wetland areas are divided into four categories.

13.7.3.2 Chemistry of Rice Soils

When the soil is submerged, the normal gas exchange that takes place between soil and air is drastically curtailed. Oxygen can enter the soil, and soil gases escape into the atmosphere by a process of molecular diffusion through water which is 10,000 times slower than normal gas exchange. The net result is that the concentration of oxygen in a rice soil is reduced to a very low value. The depletion of oxygen in the soil has important chemical and biological consequences (Ponnamperuma 1955).

The chemistry of flooded rice soil is characterized by (1) a deficiency of O_2; (2) an excess of CO_2; (3) an increase in the concentration of Fe^{2+} and Mn^{2+} due to reduction of Fe^{3+} and Mn^{4+}; (4) the decrease in nitrates and increase in denitrification; (5) the presence of sulfides; (6) the anaerobic decomposition of organic matter; (7) production of CH_4, C_2H_4, and H_2S; (8) an increase in the concentration of phosphate, calcium, magnesium, molybdenum, and silica; (9) an increase in pH of acid soils and a decrease in pH of alkaline soils; (10) a decrease in redox potential; (11) an increase in electrical conductivity; and (12) decrease in the concentration and availability of Zn, Cu, and S (Ponnamperuma 1984). These characteristics are common to almost all submerged soils whether rice is grown or not. What is uncommon to rice soils is the oxygenation of the root zone soil by internal ventilation system of rice plants. According to Ando et al. (1983), the oxidizing power of rice roots comprises two components—oxygen release and enzymatic oxidation. Jackson and Armstrong (1999) suggested that internal aeration is crucial for plant growth in waterlogged soils. Roots of rice contain large volumes of aerenchyma (Clark and Harris 1981), which provide a low-resistance pathway for diffusion of O_2 within the roots. According to Colmer (2003a), rice roots also contain a barrier against radial O_2 loss from the basal zones. The aerenchyma and barrier to radial oxygen loss enhance longitudinal diffusion of O_2 toward the root tip and thus root elongation into anoxic substrates (Colmer et al. 2006). These two traits are regarded as key features contributing to adaptation of rice to waterlogging (Colmer 2003b). Dissolved O_2 in floodwater also contributes to the oxidation of root zone soil. Thus, the immediate surface soil below the water level and the rice rhizosphere become oxidized by these processes. This is evidenced by the red coloration of iron oxide precipitates around rice roots. Below this oxidized thin layer, often less than 1 cm of upper soil, there is the reduced soil where anaerobic transformations prevail. The soil below the oxidized uppermost thin layer remains in a reduced condition during the period of submergence. The subsoil becomes dark gray, and iron, manganese, silica, and phosphate become more soluble. They move toward surface by diffusion and mass flow and are oxidized again and precipitated. Sandwiched between the oxidized surface layer and the zone of iron and manganese illuviation is the root zone of rice with reddish-brown streaks along root channels. When the land is drained after harvest of rice, almost the entire profile above the water table is reoxidized, giving it a highly mottled appearance (Ponnamperuma 1972).

On submergence the pH of an acid soil gradually increases. The probable causes are the increased concentrations of ammonia, sodium, magnesium, calcium, and ferrous ions. On the other hand, the pH of alkaline soils decreases for the production of organic acids and increased concentration of reduced iron and manganese. The entire profile of a submerged soil is not uniformly reduced. There are zones of oxidation and reduction, and there are concomitant irregularities in the distribution of E_h values within the profile. The upper yellowish-brown surface layer has an Eh value greater than 350 mV, and the subsoil has usually less than 350 mV. Along with O_2 deficiency, there is an abundance of CO_2, CH_4, and other end products of anaerobic decomposition of organic matter.

High concentrations of sulfides are formed in a submerged soil due to reduction of sulfates and anaerobic mineralization of proteins. Under low redox potential and a pH range between 5.5 and 8.5, reduction of sulfate produces hydrogen sulfide. In soils with high iron and manganese, all the sulfate in soil solution is almost completely removed as sulfides, and rice plants may suffer from sulfur deficiency (Thenabadu 1967).

13.7.4 A Sequence of Reduction Reactions Occur After Flooding a Soil

As mentioned earlier, the soil largely becomes anaerobic when they are flooded. In the complete absence of molecular oxygen (as is found in the reduced layer), microorganisms find other alternative electron acceptors than O_2. They use NO_3^-, MnO_2, $Fe(OH)_3$, SO_4^{2-}, and CO_2 as electron acceptors and reduce them. The series of reduction reactions may be shown below:

1. $O_2 + 4e^- + 4H^+ \rightarrow 2H_2O$ (Aerobic respiration: facultative anaerobes and aerobes)
2. $2NO_3^- + 10e^- + 12H^+ \rightarrow N_2 + 6H_2O$ (Denitrification)
3. $MnO_2 + 2e^- + 4H^+ \rightarrow Mn^{2+} + 2H_2O$ (Mn reduction by Mn-reducing bacteria)
4. $Fe(OH)_3 + e^- + 3H^+ \rightarrow Fe^{2+} + 3H_2O$ (Fe reduction by Fe-reducing bacteria)
5. $SO_4^{2-} + 8e^- + 8H^+ \rightarrow S^{2-} + 4H_2O$ (SO_4^{2-} reduction by SO_4^{2-}-reducing bacteria)
6. $CO_2 + 8e^- + 8H^+ \rightarrow CH_4 + 2H_2O$ (CH_4 production by methane bacteria)

There are concomitant changes in the redox potential (E_h, Chap. 7) and pH of the soils. The pH of acid soils gradually rises to 7.0 after submergence of a soil. The sequential changes in the E_h are shown below:

Observation	E_h (mV)
Disappearance of oxygen	+330
Disappearance of nitrate	+220
Appearance of manganese Mn (II) ions	+200
Appearance of ferrous iron (Fe^{2+}) ions	+120
Disappearance of sulfate	−150
Appearance of methane	−250

In the period when the soil is drained, the groundwater is lowered and the soil is exposed to free oxygen again. Reduced substances such as Fe^{2+} and Mn^{2+} are oxidized to Fe^{3+} and Mn^{4+}, respectively. Thus, in seasonally flooded soil with fluctuating groundwater table, alternate oxidation and reduction creates iron/manganese mottling in the B horizon. After prolonged submergence, the soil has a low redox potential and a neutral pH; absence of oxygen, nitrate, and sulfate; presence of carbon dioxide, ammonium, and sulfide; solubility of iron, manganese, and phosphorus; and production of an array of transitory substances ending as carbon dioxide, methane, and humus by decomposition of organic matter (Ponnamperuma 1972).

13.7.4.1 Micronutrient Deficiencies and Iron Toxicity in Wetland Rice Soils

Micronutrient deficiencies in predominantly rice-growing soils of the Indo-Gangetic Plains emerged with the adoption and spread of intensive agriculture in the region. Continuous cropping, lack of micronutrient fertilizers, decreased use of organic manure, reduced recycling of crop residues, and bumper harvests in the past three decades have induced micronutrient deficiencies in this region. Zinc deficiency has been reported from many areas. Deficiencies of Fe, Mn, B, and Mo were also recorded. Zinc deficiency is the most widespread in the four Indo-Gangetic Plain countries (Bangladesh, India, Nepal, and Pakistan; Nayar et al. 2001).

Among 2,52,000 surface soil samples collected from different parts of India, 49, 12, 4, 3, 33, and 41% showed deficiencies of Zn, Fe, Mn, Cu, B, and S, respectively. Many of these soils are under rice cultivation. Coarse-textured, calcareous, alkaline, or sodic soils having sandy texture, high pH, and low organic matter are generally low in available zinc. There were calcareous soils, Vertisols, and Inceptisols among these soils (Singh 2004). Zinc deficiency in rice soils is often linked to iron toxicity. When the amount of dissolved ferrous iron is high due to low oxygen content in the root zone, iron plaques are formed. They work as an effective adsorbent of zinc making it unavailable for the plant. The soluble iron is absorbed by the roots and accumulated in the leaves resulting in brown spots and reduced growth (Zhang et al. 1998). Iron toxicity and zinc deficiency are common problems in rice cultivation in Africa. More than half of the soils in Mali are deficient in zinc. Zinc deficiency does not only result in reduced yield but also affects human health.

13.7.5 Subaqueous Soils Lie Underwater

Subaqueous soils are soil materials (mud/sediments) found underwater in bottom of rivers, lakes, and oceans. Kubiena (1953) included two types of subaqueous soils in his natural system of soil classification: (a) subaqueous soils not forming peat and (b) peat-forming subaqueous soils. According to Ponnamperuma (1972), the unconsolidated aqueous sediments are considered as "soils" because of the following reasons: These sediments are derived from soil components; soil-forming processes operate in them; they are high in organic matter; they are biologically active; their surface layers show distinct horizons, and they are capable of supporting rooted plants and supply them nutrients. Subaqueous soils are permanently flooded soils that occur immediately below a water depth of <2.5 m. Areas permanently covered by water too deep (>2.5 m) for the growth of rooted plants are excluded.

Pedogenic processes in subaqueous soils are similar to those occurring in subaerial soils except that they form under anaerobic conditions. However, they vary in bottom type, redox substances, presence of sulfidic materials, potential for submerged aquatic vegetation restoration, etc. The fragility of such soils for providing vital functions to support life needs their study and monitoring (Erich et al. 2010).

13.8 Nutrient Transformations in Wetland Soils Occur in Low Redox Potentials

13.8.1 Nitrogen Undergoes Mineralization, Immobilization, Nitrification, Fixation, and Denitrification

Like arable soils, wetland soils contain nitrogen chiefly in the organic form and ammonium, nitrite, nitrate, and molecular nitrogen. Nitrogen is transformed within these forms by microorganisms. According to Ponnamperuma (1972), the main interconversions can be shown as:

$$\text{Proteins} \rightleftarrows \text{amino acids} \rightleftarrows \underset{\uparrow\ N_2}{NH_4^+} \rightleftarrows \underset{\downarrow\ N_2}{NO_2^-} \rightleftarrows NO_3^-$$

The biological conversion of proteins to ammonium is termed as the mineralization, which goes hand in hand with organic matter decomposition. The step is also known as ammonification. Decomposition of organic matter and ammonification are faster in the upper oxidized layer of soil than in the reduced subsoil. Ammonification can occur aerobically or anaerobically, but aerobic ammonification is much faster (Vepraskas and Faulkner 2001).

Ammonium may be absorbed by roots of plants and may be oxidized to nitrites and nitrates in the oxidized soil zone by the process of nitrification. By this process, the ammonium is first converted to nitrite (NO_2^-) and then to nitrate (NO_3^-), a biological transformation conducted by bacteria of the genera *Nitrosomonas*, *Nitrosococcus*, and *Nitrobacter*. Nitrification occurs in the aerobic soil zone of hydric soils and around aerated roots growing in anaerobic soil. Rates for nitrification are variable and depend on the quantities of both NH_4^+ and O_2 in the soil. It has been reported to range from 0.01 to 0.16 g N m^{-2} day^{-1}, being limited primarily by the supply of O_2 (Reddy and D'Angelo 1994). Nitrates may be absorbed by plant roots, diffuse to the underlying reduced layer, where it is denitrified, or leach to greater depths.

In anaerobic soil, ammonia is derived from anaerobic deamination of amino acids, degradation of purines, and hydrolysis of urea. Less than 1% comes from nitrate reduction. In the reduced layer, ammonium accumulates because the lack of O_2 in this layer restricts the nitrification. Almost all the mineralizable nitrogen in a soil is converted to ammonia within 2 weeks of submergence if the temperature is favorable; the soil is not strongly acid or very deficient in available phosphorus. The amount of ammonia produced during the first 2 weeks of submergence may range from 50 to 350 mg kg^{-1} N on the basis of the dry soil (Ponnamperuma 1972). For this accumulation of ammonium in the reduced layer, a concentration gradient will be established between high concentration in lower reduced layer and low concentration in the oxidized top layer. This may cause a passive flow of ammonium from the anaerobic to the aerobic layer, where microbiological processes convert the ion into other forms of nitrogen (Scholz 2011).

Immobilization of nitrogen takes place in submerged soil by the assimilatory reduction of nitrate into NH_3 and incorporation of the product in cell substances. Ammonia formed by ammonification is also immobilized. Assimilatory nitrate reduction is more rapid in the oxidized layer. A substantial proportion of added nitrate may, in some submerged soils, be assimilated, and enter the pool of soil organic matter.

Nitrate is not stable in the reduced soil layer. It undergoes dissimilatory reduction or nitrate respiration, a process where NO_3^- acts as an alternative electron acceptor to O_2. The process is also known as denitrification, because nitrate is converted to N_2 and lost to the atmosphere through this process. Nicholas (1963) defines denitrification as a special case of nitrate respiration in which nitrate, nitrite, or some intermediates are converted to nitrogen or its oxides. Denitrification is brought about by a large number of bacteria and fungi which include heterotrophic and autotrophic species. These organisms can use O_2 as electron acceptor in an aerated soil, but they utilize NO_3^- as the alternative electron acceptor when O_2 is not available. Denitrification in cultivated soils is undesirable because it causes the loss of a valuable plant nutrient. But in effluents from sewage works, in groundwater, streams, lakes, and estuaries, loss of nitrogen by denitrification is highly desirable because it helps to prevent the contamination of drinking water by nitrate and the pollution of natural bodies of water by excessive growth of aquatic plants (Ponnamperuma 1972).

Some microorganisms can reduce molecular nitrogen to ammonium and incorporate it into organic tissues. They are the nitrogen fixers which include some bacteria and cyanobacteria. Nitrogen-fixing bacteria can fix nitrogen anaerobically and are most abundant in the upper 5 cm of anaerobic soil. Cyanobacteria are abundant in paddy soils and other seasonally flooded soils. Members of the genera *Nostoc*, *Anabaena*, *Oscillatoria*, *Tolypothrix*, *Calothrix*, *Phormidium*, and *Aulosira* are familiar nitrogen fixers. Diffusion of N_2 through water is slow, and therefore, nitrogen fixation in wetland soils is also low. Hydrophytic plants can transport N_2 to their roots as they transport O_2. N_2-fixation rates in wetlands vary from 0.02 to 90 g N m^{-2} $year^{-1}$ around the world. N_2 fixation supplies the majority of N to some wetlands, but adds <5% of the N input to others (Vepraskas and Faulkner 2001).

13.8.2 Phosphorus Undergoes Mineralization, Immobilization, Solubilization, and Sorption

Phosphorus finds its way into the wetlands from parent materials, sediments brought in from upland watersheds, and effluents and flows out of the wetland with water; water as it flows in runoff, surface flow, and groundwater flow. Phosphorus moves within the wetlands system among soil, water, and organisms. Phosphorus has no role in redox systems because it has a constant valence state of 5. But its solubility is greatly affected by E_h. As the E_h of soil is lowered by submergence, iron and manganese are solubilized and sorbed phosphates by them are released.

There are three major forms of phosphorus in wetland soils: organic P, fixed mineral P, and soluble inorganic P. Soluble inorganic P includes anions $H_2PO_4^-$, HPO_4^{2-}, and PO_4^{3-}; their prevalence occurs at pHs of 2–7, 8–12, and >13, respectively (Vepraskas and Faulkner 2001). Fixed mineral P consists of P bound to oxides or hydroxides of Al^{3+}, Fe^{3+}, Ca^{2+}, or Mg^{2+}. Organic phosphorus is found in soil organic matter, and plant and animal residues, which contain such compounds as inositol phosphates, phospholipids, and nucleic acids.

Organic P and fixed mineral P comprise approximately 80–90% of the P in a wetland. Living plants store most of the remaining P, leaving very little in the water column (Richardson 1999). Most soil P (>95%) in organic soils is in the organic form with cycling among P forms controlled by biological forces (i.e., microbes and plants). In mineral soils, most of the soil P may be bound to minerals containing Al, Fe, Ca, or Mg. Organic P is mineralized by heterotrophic microorganisms, but the process is very slow in the reduced soil, and most P remains in the organic form in wetland soils. When P is loaded in wetland waters from wastes, fertilizer residues, and effluents, the wetlands may undergo eutrophication.

Macrophytes absorb a significant amount of phosphorus from the soil. Most phosphorus taken up from soils by macrophytes is reincorporated into the sediment as dead plant material and therefore remains in the wetland indefinitely. A means of removing excess load of phosphorus may be the harvest of macrophytes at the end of the growing season (Scholz 2006). Harvesting would reduce phosphorus levels in upper sediment layers and drive phosphorus movement into deeper layers, particularly the root zone. In deep layers of sediment, the phosphorus sorption capacity increases along with a lower desorption rate (Scholz 2011).

13.8.3 Sulfur Undergoes Mineralization, Immobilization, Oxidation, and Reduction

The transformations of sulfur in wetlands include (1) mineralization of organic sulfur, (2) oxidation of elemental sulfur and sulfides, (3) reduction of sulfates into sulfides, and (4) immobilization. Mineralization is carried out by heterotrophic microorganisms, and it is generally faster in the oxidized layer. Dissimilation of the amino acids cysteine, cystine, and methionine produces H_2S, thiol, ammonia, and fatty acids. The putrefying odor of cyanobacteria is attributed to methyl, butyl, and isobutyl thiols, and the bad egg odor is attributed to H_2S.

Hydrogen sulfide is largely produced in wetland soils by reduction of SO_4^{2-}. This reduction is brought about by the bacteria of the genus *Desulfovibrio* which use SO_4^{2-} as the terminal electron acceptor in an anaerobic environment. The H_2S, so formed, may react with metals to produce insoluble metal sulfides; it may act as hydrogen donor to photosynthetic green and purple bacteria, and it may be oxidized chemically and bacterially at aerobic/anaerobic boundaries (Ponnamperuma 1972). Sulfate reduction occurs very fast in neutral to alkaline soils in wetland conditions. All sulfates may be reduced within some weeks of submergence of the soil. In tidal and coastal wetlands, there is plenty of supply of SO_4^{2-} with seawater. In acid soils, however, sulfate reduction may be slow. In any way, loss of sulfur from wetland systems occurs through the volatilization of H_2S formed by mineralization or sulfate reduction.

Oxidation of sulfides to elemental sulfur and sulfates can occur in the aerobic layer of some soils and is carried out by chemoautotrophic (*e.g.*, *Thiobacillus* spp.) and photosynthetic microorganisms. *Thiobacillus* spp. may gain energy from the oxidation of hydrogen sulfide to sulfur and, further, by certain other species of the genus, from sulfur to sulfate. The sulfate ions may diffuse from the aerobic zone to the anaerobic zone and get reduced there.

Study Questions

1. Define wetland. Discuss the ecological significance of wetlands. How have many wetlands been lost?
2. What are the different types of wetlands? Distinguish between bog and marsh. What is the difference between peat and muck? How is peat formed? Discuss characteristics of organic wetlands.
3. Discuss characteristics of mangrove wetlands, freshwater swamps, and freshwater marshes. What do you mean by subaqueous soils?
4. Write the importance of rice as a cereal crop. Explain the chemical behavior of rice-growing wetland soils.
5. What are the differences between aerobic and anaerobic transformations in soil? Discuss the transformations of nitrogen in wetland soils.

References

Ando T, Yoshida S, Nishiyama I (1983) Nature of oxidizing power of rice roots. Plant Soil 72:57–71

Bhuiyan NI, Islam MM (1989) Sulphur deficiency problem of wet land rice soils in Bangladesh agriculture. J Soil Fert 53(1):103

Clark LH, Harris WH (1981) Observations on the root anatomy of rice (*Oryza sativa* L.). Am J Bot 68:154–161

Collins ME, Kuehl RJ (2001) Organic matter accumulation and organic soils. In: Richardson JL, Vepraskas MJ (eds) Wetland soils, genesis, hydrology, landscapes, and classification. Lewis Publishers, Boca Raton

Colmer TD (2003a) Aerenchyma and an inducible barrier to radial oxygen loss facilitate root aeration in upland, paddy and deep-water rice (*Oryza sativa* L.). Ann Bot 91:301–309

Colmer TD (2003b) Long-distance transport of gases in plants: a perspective on internal aeration and radial oxygen loss from roots. Plant Cell Environ 26:17–36

Colmer TD, Cox MCH, Voesenek LACJ (2006) Root aeration in rice (Oryza sativa): evaluation of oxygen, carbon dioxide, and ethylene as possible regulators of root acclimatizations. New Phytol 170:767–778

Cowardin LM, Carter V, Golet FC, LaRoe ET (1979) Classification of wetlands and deepwater habitats of the United States. U.S. Department of the Interior, Fish and Wildlife Service, Washington, DC

Erich E, Drohan PJ, Ellis LR, Collins ME, Payne M, Surabian D (2010) Subaqueous soils: their genesis and importance in ecosystem management. Soil Use Manage 26(3):245–252

FAO (2004) Rice is life. Proceedings of the FAO Rice Conference, 12-13 February 2004. Food and Agricultural Organization of the United Nations, Rome, Italy

Federal Register (1995) Hydric soils of the United States. Federal Register, Washington, DC

IRRI (1984) Terminology for rice growing environments. International Rice Research Institute, Manila

Jackson MB, Armstrong W (1999) Formation of aerenchyma and the processes of plant ventilation in relation to soil flooding and submergence. Plant Biol 1:274–287

Jones US, Katyal JC, Mamaril CP, Park CS (1982) Wetland rice nutrient deficiencies other than nitrogen. Rice research strategies of the future. IRRI, Los Banos

Keddy PA (2000) Wetland ecology: principles and conservation. Cambridge University Press, Cambridge

Kirk G (2004) The biogeochemistry of submerged soils. Wiley, Chichester

Kubiena WL (1953) The soils of Europe. Thomas Murray and Company, London

Mitsch WJ, Gosselink JG (1993) Wetlands, 2nd edn. Van Nostrand Reinhold Co., New York

National Technical Committee for Hydric Soils (NTCHS) (1985) Hydric soils of the United States. USDA Soil Conservation Service, Washington, DC

Nayar VK, Arora CL, Kataki PK (2001) Management of soil micronutrient deficiencies in the rice-wheat cropping system. In: Kataki PK (ed) The rice-wheat cropping systems of south Asia: efficient production management. Food Products Press, New York

Nicholas DJD (1963) The metabolism of inorganic nitrogen and its compounds in micro-organisms. Biol Rev 38:530–568

Ponnamperuma FN (1955) Some aspects of the chemistry of rice soils. Trop Agric 111:92–101

Ponnamperuma FN (1972) The chemistry of submerged soils. Academic, New York

Ponnamperuma EN (1984) Effects of flooding on soils. In: Koslowski TT (ed) Flooding and plant growth. Academic, New York

Ramsar (1987) Convention on wetlands (Ramsar, Iran, 1971), as amended in 1982 and 1987. Ramsar Convention Bureau, Gland

Reddy KR, DeLaune RD (2008) Biogeochemistry of wetlands: science and applications. CRC Press/Taylor & Francis, Boca Raton

Reddy KR, D'Angelo EM (1994) Soil processes regulating water quality in wetlands. In: Mitsch WJ (ed) Global wetlands. Old world and new. Elsevier, Amsterdam

Richardson CJ (1999) The role of wetlands in storage, release, and cycling of phosphorus on the landscape: a 25-year retrospective. In: Reddy KR, O'Connor GA, Schelske CL (eds) Phosphorus biogeochemistry in subtropical ecosystems. Lewis Publishers, Boca Raton

Richardson JL, Vepraskas MJ (2001) Wetland soils, genesis, hydrology, landscapes, and classification. Lewis Publishers, Boca Raton

Roth RA (2009) Freshwater aquatic biomes. Greenwood Press, Westport/London

Rydin H, Jeglum JK (2006) The biology of peatlands. Oxford University Press, Oxford

Schaetzl RJ, Anderson S (2005) Soils: genesis and geomorphology. Cambridge University Press, Cambridge

Scholz M (2006) Wetland systems to control urban runoff. Elsevier, Amsterdam

Scholz M (2011) Wetland systems: storm water management control. Springer, London

Singh MV (2004) Micronutrient deficiencies in Indian soils and field usable practices for their correction. IFA International Conference on Micronutrients, New Delhi, 23–24 Feb 2004

Soil Survey Staff (1998) Keys to soil taxonomy, 8th edn. USDA, Natural Resources Conservation Service, US Government Printing Office, Washington, DC

Soil Survey Staff (2003) Keys to soil taxonomy, 9th edn. USDA Natural Resources Conservation Service, Washington, DC

Thenabadu MW (1967) Chemistry of rice soils and the principles of fertilizer use. Paper presented at the symposium of the Chemical Society of Ceylon and the Royal Institute of Chemistry, Ceylon

USDA/NRCS (2006) Field indicators of hydric soils in the United States, version 6.0. In: Hurt GW, Vasilas LM (eds) USDA, NRCS, in cooperation with the National Technical Committee for Hydric Soils

USEPA (1993) (40 CFR 230.3(t)) [45 FR 85344, December 24, 1980, as amended at 58 FR 45037, August 25, 1993]. United States Environmental Protection Agency, Washington, DC

USEPA (1994) Water quality standards handbook, 2nd edn. United States Environmental Protection Agency, Washington, DC

Vepraskas MJ, Faulkner SP (2001) Redox chemistry of hydric soils. In: Richardson JL, Vepraskas MJ (eds) Wetland soils, genesis, hydrology, landscapes, and classification. Lewis Publishers, Boca Raton

Vymazal J (2010) Water and nutrient management in natural and constructed wetlands. Springer, Dordrecht

Zhang X, Zhang F, Mao D (1998) Effect of iron plaque outside roots on nutrient uptake by rice (Oryza sativa L.). Zinc uptake by Fe-deficient rice. Plant Soil 202:33–39

Forest Soils 14

In nature, forest vegetation and soil have evolved as coupled ecosystems over a long period of time by natural succession in a place. Some abandoned croplands have also been afforested so long ago that some of their original characteristics have changed. These soils are also considered as forest soils. Forest soils have some characteristics different from agricultural soils. Forest soils include usually porous, aggregated, well-differentiated soils with plenty of organic matter, often forming a forest floor on the surface soil, and high biotic activity. Most present-day agricultural soils developed under forest in the past. The upper part of many of these soils have changed significantly due to prolonged cropping practices, but their subsoils still bear the signs of a past forest vegetation. Soil properties have a bearing on selecting tree species for plantation forestry. Tree species by occupancy in a soil for long period have an effect on the soil. Some tree species may be used for the improvement of degraded lands. Forests are efficient nutrient recyclers. Very dense forests may develop in relatively poorly fertile soils due to this nutrient recycling. Nutrients recycle within the system several times and the same nutrient atom performs the same physiological functions each time. During nutrient recycling nutrients are taken up, some proportion is retained in biomass, some proportion is returned to the soil as litter, and nutrients are released by decomposition of litter materials. Too fast and too slow nutrient recycling are both undesirable in the context of soil fertility restoration and plant nutrition. Forests are also efficient carbon sequesters. Forest and forest soil management can be used for climate change mitigation.

14.1 What Is a Forest and Where Is It Found?

A forest is an ecosystem dominated by trees. A forest may more precisely be defined as "a large tract of land covered with trees and other vegetation such as herbs and shrubs, growing more or less closely together along with animals, large or minute in size." A forest may also be defined as "a dense community of trees, shrubs, herbs, microorganisms, and animals interacting among themselves and with their abiotic environment." Forests occur in almost 30% land area of the world and perform important ecological functions. The United States and Canada share 16% of the world's forests; the former Soviet Union contains 21%; Africa has 20%, and Latin America has 24%.

14.2 Forests Perform Many Important Ecological Functions

Important ecological functions of forests include (1) primary production; (2) carbon fixation, sequestration, and cycling; (3) transformation of radiant and chemical energy; (4) conservation of biodiversity; (5) regulation of hydrological cycle; (6) modification of climate; (7) biogeochemical cycling and nutrient cycling; (8) formation and conservation of soil; (9) maintenance of the food chain; (10) production of recreational opportunities; and (11) protection of wildlife.

14.3 There Are Many Different Types of Forests

Forests may be natural or artificial. Natural forests have evolved in a place spontaneously by natural succession. On the other hand, artificial forests have been created by human interventions—by sowing and planting. The practice of establishing a forest in place of an earlier forest is known as reforestation, and developing a forest in a place where there was no forest earlier is called afforestation. Natural forests are always indigenous (or native forests) consisting of indigenous trees. Indigenous trees are those trees that grow naturally in the ecological zone where there is the forest. Artificial forests may consist of indigenous or exotic trees. Exotic trees are those trees that do not naturally grow in the area and that have been imported from another ecological zone. Forests consisting of exotic trees are called exotic forests.

K.T. Osman, *Soils: Principles, Properties and Management*,
DOI 10.1007/978-94-007-5663-2_14, © Springer Science+Business Media Dordrecht 2013

Forests may be evergreen or deciduous depending on leaf persistence of dominant trees. An evergreen tree is a tree that retains leaves all the year round. It is green all the time. Forests consisting entirely or mainly of evergreen trees are evergreen forests. Examples of evergreen tress include *Artocarpus heterophyllus*, *Bischofia javanica*, *Calophyllum elatum*, *Euvodia lunuankenda*, *Hopea ponga*, *Mesua ferrea*, *and Myristica dactyloides*, which are familiar trees of tropical forests. A deciduous tree is a tree that sheds all its leaves in a season of the year. Deciduous forests consist mainly of deciduous trees. Deciduous forests remain leafless for a considerable period of the year. The deciduous trees include *Acrocarpus fraxinifolius*, *Bombax ceiba*, *Chukrasia tabularis*, *Dalbergia latifolia*, *Grewia tiliaefolia*, *Lagerstroemia microcarpa*, *Pterospermum* sp., *Tectona grandis*, *Terminalia bellirica*, and *Toona ciliate* found in tropical areas. A forest may contain evergreen and deciduous trees, both being dominant. In that case, the forest is known as either mixed evergreen forest or mixed deciduous forest, depending on their proportion. It is called mixed evergreen forest or semievergreen forest if the proportion of the evergreen trees is higher than the deciduous and it is mixed deciduous if the proportion of the deciduous trees is higher. Semievergreen forests are generally considered as a transitional stage between evergreen and moist deciduous forests.

Forests are sometimes classified on the moistness of the habitat or the amount of rainfall they receive into moist forests and dry forests. There are moist evergreen, moist deciduous, and dry deciduous forests. Tree species may have either broad, flat, and expanded leaves or narrow leaves, that is, needles. On this basis, forests may be broad-leaved forests consisting of broadleaf tree species such as mahogany, oak, and maple and coniferous forests containing cone-bearing trees having needlelike leaves such as pines. Coniferous forests are mainly found in the continental and upland regions of North America, northeast Europe, and Russia. Broadleaf forests are generally hardwood forests while coniferous forests are softwood forests, thus termed on the relative ease of sawing. There are broad-leaved evergreen and broad-leaved deciduous forests; similarly there are coniferous evergreen and coniferous deciduous forests. Forests are often classified on the basis of the habitat of their occurrence such as montane forests, swamp forests, and mangrove forests. Forests that have not yet experienced the human axe are virgin forests, but most forests are disturbed forests.

14.4 There Are Three Main Ecological Types of Natural Forests in the World

The main ecological types of world forests are tropical forests, temperate forests, and boreal forests.

14.4.1 Tropical Forests Are Rich and Diverse

Tropical forests occur on both sides of the equator in areas bounded by 23.5° N and S latitude. Tropical forest covers approximately 11.5 million km^2 area, which is about 7% of the world's total land area. Tropical forests are found in four regions: America and the Caribbean; Africa and eastern Madagascar; Asia, including India, Malaysia, Sri Lanka, Myanmar, Thailand, and Indonesia; and Australia. Approximately 57% of the total area of tropical forest is in the Americas, 25% is in Asia, and 18% is in Africa (Allaby 2006). Tropical natural forests are dense, multilayered, and highly diverse in floral and faunal composition. Trees are very tall and are mainly broad-leaved evergreens. Generally, three layers of trees are found along with undergrowth herbs and shrubs. Tropical forests are vast treasures of orchids, bromeliads, lianas, palms, climbers, mosses, etc.

Equatorial climates are generally rainy. However, rainfall varies considerably within the tropical zone. Annual rainfall may exceed 2,500 mm. Temperature is on an average 20–25 °C and varies little throughout the year. Many landforms, including mountains, deserts, deltas, plains, and floodplains, are encountered in this region. For these variations in climate, altitude, and topography, there are many subtypes of tropical forests, including rainforests, monsoon forests, evergreen forests, semievergreen forests, moist deciduous forests, dry deciduous forests, freshwater swamp forests, montane forests, and mangrove forests.

Tropical rainforests have broad-leaved evergreen trees. Many trees belong to the family Dipterocarpaceae. Krabak (*Anisoptera curtisii*), mersawa (*A. laevis*), Brunei teak (*Dryobalanops* species), lauan (*Parashorea malaanonan*), and red balan (several *Shorea* species) are typical dipterocarp timbers. The tallest of all tropical trees klinki pine (*Araucaria hunsteinii*) grows in forests of New Guinea. Tropical South America has the world's largest continuous area of lowland rainforest. Natural or Para rubber is obtained from *Hevea brasiliensis*, a tree belonging to the family Euphorbiaceae, and Brazil nuts are obtained from *Bertholletia excelsa*. The Amazon forest contains many tree species. There may be 40–300 different species growing in a single hectare of forest.

Brachystegia, *Julbernardia*, or *Isoberlinia* species dominate large areas of tropical seasonal forests. The forest consists mainly of teak (*Tectona grandis*) in Asia and *Oldfieldia africana* and mountain ebony (*Bauhinia* species) in Africa. Beneath these deciduous trees, the understory contains evergreens such as *Eucalyptus* species in parts of New Guinea and the banyan tree (*Ficus benghalensis*) in Asia (Allaby 2006). Some important tree species of tropical mangrove forests are *Avicennia alba*, *Avicennia officinalis*, *Bruguiera gymnorrhiza*, *Ceriops tagal*, *Heritiera fomes*, *Rhizophora apiculata*, *Sonneratia alba*, *Excoecaria agallocha*, etc.

14.4.2 Temperate Forests Are Relatively Low-Diversity Forests

Temperate forests occur in Western and Eastern Europe, the Urals, Western Siberia, Central Siberia, Northern China, Korea, and some parts of Japan. In North America, the forest zones stretch from the Pacific to the Atlantic, extending south along both coasts to 36°–40° N latitude. In South America, the temperate forest zones are found only in the west in the Andes, south of 38° S latitude. Temperate forests are also found on New Zealand's South Island. Annual precipitation in the temperate region varies from 750 to 1,500 mm. In some mountainous areas, rainfall may be as high as 3,000 mm. Temperature ranges from −30 to 30 °C, with an annual average of 10 °C. Climate is characterized by hot summers and cold winters. Temperate forests may contain evergreen and deciduous, broadleaf and coniferous, hard- and softwood trees with shrubs, perennial herbs, and mosses. Trees like maples (*Acer*), oaks (*Quercus*), chestnuts (*Castanea*), beeches (*Fagus*), and elms (*Ulmus*) are examples of deciduous trees. Pines (*Pinus*), firs (*Abies*), and cedars (*Cedrus*) are examples of coniferous trees. Olive (*Olea*), holly (*Ilex*), tea (*Camellia*), and eucalyptus (*Eucalyptus*) trees are all broad-leaved evergreens. Subtypes of temperate forests are moist conifer and broad-leaved forests, dry conifer forests, rainforests, Mediterranean forests, coniferous forests, and broad-leaved rainforests.

14.4.3 Boreal Forests Occur in the Cold Zone

The boreal forest or taiga is the largest forest biome in the world. It occurs between 50 and 60° N latitude in the broad belt of Eurasia and North America, two-thirds in Siberia, with the rest in Scandinavia, Alaska, and Canada. It constitutes an area of more than 13 million km^2. The summer is warm, moist, and short, while the winter is cold, dry, and long. Precipitation is mainly in the form of snow, and it ranges from 250 to 1,000 mm with an average of 500 mm $year^{-1}$. Available soil moisture is high as a result of cool temperature and low evapotranspiration rates. Boreal forests consist mostly of cold-tolerant evergreen conifers. The boreal forest is dominated by cold-tolerant, evergreen, cone-bearing, and needle-leaved trees. Four genera of conifers cover the bulk of the region: evergreen fir (*Abies*), spruce (*Picea*), and pine (*Pinus*) with some deciduous larch (*Larix*). Some broad-leaved deciduous species are of secondary importance, including alder (*Alnus*) and especially birch (*Betula*) and poplar (*Populus*). Siberian dwarf pine *Pinus pumila* covers large areas near the eastern arctic and alpine belts; its North American equivalent is white bark pine *P. albicaulis*.

14.5 Forest Soils Are Soils That Have Developed Under Forest Vegetation

Soils that have developed under the influence of forest vegetation are forest soils. This definition includes soils of natural forests only, where vegetation and soil have evolved simultaneously together by natural succession. Forests provide permanent cover, shade, roots, flora and fauna, organic matter, and nutrients; so the forest soil attains characteristics different from grassland and agricultural soils. However, most agricultural soils developed under a past forest vegetation. Under cultivation for centuries, topsoil characteristics have considerably changed. Subsoil characteristics still bear the signs of a past forest vegetation. In tropical regions, many unproductive croplands have been afforested. Moreover, many degraded forests have been reforested for some time. They are acquiring some properties due to the present forest vegetation. To accommodate all these soils, a forest soil is now defined as a soil under a forest cover.

14.6 Characteristics of Forest Soils That Distinguish Them from Agricultural Soils

Differences between agricultural and forest soils derive, in part, from the fact that often the most "desirable" soils have been selected for agricultural use and the remainder left for native vegetation such as forest (Vanmechelen et al. 1997). Forest soils are natural soils, developed undisturbed for a very long period of time under a semipermanent to permanent, thick, and dense vegetation. So, they have some characteristics that distinguish them from agricultural soils.

14.6.1 Forest Soils Have Some Limitations in Agricultural Standards

Forest soils are characterized by certain limitations in agricultural standards. These limitations may be physical or chemical. A forest soil often has one or more of these limitations. However, these limitations have become blessings for forests because they have prevented forest soils from being cleared for agricultural use. Physical limitations include (1) extreme texture (stony, gravelly, sandy, and clayey), (2) extreme moisture (arid, waterlogged, and submerged), and (3) steeply sloping and erodible. Chemical limitations include (1) acidity, (2) alkalinity, (3) salinity, (4) sodicity, (5) acid sulfate conditions, and (6) poor fertility.

Stony, gravelly, excessively sandy, and stiff clay soils are not suitable for cropping. Therefore, such areas have retained their original cover, although some have been degraded by overexploitation. These areas are now largely occupied by degraded forests and secondary forests. Wetland forests have also been disturbed but many still retain some cover. In acid, alkaline, saline, sodic, and acid sulfate conditions, plenty of luxuriant forests have developed worldwide. These soils are not good agricultural lands. So, they have retained forests.

14.6.2 Forest Soils Bear Permanent Vegetation

Forest soils develop under the influence of a permanent or a long-duration vegetation. For example, a natural forest may occupy the space for more than a millennium. On the other hand, rotations of planted forests are usually aimed at 40–50 years for timber plants and 10–15 years for fast-growing energy plantations. The occupancy of a land for so long a period by a living body exerts some effects on it. The vegetation produces biomass that adds residues and their decay products into the soil, removes water and nutrients for a long period, adds organic matter, and harbors soil organisms that have tremendous modifying effects on soil.

14.6.3 Forest Soil Has Thick Vegetative Covers

A forest has a thick and dense canopy cover, the thickness of which may be more than 30 m. It acts as an umbrella that intercepts solar radiation and rainfall. The cover regulates evaporation, transpiration, and heat transmission. As a result, a microclimate develops in the ground of the forest which is entirely different from adjacent nonforest areas. Under this cover and on the surface of the soil, the air is damp and humid. A well-developed forest canopy allows only 0–2% light to reach the ground (Thomas and Packham 2007). Significant undergrowth cannot develop at these poor light conditions. The vegetative cover keeps the soil cooler in the warm season and warmer in the cool season than the adjacent cultivated soils. Forest canopy controls the general hydrology of the forest soil. It is a kind of live mulch which intercepts almost 100% of rain. The beating action of rain is counteracted and water reaches the ground slowly as stemflow. This reduces the runoff and erosion and increases the infiltration.

14.6.4 Forest Trees Have Deep Root Systems

Most roots of agricultural crops are limited to the furrow slice (15 cm on an average), but forest tree roots usually extend to depths far more than a meter. These deep root systems penetrate greater depths unless restricted by a compacted layer or the groundwater table near the surface. Deep and extensive roots draw water and nutrients from a larger soil volume. They leave channels within this soil volume after their death and decay. These channels act as drainage and aeration channels. These channels also contribute to circulation of materials within the soil. Roots distribute organic matter and nutrients throughout the soil profile; they exude metabolites; they attract microorganisms at their vicinity, and with all these they weather and extend soil formation to greater depths. Forest soils have more roots, deeper roots, and large roots. So, forest soils have deep and well-differentiated soil profile.

14.6.5 Forest Soils Develop Through Natural Succession

In a given place, a soil and associated vegetation develop simultaneously going through stages of birth, growth, and maturity. Thus, both soil and vegetation have a life history. The series of development is called natural succession. Natural succession on a bare rock is called lithosere and in a wetland is hydrosere.

Let us see what happens in lithosere. We all know plants cannot grow on a bare hard rock. But bacteria can thrive inside the cracks and obtain nutrients from the substances dissolved from rock. Bacterial colonies secrete organic substances that enhance their own growth and dissolve substances from the rock. In this way, weathering advances and at one stage lichens (association of fungi and algae) gradually invade. As materials continue to accumulate from the wastes and dead remains of bacteria and lichens and underlying rock mass has considerably fragmented, the first true plants can gain a hold. These are often mosses, and within deeper cracks, where more organic material has collected, there may be some grass or a small flowering herb. These plants add more organic material and slowly, as this mixes with the tiny mineral fragments from the weathered rock, soil starts to develop. Soil genesis is a slow process, but after substantial parent material has accumulated, soil formation is relatively quick. At this stage, there is succession of a series of vegetation types. Herbs and shrubs come and stay for some period, after which they are replaced by pioneer trees. After soil has formed to a considerable depth, larger plants can grow in it. Their roots penetrate more deeply and leave channels after their death. These channels enhance movement of air and water. Soil materials are transformed and translocated and horizonation begins. The collection of horizons is called a soil profile. However, the horizons do not develop all at a time. As there is a succession of plants (bacteria, lichens, mosses, herbs, shrubs, pioneer trees, mature forest, and ultimately climax forest), there is also a succession of soil horizons (Fig. 14.1). At first, only the parent material, then an AC profile followed by an ABC profile, and then a mature

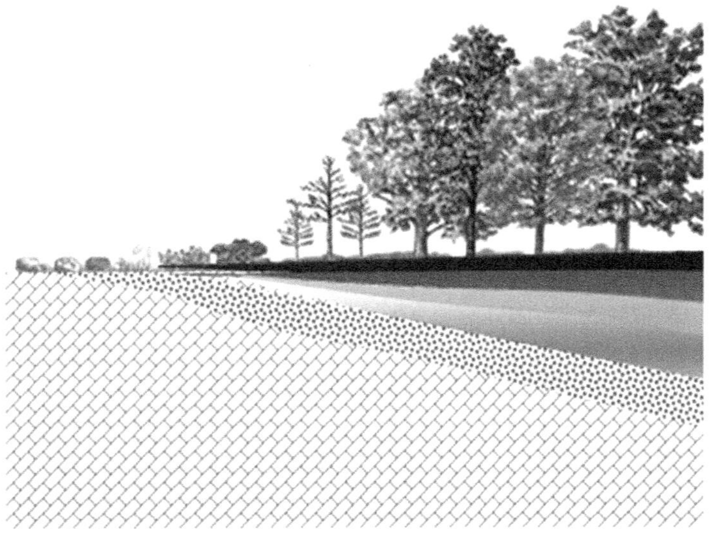

Fig. 14.1 Schematic representation of coupled evolution of forest and soil in a place

AEBC profile develop. On the surface of many forest soils, a thick O horizon develops by the accumulation of forest litters. However, for wide variations in lithology and climate, we see the development of different types of forests and associated soils (Allaby 2008). Hydrosere is a primary succession which develops in aquatic environments and results in conversion of a water body, say a lake, and its community into a land community, say a forest. In this succession, serial plant communities come and become replaced with concomitant changes in the underlying soil.

14.6.6 Forest Soils Usually Possess a Forest Floor

The most striking feature of the forest soil is the presence of a forest floor. It is a continuous layer of organic debris formed on the surface of a forest soil by the accumulation of fallen litters. Litters are dead parts of the forest vegetation, including leaves, branches, twigs, flowers, fruits, barks, and even large logs, existing in various stages of decomposition above the soil surface. Litters may also include dead bodies of soil animals and insects. Plenty of mushroom and fungi proliferate on the damp litters. As soon as litters fall on soil surface, they are attacked by soil organisms and decomposition ensues. The balance between amounts of litterfall and litter decomposition is the amount of litter accumulation in the forest floor.

Litter accumulation = Litterfall − Litter decomposition

So, development of a thick forest floor depends both on amount of litterfall and amount of litter decomposition. A thick forest floor develops where the litter production (biomass production in other words) is high or the litter decomposition rate is low. The thickness of the forest floor may range from some millimeters to more than a meter. Litter decomposition is the slowest in boreal forests, and therefore, boreal forests develop very thick forest floors. On the other hand, litter decomposition is very fast in tropical areas (all litters may be decomposed within a year). So, many tropical forests do not have a well-developed forest floor, although they can produce a large amount of litter. Temperate forests have forest floors intermediate in depth. Typical approximate figures for the accumulation of organic matter in forest floors are 15–100 t ha^{-1} in northern boreal forests, 7.5–12.5t ha^{-1} in temperate broad-leaved forest, and 1–2.5 t ha^{-1} in tropical forests (Vogt et al. 1986).

14.6.6.1 Structure of the Forest Floor

In submerged swamp forests, partially decomposed organic matter accumulates underwater as peat. There is no forest floor there. Forest floors develop in upland forests. A well-developed forest floor may be differentiated into three distinct layers, the upper undecomposed detritus called the L layer (L for litter), the intermediate partially decomposed residues called the F layer (F for fragmented or fermented), and the underlying well-decomposed layer called the H layer (H for humus). The F layer is also called "duff."

These terms about forest floor and its layers are actually coinage of forestry literature. A well-developed forest floor on the surface of mineral soils actually represents the O horizon in soil science literature. The L, F, and H layers correspond to Oi, Oe, and Oa sub-horizons, respectively (Fig. 14.2).

The L layer has freshly fallen leaves, branches, twigs, flowers, fruits, etc. They are unaltered, and their original structure is easily recognizable. The layer is loose and has a very low bulk density (even <0.2 g cm^{-3}). Bulk density increases with the stage of decomposition. Therefore, the F layer has intermediate and the H layer has the highest bulk density. The F layer is partially decomposed and the original

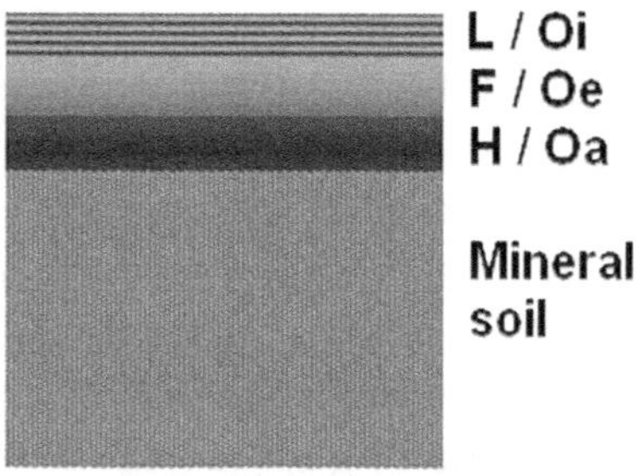

Fig. 14.2 Three zones of the forest floor

structure of materials has largely been retained. The sources of materials are identifiable. The H layer is brown to black, amorphous, and finely divided material that is relatively stable and colloidal formed by the decomposition of overlying litter materials. It may be intimately mixed with mineral matter. Forest floor materials have high water-holding capacity. Litters have high C/N ratio which decreases with advancement of decomposition. The C/N ratio of L layer may be around 60:1 and that of the H layer may be close to 15:1.

14.6.6.2 Functions of Forest Floor

1. Protection of Soil
   The forest floor is a natural organic mulch on the forest soil. It reduces raindrop impact, velocity of runoff, concentration of water in narrow channels, and detachment of soil particles. It soaks water, retains water, and enhances infiltration. It improves soil structure, aeration, and drainage.
2. Conservation of Soil Moisture
   The forest floor is a spongy material and it may hold several hundred times of its own weight of moisture. Thus, forest floor acts as a storage tank of water which gets infiltrated gradually within the soil. The forest floor acts as a barrier to the diffusion of water vapor from the soil to the air. Evaporation of water through a forest floor is slow.
3. Retention and Recycling of Nutrients
   The forest vegetation absorbs nutrients from various depths of the soil and returns much of absorbed nutrients to the soil surface in litters. Huge nutrients are locked up in the standing forest floor materials. If these nutrients were released all at a time (as by accidental fires), they would all have been lost by washing and leaching. Litters are slowly and gradually decomposed, nutrients are released as available nutrients, and plant roots absorb them. So, there is little loss of nutrients. Much of the energy and carbon fixed by forests is periodically added to the forest floor through litterfall, and a substantial portion of the nutrient requirements of forest ecosystems is supplied by decomposition of organic matter in the forest floor.
4. Habitat for Soil Organisms
   Forest floor is one of the richest components of the forest ecosystem from the standpoint of biodiversity. A large number of soil-dwelling organisms belonging to invertebrates, fungi, algae, bacteria, and archaea inhabit the forest floor. Forest floor materials provide habitat for all of them. These organisms perform many different functions, including burrowing, pedoturbation, mixing, aggregating, and nutrient cycling. Organic residues in the forest floor are food for detritivores and saprovores. Soil organisms use organic residues as food and mediate their fragmentation, mixing, and decomposition. By this way, litter materials become soil materials.
5. Seed Trap
   Fallen seeds are trapped in the forest floor and are preserved in it until germination. Otherwise, seeds are washed away by runoff water or swept away by wind. This way the forest floor contributes to forest regeneration.
6. Source of Humus
   Organic residues in the forest floor are decomposed by heterotrophic soil organisms and humus is formed as a product of this decomposition process. It has already been mentioned that the H layer of the forest floor is chiefly humus.

Foresters find great differences in the humus beneath various kinds of forests and suggest that these differences influence the physical, chemical, and biological properties of forest soils. There are three major types of forest humus: mull humus, mor humus, and duff mull or moder humus (Hoover and Lunt 1952; Fisher and Binkley 2000). These humus types differ in the presence or absence of a humified layer, the degree of incorporation of organic matter into the upper mineral soil layer, and the structure, thickness, and organic matter content of the humus layer (Fisher and Binkley 2000).

Mull humus is the brown to black, amorphous, colloidal, decomposed organic matter that is intimately mixed with the mineral matter of the surface soil. Mull humus is less acid, has higher cation exchange capacity, has a crumb structure, and is rich in bases. It generally develops under broadleaf forests on fertile calcium-rich soils. It represents true humus. Mor humus is a superficial deposit of fibrous and partially decomposed organic matter that is acidic in reaction and poor in bases. It generally develops under coniferous forests in poorly fertile soils. It resembles peat. Acidic forest soils tend to have a buildup of mor humus on the surface caused by the absence of earthworms below pH 4.8 and reduction in fungal and bacterial activity. It is low in nitrogen and wide in C/N ratio, 20–40. Mor humus is also called raw humus. Duff mull or moder humus is intermediate or transitional between mor and mull, characteristic of fairly acidic soils often under trees with hard litter such as beech or oak. Here, earthworms are usually absent and the litter is mainly decomposed by fungi and arthropods (ants, mites, springtails, millipedes, woodlice/sow bugs, etc.), leaving a mix of plant fragments and mineral particles held together by arthropod fecal material. This gives a loose, crumbly texture to the humus (Thomas and Packham 2007).

14.7 Forest Soils Have Some Differences from Agricultural Soils

Agricultural soil	Forest soil
Deep productive soils on level or gently sloping land with fewer limitations. Land used for annual crops of high value. Cultivation, fertilization, and land-conservation practices frequently used	Poorly fertile, agriculturally unproductive soils with severe limitations to agricultural use. Less subjected to cultural operations
Disturbed soils whose surface organic layers have lost their identity through cultivation.	Natural soils with undisturbed organic layers and soil horizons
Incorporated organic content usually low because of decreased litter deposition, annual crop removal, and increased oxidation resulting from cultivation and increased solar insolation	Incorporated organic content is high because of continual deposition of forest litter, long intervals between harvest, and lower rates of oxidation
Slower rates of water infiltration and percolation than forest soil due to a decreased content of large pores	Faster rates of water infiltration and percolation than agricultural soils because of the surface organic layers, large numbers of root channels extending into subsoil, and a high content of large soil pores
Poorer structure than forest soil due to low organic matter, decreased activity of macro- and microorganisms, exposure to raindrop action, clogging soil pores, increased erosion, deterioration of soil aggregates, and compaction of surface soil layers	Better structure than agricultural soil due to higher organic matter, higher activity of organisms leading to aggregation, protection from raindrop action, relatively little reduction in porosity, and large numbers of roots in surface and subsoil layers

14.8 Forest Soils Develop Through Different Processes

The dynamic interaction between vegetation and soil is so strong that it remains unclear which is dominant. Soil helps secure and renew the forest; forests help secure and renew the soil. Forest vegetations direct the kinds of soil formation processes to occur under prevailing climatic conditions.

Forest soils are usually highly weathered and leached soils. There are more horizons, thicker horizons, and deeper soil profile in forest soils, although the same four general processes—addition, removal, transformation, and translocation—also operate in the formation of forest soils. Large amounts of organic matter are added on forest soils through litters, and CO_2 is formed by their decomposition, which reacts with water to produce carbonic acid. Some other organic acids are also produced by organic matter decomposition. These acids hydrolyze soil minerals and produce new minerals, soluble substances, and bases. Thus, forest soils undergo intense biogeochemical weathering.

Pedogenic processes in the humid and subhumid tropics are influenced by (1) abundant moisture, (2) high temperatures, (3) old landscapes, and (4) active biota. These factors cause rapid mineralization of organic matter and intense and nearly complete weathering of primary minerals. By this intense weathering of minerals, bases are released and leached out of the soil profile with downward movement of water. Organic residues are rapidly decomposed giving little scope to form organic acids. Moreover, bases are also leached; as a result, the surface soil becomes mildly acid ($pH > 5.5$). Under this condition, oxides of iron, aluminum, and manganese are precipitated in the surface soil giving it a red color. The predominant soil-forming processes are laterization, latosolization, and desilication.

In humid temperate regions, there is an accumulation of organic materials in the soil surface giving rise to an organic O horizon. Decomposition of organic matter produces organic acids and makes the surface soil extremely acid ($pH < 4.0$). At this level of acidity, iron, manganese, and aluminum compounds become mobile and are removed from the A and underlying E horizons by eluviation. The A horizon is darkened by the admixture of humus with the mineral matter but the E horizon takes a gray color. On the other hand, iron, aluminum, manganese, bases, humus, and clay may be accumulated in the B horizon. The predominant process is podzolization.

Since forests occur in widely varied climatic, edaphic, and topographic conditions and develop through variable time, the magnitude and kind of all these changes also differ. So, we see widely different soil-forming processes in different regions.

14.9 Different Forest Ecosystems Have Different Soils

Forests are found to occur in almost all soil orders of Soil Taxonomy. However, dominant soil orders in tropical forests are Oxisols, Ultisols, and Alfisols with some associated Inceptisols and Entisols. Oxisols, Ultisols, and Alfisols occur on highly weathered upland residual parent materials. On the other hand, Entisols and Inceptisols are found in active floodplains, estuarine tidal floodplains, and mangrove swamps. Some Histosols are found to develop in freshwater swamp forests. Temperate forest soils are characterized by the presence of a deep O horizon over the mineral soil surface. Typical soil orders are Alfisols, Ultisols, Spodosols, and locally Histosols. Temperate Alfisols, Ultisols, and Histosols have many properties common to their tropical counterparts. Typical soils of the boreal forest regions are the Gelisols, with associated Spodosols and Histosols.

14.9.1 Oxisols and Ultisols Are Typical Tropical Forest Soils

The tropics are a region of diverse landforms ranging from lowland swamps to mountains with extensive plains. Vegetation is more diverse, including tropical rainforests, tropical monsoon forests, tropical evergreen forests, tropical moist deciduous forests, tropical dry deciduous forests, tropical montane forests, tropical mangrove, and other swamp forests. There is high rainfall, usually greater than 2,500 mm annually, in most of the regions. In rainforests, there is no definite dry period and in monsoon forests there are 4–6 dry months. For these variations in environmental conditions and the immense interaction of these conditions for soil formation, we find development of widely diverse soils in the tropics. The high temperature and plenty of rainfall in tropical forests cause intensive weathering, dissolution of minerals, release and leaching of bases, eluviation of clay, and decomposition of organic matter. These conditions are ideal for the formation of Oxisols. High weathering and leaching also produce Ultisols and Alfisols. Many tropical forest soils are very old and impoverished, especially in regions like the Amazon basin. Vitousek and Sanford (1986) estimated that moist tropical forests have 35.3% Oxisols, 27.7% Ultisols, 3.6% Alfisols, about 15% Inceptisols, and 15% Entisols.

Oxisols are typical soils of tropical broadleaf evergreen and deciduous forests. Oxisols have an oxic horizon—a mineral subsurface horizon of sandy loam or finer texture, with low CEC and few weatherable minerals (Soil Survey Staff 1999). Oxisols are rich in 1:1 clays such as kaolinite and sesquioxide clays such as goethite, hematite, and gibbsite. Kaolinite is usually the only phyllosilicate mineral of any abundance. Most coarse fragments are coated with sesquioxides. A horizons are moderately to strongly acid and tend to be low in organic matter. Organic carbon contents decrease exponentially with depth in forested Oxisols (Schaetzl and Anderson 2005).

Ultisols are intensely weathered soils of warm and humid tropical climates under forest vegetation. They are typically formed on older geologic locations in parent material that is already extensively weathered. Ultisols have an argillic or a kandic horizon and base saturation percentage less than 35. Ultisols are highly weathered and poor in fertility. The presence of the iron oxides causes the A horizon of these soils to be stained red. Ultisols are deep soils and range from well-drained soils on upland ridges, rises, and knolls to very poorly drained soils. The higher clay content of Ultisols gives them a nutrient retention capacity greater than Oxisols but less than Alfisols.

Alfisols form in climatic conditions, which are thermic or warmer and mesic or cooler. Therefore, most Alfisols are found in temperate regions, but Alfisols are found in tropical and subtropical zones under broadleaf evergreen or deciduous forests. These soils are moderately leached and relatively fertile. The soil formation is characterized by eluviation of clay from A to E horizons and illuviation of clay in the B horizon. Extensive leaching often produces a light-colored E horizon below the topsoil. Alfisols are generally fertile and productive. They have more than 35% base saturation percentage. Deep-rooted trees bring nutrients to the surface soil by biocycling through litterfall and organic matter turnover. They are more weathered than Inceptisols but less weathered and leached than Ultisols.

Tropical dry forests occur on substrates ranging from nutrient-rich alluvial soils to nutrient-poor rock outcrops. They can occur on volcanic, limestone, or ultramorphic rocks, with soil textures ranging from sandy to clay, rocky, or organic. If soils are droughty and have low water-holding capacity, vegetation may acquire a greater xeromorphic aspect than expected. Conversely, in locations where soils store water well or where water is channelized, vegetation may acquire great stature and biomass (Murphy and Lugo 1986). These soils are Aridisols, Entisols, or Inceptisols. Aridisols are characterized by high evapotranspiration to precipitation ratio; the dry soil has relatively low leaching. Entisols are soils that lack a true soil profile; they are very young soils formed in recent alluviums or aeolian or glacial parent materials. Inceptisols are young soils that have initiated soil profile differentiation. They are found on active floodplains, tidal floodplains, or river alluviums. Mangrove forests may occur in Entisols and Inceptisols.

14.9.2 Alfisols and Spodosols Are Typical Temperate Forest Soils

Almost all types of landforms are found within the temperate forest biomes. Soil groups associated with the biome are limited to a few important ones, most of which are a result of the soil-forming process podzolization. They include the Spodosols of the northern forests; the highland soils (Entisols/Inceptisols) of the alpine mountain chains and other regions, where soils are extremely thin; and the Alfisols. These soils have formed largely under current climatic conditions, and they developed with a vegetation cover that is approximately the same as that found today (Kuennecke 2008). Inceptisols are young soils and leaching of soluble compounds from the A horizon is at an early stage. There is a considerable variety of Inceptisols, but because they are young, all of them lack a clearly defined B horizon. The soil is often brown all the way to the C horizon. The color depends on the parent material, however, and can be distinctly reddish if this includes sandstone or red marl, a very fine clay derived from seabed or lakebed deposits. Inceptisols are well drained and fertile. Inceptisols are the predominant soils in a belt down the Eastern United States from the Adirondack Mountains, through Pennsylvania and the Appalachian Mountains, to as far south as northern Alabama. They also occur in the northwest, in parts of Washington. Where the climate is suitable, they support broad-leaved deciduous forest (Allaby 2008).

Alfisols develop under the temperate broadleaf deciduous forests. Broadleaf trees tend to be nutrient-demanding and their leaves are rich in major nutrient elements. The litter under this forest is not as acidic as under needleleaf trees. The autumn leaf fall provides abundant organic residues. Decomposition of the litter produces humus which gives the A horizon a dark-brown color. Alfisols were considered the most fertile and most easily worked soils. Therefore, extensive forest areas with Alfisols have been cleared for agricultural crop production. Alfisols occur to the south of the Great Lakes, from Ohio to southern Wisconsin, and also in parts of Texas, Colorado, Montana, North Dakota, and Southern California.

Spodosols are a group of soils that are characterized by the presence of a spodic horizon. They are commonly found in cool, moist temperate climate under coniferous forest vegetation. They are coarse-textured acid soils that occur in New England; on the shores of Lakes Huron, Michigan; and in parts of Florida. A typical Spodosol has a distinctive profile with four master horizons: O and/or A horizons have black or dark-brown colors due to accumulation of organic materials; the E horizon is an ash-gray albic horizon developed as a result of weathering of nonresistant minerals and subsequent leaching, and the Bhs or Bs horizons have dark reddish-brown colors resulting from illuvial accumulations of organometallic complexes. Most Spodosols have little silicate clay. The particle size class is mostly sandy, sometimes loamy. These soils are porous and permeable. Spodosols are acidic and naturally poorly fertile.

The Ultisols are the most highly weathered and leached soils of the temperate zone. They are characterized by a thin A horizon, with a thick, strongly expressed B horizon. The soils are often very red or yellow red. Ultisols are acid forest soils with relatively low native fertility. They are found on older, stable landscapes. Because of the favorable climate regimes in which they are typically found, Ultisols often support productive forests. Most Ultisols support mixed coniferous and hardwood forest vegetation. Ultisols are usually moist but some are dry during the warm season of the year. Xerults are the typical Ultisols of the temperate region.

14.9.3 Gelisols Are the Typical Boreal Forest Soils

The boreal climate is characterized by long, usually very cold winters and short, cool to mild summers. Precipitation is low, usually <900 mm, with most areas receiving less than half of this amount. In winter, temperature is <−40 °C and in summer temperature may be >30 °C. Average temperature of the summer months is around 10 °C. Precipitation is mainly in the form of snow. The evapotranspiration is also low; so there is enough available soil moisture during the growing season. The growing season is regulated by the temperature. Very cold-tolerant coniferous evergreen trees constitute the vegetation.

The soils are generally moist and may be either partially or totally frozen beneath the surface in the northern latitudes close to the arctic tree line. When this occurs, permafrost forms a discontinuous layer (Kuennecke 2008). Above the permafrost layer, there is the active layer which becomes frozen during winter and thaws during the summer. The soil development is recent and possibly incomplete, as indicated by the shallowness of some of these soils. These soils are classified as Gelisols. In some boreal areas, where the climate is mild, the predominant soils are Spodosols. Boreal forest soils tend to be young and poor in nutrients. The thinness of the soil is due largely to the cold, which hinders the development of soil and the ease with which plants can use its nutrients. Fallen leaves and moss can remain on the forest floor for a long time in the cool, moist climate, which limits their organic contribution to the soil.

The forest floors are usually thick with well-developed layers of humus underlain by carpets of moss and lichens (Fisher and Binkley 2000). In these soils, a thick O horizon develops on the mineral A horizon. Plenty of organic acids are produced by decomposition of organic matter. These acids decompose the minerals and aid in leaching bases from the upper soil layer, usually the E horizon. Clay and humus are also eluviated from the A and E horizons giving them a coarser texture. Iron, manganese, and aluminum and humus and clay may be accumulated in the B horizon. As discussed earlier this process of soil formation is known as podzolization and the resulting soils are Spodosols in the Soil Taxonomy. Histosols may develop in some low-lying areas (e.g., bogs, fens) of the boreal region. These soils are generally deep acidic peat.

14.10 There Is a Long History of Plantation Forestry

Tree planting is an ancient practice, and it probably started around 4000 BC in Greece. There are records of planting olive trees in Greece around 3000 BC. People planted trees for forest products, shelter, ornamental, ceremonial, or religious purposes. Tree plantation was a popular practice in ancient China and other countries of Asia. Teak (*Tectona grandis*) was extensively planted for timber in Java during the fifteenth century. After 1895, almost all teak plantations have been regenerated by taungya method. Taungya is a system in which farmers plant tree seeds or seedlings and tend them in association with their food crops. This was evolved in Burma (Myanmar) and it was a modification of the slash-and-burn technique. The first teak plantations in India were established in the Nilambur hills in 1840. Forest plantations were started in South Africa in the later part of the nineteenth century. The first wattle trees (*Acacia mearnsii*) were planted for tan bark in 1864 and the first pine (*Pinus* spp.) plantations in 1884. *Pinus patula* was introduced to New Zealand in 1877 and probably other countries at about the same time. It was introduced into South Africa in 1907. Irrigated plantations of *Dalbergia sissoo* and other species have been raised in the Indus basin of Pakistan. Notable irrigated plantations have subsequently been established in Iraq, Egypt, and central Sudan. Species such as eucalypts, casuarinas, and poplars are commonly grown in these irrigated systems (Evans 2009).

In 2010, the global planted forest area was estimated to be 264 million hectares, of which three-quarters were planted for productive purposes (production of wood, fiber, fuel, or non-wood forest product) and one-quarter for protective purposes (mainly rehabilitation of degraded lands, combating desertification, or protection of soil and water) (FAO 2006, 2010a). In this decade planted forest area has increased in all regions of the world.

Asia has 50% of the world's planted forests, Europe 30%, North and Central America 10%, South America 4%, Africa 4%, and Oceania 2% (FAO 2006). The species varies by regions; on an average, conifers account for 52%, broadleaves 37%, and unspecified 11%. In order of importance, the main coniferous genera by area extent are *Pinus*, *Cunninghamia*, *Picea*, *Laryx*, and *Cryptomeria*, while the main broadleaf genera are *Eucalyptus*, *Acacia*, *Tectona*, and *Populus* (FAO 2010b).

14.11 Major Plantation Tree Species Vary with Regions

A very wide range of tree species are used for forest plantations in different parts of the world. *Eucalyptus* and *Acacia* species are the most common in the tropics and subtropics. They are fast-growing short-rotation tree species. *Eucalyptus* has softwood and is extensively damaged by tropical storms. *Acacia* and *Eucalyptus* are preferred in spaces where a rapid cover is needed. Teak (*Tectona grandis*) is popular in India, Myanmar, Bangladesh, Malaysia, and Sri Lanka. Teak is a slow-growing hardwood species needed for valuable timber. There are *Swietenia macrophylla*, *Gmelina arborea*, *Dipterocarpus turbinatus*, and *Shorea robusta* in these regions. *Gmelina arborea* grows fast in the early growth stage (0–10 years). Other species are slow-growing tropical hardwood species. Softwood conifers such as pines, especially *Pinus patula*, *Pinus radiata*, *Pinus elliottii*, *Pinus taeda*, and *Pinus caribaea*, are grown in many countries. *Populus tremuloides*, a deciduous broadleaf hardwood, is planted in Canada. Temperate and boreal spruces, Sitka spruce (*Picea sitchensis*), Norway spruce (*P. excelsa*), white spruce (*P. glauca*), and black spruce (*P. marina*), are planted in the UK (Evans 2009).

14.12 Forest Trees Have Their Own Soil Requirement

All plants, tree is no exception, prefer a fertile soil; but most soils available for planting forest trees are generally of low fertility. This low soil fertility might have resulted from impoverished parent materials, erosive climate, or mismanagement of the past vegetation. Forest tree species vary in their air–water and nutrient requirements. For example, there are very high nutrient-demanding species such as teak. It grows well on well-drained, deep, fertile soils of pH above 6.5. But most forest plants favor slightly to moderately acid soil conditions. At least, soils in their natural habitats are acidic. Soils beneath broad-leaved deciduous forests are less acidic than soils beneath coniferous forests. The pH preferences of some tree species are *Larix decidua* 5.0–6.5, *Picea glauca* 5.0–6.0, *P. mariana* 4.0–5.0, *Fagus grandifolia* 5.0–6.7, *Quercus palustris* 5.0–6.5, *Q. velutina* 6.0–7.0, and *Acer saccharum* 6.0–7.5. Some tree species may thrive well in impoverished sites such as nitrogen-fixing *Acacia auriculiformis* and *Acacia mangium*. Moreover, trees vary in their requirement of soil physical properties. For example, teak grows well in sandy loam to clay loam soils; it cannot thrive well on stiff clay soils. For these reasons, particular attention is needed before selecting a tree species for an available site.

The soil requirements of some popular forest tree species for plantation are given below.

Acacia (*Acacia auriculiformis*): It grows well on deep fertile soils, but it has been reported to grow satisfactorily on a wide range of inhospitable sites, including shallow soils, compacted clays, coral soils, laterites, mine spoils, sand dunes, and unstable slopes. This species may be used for soil protection in denuded and eroded areas.

Big leaf mahogany (*Swietenia macrophylla*): It is found mostly in mixed hardwood forest belts, along riverbanks, and on deep alluvial soils of considerable fertility. It has adapted to fine- and medium-textured soils, alluvial soils, and calcareous soils.

Caribbean pine (*Pinus caribaea*): This species grows best in frost-free areas up to about 700 m in more fertile sites with good subsoil drainage and annual rainfall of 2,000–3,000 mm. Soils are usually loams or sandy loams, sometimes with high amounts of gravel, and generally well drained. The pH is usually between 5.0 and 5.5. It is the only species so far successfully used to clothe barren, eroded, and denuded lands with a tree cover.

Eucalyptus (*Eucalyptus camaldulensis*): In its natural habitat, it grows on riverine sites, whether of permanent or seasonal water (Brooker et al. 2002). It is most extensive on gray heavy clay soils along riverbanks and on floodplains subject to frequent or periodic flooding, preferring deep moist subsoils with clay content. It has extensively been planted in upland areas, even on hillslopes, in the tropical countries. It has been blamed in India to be responsible for groundwater depletion.

European larch (*Larix decidua*): It is native to Europe; it prefers mountainous regions with well-drained, slightly acidic to neutral soil conditions (pH from 6 to 7). It is adapted to a wide range of soil textures ranging from clay loam to sandy through silts. It grows satisfactorily on soils low in fertility.

Gamhar (*Gmelina arborea*): It grows on different types of soils ranging from acidic laterites to calcareous loams. It performs poorly on thin nonfertile soils with hardpan, dry sands, or heavily leached acidic soils. It grows well and fast on well-drained, deep, and fertile soils, such as Alfisols.

Mahogany (*Swietenia mahagoni*): It prefers deep, fertile, and well-drained soil. It is tolerant of acidic to alkaline soils.

Pin oak (*Quercus palustris*): In its natural habitat, it grows in moist, usually rich moist, clay soils of bottomlands. It is common in floodplains along streams and at the edges of swamps. It commonly grows on sites that flood occasionally, but not during the growing season.

Radiata pine (*Pinus radiata*): It naturally occurs within a narrow coastal zone of California on Mollisols, Entisols, Alfisols, and Ultisols. Most soils are found on sloping ground and are reasonably well drained. Soil pH is generally acidic. *Pinus radiata* has been grown successfully in various types of soils in Australia, New Zealand, Spain, Argentina, Chile, Uruguay, Kenya, and the Republic of South Africa. *Pinus radiata* is one of the most widely planted pines in the world.

Sitka spruce (*Picea sitchensis*): In plantations of Ireland, the species was found to develop best on fresh to very moist sites, with rich to very rich soil nutrient regimes (Farrelly et al. 2011). Sitka spruce has a strong affinity for soils high in calcium, magnesium, and phosphorus in the soil orders Entisols, Spodosols, Inceptisols, and Histosols. These soils are usually acidic with pH typically ranging from 4.0 to 5.7.

Sugar maple (*Acer saccharum*): Sugar maple grows on sands, loamy sands, sandy loams, loams, and silt loams but it does best on well-drained loams. It does not grow well on dry, shallow soils and in swamps. Sugar maple grows on soils ranging from strongly acidic (pH 3.7) to slightly alkaline (pH 7.3), but it most commonly grows on soils with a pH from 5.5 to 7.3.

Teak (*Tectona grandis*): Teak can grow on a variety of soils. The quality of its growth, however, depends on the depth, structure, porosity, drainage, and moisture-holding capacity of the soil. It develops best on deep, well-drained, and fertile soils, especially on volcanic substrata such as igneous and metamorphic soils or on alluvial soils of various origins. The optimal soil pH is between 6.5 and 7.5. The calcium content of the soil is also an important factor; calcium deficiency in the soil results in stunted growth of teak (Kaosa-ard 1981). Teak has been successfully grown on Alfisols, Entisols, Inceptisols, Ultisols, and Vertisols.

White spruce (*Picea glauca*): It is found on a wide variety of soils on glacial, lacustrine, marine, or alluvial parent materials. It grows well on loams, silt loams, and clays, but rather poorly on sandy soils. It grows poorly on sites with high water tables and is intolerant of permafrost. It grows mostly on acid Spodosols, Inceptisols, or Alfisols, with a pH ranging from 4.0 to 5.5.

14.13 Forest Plantations Influence Soil Properties

Trees are planted in a soil for at least 10 to more than 50 years, and during this long time, these trees draw water and nutrients from the soil. They grow, live, produce and return organic matter, and through all these activities exert some influence on the soil. Their effects may have positive and negative impacts, and improvement or degradation of the soil can be seen after the rotation. In fact, there are more reports of soil improvement by planting trees on denuded/deforested sites than their degradation. Generally, when deforested or degraded sites are reforested, there is an improvement. But when natural forests are replaced by plantations, there is a deterioration, especially in the early stages. However, the change in soil properties due to forest plantations depends on the tree species, soil, and environment.

Soils are changed physically, chemically, and biologically by a forest overstory. Aweto and Dikinya (2003) observed that growing *Combretum apiculatum* and *Peltophorum africanum* in a semiarid savanna rangeland in Botswana lowered the bulk density of the soil. They attributed this change in soil to the addition of litter materials that increased organic matter content and aggregation.

Forest trees have profound influence on soil pH and exchangeable cations. The effect of different tree species on soil pH is most significant in the first few centimeters of the topsoil. Augustoa et al. (2002) observed that topsoil pH under *Picea abies* and *Pinus sylvestris* was significantly lower than under *Fagus sylvatica*, *Quercus petraea*, or *Quercus robur*. *Abies alba* and *Pseudotsuga menziesii* appeared to be intermediate. All these trees exerted an acidifying effect on soil. The order of decreasing soil acidifying ability was (*Picea abies*, *Picea sitchensis*, *Pinus sylvestris*) > (*Abies alba*, *Pseudotsuga menziesii*) > (*Betula pendula*, *Fagus sylvatica*, *Quercus petraea*, *Quercus robur*) > (*Acer platanoides*, *Carpinus betulus*, *Fraxinus excelsior*, *Tilia cordata*). The base content of litter might have a bearing on the species' ability to affect soil pH. The processes involved in creating the observed variation in pH among species are likely to be threefold: (1) interspecific differences in the production of organic acids from decomposing litter that change the relative quantities of exchangeable base (Ca, Mg) and acid (Al, Fe) cations in soils, (2) differences in cation uptake and allocation to biomass pools with differing turnover times, and (3) variation in the mineral content of the parent material. Higher organic acid production and lower soil pH have been noted in sites dominated by species whose litter is relatively recalcitrant to the decomposition process. Calcium uptake and/or allocation to biomass pools with differing turnover times may explain the high soil pH and the large quantity of exchangeable Ca beneath sugar maple. Sugar maple trees are characterized by deep crowns and relatively high leaf litter Ca concentrations (Finzi et al. 1998).

Aweto and Dikinya (2003) also observed improvement in the levels of organic matter, exchangeable potassium, calcium, magnesium, and cation exchange capacity of soils by *Combretum apiculatum* and *Peltophorum africanum*. *Acacia nilotica* trees have enriched the organic matter and nutrient levels in the soil under their canopies in central India (Pandey et al. 2000). A similar effect has been reported by Hailu et al. (2000) for the soil under the canopy of *Milletia ferruginea* in Ethiopia. The content of nutrients in litter differs greatly with the type of tree species and so does the rate of litter decomposition and release of nutrients. Foliage of hardwood species usually has higher concentrations of N, K, Ca, and Mg than coniferous species. Thus, hardwood litters can be richer in nutrients than coniferous species (Phillips and Marion 2006). Enrichment of surface soil with nutrients may also be due to absorption of nutrients by tree roots from the subsoil and recycling them to the topsoil. Roos and Allsopp (1997) reported that growing *Acacia sieberana* in South African Savanna improved organic matter content, total nitrogen, mineralizable nitrogen, resin-extractable phosphorus, total phosphorus, and mineralizable phosphorus of soil.

Development of forest floor is also influenced by the tree species. For instance, the litter weight under *Picea abies* could be up to twice that of hardwood species like *Fagus sylvatica*. The decomposition rate of litter depends on characteristics, which are tree species dependent such as hardness, morphology, lignin/N ratio, foliage longevity, or the content of hydrosoluble components. Numerous studies have provided evidence that canopy composition has an impact on nitrogen mineralization and nitrification.

Fig. 14.3 Surface soil condition of a young teak plantation (Photo courtesy of Dr. Animesh Biswas)

Soil deterioration by successive rotations of *Cunninghamia lanceolata* monocultures was reported from subtropical China. The widespread practices of whole-tree harvesting, total removal of all organic matter from a site, and intensive soil cultivation that favors bamboo and grass invasion all contribute substantially to the problem. Ding and Chen (1995) conclude that nutrient losses and soil erosion after burning of felling debris and slash were primary factors responsible for the soil deterioration and yield decline. Pure teak plantation has long been described as a problem in India (Griffith and Gupta 1948). Soil erosion and loss of organic matter are widespread under teak (*Tectona grandis*). Site deterioration by teak plantation was also described by Chundamannii (1998). Figure 14.3 shows the surface soil conditions of a young teak plantation in Rangamati, Bangladesh. The forest is devoid of any undergrowth and the surface soil is highly prone to erosion.

14.14 Agroforestry Involves Growing Trees and Crops Together

Agroforestry is a system of growing agricultural or horticultural crops or/and rearing livestock along with trees simultaneously or sequentially in the same piece of land. Growing trees along with crops and livestock is aimed at enhancing crop yields, conserving soil, and recycling nutrients while producing fuelwood, fodder, grain, fruit, and timber (Nair 1989). Agroforestry involves the integration of trees, plants, and animals in conservative, long-term, productive systems. The positive interactions among all these components are exploited in carefully designed sustainable agroforestry systems (Sanchez 1995) by (1) multiple use of land; (2) improved utilization of land, labor, and resources; (3) protection and improvement of soil by reducing water and wind erosion and providing soil organic matter; (4) production of diverse food crops such as fruits, nuts, grains, and seeds; (5) production of feed for farm animals; (6) long-term production of tree products; and (7) enhanced productivity and net economic returns. In agroforestry trees are used in these ways: (1) individual trees in home gardens and around houses, paths, and public places; (2) dispersed trees in cropland and pastures; (3) rows of trees with crops between; (4) strips of trees along contours or waterways; (5) living fences and borderlines; and (6) windbreaks. There are two functionally different types of agroforestry systems: simultaneous and sequential (ICRAF 1994). In simultaneous agroforestry system, the tree and the crop components grow at the same time and in close enough proximity for interactions to occur. Examples of this type are alley cropping, contour buffering, border planting, and different silvopastoral systems. In sequential agroforestry systems the maximum growth rates of the crop and the tree components occur at different times even though both components may have been planted at the same time and are in close proximity. Examples of this type are shifting cultivation, improved fallows, taungya, and some multistrata systems.

14.14.1 Alley Cropping

In alley cropping crops (grains, forages, vegetables, etc.) are grown between tree rows spaced widely enough to accommodate the mature size of the trees without interfering for light and moisture with the crops between the rows. When light demanding crops like corn (maize) will be grown, the alleyways need to be wide enough to let in plenty of light even when the trees have matured. Alternatively, the cropping sequence can be planned to change as the trees grow. For instance, soybeans or corn could be grown when the trees are very small; as the tree canopy closes, forages could be harvested for hay; and finally, when the trees are fully grown and the ground is more shaded, grazing livestock or shade-tolerant crops like mushrooms or ornamental ferns could occupy the alleyways. For alley cropping, generally fast-growing multipurpose leguminous tree species including white lead tree (*Leucaena leucocephala*), Kassod tree (*Senna siamea*), Erythrina (*Erythrina poeppigiana*), and Gliricidia (*Gliricidia sepium*) are selected. Between these tree rows, corn (maize), rice, soybean, etc., are grown as alley crops. Figure 14.4 shows an example of alley cropping.

Windbreaks can be used in agroforestry systems. Windbreaks are created by strategically planting trees and shrubs along property lines or in fields to protect houses, livestock, human, and crops against strong winds and storms (Chap. 12). For contour buffering, buffer lines are created by planting trees and shrubs in rows on a contour or cross slope.

Fig. 14.4 Cowpea alley cropping with Leucaena (Image courtesy of http://freeimagefinder.com/tag/leucaena.html)

Rows are often planted along with crops and leaf litter provides soil cover and fertilizer. Contour buffers reduce sheet and rill erosion and even trap soil within the rows.

14.14.2 Silvopasture

Tree and pasture combinations are called silvopastoral agroforestry. Trees are planted in single or multiple rows, and livestock graze between them. Although both the trees and the livestock must be managed for production, some systems emphasize one over the other. Usually, in the early years of establishment, crops or hay is harvested from the planting. Grazing generally begins after 2 or 3 years, when the trees are large enough that the livestock cannot damage them.

14.15 Deforestation Causes Environmental Degradation

Deforestation is the conversion of a forest into a nonforest use such as farmland, ranches, pasture, industrial complexes, and urban areas. Deforestation may also result from massive reduction of tree canopy cover to less than 10% threshold. Almost half of the forests that originally covered the Earth are already lost, and further 16 million hectares disappear annually. There are many causes of contemporary deforestation, including overpopulation, lack of governance, unplanned human settlement, industrialization, and urbanization. The processes of deforestation are industrial logging, clear felling, indiscriminate use, forest fires, shifting cultivation, encroachment, etc. The effects of deforestation include loss of forest; loss of biodiversity; climate change; natural disasters such as cyclones, flooding, and drought; disruption of water cycle; decline in water quality; soil erosion; and sedimentation.

Forests absorb and store great amounts of water quickly when there are heavy rains. Trees are regulators of the hydrologic processes, including evaporation, transpiration, infiltration, and surface flow. A large part of the water that circulates in the ecosystem of rainforests, for instance, remains inside the plants. When these trees are cut down, it results in the climate getting drier in that area. When forests are cleared, the bare soil is exposed to the scorching effect of the sun and the beating action of the rains. Rainfall washes significant amount of topsoil. Due to elevated temperature, soil organic matter is decomposed at a faster rate. Soil aggregates are broken down by raindrop impact. Infiltration rate is reduced and more water runs off. Moving water concentrates in narrow channels in absence of tree roots and trunks. For all these, large tracts of land become permanently impoverished due to soil erosion. The rate of increase for soil loss after forest clearing is astonishing; a study in Ivory Coast found that forested slope areas lost 0.03 Mg ha^{-1} $year^{-1}$; cultivated slopes annually lost 90 Mg ha^{-1} $year^{-1}$, while bare slopes lost 138 Mg ha^{-1} $year^{-1}$ (Butler 2006).

Deforestation has a direct association with carbon dioxide emission into the atmosphere. Trees absorb carbon dioxide from the air to produce their body materials. When deforestation occurs, CO_2 fixation is reduced and many of the trees are burnt or they are allowed to rot, which results in releasing the carbon dioxide to the atmosphere leading to greater concentrations of carbon dioxide in the atmosphere. Deforestation accounts for about 20% of total global carbon dioxide (CO_2) emissions, due primarily to tropical deforestation (IPCC 2007).

14.16 Shifting Cultivation Enhances Soil Erosion

Shifting cultivation or slash-and-burn agriculture is an ancient farming system often practiced by the indigenous people in the hilly areas of the humid tropics. It is also known as "jhum" in India and Bangladesh, "taungya" in Myanmar, and "Sweden agriculture" in Europe. In this system, a patch of natural forest, usually on gentle slopes or on summits of hills, is cleared; vegetation is slashed and burnt; holes are dug in soil with elementary tools; and seeds of assorted crops are sown before monsoon (Fig. 14.5). Seeds germinate after the first rains. The crop lasts in the "field" for 6–8 months. After harvest, the land is left in the hand of nature. Shifting farmers clear a new patch of forest for cultivation in the next year. They are to return to the previous land for cropping after 15–20 years.

Earlier, it could be argued that shifting cultivation might be sustainable in its conventional form, because forests should recover by natural succession during the long fallow period. Tanaka et al. (2005) suggested that the general impacts of shifting cultivation on soil properties after more than 10 years of fallow and 1 year of cultivation were limited. But, due to overpopulation, settlement of plain landers to hills, and scarcity of available land for cropping, the fallow period has now shrunk to less than 3 years. This has created deep concerns about deforestation, loss of biodiversity, and soil degradation due to shifting cultivation. On the basis of data given in FAO and other sources, it is estimated that each year approximately $1.9–3.6 \times 10^6$ ha land of primary close forests, $3.4–40 \times 10^6$ ha land of secondary close forests, and $6.9–21.9 \times 10^6$ ha land of secondary open forests are being lost due to shifting cultivation (Detwiler and Hall 1988).

Fig. 14.5 Indigenous women are sowing seeds in a "jhum" in Chittagong Hill Tracts, Bangladesh (Photo courtesy of the daily Prothom Alo)

Shifting farmers burn the slashed debris so that nutrients in its ash should increase fertility and enhance crop productivity. Increases in 234 kg Ca, 55 kg Mg, and 20 kg K ha^{-1} after burning were reported by Salacedo et al. (1997). The effect of burning is, however, temporary. Rains following burning remove bases rapidly by erosion and leaching. Gafur et al. (2000) reported loss of 61 kg Ca, 13 kg Mg, 13 kg K, 0.14 kg P, 0.20 kg S, 0.05 kg Cu, 6.7 kg Fe, 6.1 kg Mn, and 0.065 kg Zn due to shifting cultivation in Bandarban hills in Bangladesh. Growing evidence suggests that shifting cultivation in its distorted form leads to an adverse effect on soil and water. Such activities affect soil physical and chemical properties, reduce nutrient stocks, and accelerate soil erosion and sedimentation. Gafur et al. (2000) estimated soil loss of 43 Mg ha^{-1} $year^{-1}$ due to shifting cultivation in Bandarban, Bangladesh. There are reports of erosion loss from many other countries. Values vary due to differences in soil, slope, and climate. Alake Biswas et al. (2011) carried out an investigation on soil properties under different land-use systems including shifting cultivation in Chittagong Hill Tracts, Bangladesh. They observed significantly higher nutrient levels in soils under natural forests and bushy lands than slashed, burnt, and shifting cultivated sites. In this region, vast areas have been denuded by repeated slashing and burning.

14.17 Nutrient Cycling Maintains Productivity in Forest Ecosystems

14.17.1 Nutrient Cycling Involves Pools and Fluxes

Nutrient cycling is the cyclic movement of nutrients through the lithosphere–biosphere–hydrosphere–atmosphere systems. This movement is also known as biogeochemical cycling. The biogeochemical cycle is a relatively open cycle and has several subcycles: the geochemical cycle (e.g., atmospheric deposition and volatilization), the biological cycle (e.g., biological nitrogen fixation and denitrification), and the biochemical cycle (e.g., litterfall and litter decomposition). In nutrient cycling, there are nutrient inputs and outputs (Fig. 14.6).

Nutrient input occurs by the addition of nutrients to the ecosystem through precipitation, dryfall, weathering, N_2 fixation, animal inputs, and fertilization, while output occurs by the removal of nutrients from the ecosystem through leaching, erosion, animal losses, and harvest (Turner et al. 1977). Some examples of nutrient inputs and outputs in forest ecosystems are given below.

14.17.1.1 Nutrient Inputs

Atmosphere is a very good source of nutrient inputs in forest ecosystems; it has large amounts of suspended dusts containing nutrients. These nutrients are carried to the soil as wetfall with precipitation and as dryfall. Suspended particles are an appreciable source of bases, nitrogen, phosphorus, and sulfur. In the Eastern United States, atmospheric nitrogen input to forests tends to average 5–10 kg ha^{-1} $year^{-1}$ for low-elevation forests (including precipitation, vapor, and particulates) and 15–30 kg ha^{-1} $year^{-1}$ for high-elevation forests (Johnson and Lindberg 1992). Vitousek and Sanford (1986) reported 0.3–2.3 kg P, 0.8–24 kg K, 0–34 kg Ca, and 0.3–26 kg Mg per hectare per year precipitation input in several tropical forests. Forests are efficient trappers of atmospheric dusts, but trapping efficiency varies with species. Moreover, atmospheric inputs depend on location of the forests relative to industries, farmlands, and coasts. Near these localities, atmospheric inputs are higher.

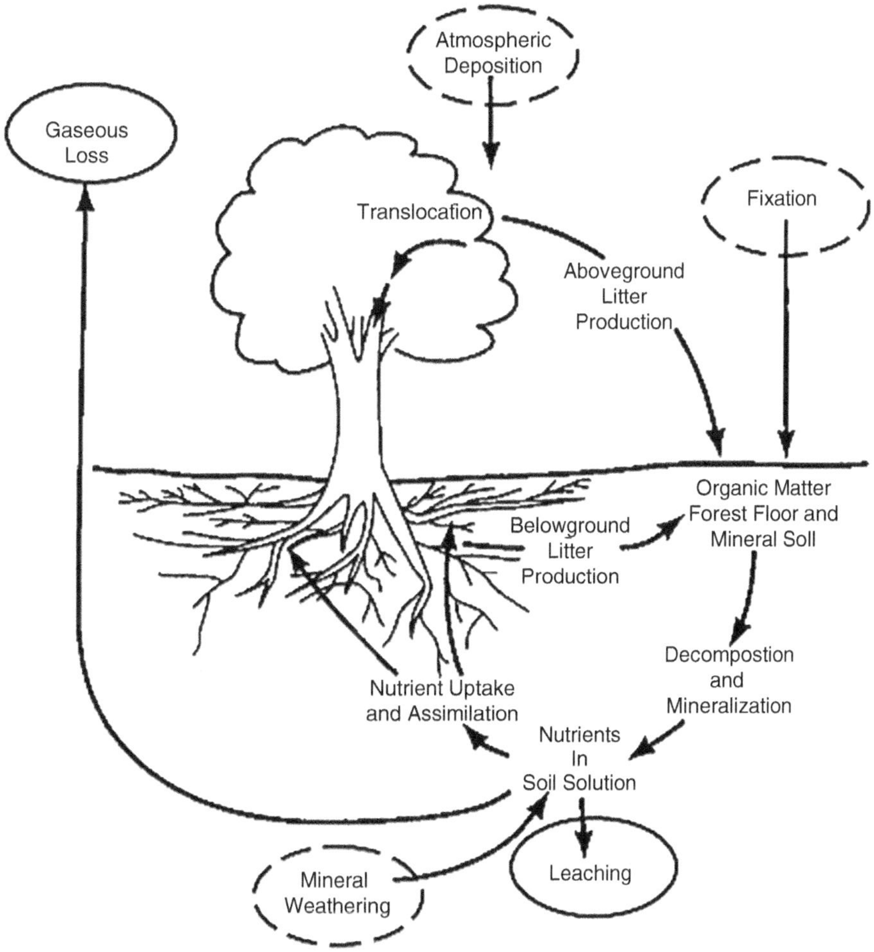

Fig. 14.6 Nutrient cycling in forest ecosystem (*dotted circles* represent inputs and *solid circles* represent outputs; *arrows* represent flux)

Weathering of soil minerals is considered to be an input because unavailable nutrients are added to the available soil nutrient pool. Weathering input depends largely on the parent materials and on conditions of weathering. Weathering inputs of 1–28.4 kg K, 0–86 kg Ca, and 1–52 kg Mg ha^{-1} $year^{-1}$ were reported from a good number of tropical and temperate forests. Forest types and species influence the magnitude of these inputs (Fisher and Binkley 2000). Sylvester-Bradley et al. (1980) estimated nitrogen input by symbiotic fixation in major forest ecosystems of Amazonia; they suggested high rates (up to 200 kg ha^{-1} $year^{-1}$) in fertile floodplain forests, intermediate rates in moderately infertile Ultisols (20 kg ha^{-1} $year^{-1}$), and very low rates (2 kg ha^{-1} $year^{-1}$) in infertile Oxisols. Nonsymbiotic fixation adds a very low amount of nitrogen, usually less than 1 kg ha^{-1} $year^{-1}$, but high amounts as 14.2 kg ha^{-1} $year^{-1}$ were also reported (Bormann et al. 1977).

14.17.1.2 Nutrient Outputs

Most undisturbed forests retain nutrients so efficiently that losses are small relative to the quantities cycled (Fisher and Binkley 2000). In some forests, nitrate leaching may be as high as 50 kg ha^{-1} $year^{-1}$ (Binkley et al. 1985); in most cases, however, it is less than 10 kg ha^{-1} $year^{-1}$. There were reports of 0.2–19.4 kg N, 0–30 kg P, 0.4–15 kg K, 1.6–163 kg Ca, and 0.7–51 kg Mg ha^{-1} $year^{-1}$ hydrologic outputs (erosion and leaching) from tropical forests. This enormous variation might have resulted from types of forests and soils. Denitrification losses were estimated to be 1–2 kg N ha^{-1} $year^{-1}$ from Amazonian forests (Vitousek and Sanford 1986). Forest harvest is a major route of nutrient output. Export of 50–150 kg N, 5–20 kg P, 20–120 kg K, 80–240 kg Ca, and 15–30 kg Mg was reported for bolewood harvest in some temperate forests (Pritchett and Fisher 1987). Whole-tree harvest would cause higher and debarked bole harvest would cause significantly less removal of nutrients. However, harvest is not an annual phenomenon.

14.17.2 Nutrient Recycling Is a Part of the Nutrient Cycling

Forests are self-nourished systems. They depend on native soil nutrients and on those retained in their own biomass. Their nutrient requirement often greatly exceeds the nutrient inputs from the atmosphere and weathering of soil minerals. The extra amount is supplied by the mineralization of litter materials. On decomposition of litter, nutrients return to the available soil nutrient pool, where they are reused by plants. It is known as nutrient recycling. Nutrient recycling has four major steps: uptake, retention, return, and release.

14.17.2.1 Uptake

Uptake is defined as the amount of annual nutrient absorption by vegetation from the soil through roots and absorption from rainfall and dryfall through foliage. Forests differ in their nutrient uptake; there are high- and low-demanding tree species. Again, tree age, soil fertility, and climate are important determinants of nutrient requirements. The annual requirement of 115.6 kg N, 12.3 kg P, 67.3 kg K, 62.2 kg Ca, and 9.5 kg Mg ha^{-1} for hardwood temperate forest at Hubbard Brook, New Hampshire, was estimated by Schlesinger (1991). On the other hand, annual uptake per hectare of 34–123 kg N, 2–12 kg P, and 5–75 kg K was reported from several other temperate forests. Thus, nitrogen uptake is ten times that of phosphorus and three times that of potassium (Pritchett and Fisher 1987). However, the greater part of the annual nutrient uptake is returned to the soil in litterfall.

14.17.2.2 Retention

Trees use a part of the nutrients absorbed annually for their physiological functions. The remaining part is returned to the soil as litterfall. Annual retention of 5–44 kg N, 1–6 kg P, and 2–20 kg K was found in some temperate forests. These values represented 20–50% of the annual nutrient uptake (Pritchett and Fisher 1987).

Nutrient accumulation is the total amount of nutrients that are held in biomass. Nutrients are retained in different biomass components such as leaves, boles, barks, and roots. Biomass production, nutrient concentration, and nutrient accumulation are all related to species (Fensham and Bowman 1995), stand age (Rytter 2002), soil fertility (Vitousek and Sanford 1986), and climate (Brubaker 1986). Vitousek et al. (1988) reported biomass accumulation in several temperate and tropical forests to range from 233 to 624 Mg ha^{-1} having, on an average, 60% in bole, 20% in roots, 15% in branch, and 5% in leaf.

The concentration of nutrients is usually the maximum in leaves among the biomass components. For the three species *Acacia polyacantha*, *Senna siamea*, and *Eucalyptus camaldulensis*, the nutrient concentrations generally graded in the following order: for N, P, K, and Mg, leaves > stembark or living branches > stemwood and, for Ca, stembark > leaves > living branches > stemwood (Harmand et al. 2004).

Foliar nutrient concentrations greatly differ among forests. Even within the same stand, there may be large variation due to developmental stage of the leaf, position of leaf in the crown, and disease incidence. Foliar concentrations of N, P, and K usually decrease and foliar Ca increases as the leaves become older. More mobile nutrients retranslocate to stems and barks as the leaves age; this retranslocation is an internal cycling. There may be a seasonal pattern for different tree species, too (Lu et al. 2007). Therefore, great care must be taken for sampling leaves and interpretation of tree nutrition based on results of foliar nutrient concentrations. Singh (1982) reported nutrient concentration in aboveground biomass of *Pinus patula* in the Darjeeling hills, India. Among the nutrients, N concentration was the highest (2.2%), followed by Ca, Mg, K, and P. In other species, different nutrient orders as Ca > N > K > Mg > P may be found (Osman and Sikder 2000). Vitousek and Sanford (1986) observed that broadleaf trees in moist tropical forests on moderately fertile Alfisols and Entisols had 2.08–2.54% N, 0.12–0.15% P, 0.85–1.92% K, 0.70–2.29% Ca, and 0.26–0.88% Mg, while those on poorly fertile Oxisols and Ultisols had 0.74–1.29% N, 0.02–0.12% P, 0.32–0.72% K, 0.37–1.03% Ca, and 0.15–0.35% Mg in leaves.

Singh (1982) observed the maximum accumulation of nutrients in 8–34 years old *P. Patula* bole, which accounted for about 58–85% of the total aerial contents of different elements. Nutrient contents in the aboveground biomass increased with stand age and were 1,911 kg N, 285 kg P, 478 kg K, 2,102 kg Ca, and 875 kg Mg ha^{-1} of the 34-year-old stand. In moist tropical forests, nutrient contents in all aboveground biomass components were 1,689–1,980 kg N, 112–290 kg P, 753–3,020 kg K, 2,370–3,900 kg Ca, and 310–403 kg Mg in moderately fertile soils and 741–2,430 kg N, 27–100 kg P, 277–600 kg K, 260–1,900 kg Ca, and 69–530 kg Mg in poorly fertile tropical soils (Vitousek and Sanford 1986).

Nutrient accumulation in roots has less been studied for difficulties in sampling. However, nutrient concentration in fine roots is generally higher than stems. In a study in China, Liu et al. (2002) obtained the following order: leaves (1.03–2.14% N) > branches (0.35–0.75% N) > roots (0.28–0.46% N) > stems (0.13–0.23% N). Root biomass production in this forest was 149.45 Mg ha^{-1}, while the aboveground biomass was 345.16 Mg ha^{-1} containing 7.08 Mg ha^{-1} in leaves. Nutrient accumulations in roots were 595 kg N, 90 kg P, 538 kg K, 665 kg Ca, and 182 kg Mg per hectare.

14.17.2.3 Return

Nutrients are returned to the soil by fall of dead parts, known as litterfall; death of roots; stemflow; throughfall; and death of animals. Among these components, litterfall has been studied the most.

14.17.2.4 Litterfall

Dead parts of vegetation, including leaves, boles, barks, branches, flowers, and fruits, constitute forest litter. Litterfall represents a major biological pathway for element transfer from vegetation to soils (Yang et al. 2005). It leads to organic matter replenishment and nutrient cycling (Bhat and Jan 2010). Amount of litterfall, distribution of litter components, concentration of nutrients in litter components, and amount of nutrients returned in litter are all very important aspects of nutrient recycling in forest ecosystems.

Litterfall varies enormously among forests due to variation in climate, species composition, stand age, and soil fertility (Yang et al. 2004). Generally, conditions that favor growth and biomass production also favor litter production. The variation in litterfall from place to place is known as spatial variation. Litterfall generally decreases with increasing latitude (from equator to pole) and follows the following order: tropical forests > temperate forests > boreal forests. Reported values of annual litterfall range from 3.5 to 5.5 Mg ha^{-1} $year^{-1}$ in temperate forests, 3.0–11.0 Mg ha^{-1} $year^{-1}$ in subtropical forests, and 5.01–14.44 Mg ha^{-1} $year^{-1}$ in tropical forests (Yang et al. 2005). Boreal forests may produce < 3 Mg ha^{-1} $year^{-1}$. Litterfall also decreases with increasing altitude. Zhou et al. (2006) observed 10.16, 10.61, 8.49, and 4.89 Mg ha^{-1} $year^{-1}$ litterfall in four forests along an altitudinal gradient in China.

In *Acacia auriculiformis*, *Casuarina equisetifolia*, *Tectona grandis*, and *Acacia mangium* plantations of Karnataka, India, leaf litter contributed 72–96% to the total litterfall (Rathod and Devar 2003). Similar results were also reported from many tropical and temperate forests (Chave et al. 2010). Thus, most nutrients are returned to the soil through leaf litterfall. There is a seasonal pattern of leaf shedding. In deciduous forests, all the leaves are shed during autumn and winter. In evergreen trees, the seasonality of litterfall depends with tree species, soil moisture content, and air temperature (Okeke and Omaliko 1994). There is also species specificity in the temporal variation of litterfall. In some species, peak litterfall occurs in one definite season, for example, autumn (*Pinus elliottii*), winter (*P. radiata*), and spring (*Quercus douglasii*). This is called unimodal litterfall. There are forests with bimodal (*Dendrocnide excelsa* in Australian rainforests, Lowman 1988) and multimodal litterfall (*Cunninghamia lanceolata* in subtropical China, Yang et al. 2004).

There is a notable difference in the concentration of nutrients between green leaves and leaf litter. More mobile nutrients such as N, P, and K move toward barks and stems and less mobile elements such as Ca and Mg accumulate in leaves before senescence. Retranslocated nutrients are utilized for new leaf growth. Nutrient resorption (or retranslocation) from senescing leaves is an important mechanism of nutrient conservation in plants (Yan et al. 2006). Huang et al. (2007) reported N and P concentrations of live and dead leaves of 14 deciduous and 20 evergreen forest tree species in a secondary evergreen broad-leaved forest in eastern China. In live leaves, there were 1.8–3.6% N and 0.10–0.16% P in deciduous trees and 1.1–2.2% N and 0.03–0.3% P in evergreen trees. On the other hand, in dead leaves, there were 0.6–2.3% N and 0.03–0.14% P in deciduous trees and 0.8–1.6% N and 0.03–0.10% P in evergreen trees. N resorption efficiency varied from 27 to 65% and 21–52% in deciduous and evergreen tree species, respectively. In another report, leaf litter contained

1.49–2.2% N, 0.06–0.14% P, and 1.0–2.2% Ca in lowland tropical forests on moderately fertile soils, whereas forests on infertile soils had 0.86–1.45% N, 0.01–0.03% P, and 0.13–1.4% Ca in leaf litter (Sayer and Tanner 2010).

For an understanding of the estimates of nutrient returns, let us consider a case of a temperate forest where total litterfall was 4,500 kg, including 3,375 kg in leaf, 675 kg in branches, 135 kg in barks, and 315 kg in miscellaneous parts. Leaf litter contains 1.2% N, 0.1% P, 0.7% K, 1.1% Ca, and 0.4% Mg. Calculate contribution of litter components to total litter and return of nutrients in leaf litter.

Percentage Contribution	
Leaf litter	3,375/4,500 × 100 = 75%
Branches	675/4,500 × 100 = 15%
Barks	135/4,500 × 100 = 3%
Miscellaneous	315/4,500 × 100 = 7%
Annual Nutrient Returns in Leaf Litter per Hectare	
Nitrogen	3,375 × 1.2/100 = 40.50 kg
Phosphorus	3,375 × 0.1/100 = 3.38 kg
Potassium	3,375 × 0.7/100 = 23.63 kg
Calcium	3,375 × 1.1/100 = 37.13 kg
Magnesium	3,375 × 0.4/100 = 13.50 kg

If leaf litter contributes 70% to total nitrogen returns, then total nitrogen return becomes 40.50 × 100/70 = 57.86 kg N h a^{-1} $year^{-1}$. Vitousek and Sanford (1986) reviewed nutrient returns in moist tropical forests. They found 110–224 kg N, 4–15 kg P, 20–68 kg K, 59–290 kg Ca, and 16–64 kg Mg returns in litter in moderately fertile soils. Nutrient returns were much less in poorly fertile soils.

14.17.2.5 Throughfall, Stemflow, and Canopy Drip

Stemflow is the flow of rainwater along plant branches and stems. The amount of stemflow is determined by leaf size and shape and stem and branch architecture. In general, broadleaf trees have more stemflow than coniferous vegetation. Some rainwater also falls as canopy drip. This is especially true of coniferous vegetation. Throughfall is the passing of precipitation through the plant canopy to the ground. This process is controlled by factors like plant leaf and stem density, type of the precipitation, intensity of the precipitation, and duration of the precipitation event. The amount of precipitation passing through varies greatly with vegetation type. Some nutrients from vegetation are leached with stemflow, canopy drip, and throughfall. These pathways generally represent a relatively minor vector for nitrogen, phosphorus, and calcium transfer in tropical forests, although it is the major pathway of potassium transfer. Reiners (1972) reported quantities of nutrients returned in throughfall in three forests of Minnesota. In oak, fen, and swamp forests, respectively, there were 5.5, 5.5, and 6 kg ha^{-1} $year^{-1}$ N; 0.6, 0.7, and 0.5 kg ha^{-1} $year^{-1}$ P; 7.6, 10.5, and 10.7 kg ha^{-1} $year^{-1}$ Ca; and 3.1, 3.8, and 3.7 kg ha^{-1} $year^{-1}$ Mg.

14.17.2.6 Nutrient Returns in Root

Very little information is available on nutrient turnover by forest tree roots. However, data of Sanford (1985) from a Venezuelan Oxisol are mentioned here for an example. Estimated turnover of roots less than 2 mm in diameter in the upper 10 cm was about 25% per month; it yielded an annual fine-root production of 15.4 Mg ha^{-1}. Nitrogen and phosphorus concentrations in roots less than 2 mm in diameter were 2.23 and 0.073%, respectively. So the quantities of nitrogen and phosphorus added to the soil by turnover of these roots were 343 and 11 kg ha^{-1} $year^{-1}$, respectively, in contrast to 61 kg ha^{-1} $year^{-1}$ N and 0.8 kg ha^{-1} $year^{-1}$ P added by fine litterfall.

14.17.2.7 Release

Nutrients are released to the soil nutrient pool through decomposition of litter by soil organisms. It is chiefly a microbial process, but some soil fauna may have significant contribution in some conditions. In coniferous forests, approximately 95% of the decay is mediated by microorganisms, and the remaining 5% is done by soil animals (Berg et al. 2003).

14.17.2.8 Litter Decomposition

As soon as litters are fallen on ground, they are attacked by soil organisms and decomposition starts. The rate of decomposition depends on litter components; wood litter takes great time for decomposition and leaf litter decomposes within relatively short time. Since litterfall contributes a significant proportion of total nutrients recycling in forest ecosystems, many studies have been done on rate of litter decomposition and factors affecting it. The rate of litter decomposition on a regional scale is governed by the climate and on a local scale by litter quality (Berg and Laskowski 2006). All the leaves fallen annually in a tropical forest decompose within a year, while in a boreal forest, decomposition is only 10–20%. For this reason, a thick forest floor develops under a boreal forest. However, a time should reach, at least theoretically, when the rate of litterfall should equal the rate of litter decomposition in an undisturbed climax forest. In this stage, the thickness of the forest floor does not change with time. It is known as the "steady state."

Two techniques are generally employed for the litter decomposition studies: One is the rate of litter disappearance from the forest floor, and the other is the mass loss of litter confined in nylon litter bags. From these studies, litter decomposition has been found to be a two-stage phenomenon (Berg and Staaf 1980): The initial rapid stage is followed by a latter slow stage (Fig. 14.7). At one time, the rate diminishes to almost zero when all decomposable materials have disappeared, recalcitrant materials have accumulated, and stable humus has formed. The time required to achieve this stage is

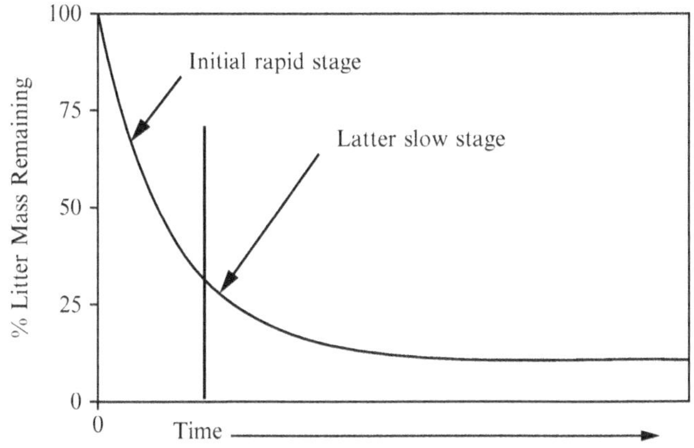

Fig. 14.7 Rate of mass loss of litter with time

highly variable from forest to forest, depending on litter quality and environment.

Litter quality is determined by the organic constituents and nutrient contents in litter. The main organic constituents of litter are celluloses, hemicelluloses, and lignins. The proportions between these fractions and also the content of mineral nutrients and other nutrients such as nitrogen, sulfur, and phosphorus vary considerably between different litters. Many of these components affect the decomposition rate in a promoting way, while others have a limiting effect. The litter decomposition process is therefore very complex, and it seems difficult to draw any general conclusions as to which parameters will control the decay dynamics (Berg et al. 1998).

In the early stage, more easily available components, such as soluble carbohydrates, are leached or degraded. Then celluloses, hemicelluloses, and proteins are lost. The early stage is regulated by nutrient levels (nitrogen, phosphorus, and sulfur) and readily available carbon. In a study from sites with a wide range of climatic conditions, Johansson (1986) reported that the initial concentrations of the nutrients nitrogen, phosphorus, potassium, and calcium and also lignin and hemicelluloses had no significant relations with litter mass loss. But carbohydrates and cellulose concentrations were however positively correlated with the litter mass loss. During the later stage, which occurs at an approximate mass loss value of 35%, most of the easily degradable materials are gone, leaving the recalcitrant lignins and other lignin-like components behind. A concomitant buildup of other recalcitrant compounds such as humic substances, from by-products, decreases the decomposition rates even further (Johansson et al. 1995).

In this stage, the rate-limiting factor is the lignin content, where a higher concentration leads to a slower decay (Andersson 2005). Soil and vegetation types are also important factors regarding decomposition rates. Not only do different types of vegetation grow on locations with different conditions, different types of vegetation also produce litters of different compositions, amplifying the differing conditions for the litter decomposition process (Laskowski et al. 2003). Wieder et al. (2009) observed a nearly fourfold variation in litter decomposition rates between tree species.

14.17.2.9 Nutrient Release from Decomposing Litter

As litters are decomposed, the mass is lost primarily by the production of CO_2 from reduced organic substrates; mass is lost at the same pace with the decomposition and the nutrients are released. However, patterns of nutrient release from decomposing litter do not always follow those for mass or C losses (Prescott 2005). Some nutrients (e.g., K, Ca, Mg) are released very fast or faster than C. On the other hand, N and P are usually retained in the litter during the initial stages of decay. When fresh C sources are added to soils, heterotrophic microorganisms grow and multiply at a fast rate. They absorb N and P from the soil nutrient pool for their growth. Thus, N and P immobilization generally take place during the early stage of litter decomposition. Net import of N or P into the litter has often been reported. A gradual increase in N concentrations narrows the C:N ratios during decomposition. When substantial carbon substrates have been lost by decomposition, microorganisms die and release the immobilized N and P. This happens usually in the latter stages of litter decomposition. However, the immobilization and release of N and P are positively correlated to their initial concentration in litter. In an experiment on nutrient dynamics in a subtropical natural oak forest in India, Pandey et al. (2007) observed 30% increase in N content of litter over the initial value during the early phase (90 days) of decomposition. After 90 days, the N was released slowly with time. On the other hand, P was released initially and then immobilized at 150 days. At the end of the study (300 days), there were 80% release of N and P. Potassium was released consistently as the decomposition progressed, and thus, the release of K followed the pattern similar to that of mass loss. The general pattern of nutrient release from decomposing leaves in temperate and boreal forests also involves the early immobilization (net accumulation) of nitrogen and often phosph orus, followed by net nutrient release (Bosatta and Staaf 1982). Litter decomposition and nutrient release in temperate forests are both slower than in tropical forests. Too slow and too fast nutrient releases are both undesired in terms of forest productivity and nutrient conservation.

14.18 Forests and Forest Soils Are Important Carbon Sequesters

There are carbon pools and carbon fluxes in the global carbon cycle. The major pools of carbon are the minerals in the Earth (limestone, dolomite, etc.), 65×10^6 Gt (1 Gt $= 10^9$ t,

1 t = 1 Mg); oceans, 39×10^3 Gt; soils, 1,580 Gt; terrestrial vegetation, 610 Gt; and atmosphere, 775 Gt (Anonymous 2008). Carbon as carbon dioxide circulates through, and accumulates in, the atmosphere, the oceans, and the land. The land and ocean pools are unable to absorb all of the anthropogenic CO_2 currently being emitted; as a result, residual carbon is accumulating in the atmosphere at a rate of about 3.2 Gt year^{-1} (NASA 2009). For this accumulation, the atmospheric CO_2 levels have risen significantly from the preindustrial 280 ppm to the present 384 ppm (Tans 2008). Evidence suggests that elevated atmospheric CO_2 concentrations are the result of a combination of expanded use of fossil fuels for energy production and transportation, deforestation, and soil cultivation. Elevated atmospheric CO_2 concentration is primarily responsible for global warming.

Carbon sequestration refers to the removal of carbon, as CO_2, from the atmosphere through photosynthesis and dissolution and the storage of C in soil as organic matter or secondary carbonates (Lal 2001). Carbon sequestration by forest is an option to mitigate climate change. Trees are efficient carbon sequesters, more efficient than crops; trees fix 10–20 times more carbon per unit area of lands than in crops or pastures. Trees lock carbon in live biomass and in wood products for a considerable length of time, say 100–200 or more years. Forest-sequestered carbon is distributed in biomass, forest floor, and forest soil. Forest vegetation and soils currently hold almost 40% of all carbon stored in terrestrial ecosystems. Much of this is stored in the great boreal forests of the Northern Hemisphere and in the tropical forests of South America and Africa.

Saving 1 Mg carbon (photosynthesis minus respiration), a forest removes 3.67 Mg CO_2 from the atmosphere. A mature pine plantation can accumulate 500 Mg biomass per hectare after 90 years of growth. If biomass contains 50% carbon, then carbon sequestration by the pine plantation stands roughly at 2.5 Mg ha^{-1} year^{-1} (Birdsey 1996). Qi et al. (2010) reported storage of 220.83 Mg C ha^{-1} in a 35-year-old eucalyptus plantation in China; the carbon sequestration rate becomes 6.3 Mg ha^{-1} year^{-1}. Growth and biomass production of trees differ with species, climate, soil, stand age, stand density, and management. So, carbon sequestration rate of forests should vary widely. Generally, longer-lived, higher-density trees store more carbon than short-lived, low-density, fast-growing trees. Respiration and losses of carbon to the atmosphere in young trees are low; therefore, most of the carbon fixed through photosynthesis is converted to biomass and sequestered. As trees age, respiration increases and carbon sequestration decreases. At a certain point in time, trees no longer sequester additional carbon but instead maintain a constant quantity. This steady state occurs when the carbon gained from photosynthesis equals the carbon lost through respiration. Different tree species reach steady state at different times.

Not all carbon present in forest floor or mineral soil has been sequestered. A greater proportion of added carbon, more than 80%, is labile which is currently being decomposed releasing CO_2 to the atmosphere. During organic matter decomposition a long-term stable fraction, humus, is formed which may resist decomposition for thousands of years. Berg et al. (2007) estimated carbon sequestration in floors of Swedish forest soils by different techniques. Direct measurements of changes in humus depth over 40 years, combined with C analyses, gave a carbon sequestration range from −20 to 730 kg ha^{-1} year^{-1} and an average rate of 180 kg ha^{-1} year^{-1}. There was also a tendency for the average depth of the forest floor to increase with time. The depth of forest floor increased from 7.47 to 9.03 mm in 40 years, and the rates of change of forest floors ranged from 0.1 to >1.2 mm year^{-1}. A long-term buildup of soil organic matter takes place in forest ecosystems at undisturbed conditions. In a boreal forest, Wardle et al. (1997) found that thick mor humus layers had accumulated for up to almost 3,000 years under growing trees in undisturbed sites. This buildup of soil organic matter based on remains from decomposing plant litter formed a C sink in the order of 65–83 kg C ha^{-1} year^{-1}. In forest soils, the average rate of soil C sequestration was reported to be 0.3 Mg C ha^{-1} year^{-1} within a range of 0–3 Mg C ha^{-1} year^{-1} across different climatic zones (Post and Kwon 2000). A synthesis of afforestation chronosequences in northwestern Europe suggested that soils can contribute about 30% of the total C sequestration in afforested ecosystems (Vesterdal et al. 2006).

Forest management practices that can be employed for enhancing carbon sequestration in forest biomass and soil include promoting sound forestry management involving residue management, tree selection, and erosion controls; managing forests for more than timber resources, increasing harvest rotation times; promoting afforestation in degraded and marginal lands; encouraging secondary succession in deforested areas; promoting growth rate through thinning, fertilizing, and irrigation; selective felling, followed by improvement planting, avoiding clear felling; restoring swamp forests; and preventing deforestation.

Study Questions

1. What is a forest? What are the different types of forests? What are the factors that determine characteristics of forest soils? Distinguish between forest soils and agricultural soils.
2. Distinguish between climate and vegetation of tropical and temperate forests. What are the typical soil orders of the tropical forests? Mention their characteristics. Discuss soil and forest as part of a coupled ecosystem.

3. Write the significance of plantation forestry. What are the major tree species for plantation and what are their soil requirements? Discuss the situation of deforestation in the tropics and its consequences.
4. What is agroforestry? What are types and benefits of agroforestry? Discuss the scope of improvement of degraded soils by forest plantations. Is shifting cultivation harmful?
5. What is nutrient cycling? How does nutrient recycling differ from nutrient cycling? Write the significance of litterfall and litter decomposition. Explain the role of forests in carbon sequestration.

References

Alake Biswas M, Alamgir M, Haque SMS, Osman KT (2011) Study on soils under shifting cultivation and other land use categories in Chittagong Hill Tracts, Bangladesh. J For Res 22(3):261–265

Allaby M (2006) Tropical forests. Chelsea House, An imprint of Infobase Publishing, New York

Allaby M (2008) Temperate forests. Facts on File Inc., An Imprint of Infobase Publishing, New York

Andersson C (2005) Litter decomposition in the forest ecosystem—influence of trace elements, nutrients and climate. PhD thesis, Environmental Science Section, Department of Biology and Environmental Science, University of Kalmar, Sweden

Anonymous (2008) The global carbon cycle. University of Michigan. http://www.globalchange.umich.edu/globalchange1/current/lectures/kling/carbon_cycle/carbon_cycle_new.html. Accessed on 5 Jan 2012

Augustoa L, Rangera J, Binkley D, Rothec A (2002) Impact of several common tree species of European temperate forests on soil fertility. Ann For Sci 59:233–253

Aweto AO, Dikinya O (2003) The beneficial effects of two tree species on soil properties in a semi-arid savanna rangeland in Botswana. Land Contam Reclam 11(3):339–344

Berg B, Staaf H (1980) Decomposition rate and chemical changes in decomposing needle litter of Scots pine. II. Influence of chemical composition. Ecol Bull 32:373–390

Berg B, Laskowski R (2006) Litter decomposition: a guide to carbon and nutrient turnover. Elsevier, Amsterdam

Berg B, Johansson MB, Meentemeyer V, Kratz W (1998) Decomposition of tree root litter in a climatic transect of coniferous forests in northern Europe: a synthesis. Scand J For Res 13:402–412

Berg B, De Santo AV, Rutigliano FA, Fierro A, Ekbohm G (2003) Limit values for plant litter decomposing in two contrasting soils—influence of litter elemental composition. Acta Oecologica 24:295–302

Berg B, Gundersen P, Akselsson C, Johansson MB, Nilsson A, Vesterdal L (2007) Carbon sequestration rates in Swedish forest soils—a comparison of three approaches. Silva Fennica 41(3):541–558

Bhat NH, Jan S (2010) Litterfall and nutrient return in *Ulmus villosa* forests of Dachigam National Park, Jammu and Kashmir. Res J Agric Sci 1(4):363–365

Binkley D, Sollins P, McGill WG (1985) Natural abundance of N-15 as tracer of alder-fixed nitrogen. Soil Sci Soc Am J 49:444–447

Birdsey RA (1996) Regional estimates of timber volume and forest carbon for fully stocked timberland, average management after final clearcut harvest. In: Sampson RN, Hair D (eds) Forests and global change: vol 2. American Forests, Washington, DC

Bormann FH, Likens GE, Melillo JM (1977) Nitrogen budget for an aggrading northern hardwood forest ecosystem. Science 196:981–983

Bosatta E, Staaf H (1982) The control of nitrogen turnover in forest litter. Oikos 39:143–151

Brooker MIH, Connors JR, Slee AV, Duffy S (2002) EUCLID: eucalypts of southern Australia (CD Rom). CSIRO Publishing, Collingwood

Brubaker LB (1986) Responses of tree populations to climatic change. Vegetatio 67:119–130

Butler RA (2006) Diversities of image—rainforest biodiversity. Mongabay.com/A place out of time: tropical rainforests and the perils they face http://rainforests.mongabay.com/0305.htm

Chave J, Navarrete D, Almeida S, lvarez EA, Araga LEOC, Bonal D, Chatelet P, Silva-Espejo JE, Goret JY, von Hildebrand P, Jimenez E, Patino S, Penuela MC, Phillips OL, Stevenson P, Malhi Y (2010) Regional and seasonal patterns of litterfall in tropical South America. Biogeosciences 7:43–55

Chundamannii M (1998) Teak plantations in Nilambur—an economic review. KFRI research report no. 144. Kerala Forest Research Institute, India

Detwiler RP, Hall CAS (1988) Tropical forests and the global carbon cycle. Science 239:42–47

Ding YX, Chen JL (1995) Effect of continuous plantation of Chinese fir on soil fertility. Pedosphere 5L:57–66

Evans J (2009) Planted forests: uses, impacts and sustainability. Food and Agriculture Organisation of the United Nations/CABI, Oxfordshire

FAO (2006) Global planted forests thematic study, results and analysis. Planted forests and trees working paper FP38E, Rome

FAO (2010a) Global forest resources assessment 2010—key findings (Brochure), Rome

FAO (2010b) Planted forests in sustainable forest management. A statement of principles (Brochure), Rome

Farrelly N, Dhubhain ÁN, Nieuwenhuis M (2011) Site index of Sitka spruce (*Picea sitchensis*) in relation to different measures of site quality in Ireland. Can J For Res 41:265–278

Fensham RJ, Bowman DMJS (1995) A comparison of foliar nutrient concentration in trees from monsoon rainforest and savanna in northern Australia. Aust J Ecol 20(2):335–339

Finzi AC, Canham CD, van Breemen N (1998) Canopy tree–soil interactions within temperate forests: species effects on pH and cations. Ecol Appl 8(2):447–454

Fisher RF, Binkley D (2000) Ecology and management of forest soils, 3rd edn. Wiley, New York

Gafur A, Borggaard OK, Jensen JR, Petersen L (2000) Changes in soil nutrient content under shifting cultivation in the Chittagong Hill Tracts of Bangladesh. Danish J Geogr 100:37–46

Griffith AL, Gupta RS (1948) Soils in relation to teak with special reference to laterisation. Indian Forest Bulletin no. 141

Hailu T, Negash L, Olsson M (2000) *Millettia ferruginea* from southern Ethiopia: impacts on soil fertility and growth of maize. Agrofor Syst 48:9–24

Harmand JM, Njiti CF, Bernhard-Reversat F, Puig H (2004) Aboveground and belowground biomass, productivity and nutrient accumulation in tree improved fallows in the dry tropics of Cameroon. For Ecol Manage 188:249–265

Hoover MD, Lunt HA (1952) A key for the classification of forest humus types. Soil Sci Soc Am Proc 14:353–358

Huang J, Wang X, Yan E (2007) Leaf nutrient concentration, nutrient resorption and litter decomposition in an evergreen broad-leaved forest in eastern China. For Ecol Manage 239:150–158

ICRAF (1994) Annual report for 1993. ICRAF, Nairobi, pp 80–87

Intergovernmental Panel on Climate Change (IPCC) (2007) Climate change 2007: the physical science basis. Cambridge University Press, Cambridge

Johansson MB (1986) Chemical composition and decomposition pattern of leaf litters from forest trees in Sweden with special reference to methodological aspects and site properties. Reports in forest ecology and forest soils. Report no. 56. Department of Forest Soils, Swedish University of Agricultural Sciences, Uppsala

Johansson MB, Berg B, Meentemeyer V (1995) Litter mass-loss rates in late stages of decomposition in a climatic transect of Pine forests—long-term decomposition in a Scots Pine forest. IX. Can J Bot 73:1509–1521

Johnson DW, Lindberg SE (1992) Atmospheric deposition and forest nutrient cycling. Springer, New York

Kaosa-ard A (1981) Teak: its natural distribution and related factors. Nat His Bull Siam Soc 29:55–74

Kuennecke BH (2008) Temperate forest biomes. Greenwood Press, London

Lal R (2001) The potential of soil carbon sequestration in forest ecosystems to mitigate the greenhouse effect. In: Soil carbon sequestration and the greenhouse effect. SSSA special publication no 57. Soil Science Society of America, USA

Laskowski R, Niklinska M, Nycz-Wasilec P, Wojtowicz M, Weiner J (2003) Variance components of the respiration rate and chemical characteristics of soil organic layers in Niepolomice Forest, Poland. Biogeochemistry 64:149–163

Liu W, Fox JED, Xu Z (2002) Biomass and nutrient accumulation in montane evergreen broadleaf forests (*Lithocarpus xylocarpus* type) in Ailao Mountains SW China. For Ecol Manage 158:223–235

Lowman MD (1988) Litterfall and leaf decay in three Australian rainforest formations. J Ecol 76:451–465

Lu EY, Liu YY, Yen CH (2007) Dynamics of foliar nutrients in major species of a broadleaf forest in the Fushan experimental forest, northeastern Taiwan. Taiwan J For Sci 22(3):307–319

Murphy PG, Lugo AE (1986) Ecology of tropical dry forest. Ann Rev Ecol Syst 17:67–88

Nair PKR (ed) (1989) Agroforestry systems in the tropics. Kluwer, London

National Air and Space Administration (NASA) (2009) The carbon cycle features. http://earthobservatory.nasa.gov/Features/CarbonCycle/carbon_cycle4.php. Accessed 11 Feb 2009

Okeke AI, Omaliko CPE (1994) Litterfall and seasonal patterns of nutrient accumulation in *Dactyladenia barteria* (Hook f ex. Oliv.) Engl. bush fallow at Ozala, Nigeria. For Ecol Manage 67:345–351

Osman KT, Sikder S (2000) Foliar nutrients of six forest tree species grown in brown hill soils of Bangladesh. Ind For 126:1279–1288

Pandey CB, Singh AK, Sharma DK (2000) Soil properties under Acacia nilotica trees in a traditional agroforestry system in central India. Agrofor Sys 49:53–61

Pandey RR, Sharma G, Tripathi SK, Singh AK (2007) Litterfall, litter decomposition and nutrient dynamics in a subtropical natural oak forest and managed plantation in northeastern India. For Ecol Manage 240:96–104

Phillips JD, Marion DA (2006) Biomechanical effects of trees on soil and regolith: beyond treethrow. Ann Assoc Am Geogr 96(2):233–247

Post W, Kwon K (2000) Soil carbon sequestration and land-use change: processes and potential. Glob Chang Biol 6:317–328

Prescott CE (2005) Decomposition and mineralization of nutrients from litter and humus. In: Bassiri Rad H (ed) Nutrient acquisition by plants: an ecological perspective. Springer, Berlin

Pritchett WL, Fisher RF (1987) Properties and management of forest soils. Wiley, New York

Qi G, Wang Q, Dai L, Wang X, Qi L (2010) Carbon sequestration in old-growth forest and plantations on Changbai mountain in northeast China. Bioinformatics and biomedical engineering 4th international conference 18–20 June 2010, Chengdu, China. pp 1–4

Rathod R, Devar KV (2003) Pattern of litter fall and contribution by various litter components to the total litter under various forest plantations. Karnataka J Agric Sci 16(3):491–493

Reiners WA (1972) Nutrient content of canopy throughfall in three Minnesota forests. Oikos 23:14–22

Roos PC, Allsopp N (1997) Soil nutrient ecology associated with *Acacia sieberana* at different tree densities in a South African savanna. Afr J Range Forage Sci 14(2):39–44

Rytter L (2002) Nutrient content in stems of hybrid aspen as affected by tree age and tree size, and nutrient removal with harvest. Biomass Bioener 23(1):13–25

Salacedo H, Tiessen H, Sampaio EVSB (1997) Nutrient availability in soil samples from shifting cultivation sites in the semi-arid Caatinga of NE Brazil. Agric Ecosyst Environ 65:177–186

Sanchez EA (1995) Science in agroforestry. Agrofor Syst 30:5–55

Sanford RL Jr (1985) Root ecology of mature and successional Amazon forests. PhD thesis. University of California

Sayer EJ, Tanner EVJ (2010) Experimental investigation of the importance of litterfall in lowland semi-evergreen tropical forest nutrient cycling. J Ecol. doi:10.1111/j.1365-2745.2010.01680.x

Schaetzl RJ, Anderson S (2005) Soils: genesis and geomorphology. Cambridge University Press, Cambridge

Schlesinger WH (1991) Biogeochemistry: an analysis of global change. Academic, San Diego

Singh B (1982) Nutrient content of standing crop and biological cycling in *Pinus patula* ecosystem. For Ecol Manage 4:317–332

Soil Survey Staff (1999) Soil taxonomy. A basic system of soil classification for making and interpreting soil surveys, 2nd edn. US Government Printing Office, Washington, DC

Sylvester-Bradley R, de Oliveira LA, de Podesta, Filho JA, John TV (1980) Nodulation of legumes, nitrogenase activity of roots and occurrence of nitrogen fixing Azospirillum spp. In: Representative soils of Central Amazonia. Agro Ecosyst 6:249–266

Tanaka S, Kendawang JJ, Yoshida N, Shibata K, Jee A, Tanaka K, Ninomiya I, Sakurai K (2005) Effects of shifting cultivation on soil ecosystems in Sarawak, Malaysia—IV. Chemical properties of the soils and runoff water at Niah and Bakam experimental sites. Soil Sci Plant Nutr 51:525–533

Tans P (2008) Trends in atmospheric carbon dioxide—Mauna Loa. National Oceanic and Atmospheric Administration, US Department of Commerce

Thomas PA, Packham JR (2007) Ecology of woodlands and forests description, dynamics and diversity. Cambridge University Press, Cambridge

Turner J, Cole DW, Gessel SP (1977) Methods employed in forest nutrient cycling studies at Cedar River. College of Forest Resources. University of Washington

Vanmechelen L, Groenemans R, Van Ranst E (1997) Forest soil condition in Europe results of a large-scale soil survey. Forest Soil Co-ordinating Centre, United Nations Economic Commission for Europe

Vesterdal L, Rosenqvist L, van der Salm C, Groenenberg BJ, Johansson MB, Hansen K (2006) Carbon sequestration in soil and biomass following afforestation: experiences from Oak and Norway spruce chronosequences in Denmark, Sweden, and the Netherlands. In: Heil G, Muys B, Hansen K (eds) Environmental effects of afforestation. Field observations, modelling and spatial decision support. Springer, Berlin

Vitousek PM, Sanford RL Jr (1986) Nutrient cycling in moist tropical forest. Ann Rev Ecol Syst 17:137–167

Vitousek PM, Fahey T, Johnson DW, Swift MJ (1988) Element interactions in forest ecosystems: succession, allometry and input-output budgets. Biogeochemistry 5:7–34

Vogt KA, Grier CC, Vogt DJ (1986) Production, turnover, and nutrient dynamics of aboveground and belowground detritus of world forests. Adv Ecol Res 15:303–377

Wardle DA, Zachrisson O, Hornberg G, Gallet C (1997) The influence of island area on ecosystem properties. Science 277:1296–1299

Wieder WR, Cleveland CC, Townsend AR (2009) Controls over leaf litter decomposition in wet tropical forests. Ecology 90: 3333–3341

Yang YS, Guo JF, Chen GS, Xie JS, Cai LP, Lin P (2004) Litterfall, nutrient return, and leaf-litter decomposition in four plantations compared with a natural forest in subtropical China. Ann For Sci 61:465–476

Yang YS, Guo JF, Chen GS, Xie JS, Gao R, Li Z, Jin Z (2005) Litter production, seasonal pattern and nutrient return in seven natural forests compared with a plantation in southern China. Forestry 78(4):1–13

Yan E, Wang X, Huang J (2006) Shifts in plant nutrient use strategies under secondary forest succession. Plant Soil 289:187–197

Zhou G, Guan L, Wei X, Zhang D, Zhang Q, Yan J, Wen D, Liu J, Liu S, Huang ZKG, Mo J, Yu Q (2006) Litterfall production along successional and altitudinal gradients of subtropical monsoon evergreen broadleaved forests in Guangdong. China Plant Ecol. doi:10.1007/s11258-006-9149-9

15 Climate Change and Soil

Emissions of greenhouse gases have alarmingly increased in the past due to land use changes, cultivation, and burning fossil fuels. Concentrations of CO_2, CH_4, and N_2O in the atmosphere have increased by 37, 156, and 19% in the atmosphere, respectively, over 1750 values. Global temperature has risen by 0.75 °C since 1850 and projected temperature rise in the twenty-first century ranges from 1.8 to 4 °C. Sea level has risen in many places, arctic ice is melting, and permafrosts are in danger of thawing. Extreme events such as floods and droughts have become more frequent. Changing climate has a crucial role on ecosystems including soil. Soil contains a large stock (1.5×10^{18} g C) of stabilized carbon in the form of soil organic matter. Rising temperature may enhance decomposition of soil organic matter and might convert soil as a source of atmospheric CO_2. Soil management practices such as no-till or minimum till, using manures and composts, mulching, cover crops, reducing summer fallow, restoration of degraded lands, afforestation, and reforestation would significantly sequester carbon, reduce greenhouse gas emissions from the soil, and help to mitigate climate change and adapt to it.

15.1 Weather Is a Short-Term and Climate Is a Long-Term Phenomenon

The sky is cloudy; it is the weather of the place today. The sky remains always cloudy here; it is the climate of the area. So, weather is a day-to-day event of the atmosphere, and climate is a general average of the conditions of precipitation, temperature, air pressure, humidity, sunshine, cloudiness, and winds of a region over a long period of time. Weather varies over short distances and within a short time such as between the morning and the evening, whereas climate remains the same in a large area over a long period of time. In fact, climate is the composite of generally prevailing weather conditions of a region with all its variations from time to time. So, the weather is what is happening to the atmosphere at any given time, whereas climate is what the statistics tell us should occur at any given time of the year (Burroughs 2007). Climate varies with geographical locations of the earth in relation to the latitude and altitude and with position relative to the mountains, oceans, deserts, etc. The climate of Antarctica is quite different from the climate of a tropical island.

15.2 Climate Has Significant Effect on Crop and Soil Management

Cultivation and management of crops require proper weather and climatic conditions. For example, we do not spray fertilizers and pesticides on wet rainy days. We do not till soil when it rains unless we prepare seedbed for a wetland crop such as rice. Climate has a deciding role in agricultural systems. Climate determines distribution of crops in different geographical regions, selection of cropping patterns and sequence, crop seasons, and management inputs. Seed germination, growth, flowering, fruiting, and ripening all are linked with climatic conditions such as temperature, moisture, and day length. Desired amount of rain does not fall in some years, or rain does not come when it is needed most. Crops may fail due to droughts in those years. Prolonged drought has been the cause of famine in some African countries. In some regions, cropping is rainfed; in many areas, ample irrigation water is not available during the growing season. Some weather conditions are favorable for specific kind of crop diseases.

Crop production is extremely susceptible to climate change. There will be major shifts in crop distribution due to climate change in the future. It has been estimated that climate changes are likely to reduce yields and/or damage crops in the twenty-first century (IPCC 2007a), although, notably, effects are expected to differ widely in different parts of the world. The warming of North America is already noticeable in the increased growing seasons and the northward movement of the limits of corn (maize) and soybean seed production growth. The Corn Belt will thus move into

K.T. Osman, *Soils: Principles, Properties and Management*,
DOI 10.1007/978-94-007-5663-2_15,

Table 15.1 Atmospheric concentration of major greenhouse gases

Greenhouse gas	Concentration[a] in 2007	Percentage increase since 1750
Carbon dioxide CO_2	383 ppmv	37
Methane CH_4	1,790 ppbv	156
Nitrous oxide N_2O	321 ppbv	19

WMO (2008), with permission
[a]*ppmv* parts per million by volume, *ppbv* parts per billion by volume

the Canadian Prairies. It was, however, observed in a study that the overall US crop production increased due to the beneficial effects of elevated CO_2 on crop yields and to marked precipitation increases. These two factors counterbalanced negative effects of warmer temperatures on crop yields. Rainfed crop production increased about 20–50%, especially benefiting winter wheat, corn (maize), soybean, and citrus crops.

15.3 Climate Is Changing

Climate has been changing due to natural and human-induced causes. Natural changes are slow and gradual, and ecosystems respond to these changes by changing themselves gradually and continually. Ecosystems tend to remain in equilibrium with their environment. This is known as ecosystem adaptation which is due to ecosystem resilience.

Human actions, including land use changes and agricultural activities, and burning fossil fuel have aggravated the climate change scenario further. Human-induced changes are rapid and abrupt, and ecosystems often fail to adjust to such changes. It has been much debated in the recent past whether climate is changing by human action and what should be the extent and consequences of such changes. Now, scientific community is virtually unanimous on this conclusion that human activities such as deforestation, biomass burning, soil tillage, drainage of wetlands, livestock farming and cropping, land clearing, and burning fossil fuel during the past 50 years have significantly increased greenhouse gases, such as carbon dioxide and methane, in the atmosphere (IPCC 2007b; CEICC 2008). Emissions of CO_2 in 2006 were about 36 billion Mg, almost 5.5 Mg for every human being (Raupach et al. 2007).

Table 15.1 presents the concentration of major greenhouse gases in the atmosphere. Carbon dioxide has increased by 37% in the atmosphere since 1750. Concentrations of other greenhouse gases have also increased. Methane and nitrous oxide have increased by 156 and 19%, respectively. Each molecule of methane and nitrous oxide absorbs 25 and 300 times more heat, respectively, than that of CO_2 (Prinn et al. 2000; Fluckiger et al. 2002). Greenhouse gas emissions have caused a rise in the temperature of the atmosphere. Projected changes indicate that some areas will get wetter and some drier and temperature will rise but at different rates in different regions. The change in climate would be exhibited by global warming, melting of arctic ice, sea level rise, and frequent occurrence of extreme events such as floods and droughts.

15.3.1 Climate Change Would Lead to Global Warming

IPCC (2007b) reported that the earth's average temperature is certainly warming. Global average surface temperature has risen some 0.75 °C since 1850. However, all parts of the planet's surface have not warmed at the same rate. Some parts are warming more rapidly than others while some other parts have slightly cooled. Projections for the twenty-first century in global temperature range from 1.8 to 4 °C. Warmer ocean waters cause sea ice to melt, result in many species shifting their geographic ranges, stress many other species that cannot move elsewhere, contribute to sea level rise, and hold less oxygen and carbon dioxide (CEICC 2008).

15.3.2 Arctic Sea Ice Would Melt Away

Sea ice in the Arctic Ocean expands in the winter and contracts in the summer. In the first half of the twentieth century, the annual minimum sea-ice area in the Arctic was usually in the range of 10–11 million km^2 (ACIA 2005). In September 2007, sea-ice area hit a single-day minimum of 4.1 million km^2, a loss of about half since the 1950s (Serreze et al. 2007). There is a decrease in thickness of the ice too. From 1975 to 2000, the average thickness of Arctic sea ice decreased by 33%, from 3.7 to 2.5 m (Rothrock et al. 2008).

15.3.3 Rising Sea Level Would Affect Coastal Environments

Sea levels are rising. Melted waters from glaciers and land ice add more volume to oceans. Warming also causes seawater to expand in volume. The global average sea level rose by 1.7 mm $year^{-1}$ in the twentieth century. Satellites measure the rate to be 3.1 mm $year^{-1}$ after 1992 (IPCC 2007a). Land is also subsiding in some coastal regions. Shoreline retreat has also been taking place as a result of sea level rise. Rising sea level inundates more coastal lands, causing changes in concerned ecosystems. Zervas (2001) indicated the rate of sea level rise at Baltimore, Maryland to be 3.12 mm $year^{-1}$ which is nearly double the present rate (1.7 mm $year^{-1}$) of global sea level rise. The higher rate might be due to land subsidence. Bangladesh, a southeast Asian country, will lose the largest amount of cultivable land due to sea level rise. A 1-m

Fig. 15.1 Flood in Bangladesh: climate change will dislodge people (Photo courtesy of the Daily Star)

rise in sea level would inundate 20% of the country's land mass (Rashid and Islam 2007).

Sea level has a profound influence on coastal environments, including beaches, barrier islands, wetlands, and estuarine systems. If the rate of sea level rise accelerates significantly, these systems will be affected adversely. If the sea level rises more rapidly than the capacity of ecosystems to accommodate, it could fundamentally change the state of the coast. Presently rising sea levels are submerging low-lying lands, eroding beaches, converting wetlands to open water, exacerbating coastal flooding, and increasing the salinity of estuaries and freshwater aquifers. Coastal installations including buildings, roads, and other infrastructures which are immobile are also vulnerable (CCSP 2009). Globally, 44% of the world's population lives within 150 km of the ocean, and more than 600 million people live in low elevation coastal zone areas that are less than 10 m above sea level (McGranahan et al. 2007), putting them at significant risk to the effects of sea level rise.

15.3.4 Hurricanes, Floods, and Droughts Would Be More Frequent

Extreme weather events including the most intense hurricanes have become more frequent (IPCC 2007b). Average annual precipitation patterns have changed in many places. At warmer temperature, moisture evaporates more quickly from land, so the amount of moisture available to plants may decline. More evaporation in oceans may lead to more cloud formation and more precipitation in some regions. However, patterns of changes in hydrology may be complex and uncertain. Many low-lying areas are flood-prone. Floods have already become frequent in many countries like Bangladesh (Fig. 15.1).

Floods and droughts are the products of patterns of wind, temperature, and precipitation that produce meteorological extremes. Floods and droughts are not isolated but are often related events driven by the same forces that shape the entire atmosphere. Drought is a sustained and regionally extensive occurrence of below average natural water availability. It is mainly caused by low precipitation and high evaporation rates. Drought is a recurring and worldwide phenomenon having spatial and temporal characteristics that vary significantly from one region to another. Drought and aridity are different environmental conditions; aridity is a long-term average feature of a dry climate. Water scarcity reflects conditions of long-term imbalances between available water resources and demands (Tallaksen and van Lanen 2004). However, the most severe consequences of drought are often found in arid or semiarid regions, where water availability is already low under normal conditions and demand is close to or exceeds natural availability. Climate change is expected to primarily affect precipitation, temperature, and potential evapotranspiration and, thus, is likely to impact the frequency and severity of meteorological droughts.

15.4 Properties and Functions of Soil Would Change in Response to Climate Change

Soils and climate are intimately linked systems. As climate is a driving variable of soil formation, it determines to a large extent the ecological functions a soil can perform. Again, soils have the potential to influence climate through greenhouse gas exchange with the atmosphere and by storing carbon. On the other hand, climate change can have a fundamental effect on functions and processes of soil. In the interplay of the soil and the atmosphere, the soil can be both a contributor to and a recipient of the impacts of climate change (Rosenzweig and Hillel 2000). According to Kardol et al. (2010), feedbacks of terrestrial ecosystems to atmospheric and climate change depend on soil ecosystem dynamics. A combination of rising atmospheric CO_2 levels and consequent changes in temperature, precipitation, windiness, and solar radiation will lead to changes in soil functioning. A rise in air temperature due to global warming would lead to concomitant rise in soil temperature in tropical, temperate, and arctic regions. Minor increases in soil temperatures in the tropics and subtropics and moderate increases with extended periods in temperate and cold climates were suggested by Emanuel et al. (1985). Increased soil temperature may enhance organic matter decomposition, soil structure deterioration, compaction and reduction in porosity, infiltration, and drainage (Lal 2004). These changes with alteration in microbial community and function may lead to soil fertility depletion. On the other hand, increased CO_2, precipitation, humidity, and biomass production may counteract some negative effects of temperature rise. A gradual improvement in soil fertility and physical conditions of soils in humid and subhumid climates has been suggested by Sombroek (1990).

15.4.1 Increased CO_2 May Enhance Biomass Production

Plant growth and carbon storage are likely to be enhanced by a "fertilization" effect of increased atmospheric CO_2. Since CO_2 is an essential ingredient of photosynthesis, as is water, photosynthesis would be faster and more efficient in enhanced atmospheric CO_2. A CO_2-enriched environment supports more growth and biomass production than would otherwise occur (Norby et al. 2005). But with limited water supplies, warming temperatures, deficiencies in other nutrients, or the influence of factors such as ozone that inhibit plant growth, the responses to increased levels of CO_2 are more complex and uncertain (Asshoff et al. 2006). Increased biomass production may lead to enhanced organic matter accumulation in soil and higher carbon sequestration.

15.4.2 Climate Change May Lead to Enhanced Decomposition of Soil Organic Matter

Primary producers (plants and other autotrophs) obtain CO_2 from the atmosphere and fix and convert carbon to biomass by photosynthesis. Decaying biomass subsequently and slowly accumulates in soils as organic matter. Meanwhile, in soil, root respiration and decomposition of organic matter return some carbon to the atmosphere as CO_2 (or as CH_4 under anaerobic conditions) and retain a part of organic matter in soil. Soil organic matter content is a fundamental property of soil because it determines the soil's capacity to facilitate many of its other functions, including retaining and transforming water, nutrients, and contaminants as well as sustaining biodiversity and storing carbon.

Organic matter is continually being added to soils and decomposed. The balance is retained in soil as soil organic matter. The carbon in soil organic matter is a significant component of the earth's carbon reservoirs, with around 1.5×10^{18} g C (Solomon et al. 1985). Significantly, more carbon is stored in the world's soils—including peatlands, wetlands, and permafrost—than is present in the atmosphere. If decomposition of soil organic matter is higher than addition, it becomes a source of atmospheric carbon dioxide. When decomposition is less than addition, soil becomes a sink of carbon dioxide.

As a result of elevated temperature, soil organic matter becomes vulnerable. The rate of biological and chemical transformations, including organic matter decomposition, increases with increasing temperature. The activity of decomposing microorganisms would also increase at elevated temperature leading to depletion of soil organic matter. An increase in temperature would deplete the soil organic carbon pool in the upper layers by 28% in the humid zone, 20% in the subhumid zone, and 15% in the arid zone (Lal 2004). However, if higher CO_2 concentration in air triggers net primary productivity and if increases of plant-derived carbon inputs to soils exceed increases in decomposition, the feedback would be negative (i.e., more carbon will be sequestered in soil). The present CO_2 concentration in the atmosphere, of 350 ppm, is suboptimal for plant growth in certain circumstances. The benefit to plants with the C_4 photosynthetic system is very small or zero because of their lack of photorespiration. In C_3 plants, there is undoubtedly the potential for increased growth of perhaps 30% if the CO_2 concentration rises to 600 ppm (Scharpenseel et al. 1990).

Despite much research, a consensus has not yet been reached on the temperature sensitivity of soil carbon decomposition. Reaching to a conclusion is difficult because the diverse soil organic compounds exhibit a wide range of kinetic properties relating to temperature sensitivity of their decomposition. Moreover, several environmental constraints obscure the intrinsic temperature sensitivity of substrate decomposition, and these constraints may, themselves, be sensitive to climate (Davidson and Janssens 2006).

15.4.3 Climate Change Would Increase Evapotranspiration

Evaporation is an important process of the hydrological cycle which involves the change of state of water from liquid to gas. Heat energy is absorbed during evaporation. So, evaporation from open water and soil is likely to increase as the temperature of the atmosphere rises. Through evaporation, water vapor enters the atmosphere, forms cloud, and falls as rain. Evaporation has a cooling effect on the earth because heat is being used for the process. But water vapor in the atmosphere acts as a greenhouse gas by trapping radiation.

The moisture holding capacity of the atmosphere increases with temperature. For every 1 °C increase in global temperatures, there is a 7% increase in the moisture holding capacity of the air. More moisture in the atmosphere ultimately leads to changes in rainfall patterns. Evaporation increases with the availability of water. More water is evaporated from a lake than from a dry soil. Moist areas like tropical rain forests have higher evaporation rates than arid regions. The amount of water that evaporates from the land surface depends on the amount that is contained in the soil.

Transpiration is the process by which terrestrial plants lose water to the atmosphere. More than 90% water that is absorbed by plants is transpired through the stomata in leaves. Leaves also intercept some rain which later evaporates to the atmosphere. As atmospheric temperature rises, so does evapotranspiration, leading to the loss of soil moisture. Plants may suffer from water stress in a warmer climate.

15.4.4 Climate Change Would Make Many Soils Saline

Soils become saline due to accumulation of soluble salts. Saline soils are problem soils, and many saline soils are not cultivable at all. Crop yield is significantly reduced, and crop failures may occur due to soil salinity.

Climate change is likely to increase all sorts of soil salinity: flooding of land by sea water in coastal areas, residual salt accumulation in arid lands, and capillary rise of salts from groundwater in humid areas. For rising sea level, flood water would make many low-lying areas saline. Sea water would intrude into inland fresh water through the river systems, and farmers would be compelled to use saline water for irrigation. The process is already advanced in many coastal riparian countries like Bangladesh. In this country, the flow of the main river system, the Ganges, has been interrupted by the Farakka Barrage in the upland watershed of Indian territory. The mighty river has dried in places, soils around have become droughty, and salt water has advanced inland through the estuary. This is an example how ill-judged human development activities cause environmental problems and lead to climate change.

Salinity in arid land soils is determined by precipitation and temperature variations. As temperature increases, evaporation and capillary rise of water also increase, resulting in soluble salt accumulation in the surface soil. Extensive observations on bore water levels and rainfall and general observations on the extent of dryland salinity have led to the conclusion that the extent of dryland salinity is related to the change in climate (Rancic et al. 2009). While a change in rainfall can alter soil salinity, the outcome depends on the condition of the soil as well as its position in the landscape (ERIC 2009).

15.4.5 Climate Change Would Alter Composition and Functions of Soil Microorganisms

Higher levels of atmospheric CO_2, warming, and changes in precipitation regimes simultaneously can have direct or indirect, positive or negative impacts on soil microbial communities. Precipitation and soil moisture changes may alter the ratio of bacteria and fungi, as well as shift their community composition (Chen et al. 2007). Rising temperatures can increase microbial activity, processes, and turnover, causing the microbial community to shift in favor of representatives adapted to higher temperatures and faster growth rates (Bradford et al. 2008). Elevated atmospheric CO_2 and precipitation changes might increase soil moisture in an ecosystem, but this increase may be counteracted by warming. Similarly, warming may increase microbial activity in an ecosystem, but this increase may be undone if changes in precipitation lead to a drier soil condition or reduced litter quantity, quality, and turnover.

Castro et al. (2010) suggested that the responses of the microbial communities to climate change that regulate ecosystem processes are less predictable. In a multifactor climate change experiment that exposed a constructed old-field ecosystem to different atmospheric CO_2 concentration (ambient, +300 ppm), temperature (ambient, +3 °C), and precipitation (wet and dry) to alter soil bacterial and fungal abundance and community structure, they found that (1) fungal abundance increased in warmed treatments; (2) bacterial abundance increased in warmed plots with elevated atmospheric CO_2 but decreased in warmed plots under ambient atmospheric CO_2; (3) the phylogenetic distribution of bacterial and fungal clones and their relative abundance varied among treatments, as indicated by changes in 16S rRNA and 28S rRNA genes; (4) changes in precipitation altered the relative abundance of *Proteobacteria* and *Acidobacteria*, where *Acidobacteria* decreased with a concomitant increase in the *Proteobacteria* in wet relative to dry treatments; and (5) changes in precipitation altered fungal community composition. They concluded that the direct and interactive impacts of climate change is likely to reshape bacterial and fungal soil communities and that changes in precipitation in particular will be important in dictating the response of microbial community composition in the future.

15.4.6 Temperature Rise May Lead to Permafrost Thawing

Permafrost is a layer of soil, in the cold zone, which remains permanently frozen. Permafrost is found in about 25% of the terrestrial ecosystems on the northern hemisphere (Zhang et al. 1999). It is found at a depth below the ground surface in some places in the boreal forest regions and in the surface

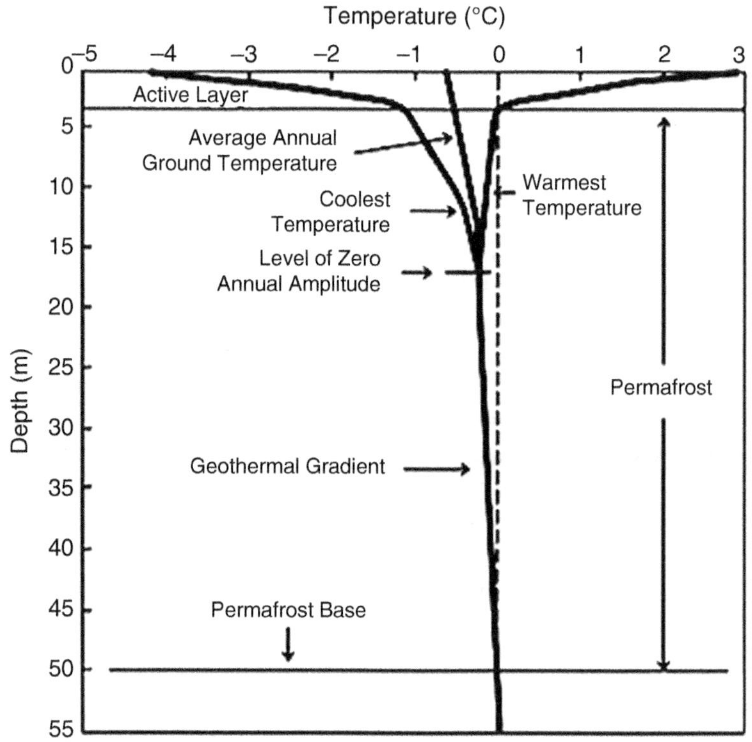

Fig. 15.2 Variation in temperature with depth in the permafrost region

of many other places in the arctic region. Where the permafrost occurs below the ground, the surface soil is frozen in the cool season and thaws in the warm season. This layer which undergoes alternate freezing and thawing is called the active layer. Thus, ground temperature of the active layer fluctuates with seasons, but the temporal variation in temperature decreases with depth. The point at which there is no discernable change in temperature is termed the "depth of zero annual amplitude" (Fig. 15.2). This depth varies from place to place.

Permafrost is a thermal condition. Environmental and anthropogenic changes that cause an alteration to the ground thermal regime determine its distribution, temperature, and thickness. However, the interaction between climate in the ground and below ground is complex and dependent on several factors influenced by climate change. Changes in climate above the ground are most often dampened below the ground due to the insulating effects of vegetation, organic material, or snow cover. There is generally a lag between a change in temperature at the ground surface and the change in permafrost at depth; for thick permafrost, this lag may be on the order of hundreds to thousands of years, for thin permafrost, years to decades.

However, in many parts of central and southern Mackenzie valley, Canada, permafrost temperatures are warm, 0 to −2 °C. Thus, small changes in ground temperatures associated with increased air temperatures will likely reduce the extent of permafrost, increase the depth of the active layer, and cause ground ice to melt (Couture et al. 2000). Numerous studies have reported permafrost degradation under climate warming in the twentieth century in the Northern Hemisphere (Camill 2005). Permafrost degradation may affect local hydrology, ecology, infrastructure, and even the climate (Zimov et al. 2006). In the period between 1989 and 1998, temperatures of upper permafrost layers have increased by 0.5–1.5 °C along a several hundred kilometer north–south transect in central Alaska and by 0.5–1 °C in the western Yamal Peninsula (Pavlov 1998). Reduction of extent of permafrost during the twentieth century has been documented for central and western Canada and Alaska (Weller and Lange 1999).

Climate models predict a mean annual temperature rise of 5 °C in the Arctic by the end of this century. A rise in temperature may have important consequences for the stability of permafrost soils. When permafrost thaws, it can cause the soil to sink or settle, damaging structures built upon or within that soil. The thickness of the active layer should increase at a warmer climate (Waelbroeck et al. 1997). Permafrost soils store twice as much carbon as is currently present in the atmosphere. If the permafrost thaws due to increased temperature, much of the carbon stored will be released to the atmosphere due to enhanced decomposition.

Thawing permafrost and the resulting mineralization of previously frozen organic carbon is considered an important future feedback from terrestrial ecosystems to the atmosphere. Hollesen et al. (2011) examined the Coup model to link surface and subsurface temperatures from a moist permafrost soil in high-arctic Greenland with observed heat production and CO_2 release rates from decomposition of previously frozen organic matter. Observations showed that the maximum thickness of the active layer at the end of the summer has increased 1 cm $year^{-1}$ since 1996.

It was found in a study that artificially elevating summer temperatures by about 2 °C on plots of arctic tundra increased the CO_2 emissions by 26–38% under normal snowfall. When snowfall on some plots was increased which is one possibility with global warming, CO_2 emissions increased 112–326%. Thus, thawing permafrost might impact further climate change and soil carbon release. However, Blok et al. (2010) suggest that permafrost temperature records do not show a general warming trend during the last decade, despite large increases in surface air temperature.

15.5 Global Circulation Models Predict Future Climate and Its Impact

A global circulation model or a general circulation model (GCM) is a computer-based model that predicts future climate patterns in a place. It simultaneously applies several mathematical equations concerning the conservation of mass, energy, and momentum. From the outputs of model calculations, predictions of a number of climate patterns including ocean and wind currents to patterns in precipitation and evaporation rates that could affect lake levels and agricultural levels can be possible.

At present, the models show wind speed, wind direction, moisture, temperature, pressure, surface hydrologic processes, and radiation. New models for cloud prediction, more detail of ground physics, vegetation, the carbon cycle, and gas emissions are being developed. The Geophysical Fluid Dynamics Laboratory (GFDL) is engaged in developing and using mathematical models and computer simulations to improve our understanding and prediction of the behavior of the atmosphere, oceans, and climate. GFDL has prepared maps of the projected increase of surface air temperature. The warming is projected to be particularly large over much of the mid-latitude continental regions, including North America and Asia. Along with this surface warming, sea ice coverage over the Arctic Ocean is projected to decrease substantially. The sea level is expected to rise due to the thermal expansion of sea water as the ocean warms. Because the deep ocean will warm much more slowly than the upper ocean, the thermally driven rise in sea level is expected to continue for centuries after atmospheric CO_2 stops increasing. The sea level rise projections are the expected changes due to thermal expansion of sea water alone and do not include the effect of melted continental ice sheets. With the effect of ice sheets included, the total rise could be larger by a substantial factor. The sea level rise is not anticipated to be uniform over all regions of the globe due to the influence of ocean circulation changes as well as land movements unrelated to global warming.

Soil moisture as simulated in climate models refers to the amount of moisture available over land areas for humidification of the atmosphere. A highly simplified parameterization of soil moisture is used in the present GFDL climate model. The model simulates many of the observed large-scale climate features related to soil moisture content, such as major desert regions and moist temperate zones. Some persistent regional problems remain with these present-day simulations, including an excessively dry southeastern United States. In response to increasing CO_2, the GFDL model projects substantial decreases in soil moisture over most mid-latitude continental areas during summer (file:///C:/Documents%20and%20Settings/personal/Desktop/climate-impact-of-quadrupling-co2.htm).

15.6 Soil Management Should Also Aim at Mitigating Climate Change and Adapting to It

Soil management involves soil water, air, and nutrient management, soil organic matter and soil structure management, and management of soil microbial dynamics and nutrient cycling. Soil management aims at restoring soil fertility and productivity, conserving soil, and maximizing yield. The conventional soil and crop management practices include tilling, harrowing, weeding, fertilizing, irrigation, drainage, and liming. Such management has resulted in marked losses in soil organic carbon (including humus) and greatly reduced diversity and abundance of microbes (algae, bacteria, fungi, nematodes, protozoa) and larger organisms (e.g., mites, ants, beetles, worms) in the soil food web (Ingham 2006). Some cultivated soils have lost one-half to two-thirds of the original soil organic carbon with a cumulative loss of 30–40 Mg C ha^{-1}. The depletion of soil C is accentuated by soil degradation and exacerbated by land misuse and soil mismanagement. A considerable part of the depleted soil organic carbon pool can be restored through conversion of marginal lands into restorative land uses, adoption of conservation tillage with cover crops and crop residue mulch, nutrient cycling including the use of compost and manure, and other systems of sustainable management of soil and water resources. Measured rates of soil C sequestration through adoption of proper management practices range from 50 to 1,000 kg ha^{-1} $year^{-1}$ (Lal 2004). Management of soil carbon is required also for soil health. Powlson et al. (2011) suggested that managing soil organic carbon is central because soil organic matter influences numerous soil properties relevant to ecosystem functioning and crop growth. Even small changes in total organic carbon content can have disproportionately large impacts on key soil physical properties.

Gain of carbon by soil ecosystems is mainly through input of biomass in the form of crop residues, compost, manure, mulch, cover crops, and alluvial or aeolian deposition. Soil and crop management practices that increase the soil carbon pool include slow-release formulations of fertilizer and use of zeolites (Oren and Kaya 2006), biofertilization via rhizobia–legume symbioses (Lugtenberg et al. 2002), increasing nitrogen fixation in legumes (Jones et al. 2007) and even in nonleguminous plants (Cheng et al. 2005), and improving soil structure.

Soil management practices at present should aim at mitigating climate change and adapting to climate change along with maintaining sustainable yield and restoring soil health. Mitigating climate change includes reducing emissions, sequestering emissions, and minimizing emissions. Adopting conservation tillage (no-till, minimum till), cover cropping, mulching, use of organic residues and composts, and biofertilization may benefit both mitigation and adaptation. Strategies to mitigate climate change also include soil restoration and woodland regeneration, nutrient management, improved grazing, water conservation and harvesting, efficient irrigation, agroforestry practices, and growing energy crops on spare lands. Numerous studies of replicated, long-term field experiments comparing conventional tillage (e.g., moldboard plow, chisel, disk) and no-tillage have demonstrated that most soils, following conversion to no-tillage, show an increase in soil organic carbon content relative to tilled soils (Ogle et al. 2005). In general, positive soil carbon responses are obtained first after several years of no-till management (Six et al. 2000), and after 20–30 years, the relative rates of C accumulation tend to decline as soil C levels approach a new equilibrium level under no-till conditions (West and Post 2002). FAO-CTIC (2008) lists the benefits of conservation tillage as (1) financial benefits to farmers; (2) greater stability in yields over varying climate years and with unfavorable weather; (3) higher ratios of outputs to inputs; (4) greater resilience to drought through better water capture and soil moisture retention; (5) reduced demands for labor and much lower costs of farm power (fossil fuels) and greenhouse gas emissions, through reduced tillage and weeding; (6) release of labor at key times, permitting diversification into new on- and off-farm enterprises; (7) better cycling of nutrients and lower losses of plant nutrients through accelerated erosion caused by inversion tillage; (8) higher profit margin because of increase in use efficiency of inputs; (9) increased land value over time because of progressive improvements in soil, water, and air quality; (10) decreased compaction; and (11) opportunities for crop diversification. The environmental benefits include (1) favorable hydrologic balance and perennial flows in rivers to withstand extreme weather events; (2) reduced intensity of desertification; (3) increased biodiversity both in the soil and the aboveground agricultural environment for nutrient cycling; (4) lower levels of soil erosion and sediments in rivers, dams, and irrigation systems; (5) greater carbon sequestration and retention in soils resulting in reduced emissions of greenhouse gases; and (6) less water pollution from pesticides and applied fertilizer nutrients.

Study Questions

1. What do you mean by weather and climate? How do you understand that the climate is changing? Why does climate change?
2. Discuss the impacts of global warming on soils sea level rise. Would there be any effect of temperature rise on emission of green house gases from soil? Why will climate change bring about a shift in the cropping patterns?
3. What is a permafrost? How would it react to climate change? What is the harm of permafrost thawing when most permafrosts are far in the arctic?
4. Discuss the impacts of climate change on vegetation and soil microorganisms. How will climate change affect biomass formation and decomposition?
5. How do GCMs predict climate change and its impacts? Discuss how soil management may help mitigation of climate change and adapting to it.

References

ACIA (Arctic Climate Impact Assessment) (2005) Arctic climate impact assessment: scientific report. Cambridge University Press, Cambridge

Asshoff R, Zots G, Korner C (2006) Phenological and growth response of mature temperate forest trees to four years of CO_2-enrichment. Glob Chang Biol 12(5):848–861

Blok D, Heijmans MMPD, Schaepman-Strub G, Kononov AV, Maximov TC, Berendse F (2010) Shrub expansion may reduce summer permafrost thaw in Siberian tundra. Glob Chang Biol 16:1296–1305

Bradford MA, Davies CA, Frey SD, Maddox TR, Melillo JM, Mohan JE, Reynolds JF, Treseder KK, Wallenstein MD (2008) Thermal adaptation of soil microbial respiration to elevated temperature. Ecol Lett 11:1316–1327

Burroughs WJ (2007) Climate change a multidisciplinary approach. Cambridge University Press, Cambridge

Camill P (2005) Permafrost thaw accelerates in boreal peatlands during late-20th century climate warming. Clim Chang 68:135–152

Castro HF, Classen AT, Austin EE, Norby RJ, Schadt CW (2010) Soil microbial community responses to multiple experimental climate change drivers. Appl Env Microbiol 76(4):999–1007

CCSP (2009) Coastal sensitivity to sea-level rise: a focus on the Mid-Atlantic region. Synthesis and assessment product 4.1. Report by the U.S. Climate Change Science Program and the Subcommittee on Global Change Research

CEICC (2008) Ecological impacts of climate change. Committee on ecological impacts of climate change. The National Academies Press, Washington, DC

Chen MM, Zhu YG, Su YH, Chen BD, Fu BJ, Marschner P (2007) Effects of soil moisture and plant interactions on the soil microbial community structure. Eur J Soil Biol 43:31–38

Cheng Q, Day A, Dowson-Day M, Shen G-F, Dixon R (2005) The Klebsiella pneumoniae nitrogenase protein gene (nifH) functionally substitutes for the chlL gene in *Chlamydomonas reinhardtii*. Biochem Biophys Res Commun 329:966–975

Couture R, Robinson SD, Burgess MM (2000) Climate change, permafrost degradation, and infrastructure adaptation: preliminary results from a pilot community case study in the Mackenzie valley. Natural Resources, Ottawa

Davidson EA, Janssens IA (2006) Temperature sensitivity of soil carbon decomposition and feedbacks to climate change. Nature 440: 165–173

Emanuel WR, Shugart HH, Stevenson MP (1985) Climatic change and the broad-scale distribution of terrestrial ecosystem complexes. Clim Chang 7:29–43

ERIC (2009) Rebuttal to the return of the RGM for dryland salinity. http://www.connectedwaters.unsw.edu.au/news/salinityrainfall.html. Accessed July 2009

FAO-CTIC (2008) Managing soil carbon to mitigate climate change: a sound investment in ecosystem services a framework for action. Food and Agriculture Organization of the United Nations, Conservation Technology Information Center

Fluckiger J, Monnin E, Stauffer B, Schwander J, Stocker TF, Chappellaz J, Raynaud D, Barnola JM (2002) High-resolution Holocene N_2O ice core record and its relationship with CH_4 and CO_2. Global Biogeochem Cycles 16(1):101–108

Hollesen J, Elberling B, Jansson PE (2011) Future active layer dynamics and carbon from thawing permafrost layers in Northeast. Glob Chang Biol 17:911–926

IPCC (Intergovernmental Panel on Climate Change) (2007a) Summary for policymakers. In: Parry ML, Canziani OF, Palutikof JP, Linden PJ, Hanson CE (eds) Climate change 2007: impacts, adaptation and vulnerability. Contribution of working group II to the fourth assessment report of the intergovernmental panel on climate change. Cambridge University Press, Cambridge

IPCC (2007b) Summary for policymakers. In: Solomon S, Qin D, Manning M, Chen Z, Marquis M, Averyt KB, Tignor M, Miller HL (eds) Climate change 2007: the physical science basis: contribution of working group I to the fourth assessment report of the intergovernmental panel on climate change. Cambridge University Press, Cambridge

Ingham ER (2006) Understanding the soil foodweb – first of twelve sub-points.http://www.soilfoodweb.com.au/index.php?pageid=274. Accessed 2 Mar 2010

Jones KM, Kobayashi H, Davies BW, Taga ME, Walker GC (2007) How rhizobial symbionts invade plants: the sinorhizobium-medicago model. Nat Rev Microbiol 5:619–633

Kardol P, Cregger MA, Company CE, Classen AT (2010) Soil ecosystem functioning under climate change: plant species and community effects. Ecology 91(3):767–81

Lal R (2004) Soil carbon sequestration to mitigate climate change. Geoderma 123:1–22

Lugtenberg BJJ, Chin-A-Woeng TFC, Bloemberg GV (2002) Microbe-plant interactions: principles and mechanisms. Antonie van Leeuwenhoek International J Gen Mol Microbiol 81:373–383

McGranahan G, Balk D, Anderson B (2007) The rising tide: assessing the risks of climate change and human settlements in low elevation coastal zones. Environ Urb 19(1):17–37

Norby RJ, DeLuciac EH, Gielend B, Calfapietrae C, Giardinaf CP, King JS, Ledforda J, McCarthyh HR, Moorei DJP, Ceulemansd R, De Angelise P, Finzij AC, Karnoskyk DF, Kubiskel ME, Lukacm M, Pregitzerk KS, Scarascia-Mugnozzan GE, Schlesinger WH, Oren R (2005) Forest response to elevated CO_2 is conserved across a broad range of productivity. Proc Natl Acad Sci 102(50):18052–18056

Ogle SM, Breidt FJ, Paustian K (2005) Agricultural management impacts on soil organic carbon storage under moist and dry climatic conditions of temperate and tropical regions. Biogeochemistry 72:87–121

Oren AH, Kaya A (2006) Factors affecting absorption characteristics of Zn^{2+} on two natural zeolites. J Hazard Mater 131:59–65

Pavlov AV (1998) Active layer monitoring in northern west Siberia. In: Lewkowicz AG, All ard M (eds) Permafrost: seventh international conference. Yellowknife, Canada

Powlson DS, Gregory PJ, Whalley WR, Quinton JN, Hopkins DW, Whitmore AP, Hirsch PR, Goulding KWT (2011) Soil management in relation to sustainable agriculture and ecosystem services. Food Policy 36:S72–S87

Prinn RG, Weiss RF, Fraser PJ, Simmonds PG, Cunnold DM, Alyea FN, O'Doherty S, Salameh P, Miller BR, Huang J, Wang RHJ, Hartley DE, Harth C, Steele LP, Sturrock G, Midgely PM, McCulloch A (2000) A history of chemically and radiatively important gases in air deduced from ALE/GAGE/AGAGE. J Geophys Res 105:17751–17792

Rancic A, Salas G, Kathuria A, Acworth I, Johnston W, Smithson A, Beale G (2009) Climatic influence on shallow fractured rock groundwater systems in the Murray-Darling Basin. NSW Department of Environment and Climate Change, NSW, Sydney

Rashid MH, Islam MS (2007) Adaptation to climate change for sustainable development of Bangladesh agriculture. Paper presented at the conference of the Technical Committee of Asian and Pacific Center for Agricultural and Machinery (APCAEM) on 20–21 Nov 2007, Beijing, China

Raupach MR, Marland G, Ciais P, Le Quere C, Canadell JG, Klepper G, Field CB (2007) Global and regional drivers of accelerating CO_2 emissions. PNAS 104:10288–10293

Rosenzweig C, Hillel D (2000) Soils and global climate change: challenges and opportunities. Soil Sci 165(1):47–56

Rothrock DA, Percival DB, Wensnahan M (2008) The decline in arctic sea-ice thickness: separating the spatial, annual, and inter annual variability in a quarter century of submarine data. J Geophys Res 113. doi:200810.1029/2007JC004252

Scharpenseel H, Schomaker WM, Ayoub A (1990) Soils on a warmer earth. Elsevier, Amsterdam

Serreze MC, Holland MM, Stroeve J (2007) Perspectives on the Arctic's shrinking sea ice cover. Science 315:1533–1536

Six J, Elliott ET, Paustian K (2000) Soil macroaggregate turnover and microaggregate formation: a mechanism for C sequestration under no-tillage agriculture. Soil Biol Biochem 32:2099–2103

Solomon AM, Trabalka JR, Reichle DE, Vorhees LD (1985) The global cycle of carbon. In Trabalka JR (ed) Atmospheric carbon dioxide and the global carbon cycle. US Department of Energy, Washington, DC

Sombroek WG (1990) Soils on a warmer earth: the tropical regions. In: Scharpenseel HW, Schomaker M, Ayoub A (eds) Effects of expected climate change on soil processes, with emphasis on the tropics and subtropics. Developments in soil science 20. Elsevier, Amsterdam

Tallaksen LM, van Lanen HAJ (eds) (2004) Hydrological drought—processes and estimation methods for streamflow and groundwater, Developments in water sciences 48. Elsevier Science BV, Dordrecht

Waelbroeck C, Monfrey Poechel WC, Hastings S, Vourlitis G (1997) The impact of permafrost thawing on the carbon dynamics of Tundra. Geophys Res Let 24:229–232

Weller G, Lange M (1999) Impact of global climate change in the Arctic regions. Reports from a Workshop on the Impacts of Global Change, Tromse

West TO, Post WM (2002) Soil organic carbon sequestration rates by tillage and crop rotation: a global data analysis. Soil Sci Soc Am J 66:1930–1946

WMO (2008) Greenhouse Gas Bulletin: the state of greenhouse gases in the atmosphere using global observations through 2007. World Meteorological Organization, Geneva

Zervas C (2001) Sea level variations of the United States 1854–1999. NOAA technical report NOS CO-OPS 36. NOAA. National Ocean Service, Silver Spring, http://tidesandcurrents.noaa.gov/publications/techrpt36doc.pdf. Accessed on 27 May 2011

Zhang T, Barry R, Knowles K (1999) Statistics and characteristics of permafrost and ground-ice distribution in the northern hemisphere. Polar Geogr 23:132–154

Zimov S, Schuur E, Chapin F (2006) Permafrost and the global carbon budget. Nature 312:1612–1613

Index

K.T. Osman, *Soils: Principles, Properties and Management*,
DOI 10.1007/978-94-007-5663-2, © Springer Science+Business Media Dordrecht 2013

GPSR Compliance

The European Union's (EU) General Product Safety Regulation (GPSR) is a set of rules that requires consumer products to be safe and our obligations to ensure this.

If you have any concerns about our products, you can contact us on ProductSafety@springernature.com

In case Publisher is established outside the EU, the EU authorized representative is:

Springer Nature Customer Service Center GmbH
Europaplatz 3
69115 Heidelberg, Germany

Batch number: 10816838

Printed by Printforce, the Netherlands